Crystal Growth Technology

Crystal Growth Technology

HANS J. SCHEEL
SCHEEL CONSULTING, Groenstrasse, CH-3803 Beatenberg BE, Switzerland
hans.scheel@bluewin.ch

TSUGUO FUKUDA
Institute of Multidisciplinary Research for Advanced Materials, Tohoku University, Sendai 980-8577, Japan
t-fukuda@tagen.tohoku.ac.jp

WILEY

Chemistry Library

Copyright © 2003   John Wiley & Sons Ltd, The Atrium, Southern Gate, Chichester, West Sussex PO19 8SQ, England

Telephone (+44) 1243 779777

Email (for orders and customer service enquiries): cs-books@wiley.co.uk
Visit our Home Page on www.wileyeurope.com or www.wiley.com

Other Wiley Editorial Offices

John Wiley & Sons Inc., 111 River Street, Hoboken, NJ 07030, USA

Jossey-Bass, 989 Market Street, San Francisco, CA 94103-1741, USA

Wiley-VCH Verlag GmbH, Boschstr. 12, D-69469 Weinheim, Germany

John Wiley & Sons Australia Ltd, 33 Park Road, Milton, Queensland 4064, Australia

John Wiley & Sons (Asia) Pte Ltd, 2 Clementi Loop #02-01, Jin Xing Distripark, Singapore 129809

John Wiley & Sons Canada Ltd, 22 Worcester Road, Etobicoke, Ontario, Canada M9W 1L1

Wiley also publishes its books in a variety of electronic formats. Some content that appears in print may not be available in electronic books.

Library of Congress Cataloging-in-Publication Data

Scheel, Hans J.
    Crystal growth technology / Hans J. Scheel, Tsuguo Fukuda.
      p. cm.
    Includes bibliographical references and index.
    ISBN 0-471-49059-8 (pbk. : alk. paper)
     1. Crystallization. 2. Crystal growth. I. Fukuda, Tsuguo. II. Title.

   TP156.C7S34 2003
   660′.284298 – dc21

                            2003050193

British Library Cataloguing in Publication Data

A catalogue record for this book is available from the British Library

ISBN 0-471-49059-8

Typeset in 10/12pt Times by Laserwords Private Limited, Chennai, India
Printed and bound in Great Britain by TJ International, Padstow, Cornwall
This book is printed on acid-free paper responsibly manufactured from sustainable forestry in which at least two trees are planted for each one used for paper production.

CONTENTS

Contents xvii

CONTRIBUTORS

Abe, Takao
Shin-Etsu Handotai, Isobe R&D Center, 2-13-1 Isobe, Annaka-shi, Gunma-ken
379-01, Japan

Asahi, T.
Central R&D Laboratory, Japan Energy Corporation, 3-17-35 Niizo-Minami,
Toda, Saitama 335-8502, Japan

Brandon, Simon
Department of Chemical Engineering, Tianjin University, Tianjin 300072, P.R
China

Chani, Valery I.
Department of Materials Science and Engineering, McMaster University, 1280
Main Street West, Hamilton, Ontario, L85 4M1, Canada

Ciszek, T. F.
National Renewable Energy Laboratory, 1617 Cole Boulevard, Golden, Colorado
80401-3393, USA

Derby, Jeffrey J.
Department of Chemical Engineering and Materials Science, Army HPC Research
Center & Minnesota Supercomputing Institute, University of Minnesota, Min-
neapolis, MN 55455-0132, USA

Falster, R.
MEMC Electronic Materials SpA, Viale Gherzi 31, 28100 Novara, Italy

Fukuda, Tsuguo
Institute of Multidisciplinary Research for Advanced Materials, Tohoku University, 2-1-1 Katahiraa, Aoba-ku, Sendai 980 8577, Japan

Gektin, A. V.
Institute for Single Crystals, Lenin Av 60, 310001 Kharkov, Ukraine

Hauser, C. (Retired)
HCT Shaping Systems SA, 1033 Cheseaux, Switzerland

Hirose, Kikuji
Department of Precision Science and Technology, Department of Electrical Engineering, Graduate School of Engineering, Osaka University, 2-1 Yamada-oka, Suita 565-0871, Osaka, Japan

Inagaki, Kohji
Dept of Precision Science and Technology, Graduate School of Engineering, Osaka University, 2-1 Yamada-oka, Suita, Osaka 565-0871, Japan

Itsumi, Manabu
NTT Lifestyle & Environmental Technology Laboratories, 3-1 Morinosato Wakamiya, Atsugi-Shi, Kanagawa, 243-0198, Japan

Jurisch, M.
Freiberger Compound Materials GmbH, Am Junger Löwe Schacht, D-09599 Freiberg, Germany

Kainosho, K.
Central R&D Laboratory, Japan Energy Corporation, 3-17-35 Niizo-Minami, Toda, Saitama 335-8502, Japan

Kakimoto, Koichi
Institute of Advanced Material Study, Kyushu University, 6-1 Kasuga-Koen, Kasuga 816-8580, Japan

Kakiuchi, Hiroaki
Department of Precision Science and Technology, Graduate School of Engineering, Osaka University, 2-1 Yamada-oka, Suita 565-0871, Osaka, Japan

Kawase, Tomohiro
Sumitomo Electric Industries, Ltd., 1-1-1, Koya-kita, Itami, Hyogo 664-0016, Japan

Kohiro, K.
Central R&D Laboratory, Japan Energy Corporation, 3-17-35 Niizo-Minami, Toda, Saitama 335-8502, Japan

Lal, Krishnan
National Physical Laboratory, Dr K. S. Krishnan Road, New Delhi 110012, India

Lebeau, Michel
CERN, 1211 Geneva 23, Switzerland

Liu, Yongcai
Department of Chemical Engineering, Tianjin University, Tianjin 300072, P.R. China

Lomonova, E. E.
Laser Materials and Technology Research Center, General Physics Institute of the Russian Academy of Sciences, Moscow, Russia

Lytvynov, Leonid A.
Institute of Single Crystals, Lenin Av., 60, Kharkov, 61001, Ukraine

Mimura, Hidekazu
Department of Precision Engineering, Osaka University, 2-1 Yamada-oka, Suita, Osaka 565-0871

Miyazawa, Shintaro
Shinkosha Co. Ltd., 3-4-1 Kosugaya, Sakae-ku, Yokohama, Kanagawa 247-0007, Japan

Mori, Yusuke
Department of Electrical Engineering, Graduate School of Engineering, Osaka University, 2-1 Yamada-oka, Suita 565-0871, Osaka, Japan

Mori, Yuzo
Department of Precision Science and Technology, Graduate School of Engineering, Osaka University, 2-1 Yamada-oka, Suita 565-0871, Osaka, Japan

Mutti, P.
MEMC Electronic Materials SpA, Via Nazionale 59, 39012 Merano, Italy

Nasch, P. M.
HCT Shaping Systems SA, Route de Genève 42, Cheseaux Sur-Lausanne CH-1033, Switzerland

Nishida, Yasuhiro
Sumitomo Electric Industries Ltd., 1-1-1, Koya-kita, Itami, Hyogo 664-0016, Japan

Noda, A.
Central R&D Laboratory, Japan Energy Corporation, 3-17-35 Niizo-Minami, Toda, Saitama 335-8502, Japan

Oda, O.
Central R&D Laboratory, Japan Energy Corporation, 3-17-35 Niizo-Minami, Toda, Saitama 335-8502, Japan

Osiko, V. V.
Laser Materials and Technology Research Center, General Physics Institute of the Russian Academy of Sciences, Vavilov str. 38, Moscow 119991, Russia

Polezhaev, V. I.
Institute for Problems in Mechanics, Russian Academy of Sciences, Prospect Vernadskogo 101, 117526 Moscow, Russia

Rudolph, Peter
Institute for Crystal Growth, Rudower Chaussee 6, D-12489 Berlin, Germany

Sano, Yasuhisa
Department of Precision Science and Technology, Graduate School of Engineering, Osaka University, 2-1 Yamada-oka, Suita, Osaka 565-0871, Japan

Sasaki, Takatomo
Department of Electrical Engineering, Graduate School of Engineering, Osaka University, 2-1 Yamada-oka, Suita 565-0871, Osaka, Japan

Sato, K.
Central R&D Laboratory, Japan Energy Corporation, 3-17-35 Niizo-Minami, Toda, Saitama 335-8502, Japan

Scheel, Hans J.
SCHEEL CONSULTING, Groenstrasse, CH-3803 Beatenberg BE, Switzerland

Shimamura, Kiyoshi,
Kagami Memorial Laboratory for Materials Science and Technology, Waseda University, 2-8-26 Nishiwaseda, Shinjuku 169-0051, Japan

Sugiyama, Kazuhisa
Department of Precision Science and Technology, Graduate School of Engineering, Osaka University, 2-1 Yamada-oka, Suita, Osaka 565-0871, Japan

Tatsumi, M.
Sumitomo Electric Industries Ltd., 1-1-1, Koya-kita, Itami, Hyogo 664-0016, Japan

Terashima, Kazutaka
Silicon Melt Advanced Project, Shonan Institute of Technology, 1-1-1 Tsujido-Nishikaigan, Fujisawa, Kanagawa 251, Japan

Triboulet, R.
Laboratoire de Physique des Solides de Bellevue CNRS, 1, Place A. Briand, F-92195 Meudon Cedex, France

Virozub, Alexander
Department of Chemical Engineering, Tianjin University, Tianjin 300072, P.R. China

Voronkov, V. V.
MEMC Electronic Materials SpA, Via Nazionale 59, 39012 Merano, Italy

Yamamura, Kazuya
Department of Precision Science and Technology, Graduate School of Engineering, Osaka University, 2-1 Yamada-oka, Suita, Osaka 565-0871, Japan

Yamauchi, Kazuto
Department of Precision Science and Technology, Graduate School of Engineering, Osaka University, 2-1 Yamada-oka, Suita, Osaka 565-0871, Japan

Yasutake, Kiyoshi
Department of Material and Life Science, Graduate School of Engineering, Osaka University, 2-1 Yamada-oka, Suita, Osaka 565-0871, Japan

Yeckel, Andrew
Department of Chemical Engineering and Materials Science, Army HPC Research Center & Minnesota Supercomputing Institute, University of Minnesota, Minneapolis, MN 55455-0132, USA

Yoshii, Kumayasu
Department of Precision Science and Technology, Graduate School of Engineering, Osaka University, 2-1 Yamada-oka, Suita, Osaka 565-0871, Japan

Yoshimura, Masashi
Department of Electrical Engineering, Graduate School of Engineering, Osaka University, 2-1 Yamada-oka, Suita 565-0871, Osaka, Japan

Zaslavsky, B. G.
Institute for Single Crystals, Lenin Av 60, 310001 Kharkov, Ukraine

PREFACE

This volume deals with the technologies of crystal fabrication, of crystal machining, and of epilayer production and is the first book on industrial aspects of crystal production. Therefore, it will be of use to all scientists, engineers, professors and students who are active in these fields, or who want to study them. Highest-quality crystals and epitaxial layers (epilayers) form the basis for many industrial technological advances, including telecommunications, computer and electrical energy technology, and those technologies based on lasers and nonlinear-optical crystals. Furthermore, automobile electronics, audiovisual equipment, infrared night-vision and detectors for medicine (tomography) and large nuclear-physics experiments (for example in CERN) are all dependent on high-quality crystals and epilayers, are as novel technologies currently in development and planned for the future. Crystals and epilayers will gain special importance in energy saving and renewable energy. Industrial crystal and epilayer production development has been driven by the above technological advances and also by the needs of the military and a multibillion-dollar industry. From the nearly 20 000 tons of crystals produced annually, the largest fraction consists of the semiconductors silicon, gallium arsenide, indium phosphide, germanium, and cadmium telluride. Other large fractions are optical and scintillator crystals, and crystals for the watch and jewellery industries.

For most applications the crystals have to be machined, i.e. sliced, lapped, polished, etched, or surface-treated. These processes have to be better understood in order to improve yields, reduce the loss of valuable crystal, and improve the performance of machined crystals and wafers.

Despite its importance, the scientific development and understanding of crystal and epilayer fabrication is not very advanced, and the education of specialized engineers and scientists has not even started. The first reason for this is the multidisciplinarity of crystal growth and epitaxial technology: neither chemical and materials engineering departments on the preparative side, nor physics and electrical engineering on the application side feel responsible, or capable of taking care of crystal technologies. Other reasons for the lack of development and recognition are the complexity of the multi-parameter growth processes, the complex phase transformation from the the mobilized liquid or gaseous phase to the solid crystal, and the scaling problem with the required growth-interface control on the nm-scale within growth systems of m-scale.

An initial workshop, named 'First International School on Crystal Growth Technology ISCGT-1' took place between September 5–14, 1998 in Beatenberg, Switzerland, and ISCGT-2 was held between August 24–29, 2000 in Mount

Zao Resort, Japan with H. J. Scheel and T. Fukuda action as the co-chairmen. Extended lectures were given by leading specialists from industries and universities, and the majority of crystal-producing factories were represented. This book contains 29 selected review papers from ISCGT-1 and discusses scientific and technological problems of production and machining of industrial crystals for the first time. Thus, it is expected that this volume will serve all scientists and engineers involved in crystal and epilayer fabrication. Furthermore, it will be useful for the users of crystals, for teachers and graduate students in materials sciences, in electronic and other functional materials, chemical and metallurgical engineering, micro- and optoelectronics including nano-technology, mechanical engineering and precision-machining, microtechnology, and in solid-state sciences. Also, consultants and specialists from funding agencies may profit from reading this book, as will all those with an interest in crystals, epilayers, and their production, and those concerned with saving energy and in renewable energy.

In Section I, general aspects of crystal growth are reviewed: the present and future of crystal growth technology, thermodynamic fundamentals of phase transitions applied to crystal-growth processes, interface and faceting effects, striations, modeling of crystal growth from melts and from solutions, and structural characterization to develop the growth of large-diameter crystals, In Section II, the problems relating to silicon are discussed: structural and chemical characteristics of octahedral void defects, intrinsic point defects and reactions in silicon, heat and mass transfer in melts under magnetic fields, silicon for photovoltaics, and slicing and novel precision-machining methods for silicon. Section III treats problems of the growth of large, rather than perfect, crystals of the compound semiconductors GaAs, InP, and CdTe. Section IV discussed oxides for surface-acoustic-wave and nonlinear-optic applications and the growth of large halogenide scintillator crystals. Section V deals with crystal machining: crystal orientation, sawing, lapping, and polishing and also includes the novel technologies EEM and CVM. Finally, Section VI treats the control of epitaxial growth modes to achieve highest-performance optoelectronic devices, and a novel, fast deposition process for silicon from high-density plasmas is presented.

The editors would like to thank the contributors for their valuable reviews, the referees (especially D. Elwell), and the sponsors of ISCGT-1. Furthermore, the editors acknowledge the competent copy-proof reading of P. Capper, and the work from J. Cossham, L. James and L. Bird of John Wiley & Sons Ltd, the publishers: also for pleasant collaboration and their patience.

It is hoped that this book may contribute to the scientific development of crystal technologies, and that it is of assistance for the necessary education in this field.

HANS J. SCHEEL and TSUGUO FUKUDA

Part 1

General Aspects of Crystal Growth Technology

1 The Development of Crystal Growth Technology

HANS J. SCHEEL
SCHEEL CONSULTING, CH-8808 Pfaeffikon SZ, Switzerland

ABSTRACT

The industrial production of crystals started with A. Verneuil with his flame-fusion growth method 1902. He can be regarded as the father of crystal growth technology as his principles of nucleation control and crystal-diameter control are adapted in most later growth methods from the melt, like Tammann, Stöber, Bridgman, Czochralski, Kyropoulos, Stockbarger, etc. The important crystal pulling from melts named after Czochralski was effectively developed by Teal, Little and Dash.

The multi-disciplinary nature of technology of crystal and epilayer fabrication, the complex multi-parameter processes – where ten or more growth parameters have to be compromised and optimized, and also the scaling problem have impeded the scientific development of this important area. Only recently has the numerical simulation of Czochralski melts started to become useful for growth technologists, the deep understanding of the striation problem allowed the experimental conditions to grow striation-free crystals to be established, and the control of epitaxial growth modes permitted the preparation of atomically flat surfaces and interfaces of importance for the performance of opto-electronic and super-conducting devices.

Despite the scale of the multi-billion dollar crystal and epilayer fabrication and crystal-machining industry and the annual need worldwide of at least 400 engineers, there is so far no formation of specialists for crystal production, epitaxy technology, crystal machining and surface preparation. A special curriculum is required due to the multi-disciplinary character of crystal-growth technology (CGT) which does not fit into a single classical university discipline like chemical, mechanical, materials, or electrical engineering, or crystallography, thermodynamics, solid-state physics, and surface physics. The education scheme for CGT has to include all these disciplines and basic sciences to such an extent that the finished engineers and scientists are capable of interacting and collaborating with specialists from the various disciplines. It is up to the interested industries to request CGT engineers from the technical universities and engineering schools.

Crystal Growth Technology, Edited by H. J. Scheel and T. Fukuda
© 2003 John Wiley & Sons, Ltd.   ISBN: 0-471-49059-8

Crystal-growth technology and epitaxy technology had developed along with the technological development in the 20th century. On the other hand, the rapid advances in microelectronics, in communication technologies, in medical instrumentation, in energy and space technology were only possible after the remarkable progress in fabrication of large, rather perfect crystals and of large-diameter epitaxial layers (epilayers). Further progress in CGT and education of CGT engineers is required for significant contributions to the energy crisis. High-efficiency white light-emitting diodes for energy-saving illumination and photovoltaic/thermo-photovoltaic devices for transforming solar and other radiation energy into electric power with high yield depend on significant advances in crystal growth and epitaxy technology. Also, the dream of laser-fusion energy and other novel technologies can only be realized after appropriate progress in the technology of crystal and epilayer fabrication.

1.1   HISTORICAL INTRODUCTION

Fundamental aspects of crystal growth had been derived from early crystalliza-tion experiments in the 18th and the 19th century (Elwell and Scheel 1975, Scheel 1993). Theoretical understanding started with the development of ther-modynamics in the late 19th century (Gibbs, Arrhenius, Van't Hoff) and with the development of nucleation and crystal growth theories and the increasing under-standing of the role of transport phenomena in the 20th century. The phenomena of undercooling and supersaturation and the heat of crystallization were already recognized in the 18th century by Fahrenheit and by Lowitz. The corresponding metastable region, the existence range of undercooled melts and solutions, was measured and defined in 1893/1897 by Ostwald and in 1906 by Miers, whereas the effect of friction on the width of this Ostwald−Miers region was described in 1911 and 1913 by Young. Although the impact of stirring on this metastable region is important in mass crystallization of salt, sugar and many chemicals, it is not yet theoretically understood.

The rates of nucleation and crystallization in glasses were the foundation to nucleation theories. The crystal surface with steps and kinks of Kossel in 1927 allowed Stranski and Kaishew in 1934 to define the work of separation of crystal units as repeatable steps as the basis of the first crystal-growth theories. With the understanding of facet formation as a function of the entropy of fusion in 1958 by Jackson, and depending on the density of bonds in the crystal structure 1955 by Hartman and Perdok, the role of screw dislocations as continuous step sources in the formation of growth hillocks (Frank 1949), and with the generalized crystal growth theory of Burton, Cabrera and Frank 1951, many growth phenomena could be explained.

In the growth of crystals from a fluid medium (melt, solution, gas phase) the heat and mass transport phenomena also play a significant role, as was observed early by Rouelle 1745 and Frankenheim 1835. The diffusion boundary layer defined by Noyes and Whitney 1897 was used in the growth-rate equation of Nernst 1904 and confirmed by interferometric measurements of concentration profiles around growing crystals by Berg 1938 and by others. Forced convection was recognized to be beneficial for diffusion-limited growth by Wulff 1886, Krüger and Finke 1910, and Johnsen 1915 for open systems with stirrers, whereas smooth stirring in sealed containers can be achieved with the accelerated crucible rotation technique ACRT of Scheel 1971/1972. The growth of inclusion-free crystals from the melt can be accomplished by observing the principles of "diffusional undercooling" of Ivantsov 1951 and "constitutional supercooling" of Tiller *et al.* 1953. Formation of inclusions, i.e. growth instability, can be prevented in growth from solutions by sufficient flow against or along the crystal facets: Carlson 1958 developed an empirical theory which was utilized by Scheel and Elwell 1972 to derive the maximum stable growth rate and optimized programming of supersaturation for obtaining large inclusion-free crystals.

Microscopic and macroscopic inhomogeneities in doped crystals and in solid solutions are caused by segregation phenomena, which are related to mass and heat transfer. Based on the derivation of effective distribution coefficients for melt growth by Burton *et al.* (1953) and by van Erk (1982) for growth from solutions, the theoretical and experimental conditions for growth of striation-free crystals could be established (Rytz and Scheel 1982, Scheel and Sommerauer 1983, Scheel and Swendsen 2001).

There have been remarkable developments with respect to size and perfection of crystals, with silicon, sapphire, alkali and earth alkali halides reaching diameters up to 0.5 m and weights of nearly 500 kg. These advances in Czochralski, Kyropoulos, heat-exchanger method, and Bridgman–Stockbarger growth were accompanied by numerical simulations which have become increasingly powerful to predict the optimized conditions. However, further advances in computer modelling and in the reliability of the used physico-chemical data are required in order to increase the efficiency and precision of computer simulations and to allow the prediction of the best crystal-growth technology including growth parameters for the growth of new large and relatively perfect crystals.

1.2 THE DEVELOPMENT OF CRYSTAL-GROWTH METHODS

Industrial crystal production started with Verneuil 1902 who with the flame-fusion growth process named after him achieved for the first time control of nucleation and thus single crystals of ruby and sapphire with melting points above 2000 °C. By formation of the neck and the following enlargement of the crystal diameter, not only was the seed selection achieved in a crucible-free process, but also the structural perfection of the growing crystal could be controlled to some extent.

Figure 1.1 shows the stages of Verneuil growth of high-melting oxides, and Figure 1.2 the relatively simple apparatus where powder is fed into the hydrogen–oxygen flame and enters the molten cap of first the cone and neck, and

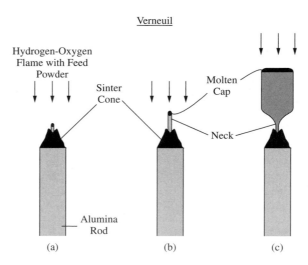

Figure 1.1 Stages of flame-fusion (Verneuil) growth of ruby, schematic: (a) formation of sinter cone and central melt droplet onto alumina rod, (b) growth of the neck by adjustment of powder supply and the hydrogen-oxygen flame, (c) Increase of the diameter without overflow of the molten cap for the growth of the single-crystal boule. (Reprinted from H. J. Scheel, *J. Cryst. Growth* **211**(2000) 1–12, copyright (2000) with permission from Elsevier Science.)

then the molten surface layer of the growing crystal. The latter is slowly moved downwards so that the surface-melting conditions remain within the observation window. Historical details of Verneuil's development were given by Nassau 1969 and Nassau and Nassau 1971. The industrialization for production of ruby watch-stones and synthetic gemstones started with Verneuil's laboratory in Paris in 1910 with 30 furnaces and with the Djevahirdjian factory at Monthey VS in Switzerland where the CIBA chemical company could provide hydrogen. The annual ruby production reached 5 million carats in 1907, and 10 million carats ruby and 6 million carats sapphire in 1913. Nowadays, the production capacity of sapphire for watch windows and other applications is 250 tons (1.25×10^9 carats), the largest fraction from the Djevahirdjian factory with 2200 Verneuil furnaces, and this with the practically unchanged process.

The principles of the Verneuil method with nucleation, growth rate and diameter control have been applied in most of the growth processes described in the following years: Tammann 1914, Stöber 1925, Bridgman 1923, Stockbarger 1936, vertical-gradient-freeze growth of Ramsberger and Malvin 1927, and in the Czochralski process 1918/1950, see Figure 1.3 and Wilke-Bohm 1988. Thus, Verneuil can really be regarded as the founder of crystal-growth technology.

However, the most important method of fabrication of semiconductor and oxide crystals consists of pulling crystals from melts contained in crucibles, a method named after Czochralski. The origin of this started with the topical area

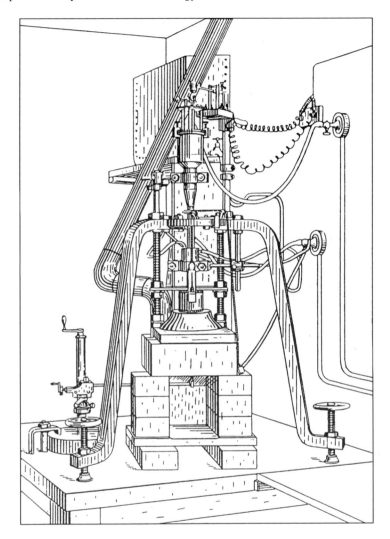

Figure 1.2 The relatively simple apparatus of Verneuil for flame-fusion growth. (Reprinted from D. T. J. Hurle (ed.) Handbook of Crystal Growth 1, Fundamentals, Part A, Chapter 1, 1993, copyright (1993) with permission from Elsevier Science.)

of physical chemistry for 50 years, the measurement or crystallization velocities, which was initiated by Gernez 1882, who had found a high crystallization rate of molten phosphorus. After Tammann's studies of glass crystallization, Czochralski presented in 1918 "a new method for measurement of the crystallization velocity of metals" where he pulled fibers of low-melting metals from melts. Soon after this, von Gomperz used floating dies to control the fiber diameter

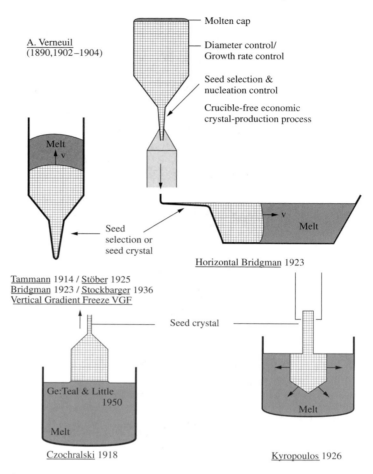

Figure 1.3 Modification of Verneuil's principles of nucleation control and increasing crystal diameters in other crystal-growth techniques. (Reprinted from H. J. Scheel, *J. Cryst. Growth* **211**(2000) 1–12, copyright (2000) with permission from Elsevier Science.)

(an early Stepanov or EFG process) 1921, and in 1928 Kapitza measured crystallization rates and prepared single-crystalline bismuth in a vertical glass tube (early zone melting).

Czochralski never considered pulling a crystal for research, although he could have adapted the inverse Verneuil principle, which was well known at that time. The metal crystals for research in Czochralski's laboratory were prepared by the Bridgman method. Therefore, it is no wonder that Zerfoss, Johnson and Egli in their review on growth methods at the International Crystal Growth Meeting in Bristol UK 1949 mentioned Verneuil, Bridgman–Stockbarger, Stöber and Kyropoulos techniques, but not Czochralski.

With the discovery of the transistor at Bell Laboratories there arose a need for single crystals of semiconductors, first of germanium. Teal and Little invented and developed crystal pulling of relatively large Ge crystals in 1950. Then it was the book of Buckley in 1951, who erroneously named pulling of large crystals after Czochralski, and this name became widespread with the fast-growing importance of crystal pulling. Possibly it is too late to change the name to Teal–Little–Dash, the latter due to his contribution to dislocation-free crystals by optimized seed and neck formation. The development of Czochralski and liquid-encapsulated Czochralski growth was discussed by Hurle 1987.

The majority of crystals is produced by pulling from the melt. Crystal dimensions have reached 10 cm for InP, $LiTaO_3$, sapphire and other oxide crystals, 15 cm for $LiNbO_3$ and GaAs, and 30 to 40 cm for silicon and for halide scintillator crystals with melt dimensions up to 1 m. Numerical modelling has assisted these developments, for instance for the optimization of the crystal and crucible rotation rates and of the temperature distribution and the design of heaters, afterheaters and heat shields. However, for the axial segregation and the striation problems the final general solution has not yet been established. The following approaches to reduce convection have been partially introduced in crystal production: double crucible for GaAs and stoichiometric $LiNbO_3$, magnetic fields in various configurations for semiconductors, and shallow melts or small melt volumes with continuous feeding for silicon and for halide scintillator crystals: these approaches are discussed in this volume. An alternative approach to solve, or at least minimize, the axial segregation and the striation problem is the co-rotating ring Czochralski (CRCZ) method which induces a hydrodynamic double-crucible effect: by optimized rotation rate of crystal and co-rotating ring (inserted into the melt) and counter-rotation rate of the crucible, a nearly convection-free melt fraction is achieved below the growing crystal, whereas the larger fraction of the melt is homogenized by combined forced and natural convection (Scheel 1995).

Only a few Czochralski-grown elements like silicon, germanium, and copper are produced dislocation-free by application of necking as developed by Dash in 1959. Most crystals have dislocations of typically 10^3 to 10^6 cm^{-2} due to the generally large temperature gradients at the growing interface and due to the post-growth defect agglomeration behind the growth front (Völkl and Müller 1989). Attempts to grow with low-temperature gradients at the Institute of Inorganic Chemistry in Novosibirsk or to grow the crystal into the melt by the Kyropoulos method (SOI-method in St. Petersburg) have yielded low dislocation crystals of, for example, sapphire up to 40 cm diameter, but have not yet been developed for large-scale production. However, alternative technologies like Bridgman and vertical gradient freeze (VGF) have gained importance for production of semiconductor crystals of improved structural perfection like GaAs, InP, and CdTe and its solid solutions. In vertical Bridgman and in VGF growth the forced convection by the accelerated-crucible rotation technique (ACRT) of Scheel 1972 has increased homogeneity, structural perfection, and stable growth rates in the cases of halides (Horowitz et al. 1983) and doped CdTe (Capper 1994, 2000).

In silicon, the octahedral void defects (see Falster and Itsumi in this volume) and their distribution as well as oxygen and dopant homogeneity require further understanding (see Abe, Falster) and process optimization, otherwise an increasing fraction of silicon wafers have to be provided with epitaxial layers in order to achieve the homogeneity of surface properties required for highly integrated microelectronic circuit structures.

Examples of other remarkable developments in the growth of bulk crystals are briefly given: Hydrothermal growth of large high-quality low-dislocation-density quartz by VNIISIMS/Russia and Sawyer/USA, hydrothermal growth of large ZnO and emerald crystals, high-temperature solution growth of perovskites and garnets by top-seeded solution growth TSSG by Linz *et al.* and by the accelerated crucible rotation technique (ACRT) by Scheel and by Tolksdorf, TSSG growth of KTP and CLBO (the latter by Sasaki and Mori), aqueous-solution growth of KDP (Sasaki, Zaitseva), Bridgman–Stockbarger growth of alkali halides and of CaF_2, float-zone growth of TiO_2 rutile by Shindo *et al.*, skull-melting growth of zirconia by Wenckus and Osiko, and Czochralski growth of numerous oxide compounds mainly in US, French, Japanese and recently Korean and Taiwan companies.

1.3   CRYSTAL-GROWTH TECHNOLOGY NOW

The world crystal production is estimated at more than 20 000 tons per year, of which the largest fraction of about 60 % are semiconductors silicon, GaAs, InP, GaP, CdTe and its alloys. As can be seen in Figure 1.4 , optical crystals, scintillator crystals, and acousto-optic crystals have about equal shares of 10 %, whereas laser and nonlinear-optic crystals and crystals for jewelry and watch industry have shares of a few % only. This scale of crystal production and the fact that most

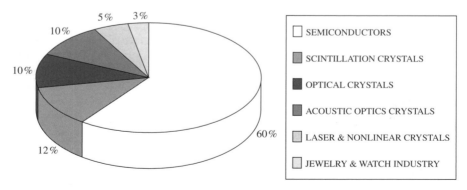

Figure 1.4   Estimated shares of world crystal production in 1999. (Reprinted from H. J. Scheel, *J. Cryst. Growth* **211**(2000) 1–12, copyright (2000) with permission from Elsevier Science.)

crystals are produced in factories specialized in silicon or GaAs or LiNbO$_3$, etc. has caused an increasing degree of **specialization.** Furthermore, crystal growth has split into five major areas so that cross-fertilization and communication are more and more reduced:

1. Fundamental theoretical and experimental crystal growth studies.
2. Laboratory crystal growth for preparing research samples.
3. Industrial fabrication of single crystals and their machining and characterization.
4. Fabrication of metallic/dendritic crystals (e.g. turbine blades).
5. Mass crystallization (salt, sugar, chemicals).

Epitaxy has similarly separated from bulk crystal growth and has split into:

1. Fundamental/theoretical and experimental epitaxy studies and surface phenomena.
2. Epitaxy growth methods (VPE, MOCVD/MOVPE, MBE, ALE, LPE).
3. Materials classes (GaAs, InP, GaN, II-VI compounds, high-T_c superconductors, low-dimensional/quantum-well/quantum-dot structures, etc.).

As a consequence of these separations and splits, numerous specialized conferences, workshops and schools are established and regularly organized, new specialized journals appear, and the common language gets more and more lost, even to the extent that formerly well-defined terms are misused.

The situation is critical in research and development involving crystal growth and epitaxy. Due to a worldwide lack of education in these areas, frequently growth methods are employed for problems for which they are not well-suited or even not suited at all. It is postulated (Scheel 2002) that for a specific crystal of defined size and perfection there is only

one optimum and economic growth technology

considering thermodynamics, growth kinetics, and economic factors. This aspect is schematically shown in Figure 1.5, where the crystal performance, the crystal size, the efficiency of the growth method, and the price of crystal are shown. The shape factor may play a role when, for instance, special shapes are needed in the application, for example square, rectangular or oval watch faces of sapphire. The concept of one or maximum two optimum technologies should be considered for industrial crystal fabrication, whereas in laboratories the existing equipment or experience may determine a nonoptimum growth method. This postulate holds obviously also for epitaxial methods: for example, high-T_c compounds have been prepared by >1000 groups (after the discovery of high-temperature superconductivity, HTSC) by vapour-phase epitaxy, which cannot yield the atomically flat surfaces required for reliable HTSC tunnel-device technology. Instead only liquid

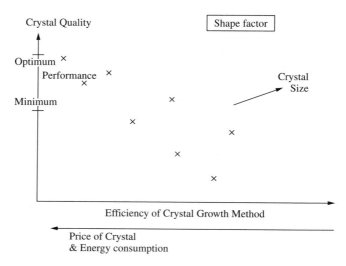

Figure 1.5 Schematic diagram of the efficiencies of various crystal-growth methods and the achieved crystal sizes and crystal quality. Note that there is normally only one optimum economic technology for the required crystal performance and size.

phase epitaxy near thermodynamic equilibrium can give atomically flat surfaces, as discussed elsewhere in this volume. Here, it could be added that the lack of recognition of the materials problems and specifically of the crystal growth and epitaxy problems have caused tremendous economic (and intellectual) losses not only in industries but also in research. As a recent prominent example, HTSC as a whole could not be properly developed, neither its physical understanding nor its many predicted applications because the chemical and structural complexity of the not very stable HTSC compounds were neglected. Most physical HTSC measurements were nonreproducible because the used crystals and layers were not sufficiently characterized for those structural and chemical defects that have an impact on these measurements or applications.

We realize that 'education' is needed for the important field of crystal-growth technology in order to achieve efficient developments in novel high technologies, in energy saving, in solar energy generation, in future laser-fusion energy and with a better approach in HTSC: in all these areas where crystal-growth technology is the progress-determining factor (Scheel 2002). However, in view of the multidisciplinary nature of crystal growth technology it will be not easy to initiate a curriculum in a classical university with the rather strict separation of disciplines in faculties or departments. In addition, the specialized education in CGT can start only after the multidisciplinary basic education and requires additional courses of 1 to 2 years. Furthermore, there are no teachers for this field as most crystal growers are specialized and experts in specific materials or specific methods only. In addition education in crystal growth technology and

Table 1.1   Complexity of crystal growth technology CGT

---

a. Multi-Disciplinary
- Chemistry (all fields) including Chemical Engineering
- Materials Science & Engineering
- Mechanical & Electrical Engineering (especially hydrodynamics, machine design, process control)
- Theoretical Physics (especially thermodynamics, non-equilibrium thermodynamics, statistical mechanics)
- Applied Crystallography and Crystal Chemistry
- Solid-State Physics

b. Complexity
- Phase Transformation from Fluid (melt, solution, vapour) to Crystal
- Scaling Problem: Control of surface on nm scale in growth system of $\sim$m size, hampers numerical simulation
- Complex Structure & Phenomena in Melts and Solutions
- Multi-Parameter Processes: Optimize and compromise $\sim$10 parameters

---

There is only one Optimum and Economic Growth Technology for a Specific Size and Performance of Crystal or Crystalline Layer!

epitaxy technology should be complemented by practical work and by periods in industries: another obstacle where, however, interested industries will collaborate. This practical work is needed in order to understand and appreciate the complexity (see Table 1.1) of crystal-growth processes in view of the involved phase transformation, of the many parameters that have to be compromised and optimized, and in view of the scaling problem, which hampers realistic numerical simulation.

1.4   CONCLUSION

Crystal-growth technology (CGT) including epilayer fabrication technology is of greatest importance for energy saving, for renewable energy, and for novel high technologies. Therefore the education of CGT engineers and CGT scientists is needed as well as significant support and publicity. Excellent students with intuition and motivation can have a great impact in crystal-growth technology and thus contribute to the energy problem and to high-technology developments of mankind.

REFERENCES

Buckley H. E. (1951), *Crystal Growth*, Wiley, New York.
Burton J. A., Prim R. C. and Slichter W. P. (1953), *J. Chem. Phys.* **21**, 1987.
Capper P. (1994), *Prog. Crystal Growth Charact.* **28**, 1.

Capper P. (2000), at second International School on Crystal Growth Technology ISCGT-2 Mount Zao, Japan August 24–29, 2000 (book in preparation).

Elwell D. and Scheel H. J. (1975), *Crystal Growth from High-Temperature Solutions*, Academic Press, London- New York, (reprint foreseen 2003).

Horowitz A., Gazit D., Makovsky J. and Ben-Dor L. (1983), *J. Cryst. Growth* **61**, 323 and Horowitz A., Goldstein M. and Horowitz Y., *J. Cryst. Growth* **61**, 317.

Hurle D. T. J. (1987), *J. Cryst. Growth* **85**, 1.

Ivantsov G. P. (1951, 1952), *Dokl. Akad. Nauk SSSR* **81**, 179; **83**, 573.

Nassau K. (1969), *J. Cryst. Growth* **5**, 338.

Nassau K. and Nassau J. (1971), *Lapidary J.* **24**, 1284, 1442, 1524.

Rytz D. and Scheel H. J. (1982), *J. Cryst. Growth* **59**, 468.

Scheel H. J. (1972), *J. Cryst. Growth* **13/14**, 560.

Scheel H. J. (1993/1994), *Historical Introduction*, Chapter 1 in *Handbook of Crystal Growth*, editor D. T. J. Hurle, Vol. 1 Elsevier, Amsterdam.

Scheel H. J. (1995), US Patent No. 5471943.

Scheel H. J. (2002), Plenary lecture at Second Asian Conference on Crystal Growth and Crystal Technology CGCT-2 August 28–31, 2002 in Seoul, Korea, (proceedings in preparation).

Scheel H. J. and Elwell D. (1972), *J. Cryst. Growth* **12**, 153.

Scheel H. J. and Sommerauer J. (1983), *J. Cryst. Growth* **62**, 291.

Scheel H. J. and Swendsen R. H. (2001), *J. Cryst. Growth* **233**, 609.

Tiller W. A., Jackson K. A., Rutter J. W. and Chalmers B. (1953), *Acta Met.* **1**, 428.

van Erk W. (1982), *J. Cryst. Growth* **57**, 71.

Völkl J. and Müller G. (1989), *J. Cryst. Growth* **97**, 136.

Wilke K.-Th. and Bohm J. (1988) *Kristallzüchtung*, VEB Deutscher Verlag der Wissenschaften, Berlin.

2 Thermodynamic Fundamentals of Phase Transitions Applied to Crystal Growth Processes

PETER RUDOLPH

Institut für Kristallzüchtung, Rudower Chaussee 6, 12489 Berlin, Germany

2.1 INTRODUCTION

Thermodynamics is an important practical tool for crystal growers. It helps to derive the most effective phase transition, i.e. growth method, and the value of the driving force of crystallization. From thermodynamic principles one can estimate the nucleation and existence conditions of a given crystalline phase, the width of compound homogeneity regions, and optimize the in-situ control of the crystal composition during the growth. In a word, no technological optimum can be found without considering thermodynamic relationships.

In general, crystal growth involves first-order phase transitions. This means there is the coexistence of two distinct uniform phases that are stable at the equilibrium point and separated by a phase boundary, i.e. an interface. Close to the equilibrium point the phases can still exist, one as thermodynamically stable, the other as the thermodynamically metastable phase, whereas the metastable phase is supersaturated (supercooled) with respect to the stable (equilibrium) phase. As a result, a thermodynamic driving force of crystallization will appear leading at a critical value of supersaturation to spontaneous nucleation of the crystalline phase within the metastable fluid phase. A controlled propagation of the solid/fluid phase boundary, however, takes place by providing a seed crystal or a substrate in contact with the fluid phase.

Classic thermodynamics (Guggenheim 1967, Wallace 1972, Lupis 1983) is concerned with macroscopic equilibrium states of quasiclosed systems. Such an approach for crystal growth is allowed due to the slow time scale of macroscopic processes compared with the kinetics of atoms, and due to the relatively small deviations from equilibrium. In order to describe nonequilibrium processes of quasiopen crystallization systems, characterized by continuous flows of heat and matter (i.e. entropy production), one uses linear nonequilibrium thermodynamics (de Groot and Mazur 1969, van der Erden 1993). In this work, however, we will treat only the fundamentals of classic thermodynamics.

Crystal Growth Technology, Edited by H. J. Scheel and T. Fukuda
© 2003 John Wiley & Sons, Ltd.   ISBN: 0-471-49059-8

2.2 PERFECT AND REAL STRUCTURE OF GROWN CRYSTALS

2.2.1 THE PRINCIPLE OF GIBBS FREE ENERGY MINIMIZATION

It is well known that all thermodynamic processes strive to minimize the free energy. Applied to the crystallization process this means that the single-crystalline state is a normal one because the free thermodynamic potential G (free potential of Gibbs) is minimal if the 'crystal growth units' (atoms, molecules) are perfectly packed in a three-dimensional ordered crystal structure, i.e. the atomic bonds are saturated regularly. Because the sum of the atomic bonds yields de facto the potential part H, the so-called enthalpy, of the internal crystal energy $U = H - PV$ (P = pressure, and V = volume) the process of ordering is characterized by the minimization of enthalpy ($H \rightarrow$ min.).

On the other side, however, an ideally ordered crystalline state would imply an impossible minimal entropy S. Thus, the opposite process of increasing entropy, i.e. disordering ($S \rightarrow$ max.) gains relevance with increasing temperature T. This is expressed by the basic equation of the thermodynamic potential of Gibbs

$$G = U + PV - TS = H(\downarrow) - TS(\uparrow) \rightarrow \min \qquad (2.1)$$

In fact, the crystallization is composed of two opposite processes – (i) regular, and (ii) defective incorporation of the 'growth units', at which the second contribution increases exponentially with T and amounts at $T = 1000$ K to about 10^{-3} % (see Section 2.2.2).

The fundamental principle of energy minimization is also used to determine the composition of the vapour phase in equilibrium with a melt and a growing crystal in multicomponent systems. This is of the highest importance for the control of the melt chemistry and for the incorporation of foreign atoms into the crystal. For instance, in a Czochralski growth chamber the inert gas atmosphere may consist of numerous additional species produced by chemical reactions of volatile melt components with sublimates of the technological materials present (heater, crucible, insulator, encapsulant, etc.). Usually, numerous complex thermochemical reactions need to be considered. It is the aim of the material scientist, however, to determine the most stable reaction products having the lowest formation energy at given growth conditions and, thus, the highest probability of incorporation in the melt and crystal. Today, one can use powerful computer packages based on the principle of Gibbs free energy minimization (e.g. ChemSage, see Eriksson and Hack 1990, for example) in combination with thermodynamic material data bases (SGTE 1996). Examples of the thermochemical analysis of the LEC GaAs growth system are given by Oates and Wenzl (1998) and Korb et al. (1999), for example.

2.2.2   EQUILIBRIUM POINT-DEFECT CONCENTRATION

Considering the dialectics of ordering and disordering forces from Section 2.2.1 it is not possible to grow an absolutely perfect crystal. In reality 'no ideal' but only an 'optimal' crystalline state can be obtained. With other words, in thermodynamic equilibrium the crystal perfection is limited by incorporation of a given concentration of point defects n (as is well known, only point defects are able to exist in an equilibrium state).

Neglecting any effects of volume change, defect type and defect interplay at constant pressure the equilibrium defect concentration n can be determined from the change of thermodynamic potential by introducing the defect as

$$\Delta G = \Delta H_d - \Delta S_d T \rightarrow \min \qquad (2.2)$$

with $\Delta H_d = n E_d$ the change of internal energy at incorporation of n defects, having a defect formation energy E_d, and $\Delta S_d = k \ln(N!)/[n!(N-n)!]$ the accompanying change of entropy (configurational entropy) with k the Boltzmann constant and N the total number of possible sites. After substitution and application of Stirling's approximation for multiparticle systems like a crystal [$\ln N! \approx N \ln N, \ln n! \approx n \ln n, \ln(N-n)! \approx (N-n) \ln(N-n)$] Equation (2.2) becomes

$$\Delta G = n E_d - kT[N \ln N - n \ln n - (N-n) \ln(N-n)] \qquad (2.3)$$

Setting the 1st derivation of Equation (2.3) $\partial \Delta G/\partial n = 0$, which equals the energetical minimum defect concentration $n_{\min}$, and considering $N \gg n$ the 'perfection limit' of a crystal is

$$n_{\min} = N \exp(-E_d/kT) \qquad (2.4)$$

exponentially increasing with temperature.

Using $N \approx N_A$(Avogadro's constant) $= 6 \times 10^{23}\,\mathrm{mol}^{-1}$ and $E_d = 1\,\mathrm{eV}$ (vacancy formation energy in copper, for example) the minimum defect concentrations $n_{\min}$ at 1000 K and 300 K are about $6 \times 10^{18}\,\mathrm{mol}^{-1}$ and $10^7\,\mathrm{mol}^{-1}$, respectively.

Figures 2.1 and 2.2 show the $G(n)$ and $n(T)$ functions schematically. Due to the limitation of diffusion rate a certain fraction of high-temperature defects freezes in during the cooling-down process of as-grown crystals (broken lines) and exceeds the equilibrium concentration at room temperature markedly (Figure 2.2). In other words, in practical cases the intrinsic point-defect concentration is still far from thermodynamic equilibrium. Note, in the case of formation of vacancy-interstitial complexes (Frenkel defects) the value of $n_{\min}^{(F)}$ is somewhat modified and yields $\sqrt{n_{is}N} \exp(-E_d^{(F)}/kT)$ with n_{is} the total number of interstitial positions depending on the given crystal structure.

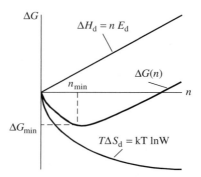

Figure 2.1 Schematic illustration of the equilibrium defect concentration ('perfection limit') n_{min} obtained by superposition of defect enthalpy ΔH_d and entropy ΔS_d using Equations (2.2)–(2.4). (Reprinted from P. Rudolph, Theoretical and Technological Aspects of Crystal Growth, 1998, copyright (1998) reprinted with permission from Trans. Tech. Publications Ltd.)

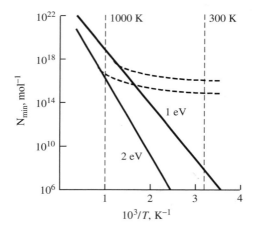

Figure 2.2 Minimum defect concentration vs. temperature at $E_d = 1\,eV$ (1) and $2\,eV$ (2) after Equation (2.4). Dashed lines: the 'freezing in' courses of high-temperature defects.

Hence, depending on the device requirements the technologist has to decide the necessary purification and equilibration of a given material. For instance, semiconductor laser diodes need high-doped material ($>10^{17}$ dopants per cm^3) allowing a residual point defect level of less than $10^{16}\,cm^{-3}$, but semi-insulating (SI) material for high-frequency devices requires the highest possible purity. Today, the standard level of residual contaminations in SI GaAs does not exceed $10^{14}\,cm^{-3}$, for example.

2.3   THERMODYNAMICS OF PHASE EQUILIBRIUM

2.3.1   THE PHASE TRANSITION

After Landau and Lifschitz (1962) crystal growth is characterized by a first-order phase transition involving two contacting phases (1 and 2) separated by an interface. Most growth methods are carried out near equilibrium (exceptions are vapour growth processes, especially MBE and MOCVD) so that the equilibrium state, as described by phase diagrams, should serve as a useful guide to the temperature and pressure conditions.

Two phases are in equilibrium when the Gibbs potentials of the phases (G_1 and G_2) are equal so that the potential difference between the phases is zero ($\Delta G = 0$). For a single component (e.g. Si, GaAs, LiNbO$_3$, where stoichiometric compounds are recorded as single components), Equation (2.1) at uniform temperature and pressure becomes

$$G_1 = U_1 - TS_1 + PV_1$$
$$G_2 = U_2 - TS_2 + PV_2 \qquad (2.5)$$
$$G_1 - G_2 = \Delta G = \Delta U - T\Delta S + P\Delta V = 0$$

where the change of internal energy can be expressed by $\Delta U = \Delta H - P\Delta V$. Figure 2.3 shows a three-dimensional sketch of two contacting phase planes with $G(T)$, $P(T)$ and $G(P)$ projections. Usually, a melt–solid phase transition involves relatively small volume change and hence $\Delta V \approx 0$. Then the free

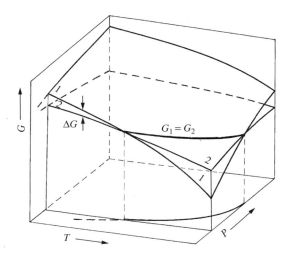

Figure 2.3   Two phase planes (1 and 2) crossing at the phase equilibrium line $G_1 = G_2$. Resulting G–T-, G–P-, and T–P-projections are demonstrated.

Helmholtz potential F can be used and at the phase equilibrium (index e) is

$$\Delta F = \Delta G_{V,P=\text{const}} = \Delta H - T_e \Delta S = 0 \qquad (2.6)$$

with ΔH the enthalpy released by crystallization (latent heat of fusion), ΔS the transition (melting) entropy and T_e the equilibrium temperature (melting point). Thus, at the phase transition the enthalpy, entropy, internal energy and volume change abruptly (see Equation (2.5)) and Equation (2.6) becomes $\Delta H = T_e \Delta S$. Table 2.1 shows the parameter changes of selected materials at vapour–solid, liquid–solid and solid–solid phase transitions.

This is of practical relevance for crystal growth. For instance, the growth-rate-dependent liberated heat participates in the thermal balance at the melt–solid phase transition according to

$$k_l \frac{dT_l}{dz} - k_s \frac{dT_s}{dz} = \frac{\Delta H}{c} v \qquad (2.7)$$

with $k_{l,s} = \lambda_{l,s}/(\rho_{l,s} c_{l,s})$ the thermal diffusivity (λ – heat conductivity, ρ – density, c – heat capacity), $dT_{l,s}/dz$ the temperature gradient along the growth axis z in the liquid (l) and solid (s) phase, and v the growth rate. In most cases, ΔH is conducted away through the growing crystal. For instance, at a typical value of $\Delta H/c \approx 10^3$ K (silicon) a crystallization velocity of $10\,\text{cm}\,\text{h}^{-1}$ releases the heat quantity $\Delta H v/c$ (right term in Equation (2.7)) of $10^4\,\text{K}\,\text{cm}\,\text{h}^{-1}$. Hence, a temperature gradient greater than $20\,\text{K}\,\text{cm}^{-1}$ is required to conduct this heat

Table 2.1 The abrupt change of thermodynamic parameters at first-order phase transitions of selected materials

Phase transition $(1 \rightarrow 2)$	Material $(T, P$ at transit. point)		Volume change $(V_1 - V_2)/V$ $= \Delta V/V (\%)$	Entropy change $S_1 - S_2 =$ ΔS (J/mol K)	Enthalpy change $H_1 - H_2 =$ ΔH (J/mol)
vapour–solid $(v \rightarrow s)$	Al	(723 K, 1 atm)	> -99.9	345	250×10^3
	Si	(1000 K, 1 atm)	> -99.9	304	304×10^3
liquid–solid $(l \rightarrow s)$	Al	(933 K)	-6.0	11	11×10^3
	Si	(1693 K)	$+9.6$	30	50×10^3
	GaAs	(1511 K)	$+10.7$	64	97×10^3
	ZnSe	(1733 K)	-19.6	7	13×10^3
	LiNbO$_3$	(1533 K)	-20.9	17	26×10^3
solid–solid $(s_1 \rightarrow s_2)$	BaTiO$_{3\,T \rightarrow C}$	(393 K)	0	0.5	0.2×10^3
	ZnSe$_{W \rightarrow ZB}$	(1698 K)	0	0.5	1.0×10^3
	CsCl$_{CsCl \rightarrow NaCl}$	(749 K)	-4	10	7.5×10^3
	C$_{GR \rightarrow D}$	(1200 K, 40 atm)	-37	4.8	1.2×10^3

through the growing crystal with thermal diffusivity of 5×10^2 cm² h⁻¹ (note no temperature gradient is assumed in the melt).

In the case of vapour growth the heat of vaporization is released during condensation (i.e. deposition) of the solid phase. For instance, in the layer-by-layer growth mode, assisted by a discontinuous two-dimensional nucleation mechanism, the liberated thermal impulse is $I_T = (\Delta H/c)\Delta z$ with Δz the crystallizing step width. Assuming a spacing between layers $\Delta z = 6.5 \times 10^{-8}$ cm and $\Delta H/c \approx 6 \times 10^3$ K (ZnSe at 1000 °C) the heat I_T of about 4×10^{-4} K cm per layer has to dissipate through the growing crystal (i.e. substrate) and ambient.

2.3.2 TWO-COMPONENT SYSTEMS WITH IDEAL AND REAL MIXING

In addition to the variables T, P, V in a material system consisting of two (or more) chemical components $i = A, B, C \ldots$ the free potential of Gibbs is also determined by the chemical phase composition, i.e. the component quantity n_i (number of gram-atomic weights, for example) so that $G = f(T, V, P, n_i)$. Hence, at constant T, V and P the change of the free Gibbs potential depends on the variation of n_i only and the partial derivative is named chemical potential μ_i

$$\frac{\partial G}{\partial n_i} = \mu_i \qquad (2.8)$$

In a solution, the composition can be expressed by the mole fraction x_i yielding, in a two-component system

$$x_A = \frac{n_A}{n_A + n_B}; \quad x_B = (1 - x_A) \qquad (2.9)$$

It proves to be convenient to write the chemical potential of a component i in a particular phase as

$$\mu_i = \mu_i^o(T, P) + RT \ln(\gamma_i x_i) \qquad (2.10)$$

with R the universal gas constant, γ_i the *activity coefficient* and μ_i^o the chemical *standard potential*. If the solution phase is a liquid, then μ_i^o is the Gibbs energy per gram-atom of pure liquid i. If the solution is a solid with a particular crystal structure α, then μ_i^o is the chemical potential of pure i with α crystal structure. For an ideal solution with completely mixing the activity coefficients are all equal to unity.

Using the dimensionless atomic fraction of the components x_i one has to translate the extensive thermodynamic functions of G, H and S in intensive (molar) values as $g = G/n$, $h = H/n$ and $s = S/n$. At fixed T, P and V Euler's Theorem can be applied to obtain the molar free Gibbs potential

$$g = \left.\frac{G}{n}\right|_{T,P,V} = \sum_{i=A}^{C} \mu_i \frac{n_i}{n} = \sum_{i=A}^{C} \mu_i x_i \qquad (2.11)$$

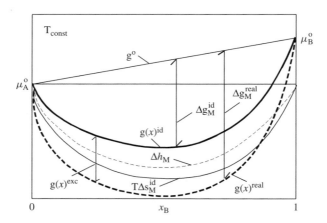

Figure 2.4 Energy composition curves for ideal $[g(x)^{id}$ with $\Delta h_M = 0]$ and real solutions $[g(x)^{real}$ with $h_M < 0]$. (Reprinted from P. Rudolph, Theoretical and Technological Aspects of Crystal Growth, 1998, copyright (1998) reprinted with permission from Trans. Tech. Publications Ltd.)

so that in a two-component phase the molar free potential depends on the composition as

$$g(x) = g^o + \Delta g_M = (\mu_A^o x_A + \mu_B^o x_B) + (\Delta h_M - T \Delta s_M) \qquad (2.12)$$

with g^o and g_M the standard and mixing term, respectively (Figure 2.4).

In the case of ideal mixing there is no interaction between the particles of the different constituting components ($\gamma_i = 1$), i.e. there is no mixing enthalpy ($\Delta h_M^{id} = 0$). However, due to the disordering by mixing, the configurational entropy can be enlarged in the mixture by the amount

$$\Delta s_M^{id} = -R(x_A \ln x_A + x_B \ln x_B) \qquad (2.13)$$

and the free molar mixing potential (bold-lined curve in Figure 2.4) is then

$$\Delta g_M^{id}(x) = -T \Delta s_M^{id}(x) = RT(x_A \ln x_A + x_B \ln x_B) \qquad (2.14)$$

However, only very few systems yields $\gamma_A \approx \gamma_B \approx 1$ over the whole composition (CCl_4-$SnCl_4$, chlorobenzene-bromobenzene, n-hexane-n-heptane; see Rosenberger 1979).

For real solutions ($\gamma_i \neq 1$) the term Δh_M is no longer more zero and deforms the $g(x)$ course as shown by the broken-lined curve in Figure 2.4. Now the real mixing potential consists of an ideal and a so-called excess term as

$$g(x)^{real} = g(x)^{id} + g(x)^{exc}$$
$$(g^o + RT \ln \gamma x) = (g^o + RT \ln x) + (RT \ln \gamma) \qquad (2.15)$$

In many cases the regular-solution model with randomly mixed alloy ($\Delta s_M^{exc} = 0$) can be applied. Then the variation of the excess potential is $\Delta g^{exc} = RT \ln \gamma = \Delta h_M^{exc} = \Omega x (1 - x)$ where Ω is the interaction parameter related in the regular solution model of a binary A–B system to the relative energies of AB, AA and BB bonds in the alloy. A positive interaction parameter indicates that AB bonds have a higher energy than the average of AA and BB and phase separation may occur. On the other side, a negative value of Ω indicates that A and B atoms are attracted to each other with a tendency of compound formation. Using the quasichemical equilibrium (QCE) model of Panish (1974) the interaction parameter can be expressed as

$$\Omega = Z N_A \left[E_{AB} - \tfrac{1}{2}(E_{AA} - E_{BB}) \right] \qquad (2.16)$$

with Z the coordination number, N_A the Avogadro number, and E_{ij} the bond energy between nearest neighbours. In the more physical delta-lattice-parameter (DLP) model of Stringfellow (1972) the enthalpy of mixing is related to the effect of composition on the total energy of the bonding electrons. In this case the interaction parameter depends on the lattice parameters of the constituents only

$$\Omega = 4.375 K (\Delta a)^2 / \overline{a}^{4.5} \qquad (2.17)$$

with Δa the lattice-parameter difference, $\overline{a}$ the average lattice parameter, and K an adjustable constant (being the same for all III-V alloys, for example). A major prediction of the DLP model is that the values of Ω are always greater than or equal to zero. This means in the regular-solution the solid alloys may cluster and phase-separate, but they will not order (Stringfellow 1997). However, ordering is also observed, especially in thin films of semiconductor alloys consisting of atoms of markedly different size. For such cases one has to apply the Hume-Rothery model (1963), which predicts in systems with large difference of atomic size the evolution of large macroscopic strain energies that drive short- and long-range ordering rather than clustering (see Section 2.3.3).

The following typical cases can be distinguished at liquid–solid crystallization, for example:

$$
\begin{array}{llll}
\gamma < 1 & \Delta h_M^{exc} < 0 & \Delta g(x)_s < \Delta g(x)_l & \text{system with } T_{max} \\
\gamma > 1 & \Delta h_M^{exc} > 0 & \Delta g(x)_s < \Delta g(x)_l & \text{system with } T_{min}
\end{array} \qquad (2.18)
$$

The extreme cases of both types lead to the appearance of intermediate phase (systems with compound; see Section 2.3.2) or eutectic composition, respectively. Further numerous transitive variants (i.e. systems with miscibility gap) are possible (see Lupis 1983 and Rosenberger 1979, for example).

2.3.3   PHASE BOUNDARIES AND SURFACES

The surface of a crystal is a cross discontinuity and has a free energy associated with it. The value of this free energy depends on the orientation of the face and on the other phase in contact (vacuum, liquid). After Brice (1987) a first approximation of the surface free energy is

$$\gamma = (1 - w/u)\Delta H N_A^{-1/3} \tag{2.19}$$

where u is the number of nearest neighbours for an atom inside the crystal and w is the number of nearest neighbours in the crystal for an atom on the surface ($w < u$). N_A is the Avogadro number and ΔH the enthalpy of fusion (i.e. vaporization). A plot of γ against face orientation is shown in Figure 2.5 schematically. Energetic minima appear at atomically smooth, so-called singular faces. Faces corresponding to less-sharp minima are not flat on the atomical scale. Some values of γ for solid/vapour, solid/liquid and liquid/vapour interfaces of selected materials are compiled in Table 2.2. The equilibrium shape of a crystal consists of a set of singular faces by using the principle of surface potential minimization (Wulff construction)

$$G = \sum_{(100)}^{hkl} \gamma_{hkl} A_{hkl} \to \min \tag{2.20}$$

with A the surface area and hkl indicating the crystallographic orientation. Hence, the change of the Gibbs potential is also determined by the surface-area variation, i.e. $\partial G/\partial A = \gamma$, and the complete thermodynamical function is now $G = f(T, P, V, n_i, A)$.

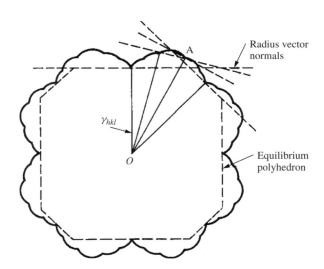

Figure 2.5 Wulff plot (bold contour) showing the variation of the free surface energy γ_{hkl} (radius vector) with orientation. The dotted contour gives the cross section of the equilibrium crystal shape.

Table 2.2 Surface tensions (γ_{lv}), solid/vapour and solid/liquid interface energies (γ_{sv}, γ_{sl}) of selected materials (in $J\,cm^{-2} \times 10^4$)

Material	γ_{sv}	(hkl)	(T)	γ_{lv}	γ_{sl}
Cu	1.70	(100)	(300 K)	1.35	0.14
Ag	1.20	(111)	(300 K)	0.93	0.11
Si	1.24	(111)	(300 K)	0.74	0.57
	1.08	(111)	(1683 K)		
Ge	0.76	(111)	(1209 K)	0.63	0.24
GaAs	0.50	(111)	(1511 K)	0.70	0.19
LiNbO$_3$	0.10	(01̄12)	(1533 K)	0.18	0.15

Such a theoretical estimation is of special interest for the overcritical nucleation state where first stable crystallites with equilibrium shape may exist. Usually, however, large real crystals contain numerous defects (dislocations, for example) and grow under specific kinetic conditions that modify the equilibrium shape.

The surface energy also plays a role for diameter control in crystal-pulling methods (Czochralski, floating zone, EFG). For instance, the equilibrium between the three interface energies γ_{sv}, γ_{sl} and γ_{lv} controls the capillary condition (stable meniscus height) by the characteristic growth angle (Tatartchenko 1994)

$$\psi_o = \arccos[(\gamma_{lv}^2 + \gamma_{sv}^2 - \gamma_{sl}^2)/2\gamma_{lv}\gamma_{sv}] \tag{2.21}$$

ψ_o values of 11° for silicon, 13° for germanium, 25° for indium antimonide and 17° for Al$_2$O$_3$ were reported (Hurle and Cockayne 1994). Hence, to pull a crystal with constant diameter in the vertical direction, the meniscus angle at the three-phase boundary (between the line tangent to the meniscus and the horizontal) of $\alpha_o = 90° - \psi_o$ needs to be held constant.

In systems of two (and more) components like solutions, the composition at the interface differs from that of the bulk (effect of adsorption). In this case the specific free interface energy depends on the concentration x and is a function of the chemical potential μ^i (Lupis 1983). If a solute is enriched at the interface then it is a surface (interface) active one ($\partial\gamma/\partial x < 0$), passive surface (interface) components ($\partial\gamma/\partial x > 0$) are rejected.

2.4 THERMODYNAMICS OF TOPICAL CRYSTAL GROWTH PROBLEMS

2.4.1 MIXED CRYSTALS WITH NEARLY IDEAL SOLID SOLUTION

Mixed bulk crystals are of increasing importance, especially as suitable substrates for epitaxial layers of pseudobinary semiconductor-alloys like Ge$_{1-x}$Si$_x$, In$_{1-x}$Ga$_x$As, Cd$_{1-x}$Zn$_x$Te, for example. In such cases, monocrystals with high compositional homogeneity (x uniformity) are required. On the other hand, mixed crystals with linear variation of the 'lattice constant' (delta-crystals) have enhanced mechanical stability (Ginzburg–Landau theory) and, after Smither (1982) are applicable as lenses in X-ray optics. The difficulty of this concept is to achieve the linear variation of the lattice constant with high accuracy and without loss of single crystallinity. In both areas, mixed crystals with nearly ideal solution behavior are under investigation.

An ideal solution means a system with complete miscibility in both the liquid and solid phase within the whole composition range $0 < x < 1$. In the low-pressure $T-x$-projection of such a phase diagram there are three fields, liquid, liquid + solid, and solid, separated by two boundaries known as the liquidus L and solidus S (Figure 2.6).

The liquidus and solidus lines of an ideal system can be calculated analytically. With respect to the component $i = B$ the equilibrium between solid and liquid phases is given by (compare Equations (2.8)–(2.10)

$$\mu_{Bs}(x_{Bs}, T) = \mu_{Bl}(x_{Bl}, T)$$

$$\mu_{Bs}^{o} + RT \ln x_{Bs}\gamma_{Bs} = \mu_{Bl}^{o} + RT \ln x_{Bl}\gamma_{Bl} \qquad (2.22)$$

$$RT \ln \left(\frac{x_{Bs}\gamma_{Bs}}{x_{Bl}\gamma_{Bl}} \right) = \Delta\mu_{B}^{o} = \Delta h_{B}^{o} - T\Delta s_{B}^{o} = 0$$

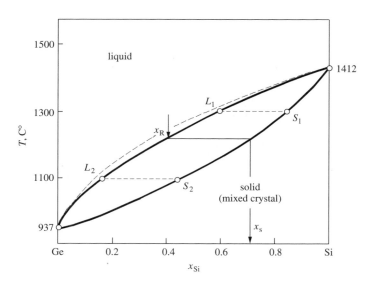

Figure 2.6 $T-x$-Projection of the near ideal system Ge–Si. Full lines were calculated, dashed line is experimental. The equilibrium distribution coefficient k_{o} is given by x_{s}/x_{l} with subscripts s, l for solid and liquid phase, respectively. (Reprinted from P. Rudolph, Theoretical and Technological Aspects of Crystal Growth, 1998, copyright (1998) reprinted with permission from Trans. Tech. Publications Ltd.)

with h_B^o and $s_B^o = h_B^o/T_{mB}$ the intensive standard enthalpy and entropy of the pure component B with melting temperature T_{mB}. Hence, Equation (2.22) becomes

$$\ln\left(\frac{x_{Bs}\gamma_{Bs}}{x_{Bl}\gamma_{Bl}}\right) = -\frac{\Delta h_B^o}{R}\left(\frac{1}{T} - \frac{1}{T_{mB}}\right) \tag{2.23}$$

Since the solutions are ideal, the activities $\gamma_{Bs,l}$ may be replaced by mole fractions and Equation (2.23) yields the *van Laar equation*

$$\frac{x_{Bs}}{x_{Bl}} = k_o = \exp\left[-\frac{\Delta h_B^o}{R}\left(\frac{1}{T} - \frac{1}{T_{mB}}\right)\right] \tag{2.24}$$

with k_o the (thermodynamic) *equilibrium distribution coefficient* of component B (in the case of Ge–Si is $k_{oSi} > 1$, as demonstrated in Figure 2.6). Using $x_{As,l} + x_{Bs,l} = 1$ the equation of the *solidus* is

$$x_{As}\exp\left[-\frac{\Delta h_A^o}{R}\left(\frac{1}{T} - \frac{1}{T_{mA}}\right)\right] + x_{Bs}\exp\left[-\frac{\Delta h_B^o}{R}\left(\frac{1}{T} - \frac{1}{T_{mB}}\right)\right] = 1 \tag{2.25}$$

and, the equation of the *liquidus* is

$$x_{Al}\exp\left[\frac{\Delta h_A^o}{R}\left(\frac{1}{T} - \frac{1}{T_{mA}}\right)\right] + x_{Bl}\exp\left[\frac{\Delta h_B^o}{R}\left(\frac{1}{T} - \frac{1}{T_{mB}}\right)\right] = 1 \tag{2.26}$$

Figure 2.6 shows the T–x-projection of the Ge–Si phase diagram calculated after Equations (2.25) and (2.26) using $\Delta h_{Ge}^o = 32\,kJ\,mol^{-1}$, $\Delta h_{Si}^o = 50\,kJ\,mol^{-1}$, $T_{mGe} = 1210\,K$ and $T_{mSi} = 1685\,K$. Since the experimental and calculated curves agree quite well, one can say the Ge–Si system exhibits nearly ideal mixing behavior. This is also reflected in the near linear variation of the lattice constant with composition (Vegard's rule). At present, one is interested to obtain (i) high-quality $Ge_{1-x}Si_x$ substrates with fixed homogeneous composition for homoepitaxy of Si–Ge MQWs expanding the area of Si technology (high-speed bipolar transistors, photodetectors), (ii) high-efficiency solar cells, and (iii) crystals with linear composition gradient, i.e. linear gradient of the lattice constant, for application in X-ray focusing systems. Looking at the phase diagram one can show that for homogeneous compositions a diffusion-controlled growth regime with $k_{eff} = 1$ is favored (Figure 2.6). Further successful measures are the double-crucible Czochralski technique and conventional Czochralski growth with in-situ recharging (Abrosimov *et al.* 1997). For growth of 'delta' crystals one has to find a well-defined program of monotonous variation of k_{eff} from k_o to 1, e.g. by convection damping applying an increasing magnetic field (as was proposed by Johnson and Tiller 1961, 1962), or one uses the travelling-solvent-zone growth in a positive temperature gradient (Penzel *et al.* 1997). However, due to the concave solidus (Figure 2.6) a slightly concaver axial temperature gradient needs to be established during the zone run.

The growth of ternary bulk mixed crystals ($Ga_{1-x}As_xP$ and $Ga_{1-x}In_xAs$) by using of the liquid-encapsulated Czochralski technique was reported by Bonner (1994). These pseudobinary III-V materials form solid solutions with significant separation between liquidus and solidus. This results in strong compositional segregation, associated supercooling effects, and potential for interface break-down resulting in polycrystalline growth. The problem was solved by intensive mechanical stirring of the melt. Also, $InP_{1-x}As_x$ ternary monocrystals with diameter of 2–3 cm, length up to 8 cm and x values up to 0.14 have been grown by the same method. With improved radial composition uniformity these crystals still exhibit a moderate x variation along the length of the boule due to the separation between liquidus and solidus lines. As a result, individual wafers cut normal to the growth direction of a given crystal may be of uniform composition, but there are still wafer-to-wafer composition variations that have to be eliminated in future (Bonner 1997). The first encouraging R&D results of MQW epitaxy on lattice-matched pseudobinary III-V substrates demonstrate the potential for a number of devices, particularly for 1.3-μm laser structures grown on $InGa_{1-x}As_x$ (Shoji *et al.* 1997).

2.4.2 SYSTEMS WITH COMPOUND FORMATION

2.4.2.1 Thermodynamic fundamentals

The tendency to form intermediate phases arises from the strong attractive forces between unlike atoms ($\gamma < 1$, $\Delta h_M^{exc} < 0$, $\Delta g(x)_s < \Delta g(x)_l$; see Equation (2.18)). In intermediate phases the bonds have a much stronger ionic or covalent character than in the metallic ordered solution. Therefore, the free energy, i.e. entropy and enthalpy, of such phases are typically low leading to their high stability. The following chemical reaction takes place

$$mA_{(s)} + nB_{(s)} = A_mB_{n(s)} \qquad (2.27)$$

As example, Figure 2.7 shows the T–x-projection of the Ga–As system with GaAs compound at $m = n = 1$ after Wenzl *et al.* (1993). Due to the relatively narrow stability region the compound is shown as a single vertical line of exact stoichiometry. If the solid and liquid phases are of identical composition in equilibrium at a given melt temperature, the compound melts congruently (as in the case of GaAs with a congruent melting point at 1513 K).

In compounds, two elements A and B mix to form an ordered structure where the atoms occupy well-defined lattice sites. Such completely ordered structure exists only if the ratio of the numbers of atoms A and B is equal to the ratio of relatively small integers ν_A and ν_B. The mixture is then designated by the formula $A_{\nu A}B_{\nu B}$. Hence, the compounds AB or A_2B are related to $\nu_A = \nu_B$ or $\nu_A = 2\nu_B$, respectively. Deviation from such formulas ($\nu \pm \delta$) corresponds to nonstoichiometric compositions with misarrangements or vacancies. As discussed

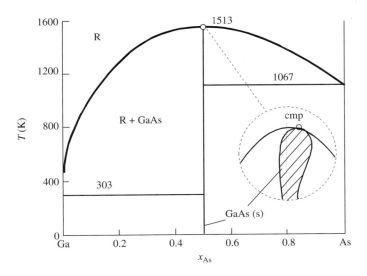

Figure 2.7 The T–x-projection of the Ga–As phase diagram with congruent melting point (cmp) on the arsenic-rich side after Wenzl and Oates (1993).

already in Section 2.2.2, at $T > 0$ K, in thermodynamic equilibrium the materials always contain a certain number of intrinsic point defects. This follows also from the plot of the free energy curves vs. composition (Figure 2.8) where the narrowness and steepness of the compound branch g_{AB} as well as the minimum positions of the g_A and g_B branches indicate the deviation from stoichiometry by the tangent intersection points. The region between these points (bold lines in Figure 2.8) is called the *range of existence* or *region of homogeneity*.

If its width is very small, i.e. the two tangents $A - A_A B_B$ and $A_A B_B - B$ touch the AB curve close to the minimum (Figure 2.8(a)), then its structure is referred to as a stoichiometric compound. If the range of stability is not negligible, i.e. the tangent contact points are markedly separated, the phase corresponds to a non-stoichiometric compound (Figure 2.8(b)). The width of the homogeneity region increases with T and depends on the formation enthalpies of nonstoichiometry-related defects.

Table 2.3 compiles some $\Delta\delta$ of selected compounds. Near stoichiometric situations with widths between 10^{-5} to 10^{-2} mole fraction are typical for semiconductor compounds with typical enthalpies of point defects between 1 and 4 eV.

2.4.2.2 Nonstoichiometry-correlated growth defects

Today, the deviation from stoichiometry is a key problem in growth of compound crystals. The side and degree of deviation influences the intrinsic point

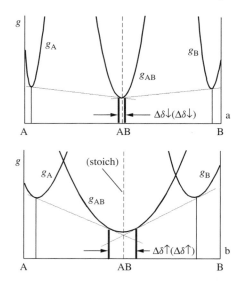

Figure 2.8 Thermodynamic derivation of the width of existence region ($\Delta\delta$) using the tangent construction of (a) near-stoichiometric compound AB and (b) nonstoichiometric compound AB.

Table 2.3 Maximum widths of the existence region $\Delta\delta$ of selected compound materials

Compound	GaAs	Cr$_3$S$_4$	CdTe	PbTe	MnSb	Fe$_3$O$_4$	Cu$_2$S	LiNbO$_3$
$\Delta\delta$ (mole fraction)	~10^{-4}	10^{-4}	3×10^{-4}	10^{-3}	10^{-3}	3×10^{-3}	3×10^{-2}	5×10^{-2}

defect concentration, i.e. electrical and optical parameters of such crystals, very sensitively. Thus, the control of stoichiometry during the growth process or by post-growth annealing is important. But in certain cases nonstoichiometry is required. For instance, the prerequisite for semi-insulating behavior of GaAs is a critical As excess. Therefore, such crystals are exclusively obtained from As-rich melt resulting in off-stoichiometric, i.e. arsenic-rich compositions and As-antisite defects known as double donors EL2, which are of crucial importance for compensation by doping control.

The generation of stoichiometry-related structural defects will be explained by the $T-x$ phase projection of the system Cd–Te. CdTe single crystals are of interest for photo-refractive devices and X-ray detectors. Figure 2.9 shows the region of homogeneity of this compound after Rudolph (1994). Three coupled growth problems can arise if the melt composition is not controlled: (i) the incongruent evaporation of the melt at the congruent melting point, therefore, (ii) segregation effects (i.e. constitutional supercooling) from the excess component on both sides

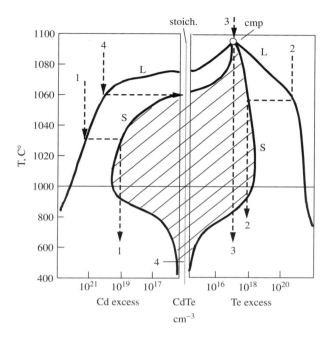

Figure 2.9 Sketched $T-x$-phase projection with region of homogeneity (hatched) of CdTe demonstrating the genesis paths of inclusions and precipitates in dependence on deviation from stoichiometry in the melt. 1,2 – formation of Cd (1) or Te (2) inclusions and precipitates; 3,4, – growth conditions of near inclusion- and precipitate-free solid composition, respectively (after Rudolph 1994).

of the congruent melting point, and (iii) the precipitation of the excess component during cooling of the as-grown crystal due to the retrograde behavior of the solidus lines. Here, one should distinguish between precipitates, generated by point-defect condensation in the solid, and inclusions that are formed by melt droplet capture from the diffusion boundary layer in front of the growing interface and enriched by the excess component (case (i)).

Second-phase precipitates and inclusions are common defects in melt-grown compound semiconductors (II-VI, III-V, IV-VI) and in several oxide systems. A sensitive dependence of the density of inclusions and also of the carrier concentration on deviation from stoichiometry was detected by Rudolph et al. (1995a) and is demonstrated in Figure 2.10. Under microgravity conditions an increased content of Te inclusions has been observed in CdTe due to the absence of convection in the melt. Variable magnetic fields (Salk et al. 1994) or accelerated crucible rotation techniques (Capper 1994, adapted from Scheel and Schulz-DuBois 1971) were used to prevent the enrichment at the interface and inclusion capture. Also post-growth annealing of bulk and wafers in active gas atmosphere (Cd saturated) was used by Ard et al. (1988) to remove inclusions. However, in this case

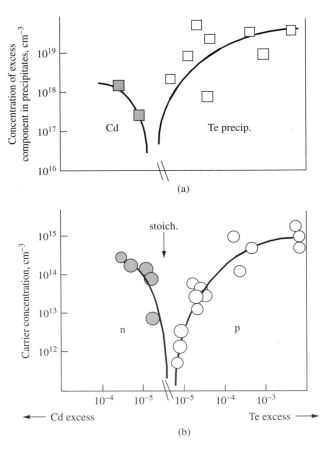

Figure 2.10 Concentration of excess component measured in (a) second-phase particles and (b) as carrier concentration vs. deviation from stoichiometry in CdTe. The results were obtained by vertical and horizontal Bridgman growth (after Rudolph *et al.* 1995).

the dissolution process of inclusions during annealing can promote the release of residual impurities by diffusion. Moreover, mechanical stress and dislocation multiplication at the site of inclusions were reported. Therefore, the best way is the in-situ control of stoichiometry during the crystal growth process.

2.4.2.3 The use of the $P-T-x$ phase diagram to control the stoichiometry

As can be seen from Figure 2.9 (path 4) the growth of stoichiometric crystals requires an off-stoichiometric melt composition (in CdTe and GaAs a Cd- or Ga-rich melt, respectively). In order to achieve constant conditions the melt composition with a given excess component must be kept constant during the whole

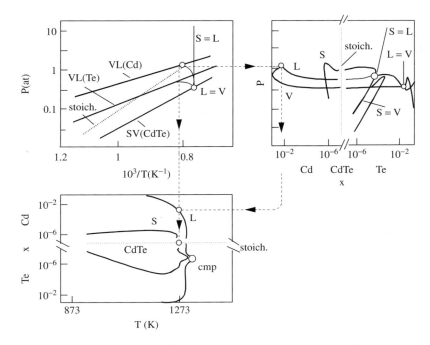

Figure 2.11 $P-T$-, $P-x$-, and $T-x$-projections of the system Cd–Te in three-phase equilibrium. Dashed lines mark the $P_{Cd} = P_{CdTe}$ (melt) equilibrium pressure for the growth of near stoichiometric crystals from the melt. As can be seen a marked Cd-excess in the melt is required to fix stoichiometric growth. (Reprinted from P. Rudolph, Theoretical and Technological Aspects of Crystal Growth, 1998, copyright (1998) reprinted with permission from Trans. Tech. Publications Ltd.)

growth run by accurate control of the gas atmosphere (Nishizawa 1990). Hence, the use of the $p-T$ and $P-x$-projections in combination with the traditional $T-x$-projection of the phase diagram is necessary (Figure 2.11).

If both elements A and B are volatile, like in II-VI compounds, and the melt evaporates incongruently with vapour species A and B_2 (whereas $P_A > P_{B2}$), the expression for the temperature of an additional source of the A component is approximately given by Rudolph $et\ al.$ (1994)

$$T_A[K] = \frac{c_1}{c_2 - \log\left[\dfrac{a\,P_{AB}}{(1/x_1) + (a-1)}\right]} \qquad (2.28)$$

with P_{AB} – total pressure above the melt, $a = P_A/P_{B2}$ – relative volatility (ratio of partial pressures of components), x_1 – melt composition for stoichiometric growth, and $c_{1,2}$ – constants of the partial function $\log P_A(\text{atm}) = -c_1/T + c_2$. Rudolph (1994) and Rudolph $et\ al.$ (1994) introduced this formula to estimate the source temperature of Zn ($T_{Zn} = 1000\,^\circ$C) and Cd ($T_{Cd} = 850\,^\circ$C) for stoichiometry-controlled Bridgman growth of ZnSe ($a = 5$, $P_{ZnSe} = 2$ atm, $c_1 = -6678.4$, $c_2 = 5.491$) and CdTe ($a = 35$, $P_{CdTe} = 2$ atm, $c_1 = 5317$, $c_2 = 5.119$), respectively. A good agreement with experimental values ($T_{Zn}^{exp} = 986\,^\circ$C and $T_{Cd}^{exp} = 848\,^\circ$C) was found.

Today it is standard to grow well-controlled semiconductor crystals of III-V and II-VI compounds by using the vertical or horizontal Bridgman method with an in situ source of the volatile component (As at GaAs, Cd at CdTe, for example) the partial pressure (i.e. temperature) of which is selected after the $P-T$ and $P-x$ projections of the phase diagram.

2.4.3   COMPOSITIONAL MODULATION AND ORDERING IN MIXED SEMICONDUCTOR THIN FILMS

In the following, a phenomenon of surface thermodynamics occurring in modern thin-film technology will be discussed (see MRS Bulletin Vol. 22, No. 7, 1997). Almost all semiconductor devices are based on the control and manipulation of electrons and holes in thin-film structures. These structures are often produced in mixed semiconductor alloys ($Ga_{1-x}In_xAs$, $Ga_{1-x}In_xP$), which become increasingly important because the electronic properties can be tailored by varying the composition. In addition, the use of very thin alloys allows the production of special structures such as quantum wells with abrupt changes in bandgap energy. Due to the microscopic scale of the layer thickness the energetical surface, i.e. the interface state becomes dominant for atom arrangement within the film "volume". In fact, recent studies of epitaxial growth of mixed semiconductor alloys showed interesting materials-growth phenomena that are driven mainly by surface thermodynamics. The well-known effect of surface reconstruction causes an atomistic ordering of the alloy components within the thin layer. First, the discovery of this phenomenon was surprising because for many years it was believed that when two isovalent semiconductors are mixed, they will form a solid solution at high temperature or separated phases at low temperature, but they will never produce ordered atomic arrangements. It was assumed that the mixing enthalpy $\Delta h(x)$ of an alloy depends on its global composition x only and not on the microscopic arrangement of atoms. However, thermodynamic analysis (Stringfellow 1997, Zunger 1997) has shown that certain ordered three-dimensional atomic arrangements within the layer minimize the strain energy resulting from the large lattice-constant mismatch between the constituents, while random arrangements do not (see also Section 2.3.2). In particular, there are special ordered structures α that have lower energies than the random alloy of the same composition x, that is

$$\Delta h(\alpha, \text{ordered}) < \Delta h(x, \text{random}) \qquad (2.29)$$

where $\Delta h(\alpha, \text{ordered}) = E(\alpha, \text{ordered}) - xE(A) - (1-x)E(B)$ and $\Delta h(x, \text{random}) = E(x, \text{random}) - xE(A) - (1-x)E(B)$ with $E(\alpha, \text{ordered})$ the total energy of a given arrangement α of A and B atoms on a lattice with N sites, $x = N_B/N$, $E(A)$ and $E(B)$ the total energies of the constituent solids.

Thus, ordered and disordered configurations at the same composition x can have different excess enthalpy $\Delta h^{exc}(x)$. In particular, the thermodynamically stable arrangement near the surface will be qualitatively different from the thermodynamically stable arrangement in an infinite bulk solid. Obviously, the surface has an important role in ordering of very thin films.

The most frequently observed form for these alloys grown epitaxially on (001) oriented substrates is the rhombohedral CuPt structure, with ordering on one or two of the set of four {111} planes (MRS Bulletin Vol. 22, No. 7, 1997). In its turn the CuPt structure is stable only for the surface reconstruction that forms [110] rows of [110] group-V dimers on the (001) surface. These group-V-dimer rows form to reduce the energy due to the large number of dangling bonds at the surface.

Figure 2.12 shows the lowest-energy configuration of a (100) film of $Ga_{0.5}In_{0.5}P$ on a GaAs (100) substrate illustrating the relation between surface reconstruction, surface segregation, and subsurface ordering (MRS Bulletin Vol. 22, No. 7, 1997). When the reconstruction is $\beta(2 \times 4)$, In segregation and CuPt-B ordering (alternating {111} planes of In and Ga) exist. CuPt-B ordering implies that there is Ga under the dimer (site A) and In between dimer rows (site C).

This thermodynamically driven effect of ordering has extremely important practical consequences. An interesting order-induced property of thin-film alloys is the reduction of the bandgap energy compared to that of bulk alloys of the same composition (Stringfellow 1997), thus moving the wavelength further into

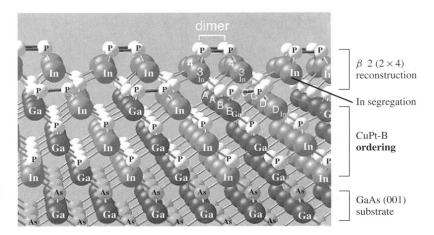

Figure 2.12   The lowest energy configuration of a (100)-oriented thin layer of $Ga_{0.5}In_{0.5}P$ on a GaAs (100)-substrate, illustrating the relation between surface reconstruction, surface segregation and subsurface ordering. (Reprinted from A. Zunger, MRS Bull. **22/7** (1997) cover image, copyright (1997) reprinted with permission from MRS Bulletin.)

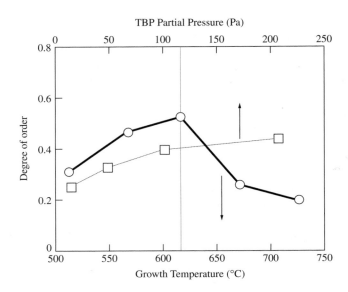

Figure 2.13 Degree of order as function of the growth temperature (for growth at a constant tertiarybutylphosphine (TBP) partial pressure of 50 Pa) and TBP partial pressure (for growth at constant temperature of 670 °C) in OMCVD-grown (Ga, In)P layers (after Stringfellow 1997).

the infrared. Hence, ordered III-V structures may become useful for IR detectors for which, so far, complicated II-VI materials ($Cd_{1-x}Hg_xTe$) were used. But for that the materials engineer must be able to control ordering over the entire surface of a wafer, to replace the spontaneous ordering by a precisely controlled one. Figure 2.13 shows a successful result to control the ordering in MOCVD grown $Ga_{1-x}In_xP$ layers as a function of the growth temperature after Stringfellow (1997).

2.5 DEVIATION FROM EQUILIBRIUM

2.5.1 DRIVING FORCE OF CRYSTALLIZATION

The prerequisite for crystallization of a stable solid phase within a metastable fluid phase is the deviation from the thermodynamic equilibrium, i.e. the crossing of coexistence conditions between the phases. In Figure 2.14 the $\mu(T)$ function in the neighborhood of a first-order phase transition is sketched. It can be seen, that μ (i.e. the Gibbs potential G) has a different dependence from the intensive variables in the two phases. Only at equilibrium are these potentials equal (Equation (2.5)). But away from equilibrium the potentials of the solid μ^s and fluid (mother phase) μ^f are different. The difference in temperatures

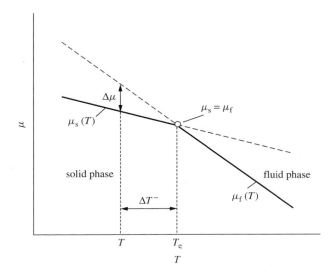

Figure 2.14 $\mu(T)$ function near the two-phase equilibrium. Solid lines: stable phases, dashed lines: metastable phases, T_e – equilibrium temperature, $\Delta\mu$ – driving force of crystallization. (Reprinted from P. Rudolph, Theoretical and Technological Aspects of Crystal Growth, 1998, copyright (1998) reprinted with permission from Trans. Tech. Publications Ltd.)

$\Delta T = T_e - T$ (Figure 2.14) denotes the *driving force of crystallization* or growth affinity as

$$\Delta\mu = \mu_f - \mu_s \qquad (2.30)$$

In practical operations, several terms are used. One differs between *supersaturation*, expressing the pressure or concentration relationships at vapour–solid or solution–solid transition, and temperature-related *supercooling* used mainly in the case of crystallization from melt. Most convenient expressions are

$$\Delta P = P - P_e \qquad \text{total supersaturation}$$
$$\Delta P/P_e = (S - 1) = \sigma \qquad \text{relative supersaturation} \qquad (2.31)$$
$$\Delta P/P_e \cdot 100 \ (\%) \qquad \text{percentage of relative supersaturation}$$

with P and P_e the actual and equilibrium pressure, respectively, and $S = P/P_e$. For small values of σ the energetic relation $\Delta\mu/kT = \ln(P/P_e) \approx \sigma$, i.e. $\ln S = \ln(1 + \sigma) \approx \sigma$, is used.

The supercooling at liquid–solid transition (l → s)

$$\Delta T = (T_m - T) \qquad (2.32)$$

is related to the driving force by

$$\mu_{l \to s} = \Delta h - T\Delta s = \Delta h - T(\Delta h / T_m) = (\Delta h / T_m)(T_m - T) = \Delta h \Delta T / T_m$$
(2.33)

where T – actual temperature, T_m – melting point, Δh, Δs – enthalpy and entropy of melting, respectively. The melt-growth processes are characterized by relatively low driving forces. Values of $\Delta h / T_m$ usually fall within the range of 10 to $100\,\text{J K}^{-1}\,\text{mol}^{-1}$, and values of ΔT are typically $1-5\,\text{K}$. Hence, $\Delta\mu_{l \to s}$ is about $10-100\,\text{J mol}^{-1}$.

In Figure 2.15 the metastable regions of supersaturation and supercooling in a sketched $P-T$-projection are demonstrated. As can be seen, below the triple point T_3 the v–s phase transition can begin with the unstable intermediate stage of fluid nuclei due to the smaller supersaturation (lower nucleation work) for generation of molten droplets than for solid nuclei. Only after that do the droplets translate into the stable crystalline phase. This phenomenon has been observed in various vapour-growth situations and is an example of *Ostwald's step rule*.

Comparing the various methods of epitaxy the relative supersaturation σ differs considerably. As 'rule of thumb', we have

liquid phase epitaxy (LPE):	σ_{LPE}	$\approx 0.02-0.1$
vapour phase epitaxy (VPE):	σ_{VPE}	$\approx 0.5-2$
molecular beam epitaxy (MBE):	σ_{MBE}	$\approx 10-100$
metal organic chemical vapour deposition (MOCVD):	σ_{MOCVD}	≈ 40

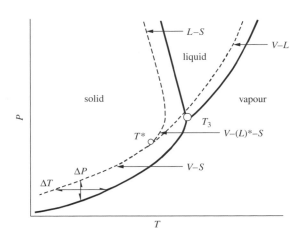

Figure 2.15 Schematic drawing of a $P-T$-projection with metastable regions of supersaturation ΔP and supercooling ΔT (dashed lines). T_3 – equilibrium triple point, T^* – intersection point between the metastable regions of vapour–solid (V $\to$ S) and vapour–liquid (V $\to$ L) transitions, i.e. non-equilibrium triple point.

Because at melt growth and LPE the crystals and layers are grown in near-equilibrium situations, the actual equilibrium states as described by the phase diagrams can be used as good approximation. At high deviations from equilibrium, however, the equilibrium phase diagrams are a rough tool only. For such cases, nonequilibrium phase diagrams considering additionally the growth kinetics are under theoretical investigations (Cherepanova and Dzelme 1981).

2.5.2   GROWTH MODE WITH TWO-DIMENSIONAL NUCLEATION

In the practice of bulk and thin-film crystal growth one starts with artificial seed crystals or single-crystalline substrates, respectively. Therefore the case of homogeneous three-dimensional nucleation in a metastable fluid phase, quite important for industrial mass crystallization, here omitted. In this work the special case of two-dimensional (2D) nucleation will be discussed which plays not only an important role in thin-film deposition processes (including homoepitaxy) but also reflects the bulk crystal growth at atomically flat phase boundaries. It is well known that after the 2D-nucleus is formed, the whole interface plane is then rapidly completed by lateral growth.

The change of the free Gibbs potential at the generation of a disc-shaped nucleus is expressed by the energy balance

$$\Delta G_{2D} = \Delta G_V + \Delta G_{IF}$$
$$\Delta G_{2D} = -\pi r^2 a \Delta\mu/\Omega + 2\pi r a\gamma \tag{2.34}$$

where ΔG_V associates with increasing volume of the nucleus (being negative due to the energy gain) and ΔG_{IF} denotes the energy required for increase of nucleus surface (r – radius of the nucleus disc, a – lattice constant, Ω – molecular volume, γ – surface energy, $\Delta\mu$ – driving force). The maximum, i.e. critical value of ΔG_{2D}, at which the nucleus is stable, can be obtained by $\partial(\Delta G)/\partial r = 0$. Then the critical of the nucleus radius becomes

$$r^* = \frac{\gamma\Omega}{\Delta\mu} \tag{2.35}$$

Combining Equation (2.34) with Equation (2.35) the critical nucleation energy becomes

$$\Delta G_{2D}^* = \frac{\pi\Omega\gamma^2 a}{\Delta\mu} \tag{2.36}$$

For instance, in the melt growth of the atomically smooth and dislocation-free {111} face of Si crystals a supercooling of $\Delta T \approx 4\,K$ was ascertained experimentally (Chernov 1985). Inserting this value in Equation (2.33) together with the heat of fusion $\Delta h = 50\,kJ\,mol^{-1}$ and melting point $T_m = 1693\,K$ a driving force of $\Delta\mu = (\Delta h/T_m)\Delta T = 118\,J\,mol^{-1}$ can be deduced. Taking from

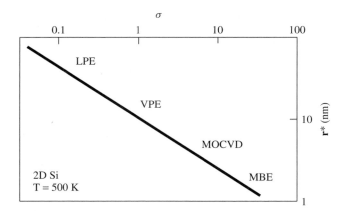

Figure 2.16   2-D nucleus radius vs. supersaturation for different thin-film deposition techniques estimated from Equation (2.35).

Table 2.2 the value of $\gamma_{ls} = 0.57 \times 10^{-4}\,\mathrm{J\,cm^{-2}}$ and for $\Omega = 12\,\mathrm{cm^3\,mol^{-1}}$ after Equation (2.35) the critical nucleus radius yields about 60 nm.

Figure 2.16 shows the critical radius of disc-shaped 2D-nucleus versus supersaturation for different methods of homoepitaxy of silicon at 500 K using Equation (2.35). As can be seen in the case of highly supersaturated growth from the vapour (MOCVD, MBE) the nucleus dimension yields only a few nm, comparable to a cluster consisting of a few atoms only.

Crystal growth from the melt occurs mainly under conditions of an atomistically rough interface (metals, semiconductors), which are growing without nucleation generation and, hence, under very small supercooling (i.e. driving force). In certain cases, however parts of the interface grow with atomically smooth interface-forming facets. On such face a discontinuous nucleation mechanism, as discussed above, takes place. As a result, in semiconductor compounds, twins can form (Hurle 1995). Furthermore typical sawtooth-like facet seams generate along the melt-grown boule surface, being a reliable indication of single crystallinity.

In epitaxial processes, where mainly atomically smooth interfaces occur, the growth mode of 2D-nucleation can be prevented by using slightly off-oriented substrates (few degrees) in order to achieve a continuous step-flow mechanism.

REFERENCES

Abrosimov, N. V., Rossolenko, S. N., Thieme, W., Gerhardt, A., Schröder, W. (1997), *J. Cryst. Growth* **174**, 182.

Ard, C. K., Jones, C. L., Clark, A. (1988), *Proc. SPIE* **639**, 123.

Bonner, W. A. (1997), *Compound Semiconductor July/August*, p. 32 [see also (1994), *Proc. SPIE 2228*, 33].

Brice, J. C. (1987), in: *Ullmann's Encyclopedia of Industrial Chemistry, Vol. A8*, VHC: Weinheim, p. 99.

Capper, P. (1994), *Prog. Cryst. Growth Charac.* **28**, 1.

Cherepanova, T. A., Dzelme, J. B. (1981), *Crystal Res. Technol* **16**, 399.

Chernov, A. A. (1985) in: *Crystal Growth of Electronic Materials*, Kaldis, E. (ed.). Elsevier: Amsterdam, p. 87.

de Groot, S. R., Mazur, P. (1969), *Non-Equilibrium Thermodynamics*, North-Holland: Amsterdam.

Eriksson, G., Hack, G. (1990), *Metall. Trans* **B12**, 1013.

Guggenheim, E. A. (1967), *Thermodynamics, 5th edition*, North-Holland: Amsterdam.

Hume-Rothery, W. (1963), *Electrons, Atoms, Metals, and Alloys, 3rd. edition*, Dover: New York.

Hurle, D. T. J., Cockayne, B. (1994), in: *Handbook of Crystal Growth, Vol. 2b*, Hurle, D. T. J. (ed.), North-Holland: Amsterdam, p. 99.

Hurle, D. T. J. (1995), *J. Cryst. Growth* **147**, 239.

Johnston, W. C., Tiller, W. A. (1961), *Trans AIME* **221**, 331 and **224**, 214.

Korb, J., Flade, T., Jurisch, M., Köhler, A., Reinhold, Th., Weinert, B. (1999), *J. Cryst. Growth* **198/199**, 343.

Landau, L. D., Lifschitz, E. M. (1962), *Statistical Physics*, Chap. 14, Pergamon Press: London.

Lupis, C. H. P. (1983), *Chemical Thermodynamics of Materials*, North-Holland: New York.

Nishizawa, J. (1990), *J. Cryst. Growth* **99**, 1.

Oates, W. A., Wenzl, H. (1998), *J. Cryst. Growth* **191**, 303.

Panish, M. B. (1974), *J. Cryst. Growth* **27**, 6.

Penzel, S., Kleessen, H., Neumann, W. (1997), *Cryst. Res. Technol*, **32**, 1137.

Rosenberger, F. (1979), *Fundamentals of Crystal Growth I*, Springer: Berlin.

Rudolph, P. (1994), *Prog. Cryst. Growth and Charac.* **29**, 275.

Rudolph, P., Rinas, U., Jacobs, K. (1994), *J. Cryst. Growth* **138**, 249.

Rudolph, P., Engel, A., Schentke, I., Grochocki, A. (1995a), *J. Cryst. Growth* **147**, 297.

Rudolph, P., Schäfer, N., Fukuda, T. (1995b), *Mater. Sci. Eng.* **R15**, 85.

Salk, M., Fiederle, M., Benz, K. W., Senchenkov, A. S., Egorov, A. V., Matioukhin, D. G. (1994), *J. Cryst. Growth* **138**, 161.

Scheel, H. J, Schulz-DuBois, E. O. (1971), *J. Cryst. Growth* **8**, 304.

SGTE *Pure Substance Database, Version 1996* and SGTE *Solution database, Version 1994*, distributor: GTT-Technologies mbH, Herzogenrath.

Shoji, H., Otsubo, K., Kusunoki, T., Suzuki, T., Uchida, T., Ishikawa, H. (1996), *Jpn. J. Appl. Phys.* **35**, L778. [see also Moore, A. (1997), *Compound Semiconductor July/August* 34].

Smither, R. K. (1982), *Rev. Sci. Instrum*, **53**, 131.

Stringfellow, G. A. (1972), *J. Phys. Chem. Solids* **33**, 665.

Stringfellow, G. B. (1997), *MRS Bull.* **22**, 27.

Tatartchenko, V. A. (1994), in: *Handbook of Crystal Growth, Vol. 2b*, Hurle, D. T. J. (ed.), North-Holland: Amsterdam, p. 1011.

van der Eerden, J. P. (1993), in: *Handbook of Crystal Growth*, *Vol. 2b*, Hurle, D. T. J. (ed.), North-Holland: Amsterdam, p. 307.

Wallace, D. C. (1972), *Thermodynamics of Crystals*, Wiley: New York.

Wenzl, H., Oates, W. A., Mika, K. (1993), in: *Handbook of Crystal Growth*, *Vol. 1a*, Hurle, D. T. J. (ed.), North-Holland: Amsterdam, p. 105.

Zunger, A. (1997), *MRS Bull.* **22**, 20.

3 Interface-kinetics-driven Facet Formation During Melt Growth of Oxide Crystals

SIMON BRANDON, ALEXANDER VIROZUB, YONGCAI LIU[1]

Department of Chemical Engineering, Technion-Israel Institute of Technology 32 000 Haifa, Israel

ABSTRACT

Facetting of crystals during their growth from the melt may lead to unwanted phenomena, adversely impacting the quality of the final crystalline product. In addition, substantial facet formation may significantly influence details of melt flow and other transport phenomena in the growth system. We present a computational method for combining the prediction of transport phenomena with interface-attachment-kinetics-driven facetting during the melt growth of single crystals; this method is applied in the analysis of vertical gradient solidification growth of oxide slabs. Calculations showing the sensitivity of facetting to changes in crystal-growth rate and furnace temperature gradient are reviewed. Results are compared with predictions based on a simple algebraic relation derived in the literature. In addition, the effects of melt convection and a transparent crystalline phase on facetting are examined. Melt flow is shown, in the bottom-seeded configuration, to enhance facetting at the center of the interface. Center facets on top-seeded crystals are, on the other hand, decreased in size due to the inverted melt flow pattern. A severely deflected, partially facetted melt/crystal interface in the case of a transparent crystalline phase demonstrates the difficulties associated with calculations of this type. Finally, semiqualitative agreement is achieved between facet positions predicted according to our calculations and those due to experiments shown in the literature. Discrepancies are most probably due to differences between the geometries studied (cylinders versus slabs) as well as inaccurate estimates of the degree of internal radiative transport within the crystalline phase.

[1] Current address: Department of Chemical Engineering, Tianjin University, Tianjin 300072, P. R. China.

Crystal Growth Technology, Edited by H. J. Scheel and T. Fukuda
© 2003 John Wiley & Sons, Ltd.   ISBN: 0-471-49059-8

3.1 INTRODUCTION

Facet formation during melt growth of single crystals is a phenomenon commonly seen in many systems. One may observe facets appearing on the melt/crystal interface as well as, in some cases (e.g. Czochralski (CZ) growth of several oxides), along the sides of the solidified crystalline material. Both anisotropy in interfacial-attachment kinetics and anisotropy in surface energy are the primary causes of facetting. In addition, transport phenomena within the growth system play a major role in controlling details of facet formation, size and position.

Understanding facetting phenomena, in particular understanding how to incorporate facetting into existing algorithms for modeling crystal-growth,[2] is of utmost importance. Achieving a capability for the prediction of facet formation is an important step towards learning how to control and minimize adverse effects resulting from facetting. Moreover, this approach is invaluable in systems involving severely facetted crystals; calculation of transport phenomena using existing algorithms may, in these systems, be erroneous due to inadequate prediction of the crystal geometry, which affects details of heat, mass and momentum transport.

Several problems related to facetting are recorded in the literature. The rapid step motion associated with growth of a facetted melt/crystal interface is often shown to be accompanied by nonequilibrium incorporation of a solute along the facet. Examples of this so, called 'facet effect' (Hurle and Cockayne, 1994) are numerous. Noteworthy is the case of Te incorporation during the CZ growth of InSb in the $\langle 111 \rangle$ direction, which was shown to exhibit an eight-fold difference in the effective segregation coefficient when comparing a (111) facet ($k_{eff} = 4.0$) to regions off the facet ($k_{eff} = 0.5$) (Hulme and Mullin, 1962). Of particular relevance to the results presented here is the facet effect in garnet crystals (Cockayne, 1968; Cockayne et al., 1973), which has been linked to the appearance of a low-quality strained inner core along the axis of certain CZ grown crystals. An additional adverse effect of facetting involves twin formation associated with facets connected to the three-phase tri-junction during CZ and encapsulated vertical bridgman (VB) growth of certain semiconductor materials (e.g. InP). Discussions and examples of this phenomenon are given in (Hurle, 1995; Amon et al., 1998; Chung et al., 1998) and references within.

In a recent publication (Liu et al., 1999) we presented a method for incorporating modeling of facet formation along the melt/crystal interface into existing algorithms for computation of transport phenomena during crystal-growth. This approach is strictly valid only in the case where facetting is caused by anisotropy of the interfacial attachment kinetics coefficient. The system studied in (Liu et al., 1999) loosely corresponds to the VB growth experiments of Nd:YAG reported in (Petrosyan and Bagdasarov, 1976). Facetting, in these experiments, was shown to exhibit a sensitivity to thermal gradients and crystal-growth velocity in a manner suggesting kinetics as the dominant cause of facet formation.

---

[2] Large-scale modeling of bulk crystal-growth (see, e.g. Dupret and van den Bogaert, 1994) has traditionally focused on transport phenomena while ignoring facetting effects.

Two classical kinetic mechanisms (see, e.g. Flemings, 1974) linking the growth rate normal to the melt/crystal interface (V_n) to the driving force for crystallization (the undercooling ΔT) are of relevance to the work presented in (Liu et al., 1999). These are growth by screw dislocations, which is characterized by a dependence of growth rate on undercooling raised to the second power ($V_n \propto \Delta T^2$), and two-dimensional nucleation limited growth, which requires a relatively large undercooling value and is associated with an exponential growth rate dependence on undercooling ($V_n \propto \exp(-A/\Delta T)$). In (Liu et al., 1999) we described calculations obtained using a model kinetic relation that is consistent both with the experimental results of Petrosyan and Bagdasarov (1976) and with a dislocation-driven growth mechanism. In addition, we showed an approach compatible with a two-dimensional nucleation growth mechanism.

In the early 1970s, Brice (1970) and Voronkov (1972) proposed simple geometric theories, relating the facet size (b) to the maximum undercooling (ΔT_0) along the facet, the radius of curvature of the melting point isotherm (R) as well as the temperature gradient at the melt/crystal interface (g). According to these the size of the facet is given by

$$b = \sqrt{8\Delta T_0 R/g} \qquad (3.1)$$

where, following (Voronkov, 1972),

$$g = \frac{k_m g_m + k_s g_s}{k_m + k_g}; \qquad (3.2)$$

here g_m and g_s are the temperature gradients normal to the interface in the melt and solid phases, respectively. Assuming facet formation does not significantly affect transport phenomena, it is possible to use results of large-scale models of crystal-growth (which do not account for facetting) as a tool for obtaining the values of R and g necessary for the prediction of facetting via Equation (3.1).

In (Liu et al., 1999) we compared results from our new approach for simultaneously calculating facetting and transport phenomena with predictions based on Equation (3.1). In this chapter we review these comparisons as well as present new calculations involving effects of melt flow, internal radiation through a transparent growing crystal, and more.

Convective heat transport in the melt of VB (or vertical gradient solidification) systems has been extensively studied in the literature (e.g. Chang and Brown, 1983; Adornato and Brown, 1987; Kim and Brown, 1991; Brandon et al., 1996; Yeckel et al., 1999). It has been shown that this mode of heat transport may or may not be of importance depending on the system studied. In the case of oxide growth (relevant to results presented here), melt convection has been demonstrated to be of importance in relatively large crucibles placed in substantial furnace gradients (Derby et al., 1994).

Internal radiation within the crystalline phase of oxides is known to be of importance, mainly due to the high growth temperatures and optical properties associated with these materials (see, e.g. Brandon *et al.*, 1996, and references within). In the results presented in (Liu *et al.*, 1999) as well as in most of the results shown here we use an enhanced thermal conductivity in the crystalline phase, therefore (in some sense) assuming the crystal to be an optically thick medium. This assumption is compatible with certain doped (see, e.g. Petrosyan, 1994) or undoped (see, e.g. Cockayne *et al.*, 1969) oxide systems. In addition, in this chapter, we briefly consider the growth of crystals whose optical properties render them almost transparent to infrared radiation.

This chapter is organized as follows. An outline of the model system and equations as well as the numerical approach is given in Section 3.2. Results of our calculations are presented in Section 3.3. These include a short review of the effects of crystal-growth rates and furnace gradient values on facet formation, calculations of convective effects and internal radiation (in a transparent crystalline phase) coupled with interface facetting, as well as an attempt at predicting the position of facets along the melt/crystal interface in the context of the experimental results of Petrosyan and Bagdasarov (1976). Finally, our conclusions are presented in Section 3.4.

3.2   MODEL DEVELOPMENT

We employ the same idealized model system, for the growth of oxide slabs, described in our previous work (Liu *et al.*, 1999). The system consists of a translationary symmetric rectangular-shaped molybdenum crucible placed in a vertical furnace whose linear temperature profile is slowly reduced with time, thus inducing directional solidification. In (Liu *et al.*, 1999) we presented both a transient and a quasi-steady-state version of our model. The transient analysis showed that after a short time the thermal field and partially facetted melt/crystal interface shape reached a steady-state form, thereby justifying the use of a quasi-steady-state approach. Therefore, most of the calculations presented in (Liu *et al.*, 1999) as well as all of those presented here are based on a quasi-steady-state formulation. Details of the transient analysis as well as further information on the model system can be found in (Liu *et al.*, 1999).

3.2.1   MATHEMATICAL FORMULATION

A two-dimensional slab-shaped system is modeled. The equations describing heat transport in this system result from a quasi-steady-state formulation involving conduction, radiation and convection heat transport mechanisms. Note, however, that convective and certain aspects of the radiative effects are not accounted for in all of the results presented here. Deviations from the model equations presented in this section are clearly described where relevant (in Section 3.3).

The temperature distribution T is governed by

$$\rho_m C_{pm} \mathbf{v} \cdot \nabla T - \nabla \cdot k_m \nabla T = 0 \quad \text{in the melt}$$

$$\nabla \cdot k_s \nabla T = 0 \quad \text{in the crystal}$$

$$\nabla \cdot k_c \nabla T = 0 \quad \text{in the crucible,} \tag{3.3}$$

where $\mathbf{v}$ describes the melt velocity field, k, ρ, C_p are the thermal conductivity, density, and heat capacity respectively, ∇ is the gradient operator and the subscripts m, s, c denote properties of melt, crystal, and crucible, respectively.

Melt convection, which was not accounted for in (Liu *et al.*, 1999), is included in the current formulation through the term on the left-hand side of Equation (3.3) (applied in the melt). This equation is coupled with the Navier–Stokes equation with the Boussinesq approximation,

$$\rho_m \mathbf{v} \cdot \nabla \mathbf{v} = \nabla \cdot \mathbf{T} + \rho_m \beta_T \mathbf{g}(T - T_{mp}) \tag{3.4}$$

as well as the continuity equation in the melt,

$$\nabla \cdot \mathbf{v} = 0. \tag{3.5}$$

In Equation (3.4) the gravity vector is denoted by $\mathbf{g}$, T_{mp} is the material's melting point temperature, β_T is the melt thermal expansion coefficient, and the stress tensor $\mathbf{T}$, which is composed of the viscous stress together with the deviation of the pressure from the hydrostatic value P (i.e. dynamic pressure), is given by

$$\mathbf{T} = -P\mathbf{I} - \boldsymbol{\tau} = -P\mathbf{I} + \mu(\nabla \mathbf{v} + \nabla \mathbf{v}^T), \tag{3.6}$$

where $\mathbf{I}$ is the unit tensor, and μ is the melt viscosity.

Assumptions of no-slip and no-penetration on the melt/crystal and crucible/melt interfaces are the basis for the boundary condition for velocity at these surfaces,

$$\mathbf{v} = \mathbf{0}; \tag{3.7}$$

note that implicit in this boundary condition is the assumption that melt and crystalline densities are identical.

External boundaries of the crucible exchange heat with an idealized furnace thermal profile according to

$$-k_c \nabla T \cdot \mathbf{n}_{ce} = h(T - T_f) + \varepsilon\sigma(T^4 - T_f^4), \tag{3.8}$$

where $\mathbf{n}_{ce}$ is the unit normal vector along the crucible's external boundary pointing away from the system, h is the convective heat transfer coefficient, ε is the crucible wall's emissivity, and σ is the Stefan–Boltzmann coefficient. The furnace temperature, T_f is given by

$$T_f = T_{co} + Gy, \qquad (3.9)$$

where G is the furnace temperature gradient, y is the spatial coordinate in the vertical direction, and T_{co}, is the cold zone temperature, positioned (in the furnace) at the same height as the bottom of the crucible.

Continuity of heat flux along the inner crucible boundaries is enforced through the boundary condition:

$$(-k_c \nabla T)_c \cdot \mathbf{n}_{ci} = \mathbf{q}_i \cdot \mathbf{n}_{ci}, \qquad (3.10)$$

where the subscript c denotes a term evaluated on the crucible side of the boundary, $\mathbf{n}_{ci}$ denotes a unit vector normal to the crucible's inner boundaries (i.e. crucible/melt and crucible/crystal interfaces), and $\mathbf{q}_i$ is the heat flux along the inner boundaries of the crucible on the melt ($i = m$) or crystalline ($i = s$) sides of the boundary.

Along the melt/crystal interface two conditions must be satisfied, the first represents a balance of thermal energy across the interface:

$$[(-k_m \nabla T)_m - \mathbf{q}_s] \cdot \mathbf{n}_{ms} = \Delta H_v V_n, \qquad (3.11)$$

and the second gives the temperature distribution along the interface:

$$T = T_{mp} - f(V_n). \qquad (3.12)$$

Here, $\mathbf{n}_{ms}$ is a unit vector normal to the melt/crystal interface (pointing into the melt), the subscript m denotes a term evaluated on the melt side of the interface, ΔH_v is the heat of solidification per unit volume of crystalline material, and $f(V_n)$ is the interfacial undercooling whose form is defined and discussed in Section 3.2.1.1 below. The growth velocity normal to the interface (V_n) is assumed (in this quasi-steady-state formulation) to correspond to

$$V_n = \frac{G^*}{G}(\mathbf{n}_{ms} \cdot \mathbf{e}_y), \qquad (3.13)$$

where $G^* = -dT_f/dt = -dT_{co}/dt$ is the rate at which the furnace temperature is reduced (thereby inducing solidification) and $\mathbf{e}_y$ is the unit vector pointing in the vertical direction.

The form of the heat flux $\mathbf{q}_i$ in Equation (3.10) depends on the dominant heat transport mechanisms acting in phase i at the interface. In the case where $i = m$ (i.e. melt), regardless of the presence or absence of convection, this flux is given by Fourier's law,[3]

$$\mathbf{q}_m = -k_m \nabla T. \qquad (3.14)$$

---

[3] This is a result of the no-penetration and no-slip conditions given by Equation (3.7).

The increase in heat transport through the crystalline phase due to thermal radiative effects is represented, in this study, either by an enhanced thermal conductivity, an approach strictly valid only for optically thick media, or by looking at the opposite limit where the crystal is transparent to infrared radiation. In the first instance the heat flux $\mathbf{q}_s$ appearing in Equations (3.10) and (3.11) is given by Fourier's law,

$$\mathbf{q}_s = -k_s \nabla T, \tag{3.15}$$

while in the case where the crystal is transparent to infrared radiation the flux is given by

$$\mathbf{q}_s = -k_s \nabla T + \mathbf{q}_R, \tag{3.16}$$

where the radiative heat flux at the surfaces of the crystal, within the crystalline phase ($\mathbf{q}_R$), is obtained as described in Section 3.2.1.2 below.

3.2.1.1 Interfacial Kinetics

Equation (3.12) can be rearranged to give the undercooling at the interface which, assuming a single-component system with a planar melt/crystal interface, is the driving force for crystal growth,

$$\Delta T = T_{mp} - T = f(V_n) \tag{3.17}$$

whose functional dependence on V_n can be inverted to give a general kinetic relation of the form

$$V_n = \beta \Delta T, \tag{3.18}$$

where the kinetic coefficient β may be a function both of ΔT and of the local crystallographic orientation.

Due to surface-energy effects, nonplanarity of the melt/crystal interface leads to modifications in the definition of the driving force for crystal-growth. In the problem studied here, surface-energy effects are assumed unimportant. However, as described in (Liu et al., 1999), including a small amount of artificial surface energy in the formulation is numerically beneficial. Results presented here were obtained with this slightly modified formulation that, in the interest of brevity, is not described here.

The undercooling is typically negligible on nonfacetted regions of the melt/crystal interface. Therefore, the dependence of β on θ (the angle denoting the crystallographic orientation in this two-dimensional model) can be ignored for all values of θ except for those that are close in magnitude to singular orientations (corresponding to surfaces that tend to facet). The most important parameters in this case are the β value corresponding to the singular orientation, the parameters determining the dependence of β on ΔT, and the range of θ values over which β significantly varies.

In this study, similar in spirit to ideas used by other authors (Yokayama and Kuroda, 1990), the kinetic coefficient (β) is modeled as a piecewise linear function of the surface slope relative to the singular orientation according to (see Figure 3.1),

$$\beta = \begin{cases} \beta^* + \dfrac{\beta_0 - \beta^*}{\tan(\Delta\theta_0)}\tan|\theta - \theta_0| & |\theta - \theta_0| < \Delta\theta_0 \\ \\ \beta_0 & |\theta - \theta_0| \geq \Delta\theta_0, \end{cases} \tag{3.19}$$

where β_0 is the temperature-independent kinetic coefficient off the facet, θ_0 is the crystallographic angle corresponding to the singular orientation along which a facet will develop, and $\Delta\theta_0$ is the maximum deviation from θ_0 beyond which the interface behaves as a nonfacetted surface. The coefficient β^*, given by

$$\beta^* = (\beta_0 - \Delta\beta_0)\Delta T_0^{n-1}, \tag{3.20}$$

is the temperature-dependent kinetic coefficient on parts of the facet where the crystallographic orientation exactly equals the singular direction ($\theta = \theta_0$). In Equation (3.20) n is the exponent of this power-law kinetic equation, ΔT_0 is the undercooling on regions of the facet where $\theta = \theta_0$, and $\beta_0 - \Delta\beta_0$ is the kinetic coefficient corresponding to these regions in the special case where $\Delta T_0 = 1$ (in units corresponding to those used for V_n/β).

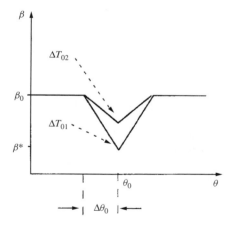

Figure 3.1 The kinetic coefficient (β) as a function of crystallographic orientation (θ) and undercooling. Here ΔT_{01} and ΔT_{02} (where $\Delta T_{01} < \Delta T_{02}$) are two different values of the maximum undercooling that correspond to melt/crystal interface orientations coinciding with the singular orientation (i.e. $\theta = \theta_0$). The lower curve corresponds to $\beta(\theta)$ for $\Delta T_0 = \Delta T_{01}$ and the upper curve corresponds to $\beta(\theta)$ for $\Delta T_0 = \Delta T_{02}$; both curves coincide for $|\theta - \theta_0| \geq \Delta\theta_0$. The β^* value appearing in this figure corresponds to $\Delta T_0 = \Delta T_{01}$. The curves are schematically plotted as piecewise linear since it is assumed, in *this figure*, that $\Delta\theta_0 \ll 1$. (Reprinted from Liu *et al.*, *J. Cryst. growth*, **205** (1999) 333–353, copyright (1999) with permission from Elsevier Science.)

Values of important parameters of the model, namely β^*, n, and $\Delta\theta_0$, were chosen as follows. The value of the kinetic coefficient β^* and the exponent n were derived to be consistent with the experimental observations in (Petrosyan and Bagdasarov, 1976), i.e. $\Delta T_0 \approx 1\,\mathrm{K}$ and $n = 2$. Estimated values of $\Delta\theta_0$ can be deduced from (Bauser and Strunk, 1984). Due to numerical considerations we employ larger values of this parameter though, as described in (Liu et al., 1999), these differences in $\Delta\theta_0$ values do not affect the results as long as $\Delta\theta_0$ is small enough. Finally, note that if β_0 is large enough it should yield negligible undercooling off the facet thereby rendering its exact value unimportant.

3.2.1.2 Radiative Flux Formulation

Formulation of the internal radiative heat flux appearing in Equation (3.16) is given by

$$\mathbf{q}_R \cdot \mathbf{n}_b = -\varepsilon_b \left[\tilde{n}^2 \sigma T_b^4 - \int_{\Omega=0}^{2\pi} i'(\Omega) \cos(\psi)\, d\Omega \right], \tag{3.21}$$

where $\mathbf{n}_b$ represents the direction normal to the crystal's surface pointing away from the crystal, ε_b is the emissivity of the surface, $\tilde{n}$ is the crystalline refractive index, σ is the Stefan–Boltzmann constant, $i'(\Omega)$ is the radiation intensity directed along the solid angle Ω, and ψ represents the angle between the direction of $i'(\Omega)$ and $\mathbf{n}_b$.

The intensity of radiation leaving the surface of the crystal towards its interior along the solid angle Ω, with the assumption of gray and diffuse boundaries, is given by

$$i'(\Omega) = \frac{1}{\pi} \left(\left\{ \tilde{n}^2 \sigma T_b^4 - \frac{1-\varepsilon_b}{\varepsilon_b} [\mathbf{q}_R \cdot (-\mathbf{n}_b)] \right\} \bigg|_{\text{surface}} \right). \tag{3.22}$$

For detailed discussions of radiative heat flux formulations in the context of crystal-growth from the melt see (Dupret et al., 1990; Brandon and Derby, 1992a; Brandon and Derby, 1992b). In the case of the present work, please see (Virozub and Brandon, 1998) for a study of a similar system, albeit without kinetic effects.

3.2.2 NUMERICAL TECHNIQUE

A full account of the techniques employed here can be found in (Brandon et al., 1996; Brandon, 1997; Virozub and Brandon, 1998; Liu et al., 1999) and references within. In particular, see (Liu et al., 1999) for details pertaining to application of Equation (3.12) including the introduction of artificial surface energy into the formulation.

Our algorithm is based on a standard Galerkin finite element method (GFEM), using biquadratic basis functions[4] for all variables except the pressure, which is

---

[4] A recent study (Virozub and Brandon, 2002) discusses the application of incomplete Hermite or complete Lagrangian cubic elements versus complete Lagrangian quadratic elements (which were used here).

represented by 2-D linear basis functions. An elliptic mesh generation technique was employed; the method is based mostly on the work described in (Christodoulou and Scriven, 1992), though in cases where the crystal was assumed transparent to infrared radiation the mesh-generation technique was modified to correspond to work presented in (Tsiveriotis and Brown, 1992).

The nonlinear Galerkin weighted residual equations associated with the temperature, velocity, pressure and nodal position unknowns were solved using the Newton–Raphson technique (Dahlquist and Björck, 1974). The mesh used in the calculations presented here included 988–1368 two-dimensional, nine-node, Lagrangian quadrilateral elements. Between 2 and 8 Newton–Raphson iterations were required for a fully convergent solution with 2–50 CPU minutes needed per iteration on a 32- processor Cray J932 supercomputer.[5] The convergence criterion was set in the L-2 norm as $\|\delta\|_2 < 10^{-6}$, where δ represents the update vector for the Newton–Raphson iteration.

3.3   RESULTS AND DISCUSSION

A number of issues are investigated in this study. First we review the effect of two important operating parameters, the crystal-growth rate and the furnace gradient, on facetting in a system in which melt convection is unimportant and internal radiative effects can be approximated by an enhanced thermal conductivity in the crystalline phase. The second set of results focuses on the effects of melt flow on facetting phenomena, while the third part of this section is concerned with modeling the growth of transparent crystals prone to facet formation along the melt/crystal interface. Finally, an attempt is made to reproduce and understand the positioning of facets as observed in the experiments of Petrosyan and Bagdasarov (1976).

Physical properties and operating parameters vary according to the subject studied. In general, the system, which is loosely based on the experiments described in (Petrosyan and Bagdasarov, 1976), was modeled using the same properties and parameters applied in our previous report (Liu et al., 1999). All deviations from the parameters given in (Liu et al., 1999) or application of additional data (e.g. necessary for modeling melt convection), are accounted for in the text.

3.3.1   EFFECT OF OPERATING PARAMETERS ON FACETTING

Two important operating parameters existing in melt growth systems are the furnace temperature gradient and the crystal axial growth rate ($V_g = G^*/G$).[6]

---

[5] The exact CPU time requirement and number of elements used depend on the transport mechanisms considered in the calculation.

[6] As indicated in Section 3.2.1, we assume that the growth rate in the vertical direction (V_g) is equal to that induced by the furnace operator (see Equation (3.13)).

Understanding how these parameters can be used to control facet size and position is most important in the event that adverse phenomena (e.g. the appearance of strain in garnets) is associated with facetting in the system under consideration.

At slow enough growth rates, the kinetic relation (Equation (3.18)) yields a negligible value for undercooling ($\Delta T = f(V_n) \approx 0$) which, when neglecting surface-energy effects, leads to an interface conforming to the shape of the melting-point isotherm. This is generally the case for regions of the interface that support a large enough value of β. However, in the vicinity of singular orientations, β may be small enough so that the undercooling $\Delta T = V_n/\beta$ (Equation (3.18)) is not negligible even for the relatively small growth rates typically employed in systems of interest. In this case, the melt/crystal interface may exhibit visible facets corresponding to singular orientations.

In Figure 3.2 we present results showing the sensitivity of interface shapes and their corresponding undercooling profiles to the axial growth rate for two different singular orientations ($\theta_0 = 0$[7] and $|\theta_0| = \pi/4$). In both cases partially facetted surfaces are predicted with facets corresponding to orientations defined as singular. In addition, an increase in facet size with increasing growth rate is obtained in these calculations. Finally, the maximum undercooling is consistent with Equations (3.18)–(3.20) (with $n = 2$); for a given singular orientation the maximum undercooling obtained is proportional to the square root of the growth rate.

Interface shapes and undercooling profiles are next shown (Figure 3.3) for different values of the furnace temperature gradient G, for the same two singular orientations considered in Figure 3.2. Here, as expected, the maximum undercooling does not vary with G (for a given orientation). However, it is evident that facet sizes (which are probably easier to estimate by looking at the undercooling profiles), grow with decreasing furnace temperature gradient.

Following Chernov (1973) we combine the Brice–Voronkov theory given by Equations (3.1) and (3.2) with the kinetic relation Equations (3.18)–(3.20) (with $\theta = \theta_0$ and $n = 2$) yielding:

$$b = [V_n/(\beta_0 - \Delta\beta_0)]^{0.25}\sqrt{8R/g} \qquad (3.23)$$

This equation predicts a linear relation between the facet size b and $V_g^{0.25}$ (for a given singular orientation $V_g \propto V_n$) as well as a linear relation between b and $g^{-0.5}$. It is conceivably possible to use crystal-growth models that ignore facetting (e.g. our model with $\beta \to \infty$) to obtain predictions of g and R that, together with Equation (3.23), can be employed in the estimation of the facet size b.

Facet sizes were obtained from the results shown in Figures 3.2 and 3.3 by relating facet edges to points along the melt/crystal interface at which $\theta = \theta_0 + \Delta\theta_0$ and $\theta = \theta_0 - \Delta\theta_0$. See Figure 3.4 for a comparison of these values of b with those calculated via the Brice–Voronkov theory Equation (3.23). It is

---

[7] Corresponding to a singular orientation coinciding with the direction of growth.

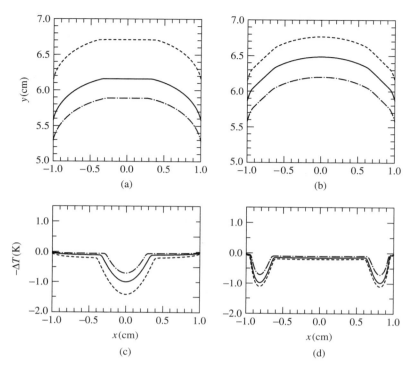

Figure 3.2 The effect of growth rate on interface shapes and interfacial undercooling for $G = 50$ K/cm. Interface shapes for (a) $\theta_0 = 0$ (singular orientation aligned with growth direction); (b) $|\theta_0| = \pi/4$ (singular orientation at an angle $\pi/4$ radians away from the growth direction). Undercooling profiles for (c) $\theta_0 = 0$; (d) $|\theta_0| = \pi/4$. In the case where $\theta_0 = 0$ (a,c) dashed curves correspond to $V_g = 0.4$ cm/h, continuous curves correspond to $V_g = 0.2$ cm/h, and dashed-dotted curves correspond to $V_g = 0.1$ cm/h. In the case where $|\theta_0| = \pi/4$ (b,d) dashed curves correspond to $V_g = 0.4$ cm/h, continuous curves correspond to $V_g = 0.3$ cm/h, and dashed-dotted curves correspond to $V_g = 0.2$ cm/h. For the sake of clear presentation, interface shapes are plotted at a vertical distance from each other that is proportional to their growth rates (assuming 2.9 h have elapsed since growth was initiated). (Reprinted from Liu *et al.*, *J. Cryst. growth*, **205** (1999) 333–353, copyright (1999) with permission from Elsevier Science.)

evident that the Brice–Voronkov theory gives predictions of b that are linear with $V_g^{0.25}$, indicating that R and g are almost unaffected by changes in V_g. A linear relation is also found between facet sizes, estimated via Equation (3.23), and $G^{-0.5}$. Consistent with this interesting observation we found that a linear relation exists between g and G. In addition, these last results also indicate that R is not significantly affected by changes in G. Finally, facet sizes obtained with (Equation (3.23)) are close to but different from those calculated by our model. Most significant are differences obtained for low values of G.

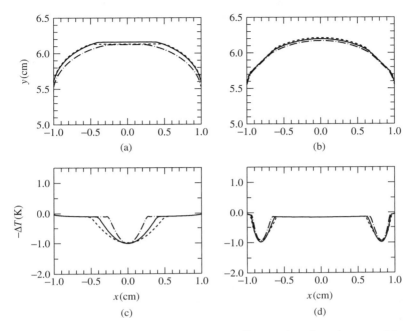

Figure 3.3 The effect of furnace temperature gradient on interface shapes and interfacial undercooling. Interface shapes for (a) $\theta_0 = 0$; (b) $|\theta_0| = \pi/4$. Undercooling profiles for (c) $|\theta_0| = 0$; (d) $|\theta_0| = \pi/4$. In the case where $\theta_0 = 0$ (a,c) $V_g = 0.2$ cm/h, dashed curves correspond to $G = 20$ K/cm, continuous curves correspond to $G = 34$ K/cm, and dashed-dotted curves correspond to $G = 90$ K/cm. In the case where $|\theta_0| = \pi/4$ (b,d) $V_g = 0.3$ cm/h, dashed curves correspond to $G = 34$ K/cm, continuous curves correspond to $G = 50$ K/cm, and dashed-dotted curves correspond to $G = 70$ K/cm. (Reprinted from Liu *et al.*, *J. Cryst. growth*, **205** (1999) 333–353, copyright (1999) with permission from Elsevier Science.)

3.3.2 INTERACTION BETWEEN MELT FLOW AND FACET FORMATION

It is reasonable to assume that heat transport in the small-scale system studied in (Liu *et al.*, 1999), as well as in Section 3.3.1 of this chapter, is only marginally affected by melt convection. However, as shown for a similar (though cylindrical) system in (Derby *et al.*, 1994) it is expected that increasing the size of the crystal will lead to increased natural-convective flows that, beyond a certain system size, will significantly affect heat transport and therefore facet formation. This enhancement in flow is due to the fact that the dimensionless driving force for flow is proportional to a characteristic length of the system (e.g. crystal width) raised to the third power (see, e.g. Landau and Lifshitz, 1959).

In this section we present calculations that account for melt flow. Properties employed here, not used in (Liu *et al.*, 1999), are the melt viscosity

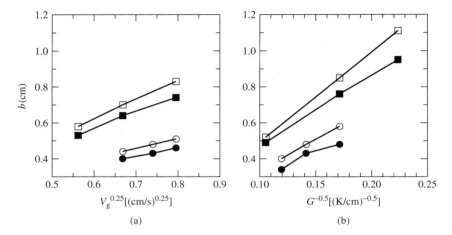

(a)   (b)

Figure 3.4 Facet size as a function of (a) growth rate and (b) furnace temperature gradient. Open symbols depict values predicted based on the Brice–Voronkov approach using calculations with $\beta \to \infty$ together with Equation (3.23). Filled symbols represent results from our calculations (shown in Figures 3.2 and 3.3) accounting for coupling between kinetics and thermal field calculations (i.e. using finite values for β). Squares and circles represent results for $\theta_0 = 0$ and $|\theta_0| = \pi/4$, respectively. (Reprinted from Liu *et al.*, *J. Cryst. growth*, **205** (1999) 333–353, copyright (1999) with permission from Elsevier Science.)

($\mu = 0.0426$ Pa s) and thermal expansion coefficient ($\beta_T = 1.8 \times 10^{-5}$ K^{-1}) both of which are based on measurements of Fratello and Brandle (1993).

Melt/crystal interface shapes as well as interfacial undercooling profiles are plotted in Figure 3.5 for a variety of crystal widths both with and without accounting for melt flow. Increasing the crystal width (B^*) leads to a strengthening of convection characterized by fluid rising along the hot crucible walls and descending along the centerline towards the melt/crystal interface. This flow pattern (as can be seen in Figure. 3.5) promotes the appearance of facet sizes larger than those calculated without taking melt convection into account.

In an attempt to better understand the results shown in Figure 3.5 we plot, in Figure 3.6 data corresponding to calculations using our model without accounting for kinetic effects (i.e. $\beta \to \infty$). This data includes values of the thermal gradient in the melt (g_m) normal to the melting-point isotherm and the curvature (R) of the melting-point isotherm, both calculated at the position on the melting-point isotherm whose orientation corresponds to θ_0. Results shown in Figure 3.6 were calculated both with and without accounting for melt convection, as a function of the width of the crystal B^*.

Considering the simple Brice–Voronkov approach Equations (3.1) and (3.23), it is apparent that the facet size is sensitive to the ratio R/g. When convection is not considered, increasing the system size leads to a decrease in g, and an increase

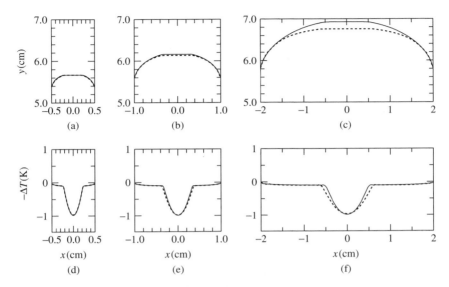

Figure 3.5 The effect of melt convection (and crystal width B^*) on interface shapes and interfacial undercooling for $\theta_0 = 0$, $G = 50$ K/cm and $V_g = 0.2$ cm/h. Interface shapes for (a) $B^* = 1$ cm, (b) $B^* = 2$ cm, (c) $B^* = 4$ cm. Undercooling profiles for (d) $B^* = 1$ cm, (e) $B^* = 2$ cm, (f) $B^* = 4$ cm. Dashed curves were obtained while accounting for melt convection and continuous curves were calculated while neglecting convection (i.e. $\beta_T = 0$).

in R, which promotes[8] an increase in facet size. In the (more realistic) event that melt convection is considered, increasing the system size leads to an increase both in g_m and in R. In this case even though R/g_m can be shown to increase with increasing system size, it is not obvious at first glance if the rapid increase in R (compared to the case where convection is ignored) is significant enough to cause widening of facets (for a given system size) due to convective effects. The convection-induced increase in g_m, compared to the case where convection is ignored, may conceivably offset the increase in R. However, looking at Figure 3.5 and at Figure 3.7 it is evident that changes in R dominate and that, for a given system size, convective effects lead to an increase in facet size. Moreover, it is interesting to note (see Figure 3.7) that for $B^* = 4$, convective effects are strong enough to cause an increase in the discrepancy between the predictions based on Equation (3.23) and calculations obtained using our approach. These discrepancies are, as in the cases shown in Figure 3.4, most serious for low values of G.

---

[8] Note that although we consider g_m rather than g, conservation of energy across the interface in this type of system (in which latent heat effects are most probably unimportant) renders $g_s \propto g_m$ and therefore $g \propto g_m$.

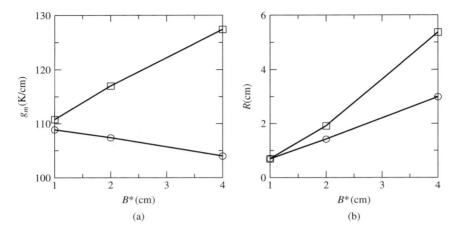

(a)   (b)

Figure 3.6 (a) The thermal gradient in the melt normal to the melting point isotherm (g_m, defined in the text) and (b) the curvature of the melting point isotherm (R), as a function of crystal width (B^*). Squares correspond to values calculated while accounting for melt convection and circles represent values calculated while neglecting melt convection (i.e. $\beta_T = 0$). All data points were evaluated at a position on the melting-point isotherm for which $\theta = \theta_0 = 0$, in the case where $G = 50$ K/cm, $V_g = 0.2$ cm/h and kinetic effects are unimportant ($\beta \rightarrow \infty$).

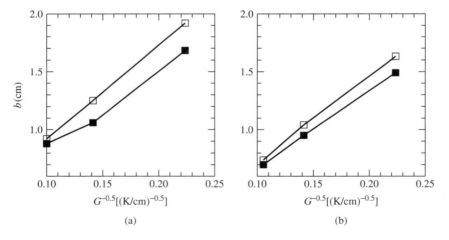

(a)   (b)

Figure 3.7 Facet size as a function of furnace temperature gradient calculated (a) with convection in the melt and (b) without convection in the melt ($\beta_T = 0$). Here $G = 50$ K/cm, $V_g = 0.2$ cm/h, $\theta_0 = 0$, $B^* = 4$ cm. Open symbols depict values predicted based on the simplified approach using calculations with $\beta \rightarrow \infty$ together with Equation (3.23). Filled symbols represent results from our calculations accounting for coupling between kinetics and thermal field calculations (i.e. a finite β value).

Our analysis is limited to the particular vertical gradient solidification system described here. It is interesting to consider many different variations of the parameters used in the calculations shown in Figure 3.5. Unfortunately, a comprehensive study of these is beyond the scope of our work. We do, however, consider one permutation of the system parameters. Results shown in Figure 3.8 exhibit melt/crystal interface shapes and undercooling profiles calculated without taking convection into account, taking convection into account and taking convection into account with the gravity vector pointing in the opposite direction. This last situation describes a top-seeded vertical growth system in which the crystal is placed above the melt into which it is growing (i.e. the gravity vector is directed from the cold zone of the furnace towards the hot zone, in the y direction).

Melt flow, in the case of top seeding, is considerably more vigorous than in the previous calculations. This is evident when observing the significant shift in the melt/crystal interface position due to flow in the case of inverted gravity compared to the usual situation. Of relevance is the direction of flow near the melt/crystal interface, which, due to inversion of gravity, is similar to the direction of natural-convective-dominated flows often observed in Czochralski systems (see Müller and Ostrogorsky, 1994, and references within). Here, melt flows towards the interface near its periphery, towards the centerline along the interface, and away from the interface near its center. This inverted flow pattern tends, in the absence of facetting, to sharpen the interface near its center thereby increasing its curvature in this region; a similar effect was observed in natural-convective dominated flow calculations during the growth of oxides via the Czochralski technique (Xiao and Derby, 1993). When facetting is considered (see Figure 3.8),

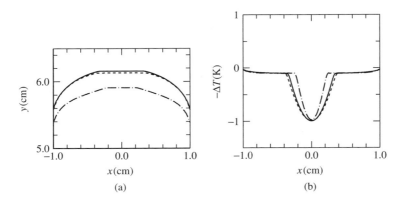

Figure 3.8 (a) Interface shapes and (b) undercooling profiles for different convective driving forces. Continuous curves correspond to calculations without a convective driving force ($\beta_T = 0$), dashed curves correspond to calculations with melt convection in the thermally stabilized, base-case configuration (bottom seeding), and dashed-dotted curves correspond to calculations with melt convection in a top-seeded configuration. Here $G = 50\,\text{K/cm}$, $V_g = 0.2\,\text{cm/h}$, and $\theta_0 = 0$.

the facet size is reduced in comparison to the calculation that does not account for melt convection. The decrease in facet size is most probably due to the reduction in R near the center of the system.

3.3.3 TRANSPARENT CRYSTALLINE PHASE

In (Liu *et al.*, 1999) and in most of the results shown here, radiative transport within the crystalline phase was modeled using an enhanced thermal conductivity. This approach, which in some sense approximates internal radiation in optically thick media, has difficulty predicting the deep interfaces associated with optically thin or transparent crystals (Brandon and Derby, 1992a). In this section the crystal is approximated as a medium completely transparent to infrared radiation. Here, system parameters remain the same as before except for: (a) the crucible height, which was set to $L = 20$ cm, thus allowing for full exploration of severe deflections in interface shapes; (b) the hot and cold zone temperatures, which were modified to allow for the same base-case gradient as before ($G = 50$ K/cm); (c) the growth rate, which was chosen to be 0.55 cm/h due to numerical limitations briefly mentioned below. Physical parameters remain the same except for the thermal conductivity in the crystalline phase, which without artificial radiative enhancement is given by $k_s = 0.08$ W/cm K. Emissivities of the crucible/crystal boundaries, the melt/crystal interface and the crystal's refractive index were chosen according to the properties used in our previous study of a similar system (Virozub and Brandon, 1998).

In (Virozub and Brandon, 1998) we showed that slab-shaped geometries, in comparison with cylindrical systems, may exhibit severely deflected melt/crystal interface shapes due to effects associated with radiative cooling through the transparent crystal. Radiative transport is, in some sense, enhanced by relatively large viewing angles between the interface and the cold end of the crucible in this slab-shaped system. The large deflection in the interface leads to difficulties in achieving closure of energy balances; this problem was solved by optimizing mesh-generation parameters. In attempting to model facetting phenomena in this system, numerical difficulties associated with the combination of radiative transport, resultant deep interfaces, and facet formation lead to the requirement of relatively intensive computational resources. For this reason we have limited the study to a case involving a relatively large value of the crystallographic orientation $|\theta_0| = 1.2$ (smaller values of $|\theta_0|$ require additional mesh refinement). However, obtaining significant facetting using $|\theta_0| = 1.2$ requires a relatively large value for V_g.[9] This is a result of the relation $V_n = V_g(\mathbf{n}_{ms} \cdot \mathbf{e}_y) = V_g \cos(|\theta|)$, which for $|\theta| = |\theta_0| = |1.2|$ yields a value of V_n significantly smaller than V_g.

A sample interface shape is shown in Figure 3.9. Notice the severe deflection of the interface (due to the crystal's transparency), which is shaped in a

---

[9] Here we use $V_g = 0.55$ cm/h.

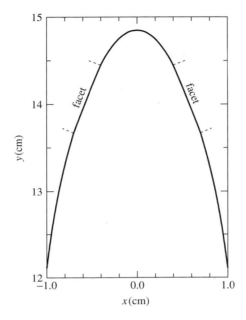

Figure 3.9 Partially facetted interface of a transparent crystal grown at $V_g = 0.55$ cm/h with $G = 50$ K/cm and $\theta_0 = 1.2$. Facet boundaries are marked by small dashed lines perpendicular to the facet.

manner where curvature is significant near the center. Facets, whose boundaries are marked by small dashed lines, appear where expected.

3.3.4 POSITIONING OF FACETS ALONG THE INTERFACE

Growth of $\langle 111 \rangle$ axis garnet crystals using the Czochralski method often yields a crystal with a strained inner core bounded by three {211} facets (see, e.g. Zydzik, 1975); {110} facets are also known to appear in this case at larger radial positions (see, e.g. Cockayne and Roslington, 1973).

Application of the vertical bridgman technique to the growth of YAG in several different crystallographic directions is shown to be characterized by a variety of facet distribution patterns depending on the growth orientation (Petrosyan *et al.*, 1978). Using this technique to grow Nd:YAG along the $\langle 221 \rangle$ direction yields {211} and {110} facets positioned near the periphery of the melt/crystal interface (Petrosyan and Bagdasarov, 1976).

In Figure 3.10 we present results attempting to reproduce the growth conditions reported in (Petrosyan and Bagdasarov, 1976) with the aim of better understanding causes of facet positioning along the interface. Parameters were chosen to closely follow those used by Petrosyan and coworkers. These are the

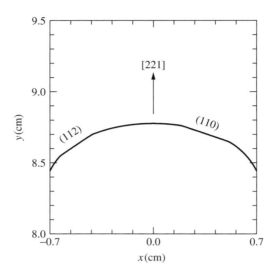

Figure 3.10 Partially facetted interface of a crystal grown under conditions (detailed in the text) closely following those used by Petrosyan and Bagdasarov (1976).

same as base-case values used in (Liu *et al.*, 1999) and in the present study except for the crucible inner half-width, the crucible inner length, and crucible wall thickness, which were estimated to be $B^* = 0.7$ cm, $L^* = 15$ cm, and $d = 0.03$ cm, respectively (Petrosyan, 1998). In addition, in these calculations, $G = 100$ K cm and $V_g = 0.26$ cm/h (Petrosyan and Bagdasarov, 1976). Here the crystal is grown along the [221] direction and the two singular surfaces considered, whose orientations are coplanar with this growth orientation, are (112) ($|\theta_0| = 0.615$) and (110) ($|\theta_0| = 0.340$).[10] Finally, note that in this small-scale system, convection was not taken into account and radiative transport within the crystalline medium was modeled using an enhanced thermal conductivity.

Facet positions shown in Figure 3.10 are semiqualitatively consistent with the experimental results of (Petrosyan and Bagdasarov, 1976). The (112) facet shown on the left is close to the crucible wall (consistent with Petrosyan and Bagdasarov, 1976), though the (110) facet shown on the right is not so obviously positioned along the periphery of the crystal surface.

3.4 CONCLUSIONS

Facetting during the growth of single crystals from the melt can cause unwanted phenomena.[11] In addition, using standard computational algorithms for the analysis

---

[10] Note that, in this asymmetric case, each singular orientation appears only on one side of the growth axis; Figure 3.10 is consistent with $\theta_0 = 0.615, -0.340$.

[11] See (Cockayne and Roslington, 1973) for a situation where facets are beneficial.

of melt growth may lead to inaccuracies in the prediction of thermal, solute and flow fields, all of which may be significantly affected by the presence of large facets along the melt/crystal interface.

Results shown here demonstrate current capabilities of a new approach designed to couple predictions of facet formation with transport phenomena during directional melt growth. Important information gleaned from our calculations can be itemized as follows:

- Predictions of facet size and undercooling profiles are qualitatively consistent with the expected sensitivity to operating parameters. Facets increase in size with increasing crystal-growth rate and decreasing furnace temperature gradient. When melt convection is unimportant and internal radiation is modeled using an enhanced crystalline thermal conductivity, semiquantitative agreement is found between facet sizes predicted here and those calculated using the Brice–Voronkov theory. As explained in (Liu *et al.*, 1999), discrepancies can partially be explained by the small level of undercooling appearing on the off-facet region in some of our calculations. However, in a number of cases (particularly for small G values), discrepancies are probably due to nonuniformity in R^{12} (in the case of $|\theta_0| = \pi/4$) as well as deviations of the thermal field near the facet from that assumed in (Voronkov, 1972).

- Including melt convection in the analysis is shown to be important for the largest crucible considered here ($B^* = 4$ cm). The increase in flow intensity with system size, the direction of the resultant flow (towards the interface at its center), and the overall 'flattening' of the melt/crystal interface shape due to flow, are all similar to observations made for a cylindrical geometry (Derby *et al.*, 1994). Following the ideas of Brice (1970) and Voronkov (1972), the increase in R due to convection should promote larger facet sizes, while the increase in the thermal gradient at the interface (also due to convection) should favor smaller facet sizes. Our calculations show that the increase in R as a result of convection dominates and leads to an increase in facet size. Interestingly, facet sizes calculated in this case significantly deviate (for small G values) from values predicted using the Brice–Voronkov approach. This is probably due to an increase in the deviation between thermal fields assumed in Voronkov (1972) and those calculated here.

- An increase in facet size due to melt convection was observed for the standard bottom-seeded configuration and specific singular orientation $|\theta_0| = 0$ studied here. Looking at Figure 3.5 it is evident that for large enough $|\theta_0|$ values (appearing near the crucible walls) R may not change or may even decrease due to convection. A certain decrease in R near the crystal periphery due to melt convection can be observed when looking at previous results for cylindrical crystals (see, e.g. Figure 3.9 in Brandon *et al.*, 1996). A decrease in R due to convection could conceivably lead to reduction in facet sizes

---

[12] R is assumed uniform in the analysis of Brice (1970) and Voronkov (1972).

near the crucible wall. Finally, considering top-seeded crystal-growth with $|\theta_0| = 0$, demonstrates that the associated inverted flow pattern (i.e. flow away from the interface at its center) acts to increase curvature (decrease R) near the system center, thereby causing a reduction in facet size compared to the situation where flow is not considered. Note that this significant result, which has possible implications for natural-convective-flow-dominated CZ systems, was obtained for a relatively small scale system ($B^* = 2$ cm) in which bottom seeding does not lead to significant flow.

- Taking internal radiation effects into account for the case of optically thin (or transparent) crystalline phases is a nontrivial task. The combination of extremely deep interfaces due to radiation in slabs with numerical difficulties associated with modeling kinetics (see Liu *et al.*, 1999), render calculations quite challenging. We demonstrate a certain capability for calculating partially facetted interfaces in this case. Ongoing work includes implementing new numerical meshing strategies aimed at giving us the capability to accurately and thoroughly study effects associated with transparent crystalline phases of the type studied here.

- Using parameters close to those used by Petrosyan and Bagdasarov (1976) we attempt to reproduce their experimental result concerning positioning of facets (near the crystal periphery). Two singular surfaces coplanar with the growth direction were chosen in this study of a slab-shaped crystal. One of the resultant facets appears near the crucible wall while the other is not as close to the wall as shown in (Petrosyan and Bagdasarov, 1976). This is not very surprising since facet positioning is very sensitive to the distribution of the crystallographic orientation angle along the interface. Effects of melt convection discussed above, variations in optical properties in semi-transparent crystals (Brandon and Derby, 1992a), and the effect of differences in geometry (slab versus cylinder, Virozub and Brandon, 1998) may lead to changes in the distribution of θ causing facets (of a particular surface orientation) to be close to or far from the crucible walls. An accurate quantitative prediction of facet positioning along the interface requires, in the case of (Petrosyan and Bagdasarov, 1976), the analysis of a cylindrical system[13] with a good knowledge of the degree of transparency of crystals grown.

ACKNOWLEDGMENTS

This Research was supported by The Israel Science Foundation administered by The Israel Academy of Sciences and Humanities. AV acknowledges partial support by the Israel Absorption Ministry.

---

[13] Note that the computation of facetting in a cylindrical system with $\theta_0 \neq 0$ requires a three-dimensional analysis.

NOTE ADDED IN PROOF

Since the completion of this manuscript an important contribution to this field has appeared in the literature. In (Lan and Tu, 2001) a slightly different approach is used to extend the ideas discussed here to three-dimensional geometries.

REFERENCES

Adornato, P. M. and Brown, R. A. (1987) 'Convection and segregation in directional solidification of dilute and non-dilute binary alloys: Effects of ampoule and furnace design' *J. Cryst. Growth*, **80**, 155–190.

Amon, J., Dumke, F. and Müller, G. (1998) 'Influence of the crucible shape on the formation of facets and twins in the growth of GaAs by the vertical gradient freeze technique' *J. Cryst. Growth*, **187**, 1–8.

Bauser, E. and Strunk, H. P. (1984) 'Microscopic growth mechanisms of semiconductors: Experiments and models' *J. Cryst. Growth*, **69**, 561–580.

Brandon, S. (1997) 'Flow fields and interface shapes during horizontal Bridgman growth of fluorides' *Model. Simul. Mater. Sci. Eng.*, **5**, 259–274.

Brandon, S. and Derby, J. J. 1992a 'Heat transfer in vertical Bridgman growth of oxides: effects of conduction, convection and internal radiation' *J. Cryst. Growth*, **121**, No. 3, 473–494.

Brandon, S. and Derby, J. J. (1992b) 'A finite element method for conduction, internal radiation and solidification in a finite axisymmetric enclosure' *Int. J. Num. Meth. Heat Fluid Flow*, **2**, 299–333.

Brandon, S., Gazit, D. and Horowitz, A. (1996) 'Interface shapes and thermal fields during the gradient solidification method growth of sapphire single crystals' *J. Cryst. Growth*, **167**, No. 1–2, 190–207.

Brice, J. C. (1970) 'Facet formation during crystal pulling' *J. Cryst. Growth*, **6**, 205–206.

Chang, C. J. and Brown, R. A. (1983) 'Radial segregation induced by natural convection and melt/solid interface shape in Vertical Bridgman growth' *J. Cryst. Growth*, **63**, 343–364.

Chernov, A. A. (1973) 'Crystallization' *Ann. Rev. Mater. Sci.*, **3**, 397–454.

Christodoulou, K. N. and Scriven, L. E. (1992) 'Discretization of free surface flows and other moving boundary problems' *J. Comp. Phys.*, **99**, No. 1, 39–55.

Chung, H., Dudley, M., Larson Jr., D. J., Hurle, D. T. J., Bliss, D. F. and Prasad, V. (1998) 'The mechanism of growth-twin formation in zincblende crystals: new insights from a study of magnetic liquid encapsulated Czochralski-grown InP single crystals' *J. Cryst. Growth*, **187**, 9–17.

Cockayne, B. (1968) 'Developments in melt-growth oxide crystals' *J. Cryst. Growth*, **3**, 60–70.

Cockayne, B., Chesswas, M. and Gasson, D. B. (1969) 'Facetting and optical perfection in Czochralski grown garnets and ruby' *J. Mater. Sci.*, **4**, 450–456.

Cockayne, B. and Roslington, J. M. (1973) 'The dislocation-free growth of gadolinium gallium garnet single crystals' *J. Mater. Sci.*, **8**, 601–605.

Cockayne, B., Roslington, J. M. and Vere, A. W. (1973) 'Macroscopic strain in facetted regions of garnet crystals' *J. Mater. Sci.*, **8**, 382–384.

Dahlquist, G. and Björck, Å. A. (1974), *Numerical Methods*, Prentice-Hall, Inc., Englewood Cliffs, N.J.

Derby, J. J., Brandon, S., Salinger, A. G. and Xiao Q. (1994) 'Large-scale numerical analysis of materials processing systems: High-temperature crystal-growth and molten glass flows' *Comp. Meth. Appl. Mech. Eng.*, **112**, 69–89.

Dupret, F., Nicodème, P., Ryckmans, Y., Wouters, P. and Crochet, M. J. (1990), 'Global modelling of heat transfer in crystal-growth furnaces' *Int. J. Heat Mass Transfer*, **33**, No. 9, 1849–1871.

Dupret, F. and van den Bogaert, N. (1994), 'Modelling Bridgman and Czochralski growth'. *Handbook of Crystal Growth Vol. 2b: Bulk Crystal Growth, Growth Mechanisms and Dynamics*, ed. D. T. J. Hurle, North-Holland, Amsterdam, 875–1010.

Flemings, M. C. (1974), *Solidification Processing*, Mc-Graw Hill, New York.

Fratello, V. J. and Brandle, C. D. (1993) 'Physical properties of a $Y_3Al_5O_{12}$ melt' *J. Cryst. Growth*, **128**, 1006–1010.

Hulme, K. F. and Mullin, J. B. (1962) 'Indium-antimonide a review of its preparation, properties and device applications' *Solid State Electron.*, **5**, 211–247.

Hurle, D. T. J. and Cockayne, B. (1994), 'Czochralski growth'. *Handbook of Crystal Growth Vol. 2a: Bulk Crystal Growth, Basic Techniques*, ed. D. T. J. Hurle, North-Holland, Amsterdam, 99–211.

Hurle, D. T. J. (1995) 'A mechanism for twin formation during Czochralski and encapsulated vertical Bridgman growth of III-V compound semiconductors' *J. Cryst. Growth*, **147**. No. 3–4, 239–250.

Kim, D. H. and Brown, R. A. (1991) 'Modelling of the dynamics of HgCdTe growth by the Vertical Bridgman method' *J. Cryst. Growth*, **114**, 411–434.

Lan, C. W. and Tu, C. Y. (2001) 'Three dimensional simulation of facet formation and the coupled heat flow and segregation in Bridgman growth of oxide crystals' *J. Cryst. Growth*, **233**, 523–536.

Landau, L. D. and Lifshitz, E. M. (1959), *Fluid Mechanics*, 1st edn. Permagon Press, Oxford.

Liu, Y., Virozub, A. and Brandon, S. (1999) 'Facetting during directional growth of oxides from the melt: coupling between thermal fields, kinetics and melt/crystal interface shapes' *J. Cryst. Growth*, **205**, No. 3, 333–353.

Müller, G. and Ostrogorsky, A. (1994), 'Convection in melt growth'. *Handbook of Crystal Growth Vol. 2b: Bulk Crystal Growth, Growth Mechanisms and Dynamics*, ed. D. T. J. Hurle, North-Holland, Amsterdam, 709–819.

Petrosyan, A. G. (1994) 'Crystal-growth of laser oxides in the vertical Bridgman configuration' *J. Cryst. Growth*, **139**, 372–392.

Petrosyan, A. G. (1998) *Personal communication*.

Petrosyan, A. G. and Bagdasarov, Kh. S. (1976) 'Faceting in Stockbarger grown garnets' *J. Cryst. Growth*, **34**, 110–112.

Petrosyan, A. G., Shirinyan, G. O., Ovanesyan, K. L. and Avetisyan, A. A. (1978) 'Facet formation in garnet crystals' *Kristall und Tecchnik*, **13**, No. 1, 43–46.

Tsiveriotis, K. and Brown, R. A. (1992) 'Boundary-conforming mapping applied to computations of highly deformed solidification interfaces' *Int. J. Num. Meth. Fluids*, **14**, No. 8, 981–1003.

Virozub, A. and Brandon, S. (1998) 'Radiative heat transport during the vertical Bridgman growth of oxide single crystals: slabs versus cylinders' *J. Cryst. Growth*, **193**, 592–596.

Virozub, A. and Brandon, S. (2002) 'Selecting finite element basis functions for computation of partially facetted melt/crystal interfaces appearing during the directional growth of large-scale single crystals' *Model. Simul. Mater. Sci. Eng.* **10**, 57–72.

Voronkov, V. V. (1972) 'Supercooling at the face developing on a rounded crystallization front' *Krystallografiya*, **17**, No. 5, 909–917.

Xiao, Q. and Derby, J. J. (1993) 'The role of internal radiation and melt convection in Czochralski oxide growth: deep interfaces, interface inversion, and spiraling' *J. Cryst. Growth*, **128**, 188–194.

Yeckel, A., Doty, F. P. and Derby, J. J. (1999) 'Effect of steady crucible rotation on segregation in high-pressure vertical Bridgman growth of cadmium zinc telluride' *J. Cryst. Growth*, **203**, 87–102.

Yokayama, E. and Kuroda, T. (1990) 'Pattern formation in growth of snow crystals occurring in the surface kinetic process and the diffusion process' *Phys. Rev. A.*, **41**, No. 4, 2038–2049.

Zydzik, G. (1975) 'Interface transitions in Czochralski growth of garnets' *Mater. Res. Bull.*, **10**, 701–708.

4 Theoretical and Experimental Solutions of the Striation Problem

HANS J. SCHEEL

SCHEEL CONSULTING, Sonnenhof 13, CH-8808 Pfaeffikon, Switzerland

ABSTRACT

Striations are growth-induced inhomogeneities that hamper the applications of solid-solution crystals and of doped crystals in numerous technologies. Thus the optimized performance of solid solutions often can not be exploited. It is commonly assumed that striations are caused by convective instabilities so that reduced convection by microgravity or by damping magnetic fields was and is widely attempted to reduce inhomogeneities.

In this chapter it will be shown that temperature fluctuations at the growth interface cause striations, and that hydrodynamic fluctuations in a quasi-isothermal growth system do not cause striations. The theoretically derived conditions were experimentally established and allowed the growth of striation-free crystals of $KTa_{1-x}Nb_xO_3$ 'KTN' solid solutions for the first time.

Hydrodynamic variations from the accelerated crucible rotation technique (ACRT) did not cause striations as long as the temperature was controlled within $0.03\,^\circ C$ at $1200\,^\circ C$ growth temperature. Alternative approaches to solve the segregation and striation problems are discussed as well.

4.1 INTRODUCTION

Solid solutions or mixed crystals are special crystals or alloys in which one or more lattice sites of the structure are occupied by two or more types of atoms, ions or molecules. By varying the concentration of the constituents, the physical or chemical properties of solid solutions can be optimized for specific applications, so that solid solutions play an increasing role in research and technology. One example are III-V semiconductors where the bandgap and thus emission and absorption wavelengths can be adjusted, along with the lattice constant to match the available substrates for epitaxial growth, for optical communication systems. Another example are III-V compounds for photovoltaic devices where

Crystal Growth Technology, Edited by H. J. Scheel and T. Fukuda
© 2003 John Wiley & Sons, Ltd. ISBN: 0-471-49059-8

the composition can be adjusted to optimize the solar-cell efficiency and to maximize radiation resistance (Loo *et al.* 1980). In certain cases, properties and effects may be obtained in solid solutions that are not observed in the constituents: The phase transition temperature and the related anomaly of a high dielectric constant may be shifted to the application temperature for electro-optic, nonlinear-optic and acousto-optic applications (Chen *et al.* 1966, Scheel and Günter 1985). The hardening effect of solid solutions is often used to improve the mechanical properties (Nabarro 1975).

The statistical distribution of species in a given lattice site normally is at random, but can deviate in the direction of ordering (with the extreme case of a superlattice) or in the direction of clustering (with the extreme case of immiscibility or phase separation) as it was shown by Laves 1944. This site distribution has an impact on the physical properties and can be controlled to some extent in metallic alloys, with their high diffusivities, by preparation or annealing conditions. However, in oxide systems with low diffusivities, the control of the distribution of species on lattice sites during crystal growth experiments has not been reported and will not be considered here.

In this chapter the bulk fluctuations of concentration will be treated with respect to the application-dependent homogeneity requirements, the types of inhomogeneities and their origin, and how inhomogeneities can be reduced or completely suppressed. There is a strong tendency for fluctuations of the growth conditions causing concentration fluctuations along the growth direction known as striations or growth bands. The suppression of striations is an old problem in crystal-growth technology, so that several authors had described striations as an 'intrinsic', 'inherent' or 'unavoidable' phenomenon in crystal growth, see Rytz and Scheel 1982, Scheel and Günter 1985.

Based on the segregation analyses for melt growth by Burton *et al.* (1953) and for growth from diluted solutions by Van Erk 1982, the role of hydrodynamics will be discussed, and the experimental conditions for growth of striation-free crystals derived. This theoretical result will be confirmed by the growth of quasi striation-free crystals of solid solutions. Finally, alternative approaches to reduce or eliminate striations are discussed, also novel approaches that require to be tested.

4.2   ORIGIN AND DEFINITIONS OF STRIATIONS

Nearly all crystals have inhomogeneities and growth bands called striations. Early observations of striations in semiconductor crystals have been reported by Goss *et al.* (1956) and by Bardsley *et al.* (1962) and for CaF_2 by Wilcox and Fullmer (1965). Hurle (1966) has shown for semiconductor crystals and Cockayne and Gates (1967) for Czochralski-grown oxide and fluoride crystals, that striations are caused by temperature fluctuations that may be correlated with thermal unsymmetry or with convective oscillations when a critical Rayleigh number is surpassed.

Many compounds like GaAs and LiNbO$_3$ are not line compounds, but have an existence range, that is they have a certain degree of solid solubility with one or both of the constituents. This causes a difference of the congruent melting composition from the stoichiometric composition. Therefore the grown crystals show variations of composition, often in the form of striations, depending on the exact growth conditions (melt composition and growth temperature and their fluctuations). This topic is discussed by Miyazawa in this volume for oxides and by Wenzl *et al.* (1993) for GaAs and compound semiconductors.

Also, melt-grown elements like silicon and line compounds like Al$_2$O$_3$ show striations due to impurities, in the case of silicon–oxygen striations from partial dissolution of the SiO$_2$ crucible. Very pronounced striations are often found in doped crystals (semiconductors, lasers) and in crystals of solid solutions.

The homogeneity requirements depend on the material and on the application. Examples of the maximum composition (X) variations $\Delta X / X$ are given in Table 4.1.

Corresponding to these tolerance limits, analytical methods are to be applied or developed in order to assist in the achievement of 'striation-free' crystals. In addition to the well-established methods to visualize striations, Donecker *et al.* (1996) developed a colorful optical diffraction method to visualize striations and demonstrated it with oxide, doped InP and Si$_{1-x}$Ge$_x$ crystals.

It is recommended to use the term 'striation-free' in those cases where striations can not be detected or where they are not harmful for the specific application. An absolute striation-free crystal may neither be achieved nor proven, although in facetted growth or in liquid-phase epitaxy in the Frank–Van Der Merwe growth mode quasi striation-free crystals and layers could be expected when step-bunching is prevented.

Various definitions for striations had been suggested. In the following 'functional terms' are proposed that relate to the origin of the specific striations (instead of type-I, type-II, etc.). Striations are defined as growth-induced inhomogeneities in the crystal that are aligned along the facetted or non-facetted growth surface, or in the case of facetted growth with step-bunching are related to the traces of macrosteps.

Table 4.1 Homogeneity requirements of material classes

Crystals, substrates, epilayers	$\Delta X / X$
Semiconductors and solid solutions GaAs/GaP, (Ga,In)As, (Ga,In)Sb, (Cd, Hg)Te	10^{-4} to 10^{-5}
Dielectrics Piezo-electrics, pyroelectrics, electro-optic, Nonlinear-optic and laser crystals	10^{-5} to 10^{-7}
Magnetic and magneto-optic crystals and layers	10^{-4} to 10^{-5}
Metals and alloys	10^{-2} to 10^{-3}

These, often periodic, inhomogeneities are caused in the first case by growth rates which fluctuate with time due to temperature fluctuations and are schematically shown in Figure 4.1(a). Therefore they could be named 'thermal striations'. In Czochralski crystal pulling these striations are frequently linked to crystal rotation rate, since the crystal feels any unsymmetry in the heater-insulation configuration (Camp 1954): these specific thermal striations are called 'rotational striations' or in short 'rotationals'. It was shown by Witt *et al.* (1973) by using time markers and etching of crystal cross sections, that remelting may occur followed by fast growth, so that they defined microscopic (instantaneous) and macroscopic (average) growth rates and thus could explain complex striation patterns. The fluctuating temperatures at the growth interface, leading to growth-rate fluctuations in Czochralski growth, are related to convective instabilities (Hurle 1967), to interactions of several different kinds of flows, which will be discussed further below.

The second class of striations is caused by lateral growth-rate differences as shown in Figure 4.1(b) and are named 'macrostep-induced striations' or 'kinetic striations'. Facetted growth may be observed in growth from melts in small temperature gradients, and it is practically always observed in growth from solutions and from high-temperature solutions. At high densities of growth steps, i.e. at high supersaturation respectively growth rates, the bunching of steps occurs and leads to macrosteps with a large integer of the height of the monostep. For a qualitative explanation of this phenomenon Frank (1958) and Cabrera and Vermilyea (1958) applied the kinematic wave theory that was developed by Lighthill and

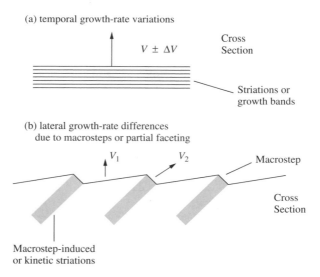

Figure 4.1 Definition of striations (a) caused by temporal growth-rate and temperature fluctuations as 'thermal striations', (b) caused by macrosteps and lateral growth-rate differences as 'macrostep-induced striations' or as 'kinetic striations'.

Whitham (1955) for the general traffic-flow problem. As shown in Figure 4.1(b) growth surfaces with macrosteps respectively with a terrace-and-riser structure with different local growth mechanisms and growth rates: The terrace grows due to lateral propagation of single or double steps with the facet growth rate v_1, whereas the macrostep has the velocity v_2. These growth-rate differences cause corresponding differences in impurity or dopant incorporation and thus lead to the striations which mark the traces of the macrosteps. These 'macrostep-induced' or 'kinetic striations' are shown in Figure 4.2 with LPE-grown multilayers, where the correspondence of the kinetic striations (visible in the angle-lapped and etched p-GaAs layers) with the marked macrosteps is clearly recognized (Scheel 1980). In that work the transition of the misoriented macrostep-surface to the facet with a continuous step propagation by the Frank–Van Der Merwe growth mode is described, a transition that of course leads to layers with excellent homogeneity, i.e. without striations. Chernov and Scheel (1995) analyzed the conditions for achieving such atomically flat surfaces and extremely homogeneous layers (and crystals), see also the epitaxy review of Scheel in this volume.

Macrosteps can be regarded as a first step towards growth instability and are formed not only at high supersaturation, but also in the presence of certain impurities, and on misoriented surfaces (Chernov 1992). When the thermodynamic driving force is further increased, then the impurity built up in front of

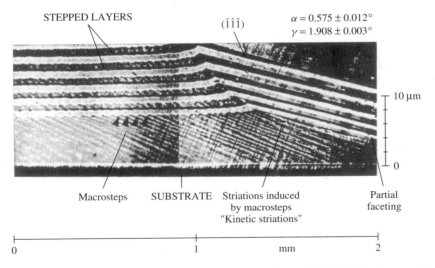

Figure 4.2 Macrostep-induced striations visible in the etched p-GaAs layers of an 11-layer p-/n-GaAs structure grown by liquid phase epitaxy (Scheel 1980). On the GaAs substrate of 0.58° misorientation first a thick layer is grown, which on the right side shows the transition to the {111} facet, whereas the macrostepped surfaces on the left side cause the striations. Angle-lapped ($\gamma = 1.9°$) and etched, composite differential interference (Nomarski) micrograph.

the growing crystal may reach a critical level and is suddenly incorporated, then crystal growth continues until again the impurity is incorporated, and so on. This oscillation of growth rate and impurity incorporation leading to striations was analyzed theoretically and experimentally by Landau (1958). This kind of inhomogeneity could be named 'instability-induced striations'.

4.3   HOMOGENEOUS CRYSTALS WITH $k_{eff} \rightarrow 1$

The segregation problem is defined by the distribution coefficient k which gives the concentration ratio of a constituent in the grown crystal to that in the growth melt or solution as will be discussed later. Normally k is not unity so that either more constituent is incorporated ($k > 1$) or less constituent than in the growth fluid is built into the crystal ($k < 1$). In these cases $k = 1$ could be achieved in quasi-diffusionless growth, at very high solidification rates. These cannot be applied in bulk crystal growth, but might be acceptable in certain cases of low-dimensional growth, for instance in one-dimensional whiskers or thin rods, or in two-dimensional structures like thin plates or in lateral growth of thin layers.

Mateika (1984) has shown that in the complex garnet structures with tetrahedral, octahedral and dodecahedral sites and by crystal-chemical considerations, cations could be introduced and combined, so that $k = 1$ was achieved. In order to obtain a garnet substrate crystal with a specific lattice constant, the ionic radii and valency of the cations had to be taken into account. For example, the garnet $Gd_3Sc_xGa_{5-x}O_{12}$ with $a = 12.543$ Å could be grown from melt with $x = 1.6$, when $k(Sc) = 1$.

A constant effective distribution coefficient of unity could also be achieved in crystal growth from high-temperature solutions. By the use of solutions an additional degree of freedom is obtained, since the distribution coefficient also depends on the properties of the solvent and on solvent–solute interactions. Systematic experiments with the growth of oxide solid solutions have shown that different solvents and solvent mixtures may cause $k_{eff} > 1$ and $k_{eff} < 1$ for a given solid solution. By proper mixing of the solvent it is then possible to obtain $k = 1$ and thus eliminate the segregation problem. Two examples are given with the perovskite solid solutions $Gd_{1-x}Y_xAlO_3$ and $Gd_{1-x}La_xAlO_3$ that are grown from solvent mixtures with the major components PbO and PbF_2 and the minor components B_2O_3 and excess Al_2O_3 (Scheel and Swendsen 2001).

In Figure 4.3 the measured effective distribution coefficients of the La and Y dopants in $GdAlO_3$ are shown as a function of the $PbO–PbF_2$ solvent composition. For the pure PbO flux, k_{eff} for La is nearly four, whereas for equal concentrations of oxide and fluoride k_{eff} is less than 0.5. At a composition of about 37 mol% PbF_2, $k_{eff} = 1$ is obtained. In the case of Y-doped $GdAlO_3$, a nearly pure PbO solvent is required to obtain $k_{eff} = 1$.

An example of garnet solid solutions grown from $PbO–PbF_2$ solvent mixtures is shown in Figure 4.4 using the data of Krishnan (1972). Here the distribution

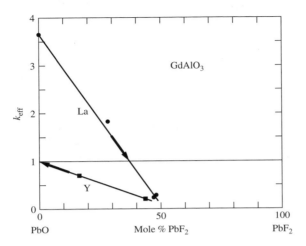

Figure 4.3 The dependence of the effective distribution coefficient on solvent composition in growth of La-doped and Y-doped gadolinium aluminate, after Scheel and Swendsen 2001.

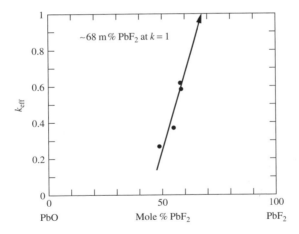

Figure 4.4 The dependence of the effective distribution coefficient on the $PbO-PbF_2$ solvent ratio in flux growth of Cr-doped yttrium iron garnet, after Scheel and Swendsen 2001.

coefficient for Cr in yttrium-iron garnet increases with increasing PbF_2 concentration, so that by extrapolation a solvent mixture with about 32 mol% PbO should give $k_{eff} = 1$. Another example of $Ga_{2-x}Fe_xO_3$ solid solutions grown from PbO-B_2O_3 and Bi_2O_3-B_2O_3 solvent mixtures shows that $k_{eff} = 1$ can be approached with the Bi_2O_3-rich flux (data of Schieber 1966).

A systematic investigation would help to understand these approaches to achieve $k_{\text{eff}} = 1$ and to establish rules for the general applicability.

4.4 SEGREGATION PHENOMENA AND THERMAL STRIATIONS

The concentration of the constituents of a solid solution generally differs from that in the liquid from which the mixed crystal is grown, a phenomenon known as segregation. In equilibrium or at very low growth rates the ratio of the concentration of component A in the solid to that in the liquid is defined as equilibrium segregation (or distribution) coefficient

$$k(A)_0 = C(A)_S/C(A)_L \tag{4.1}$$

k_0 may be derived from the equilibrium phase diagram by the ratio of solidus and liquidus concentrations of A at a given temperature.

In a crystallizing system with a limited melt volume, segregation at the growing interface leads to a continuous change of the fluid: $C(A)_L$ decreases for $k_{\text{eff}} > 1$, and $C(A)_L$ increases for $k_{\text{eff}} < 1$, so that the concentration $C(A)_S$ in the solid continuously changes. This causes an inherent concentration gradient in the crystal of which the concentration at any location of the growth front is given by

$$C(A) = k_{\text{eff}}C_0(1 - g)^{k_{\text{eff}}^{-1}} \tag{4.2}$$

where g is the fraction of crystallized material and C_0 the initial concentration as derived by Pfann (1952). Here it is assumed that k_{eff} does not vary with concentration or temperature changes. Obviously the inherent concentration gradient can be made zero by keeping the fluid concentration constant. This can be done by growth at constant temperature in combination with transporting feed material from a higher temperature, a growth technique called gradient-transport technique (Elwell and Scheel 1975). However, this approach generally involves large temperature gradients (for acceptable growth rates) and thus leads to temperature fluctuations, so that striations cannot be prevented.

In Figure 4.5 the phase diagram of the system $KTaO_3 - KNbO_3$ of Reisman et al. (1955) is presented where the gradient-transport technique, using the temperature difference T_3 to T_4, is shown. Numerous groups listed by Rytz and Scheel (1982) using this approach could not grow the required striation-free crystals of KTN ($KTa_{1-x}Nb_xO_3$) solid solutions. For electro-optic and other optical applications the variation of x must be smaller than 0.00003, and this requires temperature fluctuations smaller than 0.01 °C as is indicated in Figure 4.5. Also shown is the slow-cooling technique, where a quasiisothermal solution is cooled from T_1 to T_2. This allows to grow striation-free crystals, and the inherent concentration gradient can be kept within tolerated limits by using a large melt. For the KTN system Rytz and Scheel (1982) have calculated the mass of melt M

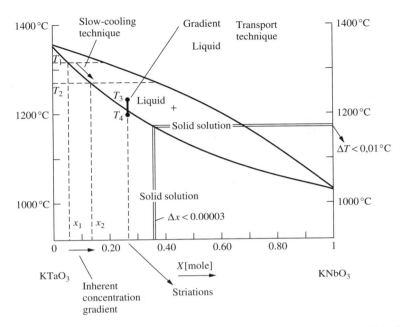

Figure 4.5 The phase diagram $KTaO_3$–$KNbO_3$ of Reisman *et al.* (1955) with indicated growth techniques and resulting inhomogeneities, after Rytz and Scheel 1982.

that is required to grow a crystal of specified volume V with a tolerated inherent concentration difference $x_1 - x_2$. These data and also the temperature cooling range can be read from the nomogram shown in Figure 4.6. For example, a KTN crystal of $1\,cm^3$ with $\Delta x < 0.02$, requires a melt of $1000\,g$ that is cooled by $12\,°C$. The results of growth experiments will be discussed further below.

In normal crystal growth we have neither the case of very fast growth and $k_{eff} = 1$ nor are we near equilibrium with very low growth rates and $k_{eff} = k_0$, as crystals should be grown at the fastest possible rate that still gives high structural perfection. Figure 4.7 shows the situations at the crystal/liquid interface for the three cases of equilibrium (which theoretically could also be achieved by 'complete' mixing), for the steady-state normal crystal growth, and for fast diffusionless solidification. Also shown are the concentrations in the solid and in the liquid with the diffusion boundary layers. In the steady-state crystal growth the effective distribution coefficient k_{eff} lies between the equilibrium distribution coefficient k_0 and 1 and is dependent on the diffusion boundary layer δ and the growth rate v as shown in Figure 4.7.

For the case of pulling a rotating crystal from the melt by the Czochralski technique, the flow analysis of Cochran (1934) towards an infinite rotating disc was applied by Burton, Prim and Slichter (BPS, 1953a) in order to derive the effective distribution coefficient. Thereby it was considered that the solute concentration

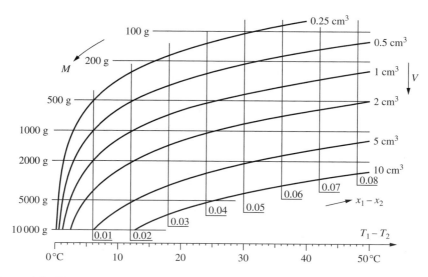

Figure 4.6 Nomogram to read the experimental parameters for growth of KTN solid solution crystals of a specified maximum inhomogeneity $x_1 - x_2$ and volume V. M = mass of the melt, $T_1 - T_2$ is the cooling interval (Rytz and Scheel 1982).

profile is virtually uniform in the radial direction, i.e. the diffusion boundary layer δ can be regarded as having constant thickness across the idealized flat growth interface. Up to a not too large growth rate, δ depends essentially on the crystal rotation rate ω (except for the rim, which is neglected), the kinematic viscosity v, and a bulk diffusion coefficient D (which includes solute diffusion and solvent counter-diffusion) by

$$\delta = 1.6 D^{1/3} v^{1/6} \omega^{1/2}. \tag{4.3}$$

BPS derived for the steady-state case, in which equilibrium prevails at the interface virtually independently of growth rate, the equilibrium distribution coefficient

$$k_{\text{eff}} = k_0 / [k_0 + (1 - k_0) \exp -(v\delta/D)] \tag{4.4}$$

with v the growth rate. In the following, this approximation can be utilized, since the growth-rate dependence of k_0 will be negligible due to the fact that in our attempt to grow striation-free crystals the effective variations of the growth rate are small.

In the accompanying experimental paper Burton *et al.* (1953b) measured the distribution coefficients of several elements in solid/liquid germanium and were able to correlate them with atomic size, respectively, the tetrahedral covalent radii: the larger the element, the smaller the distribution coefficient. They also measured striations by means of incorporated radioactive tracers and photographic

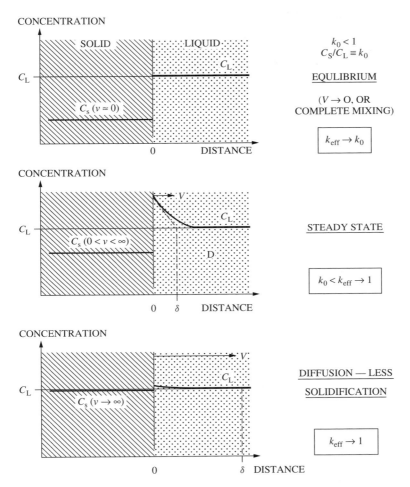

Figure 4.7 The concentration relations at the growth interface and the effective distribution coefficients for three growth situations: near equilibrium k_{eff} approaches k_0, in very fast solidification k_{eff} approaches unity, and in typical crystal growth k_{eff} is between k_0 and 1. The diffusion boundary layer thickness δ depends on the growth rate v and on the flow rate of melt or solution.

film and pointed out the importance of adequate stirring for the growth of homogeneous crystals.

For growth of mixed crystals from dilute solutions, van Erk (1982) has considered the complex solute–solvent interactions and has derived the effective distribution coefficient for diffusion-limited growth as

$$\ln k_{\text{eff}} = \ln k_0 - (k_{\text{eff}} - 1)(v\delta/D). \tag{4.5}$$

The plotted solutions of Equations (4.4) and (4.5) shown in Figures 4.8 and 4.9, respectively, look similar, but the sensitivity to fluctuations of the growth parameters is different (Scheel and Swendsen 2001).

Growth from solutions is normally limited by volume diffusion, and the relatively fast interface kinetics can be neglected. Based on the diffusion-boundary

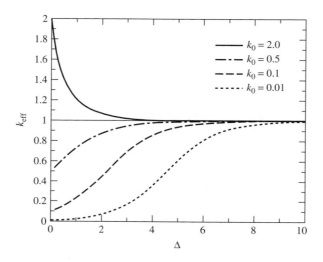

Figure 4.8 The effective distribution coefficient as a function of the exponent $(v\delta/D) = \Delta$ from the Burton-Prim-Slichter Equation (4.4) for four values of k_0 (Scheel and Swendsen 2001).

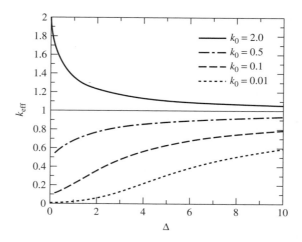

Figure 4.9 The effective distribution coefficient as a function of the exponent Δ from the Van Erk theory and Equation (4.5) for four values of k_0 (Scheel and Swendsen 2001).

layer concept, Nernst (1904) has derived the growth rate as

$$v = D(n_\infty - n_e)/\rho_c \delta \tag{4.6}$$

with n_e and n_∞ the equilibrium and effective bulk concentrations of the solute in the solution, and ρ_c is the solute density. The time constant for the effects of temperature fluctuations (and thus on growth rate v) is on the order of seconds, while that for hydrodynamic fluctuations on δ and v is on the order of minutes. In steady-state growth, within a given range of time and temperature, the diffusion coefficient D and the equilibrium distribution coefficient k_0 in Equations (4.4) and (4.5) can be taken as constants. Therefore the changes in the effective distribution coefficient k_{eff} are essentially determined by the product $(v\delta)$ of the exponent. As a first approximation this product is constant due to the inverse relation between v and δ in the growth-rate Equation (4.6). This means that hydrodynamic changes, which lead to changes of δ, are compensated by growth-rate changes. On the other hand, growth-rate changes caused by temperature fluctuations are not compensated and thus lead to changes of k_{eff} and to thermal striations.

It follows from this discussion that for growth of striation-free crystals the temperature fluctuations should be suppressed to less than about $0.01\,°C$, and therefore also the temperature gradients should be minimized to less than about $1\,°C$ per cm. On the other hand the crystal has to be cooled to remove the latent heat at practical growth rates, and to control nucleation.

The application of forced convection is recommended for efficient growth. Homogenizing the melts and solutions facilitates the achievement of above temperature conditions, and it reduces the diffusion problems leading to growth instability as discussed in Chapter 6 of Elwell and Scheel (1975). Stirring may be achieved by a continuous flow along the growth interface, by Ekman or Cochran flow towards the rotating growth surface, by periodic flow changes, as in reciprocating stirring in growth from aqueous solutions or by accelerated crucible rotation technique (ACRT) in growth from high-temperature solutions or in growth by the Bridgman–Stockbarger technique, etc.

For many years the favored approach to minimize striations was to reduce convection, to apply microgravity or in the case of semiconductors to apply convection-damping magnetic fields. But the above discussion has shown that, except for special cases requiring large temperature gradients, forced convection has many advantages, not least to achieve economic growth rates approaching the maximum stable growth rates for inclusion-free crystals.

From the above discussion we can also derive the experimental conditions to induce regular striations and superlattice structures, for instance by applying a large temperature gradient in combination with periodic hydrodynamic changes (by ACRT) and applying a constant mean supersaturation, respectively, a constant mean growth rate.

The suppression of striations represents an old problem in crystal-growth technology, and several authors have described striations as an 'inherent', 'intrinsic' or 'unavoidable' phenomenon in crystal growth, for example Byer (1974), Räuber (1978), and Reiche *et al.* (1980).

In the following, the theoretical considerations discussed above will be applied to an example of growing striation-free crystals of KTN solid solutions which could long not be achieved despite the numerous attempts listed by Rytz and Scheel (1982).

4.5   GROWTH OF STRIATION–FREE KTN CRYSTALS

The solid-solution system $KTa_{1-x}Nb_xO_3$ (KTN) is of interest due to its very large electro-optic coefficient which can be optimized for specific application temperatures by the choice of x (Scheel and Günter 1985). The composition with $x = 0.35$ is used for room-temperature applications, because the ferroelectric transition temperature is then $10\,°C$ and the very large dielectric constant and electro-optic coefficient are observed just above the transition. However, for optical applications the inhomogeneity in refractive index should be less than 10^{-6}, requiring that the crystals and layers be striation-free to this level.

As discussed above, this homogeneity can not be achieved by a gradient-transport technique, but only by slow-cooling of nearly isothermal solutions. The latter was applied by Rytz and Scheel (1982) and Scheel and Sommerauer (1983) who, furthermore, combined an ultra-precise temperature control with optimized temperature distribution in the furnace and applied stirring by the accelerated crucible rotation technique (ACRT) (Scheel 1972) and thus could obtain striation-free KTN crystals for the first time. A typical crucible arrangement with bottom cooling to provide a nucleation site is shown in Figure 4.10 along with the applied ACRT cycle. In ACRT the crucible is periodically accelerated and decelerated so that by inertia the liquid is moving relative to the crucible wall and forms a spiral when seen from top. This spiral shear flow (or spiral shearing distortion) was analyzed and simulated by Schulz-DuBois (1972). He also analyzed the Ekman layer flow in which, under optimized conditions, the liquid is always pumped through a thin Ekman layer at the bottom of the crucible when this is accelerated and decelerated. In Chapter 7 of the book of Elwell and Scheel (1975) the derivation of optimized ACRT stirring based on the kinematic viscosity of the liquid and on the crucible dimension is treated.

Temperature control with a precision of $0.03\,°C$ at around $1300\,°C$ could be achieved by using a thermopile of Pt-6%Rh versus Pt-30%Rh thermocouples (Scheel and West 1973). In Figure 4.11(a) the electromotive forces, i.e. the high-temperature sensitivities of conventional thermocouples and of the 6-fold thermopile are compared whereby the room-temperature sensitivity of the thermopile (Figure 4.11(b)) is practically zero, so that cold-junction compensation is not required. The temperature distribution in the chamber furnace was optimized by arranging heating elements and ceramic insulation according to the reading of numerous thermocouples in the system, and the control thermopile was positioned at the optimum site between heating elements and crucible. KTN crystals

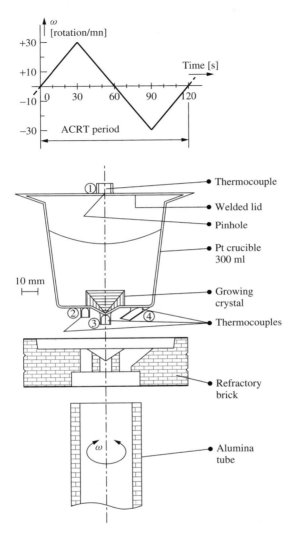

Figure 4.10 Side view of a 300-cm³ platinum crucible, with welded lid and slightly cooled spot at crucible bottom for nucleation control, used in the ACRT experiments for growth of KTN crystals (Rytz and Scheel 1982). A typical ACRT period is also shown.

with $x = 0.01$ and $x = 0.25$ and up to $33 \times 33 \times 15\,\text{mm}^3$ in size were grown, see Table 4.2 for details of the results.

In the best crystals no striations were visible in a polarizing microscope, although very faint striations could be revealed by using extremely sensitive methods (Scheel and Günter 1985).

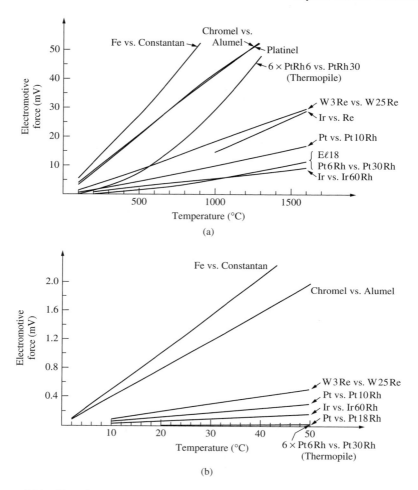

Figure 4.11 The electromotive forces of various common thermocouples (a) for high temperatures, (b) for the temperature range 0 to 50 °C. Note that the thermopile of 6 × PtRh6% versus PtRh30% has a very high sensitivity at high temperatures and practically no sensitivity around room temperature (Scheel and West 1973, from Elwell and Scheel 1975).

4.6 ALTERNATIVE APPROACHES TO REDUCE STRIATIONS

Based on the early studies of striations discussed in Section 4.2, the striations were generally connected with convective oscillations: a clear distinction between the effects of hydrodynamics and of temperature as discussed by Scheel and Swendsen (2001) and in Section 4.4 was not consequently attempted before. This explains the numerous approaches to *reduce* convection in order to solve

Table 4.2 Results of 'striation-free' $KTa_{1-x}Nb_xO_3$ crystals

with $x \sim 0.01$ and $T_c \sim 0\,K$ (Rytz and Scheel 1982)
(cooling rate $0.15\,°C$ per hour; ACRT)
Crystal size up to $3.5 \times 3.0 \times 1.8\,cm^3$ with 'perfect' regions up to $0.8 \times 0.4 \times 0.4\,cm^3$
$x = 0.0070 \pm 0.00002$ and $\Delta x < 0.0003/cm$ (detection limit of ARL electron microprobe and of acoustic resonance measurements)
Very faint striations by polarizing microscope, not detectable by electron microprobe
with $x \sim 0.25$ and $T_c \sim 300\,K$ (Scheel and Sommerauer 1983)
(cooling rate $0.13\,°C$ per hour; ACRT; growth rate $\sim 500\,Å/s$)
Crystal size up to $3.6 \times 3.3 \times 1.5\,cm^3$ with 'perfect' regions up to $0.5 \times 0.5 \times 0.25\,cm^3$
$\Delta T = 0.1\,°C \sim \Delta x = 0.0003/\,cm$ at $x = 0.25$
No striations detectable in polarizing microscope and by interferometry (no strain birefringence)
Best literature values for $x \sim 0.25$–0.3:   always with striations $\Delta x = 0.006/cm$ Fay 1967 $\Delta x = 0.002/cm$ Levy and Gashler 1968

the striation problem, whereas it was discussed above that *forced convection* and even *oscillating convection* of quasi-isothermal solutions can solve the striation problem.

In the following, first the attempts to reduce convection will be briefly discussed on the base of the Rayleigh number Ra. At a critical value of Ra one expects the onset of convection in a fluid between a warm bottom plate and a cool top plate (although in reality convection will always occur whenever there is a temperature difference in a real growth system):

$$Ra = \frac{g\alpha L^3 \Delta T}{k\nu} \tag{4.7}$$

where α is the thermal expansion coefficient, L the fluid height, ΔT the temperature difference, k thermal conductivity, ν the kinematic viscosity, and g the gravitation constant. The first approaches have been to reduce L by means of shallow melts or by baffles below the growth interface in Czochralski crystal pulling. Also double crucibles may have a certain effect on striations (Kozhemyakin 2000), but their greatest advantage is the increased axial crystal yield with the same composition, as it was shown by Benson *et al.* 1981 for silicon and later applied to GaAs and $LiNbO_3$ growth.

Edge-defined film-fed growth EFG allows $k_{eff} \sim 1$ to be approached as was shown theoretically by Kalejs (1978) and experimentally by Fukuda and Hirano (1976) and Matsumura and Fukuda (1976) for platelet growth of doped $LiNbO_3$ and $LiTaO_3$. Shaped growth by pulling downwards has been studied by Miyazawa (1982) who grew relatively homogeneous Te-doped GaSb platelets by 'shaped melt lowering'. A capillary-controlled Czochralski process for shaped crystals of $Sr_xBa_{1-x}Nb_2O_6$, which has similarity with Stepanov and EFG, was applied by Ivleva *et al.* (1987, 1995) and has led to nearly striation-free crystals due to the thin melt layer between shaping multicapillary die and the crystal. Nakajima (1991) and Kusonoki *et al.* (1991) tried to grow InGaAs solid-solution crystals by ramp-cooling and by GaAs supply at constant temperature similar to W. Bonner, but it seems that large homogeneous crystals for technological use as substrates has not yet been achieved. Ostrogorsky and Müller (1994) proposed a submerged-heater method for vertical Bridgman growth, where the combined effect of thin melt layer and stabilizing temperature gradient should minimize striations, but the complexity has hampered technological application.

Large-scale application has found the modified Czochralski process for pulling huge halide scintillator crystals (up to 700 mm diameter and 550 kg weight) from small melt volumes with liquid feeding by Gektin and Zaslavsky, see in this volume. In this case, there is a combined effect of small liquid volume and forced convection due to counterrotation of crystal and crucible.

But from the time of Skylab and with the available funding for space experiments, the interest shifted in the direction of microgravity, of reduction of g in Equation (4.5). In practical fabrication of semiconductor crystals it was recognized that microgravity would not be applicable for various reasons, so that here the parameter v of Equation (4.7) was considered: by application of magnetic fields the viscosity was increased, and convection was damped. Magnetic fields were first introduced in 1980 by SONY to silicon production, followed by Fukuda and Terashima, who in 1983 applied magnetic field to GaAs growth, which in industry is still applied in LEC production of specific GaAs products.

Alternative approaches to apply forced convection have been discussed in the Elwell–Scheel book, for instance 'sloshing' of the crucible by moving the center of the container in a horizontal circular path described first by Gunn (1972) as 'sloshing' and later as a special vibration technique by Liu *et al.* (1987). Kirgintsev and Avvakumov (1965) compared various stirring techniques including vibrators and found the latter not very effective. Hayakawa and Kumagawa (1985) and more recently Kozhemyakin (1995) and Zharikov *et al.* revived the interest in vibration stirring, although it has not yet found application in real crystal production.

An optimized technology for growth from solutions using a large rotating seed crystal (or a substrate for liquid phase epitaxy) is shown in Figure 4.12. In bulk growth the grown crystal can be cut off and, after etching to remove the surface damage from slicing, the seed crystal can be used again for the next growth cycle, a little like the large-diameter seed crystals recently developed in silicon Czochralski technology. The optimum crystallographic orientation of the seed is important firstly in view of the growth mechanism to remain at a flat growth surface and to prevent macrostep formation, and secondly with respect to the application of the crystal. In this configuration, a sufficiently high seed rotation rate provides Ekman layer flow and thus a quasiconstant diffusion boundary layer along the crystal surface.

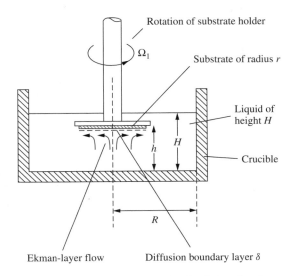

Figure 4.12 An ideal growth system for growth of striation-free large-diameter crystals and of thick and homogeneous LPE layers.

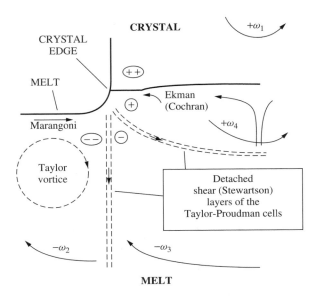

Figure 4.13 More than six types of flow, differing in direction and/or velocity, below the edge of a Czochralski-grown crystal for the case of crystal rotation $+\omega_1$ and crucible counterrotation $-\omega_2$ above the critical Rossby number (Scheel and Sielawa 1985).

For growth from melts by Czochralski crystal pulling there are six or more different kinds of flows below the rim of the crystal as shown in Figure 4.13: flows with different directions and/or different velocities. Double crucibles discussed above will help somewhat to simplify the flow pattern but have practical disadvantages in the growth process. A technologically simpler approach consists of a solid ring coaxial with the crystal rotation axis and rotating with same direction and velocity as the crystal. This ring can be introduced into the melt after complete melting and it can be easily removed at the end of the growth process. In simulation experiments it was shown that at optimized setting of the rotation rate of crystal and ring, and counterrotation of the crucible, the melt fraction inside the ring is nearly stationary and separated from the well-mixed bulk melt, see Figure 4.14. This co-rotating ring Czochralski CRCZ approach of Scheel (1995) has the positive double-crucible effects without the disadvantages of a double crucible: an increased yield of axially homogeneous crystal

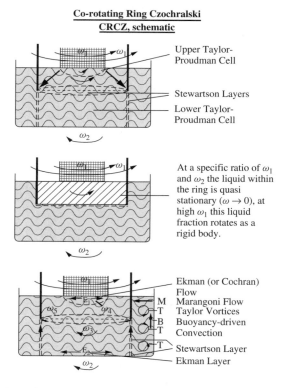

Figure 4.14   Schematic presentation of the Co-rotating ring Czochralski CRCZ concept where the complex flow region of Figure 4.13 is transferred deeper into the melt and where the liquid fraction within the ring can be separated from bulk melt flow under optimized conditions (Scheel 1995).

and reduced striations. Numerical simulations of CRCZ have started (Kakimoto 2002; Zhong Zeng 2001), but the growth technology has yet to be developed.

4.7 DISCUSSION

Several experimental and theoretical solutions were presented to overcome the as intrinsic regarded striation problem:

1. k_0, D, v, δ = constant: very difficult; chance with rotating seed (Figure 4.12) & with CRCZ (Figure 4.14)
2. k_0, D, v = constant; δ may vary   Rytz & Scheel 1982; Scheel and Sommerauer 1983; Scheel and Swendsen 2001
3. $k_{eff} = 1$ for growth from melt: Mateika 1984
4. $k_{eff} = 1$ for growth from solution: Scheel and Swendsen 2001
5. Avoid solid solutions by growth of simple stoichiometric compounds with well-defined site distribution from ultra-pure chemicals

Further conditions are

(a) Continuous flat or smooth growth surface
(b) Isothermal growth surface with $\Delta T / T < 10^{-5}$
(c) Homogeneous melt or solution with $\Delta n / n < 10^{-6}$
(d) Constant growth rate with $\Delta v / v < 10^{-5}$.

These indicated tolerances are typical values and depend on the individual system and on the tolerated inhomogeneity of the crystal. Therefore it is advisable to analyze a new growth system and the phase relations theoretically so that the required technological parameters can be established. With respect to optimized hydrodynamics, simulation experiments with a liquid of similar kinematic viscosity are very useful in the early phase and may be complemented by numerical simulation for process optimization. However, the latter require reliable material (liquid) parameters, which often are not available. In certain cases dimensionless numbers may be helpful to get a feel for the convection regime.

In conclusion, one can say that the striation problem is solvable (on earth) but requires a certain theoretical and technological effort. Hydrodynamic fluctuations are not harmful as long as the fluid is sufficiently isothermal, as long as the transport of fluid of different temperature to the growth interface is suppressed.

REFERENCES

Bardsley W., Boulton J. S. and Hurle D. T. J. (1962), *Solid State Electron.* **5**, 365.
Burton J. A., Prim R. C. and Slichter W. P. (1953a), *J. Chem. Phys.* **21**, 1987.

Burton J. A., Kolb E. D., Slichter W. P., and Struthers J. D. (1953b), *J. Chem. Phys.* **21**, 1991.

Byer R. L. (1974), *Ann. Rev. Mater. Sci.* **4**, 147.

Cabrera N. and Vermilyea D. A. (1958) in *Growth and Perfection of Crystals*, editors R. H. Doremus, B. W. Roberts and D. Turnbull, Wiley, New York and Chapman and Hall, London, p. 393.

Camp P. R. (1954), *J. Appl. Phys.* **25**, 459.

Chen F. S., Geusic J. E., Kurtz S. K., Skinner J. C. and Wemple S. H. (1966), *J. Appl. Phys.* **37**, 388.

Chernov A. A. (1992), *J. Cryst. Growth* **118**, 333.

Chernov A. A. and Scheel H. J. (1995), *J. Cryst. Growth* **149**, 187.

Cockayne B. and Gates M. P. (1967), *J. Mater. Science* **2**, 118.

Donecker J., Lux B. and Reiche P. (1996), *J. Cryst. Growth* **166**, 303.

Elwell D. and Scheel H. J. (1975) *Crystal Growth from High-Temperature Solutions*, Academic Press, London, New York, Chapter 7.

Frank F. C. (1958) in *Growth and Perfection of Crystals*, editors R. H. Doremus, B. W. Roberts and D. Turnbull, Wiley, New York and Chapman and Hall, London, p. 411.

Fukuda T. and Hirano H. (1976), *J. Cryst. Growth* **35**, 127.

Goss A. J., Benson K. E. and Pfann W. G. (1956), *Acta Met.* **4**, 332.

Gunn J. B. (1972), *IBM Tech. Discl. Bull.* **15**, 1050.

Hayakawa Y. and Kumagawa M. (1985), *Cryst. Res. Technol.* **20**, 3.

Hurle D. T. J. (1966), *Philos. Mag.* **13**, 305.

Hurle D. T. J. (1967), in *Crystal Growth*, editor H. S. Peiser, Pergamon, Oxford, 659–663.

Ivleva L. I., Yu. S. Kuz'minov V. V. Osiko and Polozkov N. M. (1987), *J. Cryst. Growth* **82**, 168.

Ivleva L. I., Bogodaev N. V., Polozkov N. M. and Osiko V. V. (1995), *Opt. Mater.* **4**, 168.

Kakimoto K. (2002), private communication.

Kalejs J. P. (1978), *J. Cryst. Growth* **44**, 329.

Kirgintsev A. N. and Avvakumov E. G. (1965), *Sov. Phys.-Cryst.* **10**, 375.

Kozhemyakin G. N. (1995), *J. Cryst. Growth* **149**, 266.

Kozhemyakin G. N. (2000), *J. Cryst. Growth* **220**, 39.

Krishnan R. (1972), *J. Cryst. Growth* **13/14**, 582.

Kusonoki T., Takenaka C. and Nakajima K. (1991), *J. Cryst. Growth* **115**, 723.

Landau A. P. (1958), *Fiz. Metal. Metalloved* **6**, 193.

Laves F. (1944), *Die Chemie* **57**, 30–33; reprinted in *Z. Krist.* **151** (1980) 21.

Lighthill M. J. and Whitham G. B. (1955), *Proc. Roy. Soc.* **229**, 281.

Liu W.-S., Wolf M. F., Elwell D. and Feigelson R. S. (1987) *J. Cryst. Growth* **82**, 589.

Loo R., Kamath G. S. and Knechtli R. C. (1980), *Radiation damage in GaAs solar cells*, 14th IEEE Photovoltaics Specialists Conference.

Mateika D. (1984) in *Current Topics in Materials Science*, editor E. Kaldis, North-Holland/Elsevier, Chapter 2.

Matsumura S. and Fukuda T. (1976), *J. Cryst. Growth* **34**, 350.

Miyazawa S. (1982), *J. Cryst. Growth* **60**, 331.

Nabarro F. R. N. (1975) in *Solution and Precipitation Hardening. The Physics of Metals, Part 2, Defects*, editor P. B. Hirsh, Cambridge University Press Cambridge, p. 152.

Nakajima K. (1991), *J. Cryst. Growth* **110**, 781.

Nernst W. (1904), *Z. Phys. Chem.* **47**, 52.

Ostrogorsky A. G. and G. Müller (1994), *J. Cryst. Growth* **137**, 66.

Pfann W. G. (1952), *J. Met.* **4**, 747.

Räuber A. (1978), in *Current Topics in Materials Science* Vol. 1, editor E. Kaldis, North-Holland, Amsterdam, p. 481.

Reiche P., Schlage R., Bohm J., and Schultze D. (1980), *Kristall U. Technik* **15**, 23.

Reisman A., Triebwasser S. and Holtzberg F. (1955), *J. Am. Chem. Soc.* **77**, 4228.

Rytz D. and Scheel H. J. (1982), *J. Cryst. Growth* **59**, 468.

Scheel H. J. (1972), *J. Cryst. Growth* **13/14**, **560**.

Scheel H. J. (1980), *Appl. Phys. Lett.* **37**, 70.

Scheel H. J. (1995), US Patent 5,471,943.

Scheel H. J. and Günter P. (1985), Chapter 12 in *Crystal Growth of Electronic Materials*, editor E. Kaldis, Elsevier, 149–157.

Scheel H. J. and Sielawa J. T. (1985), Proceedings International Symposium on High-Purity Materials, Dresden/GDR May 6–10, 1985, 232–244.

Scheel H. J. and Sommerauer J. (1983), *J. Cryst. Growth* **62**, 291.

Scheel H. J. and Swendsen R. H. (2001), *J. Cryst. Growth* **233**, 609.

Scheel H. J. and West C. H. (1973), *J. Phys E* **6**, 1178; see also Elwell and Scheel (1975) Chapter 7.

Schieber M. (1966), *J. Appl. Phys.* **37**, 4588.

Schulz-DuBois E. O. (1972), *J. Cryst. Growth* **12**, 81.

Van Erk W. (1982), *J. Cryst. Growth* **57**, 71.

Wenzl H., Oates W. A. and Mika K. (1993), Chapter 3 in *Handbook of Crystal Growth* Vol. 1, editor D. T. J. Hurle, Elsevier, Amsterdam.

Wilcox W. R. and Fullmer L. D. (1965), *J. Appl. Phys.* **36**, 2201.

Zhong Zeng (2001), private communication.

5 High-Resolution X-Ray Diffraction Techniques for Structural Characterization of Silicon and other Advanced Materials

KRISHAN LAL

National Physical Laboratory New Delhi-110 012 India

5.1 INTRODUCTION

The increasing density of devices on individual integrated circuits requires an unprecedented level of homogeneity in chemical composition and structural perfection of silicon and other semiconductor crystals (Claeys and Deferm 1996, Bullis and Huff 1993). A strict control of composition and trace impurities as well as on structural quality with particular emphasis on defects like dislocations, point defects and their clusters is needed. Therefore, the demand on material characterization technology is increasing as advances in other sectors of technology are reaching newer levels of sophistication (Laudise 1972, Lal 1991a).

Structural characterization of materials involves phase identification as well as determination of real structure. We shall be concerned with determination of real structure, which requires information about all types of defects and their concentrations in a given material (Lal 1998). The information regarding crystalline perfection may be required from the surface or from the bulk of the crystal. In the case of thin-film single-crystal substrate systems, the interface region is also important.

A wide variety of experimental techniques are available for observation and characterization of defects in bulk single crystals and thin films. These include chemical etching by suitable solvents (Sangwal 1987), decoration, birefringence (for specific materials) (Mathews *et al.* 1973), electron microscopy/diffraction (Heydenreich 1982, Spence 1984) and high-resolution X-ray diffraction techniques (Lang 1958, 1970, Kato 1975, Tanner 1976, Lal 1982, 1991a, 1991b 1998). The recently developed scanning tunneling microscopy (Binnig *et al.* 1982, Hansma and Tersoff 1987) and atomic force microscopy (Binnig *et al.* 1986, Alexander *et al.* 1989) are used to reveal atomic arrangement on solid surfaces. In this chapter, we shall consider

Crystal Growth Technology, Edited by H. J. Scheel and T. Fukuda
© 2003 John Wiley & Sons, Ltd.   ISBN: 0-471-49059-8

fundamentals of high-resolution X-ray diffraction techniques and their illustrative applications.

5.2 HIGH-RESOLUTION X-RAY DIFFRACTION TECHNIQUES

5.2.1 THEORETICAL BACKGROUND

The kinematical theory (Cullity 1967, James 1950, Guinier 1952, and Barrett and Massalski 1968) and the dynamical theory of X-ray diffraction (Laue 1960, Zachariasen 1945, Batterman and Cole 1964, Kato 1992, Pinsker 1978, Schneider and Bouchard 1992) are used to understand diffraction of X-rays from single crystals. Since, the semiconductor crystals are nearly perfect it is necessary to use the dynamical theory of X-ray diffraction. The kinematical theory can be used only when one is dealing with rather imperfect crystals or their dimensions are small. When a parallel beam of monochromatic X-rays travels through a crystal, electrons of the atoms in its path become scattering centres due to their interaction with the electric vector of the X-ray beam. An atom located in the interior of the crystal will receive not only the incoming radiation whose intensity has decreased due to scattering and absorption by layers above it but also it will receive scattered waves from all the other atoms that are irradiated. This implies that interaction between the scattered and the incoming waves as well as among the scattered waves is to be considered for determining the total wave field and the intensity of the diffracted beam and the shape of the diffraction curves or rocking curves. The diffraction curve is essentially a plot of the diffracted intensity as a function of the glancing angle θ around the exact Bragg angle θ_B. In the following, we give the most essential steps needed for calculation of diffracted intensity in the case of Bragg geometry (reflection) as well as the Laue geometry (transmission).

In the dynamical theory one solves Maxwell's equations for propagation of X-rays inside the crystal that is oriented near a diffraction maximum, so that:

$$\mathbf{K}_H = \mathbf{K}_0 + \mathbf{H} \tag{5.1}$$

Here $\mathbf{K}_0$ is the wave vector of the incident beam and $\mathbf{K}_H$ is that of the scattered beam from the crystal lattice having periodicity $\mathbf{H}$ ($H = 1/d$, or reciprocal of the lattice spacing of the diffracting planes). The X-rays propagating in the crystal will satisfy Maxwell's equations (e.g. Batterman and Cole 1964):

$$\nabla \times \mathbf{E} = -\partial \mathbf{B}/\partial t = -\mu_0 \partial \mathbf{H}/\partial t \tag{5.2}$$

$$\nabla \times \mathbf{H} = \partial \mathbf{D}/\partial t = \varepsilon_0 \partial (\kappa \mathbf{E})/\partial t \tag{5.3}$$

Here $\mathbf{E}$, $\mathbf{D}$, $\mathbf{B}$ and $\mathbf{H}$ are the four vector fields: the electric vector, the displacement vector, the magnetic vector and the magnetic induction vector, respectively. The conductivity σ of the crystal is taken as zero at X-ray frequencies. Also, the crystal is considered as nonmagnetic and it behaves like empty space so that $\mu = \mu_0$. The structure of the crystal is introduced by the dielectric constant κ, which is related to the electron density ρ through

$$\kappa(\mathbf{r}) = 1 - r_e(\lambda^2/\pi)\rho(\mathbf{r}) \tag{5.4}$$

Here, r_e is the classical radius of electron ($= e^2/4\pi\varepsilon_0 mc^2 = 2.818 \times 10^{-13}$ cm) and λ is the wavelength of the X-rays. The electron density $\rho(\mathbf{r})$ at any point $\mathbf{r}$ inside the crystal can be expressed as a Fourier sum over the reciprocal lattice. Hence, we can express Equation (5.4) as:

$$\kappa(\mathbf{r}) = 1 - \Gamma\Sigma_H F_H\exp(-2\pi i\mathbf{H}\cdot\mathbf{r}) \tag{5.5}$$

where F_H is the structure factor and $\Gamma = r_e\lambda^2/\pi V$. The dielectric constant $\kappa(\mathbf{r})$ is very close to unity, differing from it by a few parts per million or so. A sum of plane waves is assumed as the solution of Maxwell's equations, and can be expressed in the form:

$$\mathbf{A} = \exp(2\pi i\nu l)\Sigma_H\mathbf{A}_H\exp(-2\pi i\mathbf{K}_H\cdot\mathbf{r}) \tag{5.6}$$

Here, $\mathbf{A}$ stands for $\mathbf{E}$, $\mathbf{D}$, $\mathbf{B}$ or $\mathbf{H}$. When only one reciprocal lattice point is near the Ewald sphere, one obtains the following fundamental set of equations for the wave field inside the crystal after some mathematical analysis:

$$[k^2(1 - \Gamma F_0) - (\mathbf{K}_0.\mathbf{K}_0)]E_0 - k^2 P\Gamma F_{\overline{H}}E_H = 0 \tag{5.7a}$$

$$-k^2 P\Gamma F_H E_0 + [k^2(1 - \Gamma F_0) - (\mathbf{K}_H.\mathbf{K}_H)]E_H = 0 \tag{5.7b}$$

Here, the parameter P represents both states of polarization and equals 1 for σ polarization and is $\cos 2\theta$ for the π state. k is the vacuum value of the wave vector. From these basic equations one can, after some analysis, obtain the wave vectors as well as expressions for the wave amplitudes. The relationship between different wave vectors is elegantly represented by using the dispersion-surface concept. It is not possible to cover all aspects here. We shall consider the most essential part, which helps in calculating diffraction curves.

5.2.1.1 Laue Case

For symmetrical Laue geometry we finally get the following expressions for intensities of the diffracted ($I_{H\omega}{}^e$) and the forward diffracted beams ($I_{0\omega}{}^e$)

$$\frac{I_{0\omega}{}^e}{I_0} = \frac{1}{4}\left[1 \mp \frac{\eta'}{|1 + (\eta')^2|^{1/2}}\right]^2 \exp\left[-\frac{\mu_0 t_0}{\gamma_0}1 \mp \frac{|P|\varepsilon}{|1 + (\eta')^2|^{1/2}}\right]$$

$$\frac{I_{H\omega}{}^e}{I_0} = \frac{1}{4}\frac{1}{1 + (\eta')^2}\exp\left[-\frac{\mu_0 t_0}{\gamma_0}\left\{1 \mp \frac{|P|\varepsilon}{[1 + (\eta')^2]^{1/2}}\right\}\right] \tag{5.8}$$

Here, ω refers to a particular polarization state of either the α or the β branch of the dispersion surface, I_0 is the intensity of the exploring beam and t_0 is the thickness of the crystal. The upper negative sign is for the α branch of the dispersion surface, whereas for β branch positive sign has to be used. Also, it has been assumed that the crystal is centrosymmetric. The orientation of the crystal is defined with respect to the exact diffraction condition by $\Delta\theta$. It is related to a parameter η' by the expression:

$$\eta' = (\Delta\theta \sin 2\theta)/|P||\Gamma||F_H'| \qquad (5.9)$$

Theoretical diffraction curves can be obtained from Equations (5.8) and (5.9). The absolute value of integrated intensity can be calculated from the area under the diffraction curve. The half-width can be obtained from Equation (5.8) as the angular spread corresponding to $\Delta\eta' = 2$, i.e.

$$[\Delta\theta]_{HW} = 2|P||\Gamma||F_H'|/\sin 2\theta \qquad (5.10)$$

5.2.1.2 Bragg Case

In the case of Bragg diffraction the diffracted beam emerges from the same surface on which it is incident. It shows interesting features like total reflection for nonabsorbing crystals. For a centrosymmetric crystal, the reflection coefficient defined as $|E_H^e/E_0^i|^2$ is given by the following expression:

$$\frac{|E_H^e|^2}{|E_0^i|^2} = |b||\eta \pm (\eta^2 - 1)^{1/2}|^2 \qquad (5.11)$$

For symmetrical Bragg geometry $b = -1$.

All real crystals have finite absorption coefficients for X-rays and therefore the structure factor, the wave vectors, the dielectric and η are complex quantities. For symmetrical Bragg geometry $\eta(= (\eta' + i\eta''))$ is defined by:

$$\eta' = (-\Delta\theta \sin 2\theta + \Gamma F_0')/|P|\Gamma F_H'$$
$$\eta'' = -(F_H''/F_H')(\eta' - 1/|P|\varepsilon) \qquad (5.12)$$

Figure 5.1 shows a diffraction curve of a silicon crystal calculated for 111 reflection and $MoK\alpha_1$ radiation for symmetrical Bragg geometry. This curve shows very high reflectivity at the peak and its half-width is only $\sim$3 arcsec.

5.2.2   HIGH-RESOLUTION X-RAY DIFFRACTION EXPERIMENTS: A FIVE-CRYSTAL X-RAY DIFFRACTOMETER

To obtain diffraction curves of real crystals, one would ideally require an absolutely parallel and monochromatic X-ray beam, which obviously one cannot

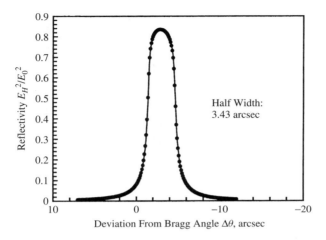

Figure 5.1 A theoretical diffraction curve of a perfect silicon crystal for (111) diffracting planes calculated on the basis of the plane-wave dynamical theory for symmetrical Bragg geometry and MoKα₁ radiation. (Reprinted from Krishan Lal, PINSA **64A(5)** (1998) 609–635, copyright (1998) with permission from PINSA.)

achieve. One has to work with a beam that is nearly parallel and monochromatic. The divergence and the wavelength spread are to be reduced sufficiently so that the broadening produced by these two sources is a small fraction of the theoretical half-width of the diffraction curve of interest. This problem is quite challenging and several approaches are used.

Conventionally, the divergence of X-ray beams can be reduced by using long collimators fitted with fine slits, which is not practical to reduce divergence to the arcsec level. Nearly perfect single crystals are used as monochromators and collimators due to their small angular range of diffraction. By successive diffraction from nearly perfect crystals, it is possible to obtain nearly parallel and monochromatic X-ray beams. The use of channel-cut monolithic monochromators of the Bonse–Hart type (Bonse and Hart 1965) has made it possible to use a single unit for two or even more reflections. It is profitable to combine X-ray sources having low dimensions with long collimators and one or more monochromator crystals to achieve high-quality X-ray beams. Several types of multicrystal X-ray diffractometers have been developed. Some of these are now commercially available. The author and his coworkers have developed different types of multicrystal X-ray diffractometers (Lal and Singh 1977, Lal and Bhagavannarayana 1989, Lal et al 1990a, Lal et al 1996a and Lal 1993). A five-crystal X-ray diffractometer with state-of-the-art level resolution is briefly described in the following (Lal 1993, 1998).

Figure 5.2 shows a schematic line diagram of the five-crystal X-ray diffractometer designed, developed and fabricated in the author's laboratory. It can be used with a fine-focus sealed X-ray tube or a rotating-anode X-ray generator.

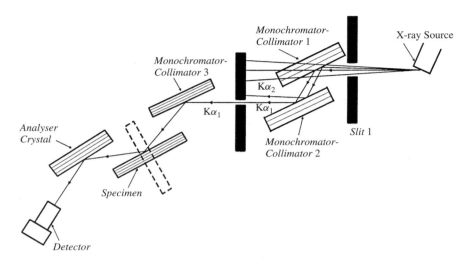

Figure 5.2 A schematic line diagram of the five-crystal X-ray diffractometer designed, developed and fabricated in the author's laboratory. (Reprinted from Krishan Lal, PINSA **64A(5)** (1998) 609–635, copyright (1998) with permission from PINSA.)

A long collimator is combined with a set of two plane silicon monochromator crystals of Bonse–Hart type (Bonse and Hart 1965). The beam diffracted from the monochromator crystals consists of $K\alpha_1$ (for $0.1 \times 0.1 \, \text{mm}^2$ source) or well-resolved $K\alpha_1$ as well as $K\alpha_2$ components of the $K\alpha$ doublet (with a focal spot of $0.4 \times 0.4 \, \text{mm}^2$). The $K\alpha_1$ beam is isolated with the help of a fine slit and is further diffracted from a third monochromator-collimator crystal as shown in Figure 5.2. The third crystal is oriented for diffraction from (800) planes in $(+, -, +)$ geometry. The small intrinsic angular width of its diffraction curve leads to a substantial reduction in divergence as well as the wavelength spread of the beam (about 1/50th of the intrinsic spread of the $K\alpha_1$ beam). The specimen generally occupies the fourth crystal position and can be oriented in Bragg or Laue geometries. The X-ray beam diffracted from the fourth crystal stage is further diffracted from the fifth crystal, which acts as an analyzer. The analyzer crystal is very effective in determining spatial variations in lattice parameters of the specimen located at the fourth crystal position. It also enables accurate determination of changes in lattice parameters produced by external influences such as high electric fields (Lal and Thoma 1983 and Lal and Goswami 1987). It has been found that in the case of specimens with epitaxial films, there is a finite mismatch between crystallographic orientations of the film and the substrate in addition to the lattice mismatch between the two (Lal *et al.* 1996b and Goswami *et al.* 1999). With the use of the analyzer crystal, the two mismatches can be conveniently isolated and measured with low level of uncertainty.

The diffractometer is generally oriented in $(+, -, +, -, +)$ geometry. There is considerable flexibility in fixing the geometry to meet a given requirement. The

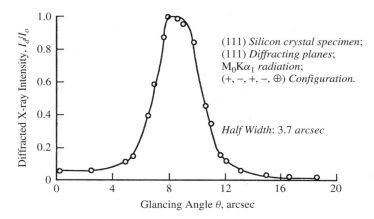

Figure 5.3 A typical high-resolution X-ray diffraction curve of a nearly perfect silicon single crystal recorded with (111) planes on the five-crystal X-ray diffractometer in symmetrical Bragg geometry. MoKα₁ radiation and $(+, -, +, -, +)$ configuration were employed. (Reprinted from Krishan Lal, PINSA **64A(5)** (1998) 609–635, copyright (1998) with permission from PINSA.)

specimen crystal can occupy any of the positions, from first to fifth. Figure 5.3 shows a typical diffraction curve of a silicon single crystal located at the fifth crystal position in $(+, -, +, -, +)$ configuration of the diffractometer (Lal 1993). This is an as-recorded curve without any corrections being applied. The intensity at the peak is nearly equal to that of the exploring beam and its half-width is only 3.7 arcsec as compared to 3.3 arcsec of the theoretically calculated curve under ideal experimental conditions (Figure 5.1). Even when one uses relatively higher-order reflections, the dispersion broadening does not significantly affect the shape of experimental diffraction curves. For example, the experimentally recorded diffraction curve of (400) planes is only 1.7 arcsec in half-width (Lal 1998). The theoretical value of half-width for this set of planes is ∼1.4 arcsec. Due to these features, the diffractometer allows one to make high-resolution X-ray diffraction experiments with low experimental uncertainty.

The diffractometer can be used for making high-resolution X-ray diffraction experiments of a wide variety, including (i) diffractometry; (ii) topography; (iii) determination of absolute value of integrated X-ray intensity; (iv) diffuse X-ray scattering measurements; (v) measurement of curvature of wafers and determination of biaxial stress in thin-film-single-crystal substrate systems; (vi) accurate determination of lattice parameters of crystals; (vii) accurate determination of lattice and orientational mismatches between epitaxial films and their substrates; (viii) accurate determination of orientation of crystal surfaces as well as straight edges on wafers; and (ix) determination of absorption coefficient of single crystals for X-rays.

5.3 EVALUATION OF CRYSTALLINE PERFECTION AND CHARACTERIZATION OF CRYSTAL DEFECTS

Evaluation of crystalline perfection is concerned with identification and characterization of all types of crystal defects and determination of their densities. Since the presence of grain boundaries means that the material is not a single crystal, we will not discuss characterization of grain boundaries in this paper. Low-angle boundaries produce tilts of ≥ 1 arcmin between adjoining subgrains. Recent experimental investigations have shown that some crystals contain very low-angle boundaries with tilt angles of less than 1 arcmin (Murthy *et al.* 1999, Kumar *et al.* 1999). Figure 5.4(a) shows a typical diffraction curve of a very low-angle boundary in a bismuth germanate crystal (Murthy *et al.* 1999). In this case, we observe two well-resolved peaks, labelled as a and b, representing two regions of the crystal that are tilted with respect to each other by an angle of only 33 arcsec. Traverse topographs shown in Figure 5.4(b) were recorded by orienting the crystal on each of the peaks. It is seen that in the topograph on the left, recorded with peak a, the entire illuminated region is diffracting except for a small portion near the top. The other topograph recorded with peak b is essentially the small missing part in the topograph recorded on peak a. The subgrains on

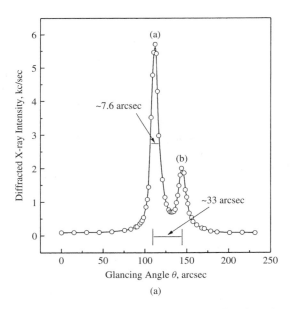

(a)

Figure 5.4   (a) A high-resolution X-ray diffraction curve of a bismuth germanate crystal showing a very low-angle boundary. A double-crystal X-ray diffractometer was used in $(+, -)$ geometry with MoKα_1 radiation, (b) A set of two traverse topographs recorded by orienting the bismuth germanate specimen of (a) on the two peaks. (Reprinted from Murthy *et al.*, *J. Cryst. Growth* **197/4** (1999) 865–873, copyright (1999) with permission from Elsevier Science.)

Peak a   Peak b

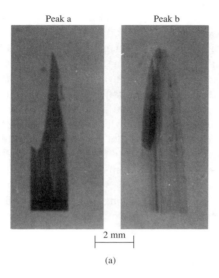

2 mm

(a)

Figure 5.4 (*continued*)

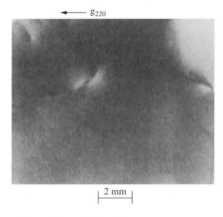

g_{220}

2 mm

Figure 5.5 A high-resolution X-ray diffraction traverse topograph of a nearly perfect semi-insulating gallium arsenide crystal recorded in symmetrical Laue geometry and (+, −, +) setting of the diffractometer. Images of only three dislocations are seen. (Reprinted from Krishan Lal *et al.*, *J. Appl. Phys.* **69(12)** (1991) 8092−8095, copyright (1991) with permission from American Institute of Physics.)

the two sides of the very low-angle boundary are directly imaged in this manner. The intensity in these topographs is not uniformally distributed. The changes in contrast are due to deformations present in the subgrains.

We shall now consider an example of direct observation and characterization of dislocations. Figure 5.5 shows a traverse topograph of a nearly perfect gallium arsenide crystal (Lal *et al.* 1990a). It is seen that the intensity of diffraction is nearly uniform in most parts of the topograph, except near the right-hand top showing a small misoriented region. Images of only three dislocations are prominently seen in an area of about $122 \, \text{mm}^2$. The image of each dislocation is quite large (linear dimension of almost $2 \, \text{mm}$). This means that the disturbance produced in regions located about a million atoms away from the dislocation core can be directly photographed. For detailed characterization regarding the nature of defect, traverse topographs are recorded with different diffraction vectors. The strain at the core of dislocations is not isotropic. Therefore, contrast changes with diffraction vector and the defects can be identified from observed variations in contrast (Lang 1958, 1970, Tanner 1976, Lal 1982, 1985). It is possible to determine the magnitude of Burgers vector from the intensity distribution in section topographs (Authier 1977). It may be mentioned that device-quality silicon crystals are expected to be free from dislocations, and in their topographs one observes nearly uniform intensity.

The absolute values of integrated X-ray intensities ρ are also a good measure of gross crystalline perfection. In high-resolution X-ray diffraction experiments, it is possible to make measurements of ρ with a low level of uncertainty. For ideally perfect crystals with moderate values of absorption coefficient μ, ρ is given by (James 1950):

$$\rho = \frac{8}{3\pi} N\lambda^2 |F| \frac{e^2}{mc^2} \frac{1 + |\cos 2\theta|}{2 \sin 2\theta} \tag{5.13}$$

Here N = number of unit cells per unit volume, λ = wavelength of the exploring X-ray beam, F = structure factor of the reflection under consideration, θ = glancing angle and e, m and c are well-known symbols. Integrated intensity expressions for ideally imperfect or mosaic crystals are well known (James 1950). A real crystal is expected to have its integrated intensity in the range lying between that for a perfect and an imperfect crystal. The experimentally measured values of ρ for nearly perfect silicon crystal lie close to that obtained from Equation (5.13) (Lal *et al.* 1990b, Ramanan *et al.* 1995).

Single crystals free from boundaries and dislocations can not be considered as ideally perfect. These may contain varying concentrations of point-defect aggregates and isolated point defects far in excess of those expected on the basis of thermodynamics. Point defects in excess of equilibrium concentrations, caused for instance by thermal treatment are generally clustered and form aggregates. Many of these aggregates are small dislocations loops on slip planes like {111} in silicon crystals.

The strain field produced by the clusters and isolated point defects is very small as compared to that produced by other defects. These can not produce sufficient contrast in X-ray diffraction topographs to be directly observable. However, these induce measurable changes in scattered intensity in the wings of the diffraction curves or produce appreciable scattering from the regions of reciprocal space surrounding a reciprocal lattice point. The intensity of such scattered X-rays is diffusely distributed without any specular reflection. Therefore, it is known as diffuse X-ray scattering (DXS). Several theoretical formulations are available to compute intensity of scattering due to point defects and their clusters (Krivoglaz 1969, Trinkaus 1972, Dederichs 1973, Larson and Schmatz 1980, Lal 1989). The DXS intensity in reciprocal space in different directions with respect to the recip-rocal lattice point under consideration gives important information about point defects and their clusters. It can be determined whether point defects are clustered and if so, one can obtain their size and shape by using model calculations. One particularly useful model that can be used for isolated point defects as well as point defect aggregates is that given by Dederichs (Dederichs 1973). According to this model the DXS intensity is given by:

$$S(\mathbf{R}^* + \mathbf{K}^*) = C_{cl}|F_{pT}|^2 \left| -4\pi \frac{A_{cl}}{V_c} \left[\frac{\mathbf{R}^*.\mathbf{K}^*}{K^{*2}} \right] \left[\frac{\sin(K^*R_{cl})}{K^*R_{cl}} \right] \right.$$

$$\left. -\frac{4\pi R_{cl}^3}{V_c} \left[\frac{\sin(K^*R_{cl}) - K^*R_{cl}\cos(K^*R_{cl})}{(K^*R_{cl})^3} \right] \right|^2 \quad (5.14)$$

Here, C_{cl} is the cluster density, $F_{p,T}$ is the structure factor corrected for polar-ization and temperature, V_c is the volume of the unit cell, R_{cl} = cluster radius, $A_{cl} = R_{cl}^2/R^*$, $\mathbf{K}^*$ is a vector that joins the elemental volume of reciprocal space under investigation to the nearest reciprocal lattice point, and $\mathbf{R}^*$ is the reciprocal lattice vector. It may be mentioned that phonons or elastic thermal waves also produce diffuse X-ray scattering. A series of systematic experiments performed in the author's laboratory established clearly that at or near room temperature, the contribution of defects to the observed DXS intensity is predominant (Lal *et al.* 1979, Lal 1989).

In recent times, the volume of reciprocal space being explored has been enhanced, particularly by employing high-brilliance synchrotron sources of X-rays. This tech-nique is referred to as reciprocal lattice mapping. As an illustration, we shall consider here results of a recent study on characterization of point defects and their clus-ters in silicon crystals grown by Czchoralski (Cz) method and the floatzone (FZ) method. High-purity crystals characterized by high electrical resistivity of about $10\,k\Omega$ (Ramanan *et al.* 1995) were used as specimens. Figure 5.6 shows a set of DXS intensity versus K^* plots for the FZ and Cz crystals. It is seen that both curves have nearly the same slope of -2 at low values of K^*. At higher values, the slope increases to -4 in both the cases. It is known that isolated point defects give a slope of -2 in DXS intensity versus K^* plots (Huang scattering). The nature of point defects, whether vacancies or interstitials, is obtained from the observed anisotropy in the distribution of DXS intensity. The anisotropy for any value of K^* is defined as: anisotropy = $DXSI_{(\theta > \theta B)} - DXSI_{(\theta < \theta B)}$. A positive value of anisotropy indicates the presence of interstitial defects that extend the surrounding lattice. Vacancies, on the other hand, produce negative anisotropy in the DXS distribution. In the present set of experiments, it was found that the FZ crystals show negative anisotropy, and for the Cz crystals the anisotropy was positive. The presence of a region of slope

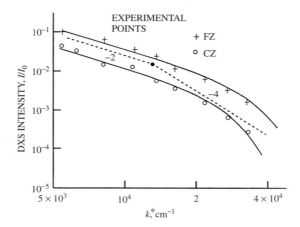

Figure 5.6 A set of log DXS intensity versus Log K^* plots for highly pure silicon crystals grown by the float zone and the Czochralski techniques. (Reprinted from Murthy *et al.*, *J. Cryst. Growth* **197/4** (1999) 865–873, copyright (1999) with permission from Elsevier Science.)

Table 5.1 Values of point defect cluster radius R_{cl}, cluster volume A_{cl} and number of defects per cluster N_{cl}

Specimen	Cluster radius μm	Cluster volume A_{cl} cm^3	Number of defects per cluster N_{cl}	Nature of defects
Cz	0.8	2.02×10^{-16}	1.26×10^6	Interstitial
FZ	0.6	1.32×10^{-16}	8.3×10^5	Vacancy

-4 at higher values of K^* in Figure 5.6 (Stokes–Wilson) indicates the presence of clustered point defects. To analyze the experimental data, we have used Dederich's model (Equation. (5.14)). The defect parameters obtained as a result of this analysis are given in Table 5.1. It is seen that sub-micrometer clusters are present in these crystals. The distribution of the DXS intensity is such that the clusters appear to be dislocation loops lying on (111) planes.

In a recent investigation, oxygen has been deliberately introduced in FZ crystals in a controlled manner. It has been observed that when the oxygen concentration approaches values generally observed in Cz crystals interstitial defects provide a dominant contribution to DXS even in FZ (Lal *et al.* 1999).

5.4 ACCURATE DETERMINATION OF CRYSTALLOGRAPHIC ORIENTATION

For all main applications of semiconductor crystals, it is necessary to restrict the orientation of their visible surfaces with respect to the nearest crystallographic planes to a small fraction of a degree of arc. Also, most wafers have one or two flats or straight edges ground on them. Their orientation is also well specified. The main techniques to determine crystallographic orientations of crystal surfaces and straight edges are (a) optical examination of wafer surfaces after chemical etching (ASTM 1981a); (b) X-ray Laue method (Christiansen *et al.* 1975 and ASTM 1981b); (c) conventional X-ray diffractometry technique (Cullity 1956, ASTM 1981a, Kholodnyi *et al.* 1984); and (d) high-resolution X-ray diffraction techniques (Lal and Goswami 1988, Lal *et al.* 1990c). In this chapter, we shall briefly describe the recently developed high-resolution X-ray diffraction techniques that give unprecedented accuracy and convenience of measurement.

We have utilized very narrow diffraction curves, obtained with nearly perfect crystals on multicrystal X-ray diffractometers, for measurement of crystallographic orientation (Lal *et al.* 1990c). Let us consider a wafer whose surface normal N_s makes an angle α with the normal to the nearest crystallographic set of planes N_p as shown in Figure 5.7. If the crystal is rotated azimuthally around the surface normal N_s, the normal to the lattice planes N_p will rotate on a conical surface as shown in Figure 5.7. If this crystal is oriented for diffraction on a multicrystal X-ray diffractometer, its peak position will depend upon its azimuthal orientation. The glancing angle θ has to be readjusted by say $\Delta\theta$ to bring it back on the peak as the azimuthal orientation is changed. If these rotations for different azimuthal positions are plotted as a function of azimuthal orientation of the specimen one would get a sinusoidal curve, from which one can get the value of α that defines the orientation of the specimen. In practice, this is realized by a device (Lal *et al.* 1990c) that provides azimuthal rotations to the specimen. In general, the axis of rotation of the device and the normal to its flat surface on which the specimen crystal is mounted are not exactly collinear. There is a finite

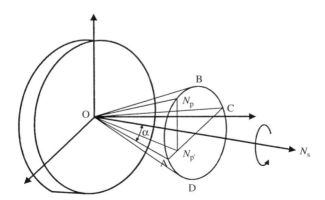

Figure 5.7 A schematic diagram showing movement of normal to the diffracting planes N_p on a conical surface as the crystal is rotated azimuthally around surface normal N_s. (Reprinted from Krishan Lal *et al.*, *Meas. Sci. Technol.* (1990) 793–800, copyright (1990) with permission of Institute of Physics.)

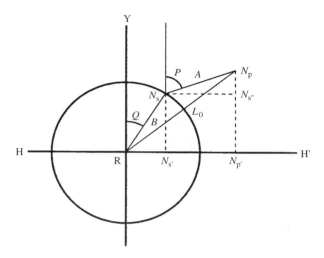

Figure 5.8 A right-angle cross-section of the cones of rotations of ON_s and ON_p of Figure 5.7. (Reprinted from Krishan Lal *et al.*, *Meas. Sci. Technol.* (1990) 793–800, copyright (1990) with permission of Institute of Physics.)

angle between these and it can vary from device to device. Therefore, the device should be examined in detail and this angle should be measured experimentally. In general, one is dealing with two cones, one due to specimen (Figure 5.7) and the other due to the device. The cross sections of the two cones would be as shown in Figure 5.8. For the sake of clarity only the inner circle representing the device is shown. The normal to the specimen surface intersects it at N_s, whereas N_p is the extremity of the normal to the crystallographic plane. Experimentally, one would measure the projection of N_p on the horizontal plane along RH'. The value of RN_p' will depend upon the relative orientations of the device normal and that of the wafer in the vertical plane. The measured value of RN_p' will be a function of angles P and Q and those corresponding to A and B. The lengths A and B correspond to the angles between the normal to the device surface and its rotation axis and that between the crystal surface normal and the lattice planes, respectively.

It can be shown from Figure 5.8 that:

$$RN_p^{\,2} = L_0 = A^2 + B^2 + 2AB \cos K \tag{5.15}$$

where $K = P - Q$

If angle P is changed to $P + \Delta P_1$, where ΔP_1 is a known angular increment, we get:

$$A^2 + B^2 + 2AB \cos(K + \Delta P_1) = L_1 \tag{5.16}$$

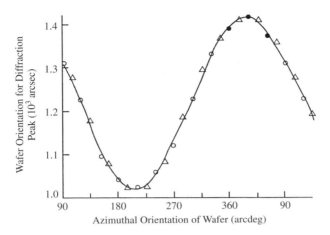

Figure 5.9   A plot of the angle of reorientation of the specimen at the diffraction peak as a function of the azimuthal orientation of a 100-mm diameter (100) silicon wafer. O represents experimental points; experimental points chosen for calculation of A, B, P and Q; and $\triangle$ are the theoretically computed points. (Reprinted from Krishan Lal *et al.*, *Meas. Sci. Technol.* (1990) 793–800, copyright (1990) with permission of Institute of Physics.)

By giving different angular increments ΔP_1, ΔP_2, etc., one can get a set of equations whose solution yields values of A, B, P and Q (Lal *et al.* 1990c) and therefore, the orientation is determined.

Figure 5.9 shows a typical plot of the angle of reorientations of a 100 mm diameter silicon single-crystal specimen to bring it at diffraction peak as a function of azimuthal orientations. In this case, the angle between the (100) planes and the surface was determined as 20.26 arcmin. For the device the value of B was found to be 3.36 arcmin. The value of P was zero. Once the values of A, B, P and Q were known values of RN_p were calculated at different azimuthal orientations of the wafer with respect to the device. These are shown as triangles on the curve (Figure 5.9). The fit between the experimental curve and the calculated points is good. The experiments performed on different wafers and on different diffractometers confirmed that the overall uncertainty is less than ±6 arcsec.

For determination of orientation of flats or straight edges on semiconductor wafers, one can use X-ray Laue techniques (Cullity 1956, ASTM 1987), diffractometric methods (Bond 1976, ASTM 1987) and a recently developed technique based on X-ray diffraction topography (Lal and Goswami 1988). The last technique is very convenient and is a significant improvement over all the other known methods. The specimen crystal is oriented for diffraction on an X-ray diffraction topography set up. Figure 5.10 shows a schematic diagram of the basic principle. A traverse topograph is recorded covering the straight edge at one extreme and a small adjoining region. After the exposure, the specimen, together with the film, is translated by a short distance of about 1 mm or so and a stationary topograph

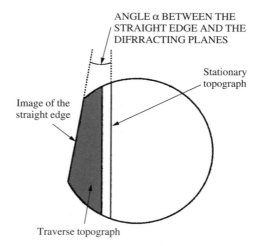

Figure 5.10   A schematic diagram illustrating essential features of the x-ray diffraction topographic method for determining orientations of the straight edges on semiconductor wafers. (Reprinted from Krishan Lal and Niranjana N. Goswami, *J. Rev. Sci. Instrum.* **59(8)** (1998) 1409–1411, copyright (1998) with permission of American Institute of Physics.)

is recorded. The traverse topograph has the image of the straight edge at one extreme. The stationary topograph is the projection of the lattice planes that are nominally parallel to the straight edge. One can conveniently measure the angle between the stationary topograph and the image of the straight edge that determines the orientation. The uncertainty is ± 3 arcmin. It may be mentioned that it is not necessary to use ground, lapped and polished wafers. Raw cut wafers can be used and corrective measures can be taken if necessary.

5.5   MEASUREMENT OF CURVATURE OR BENDING OF SINGLE-CRYSTAL WAFERS

Semiconductor single-crystal wafers used for fabrication of microelectronic devices have to conform to strict specifications in terms of warpage or bending. The bending is restricted to about $30\,\mu$m for large size wafers. The site flatness is restricted to $\sim 0.2\,\mu$m only for a site area of $26\,$mm $\times\,26\,$mm (Huff and Goodall 1996). The bending measurements are also required for determining the biaxial stress in thin-film single-crystal substrates systems. Curvature of wafers can be measured optically as well as by using high-resolution X-ray diffraction methods (Segmuller *et al.* 1989, Lal *et al.* 1990a). Figure 5.11 shows a schematic diagram of the bending of a typical semiconductor wafer (with a thin film). If the wafer is not flat, the central region (B) will not touch the flat surface and would

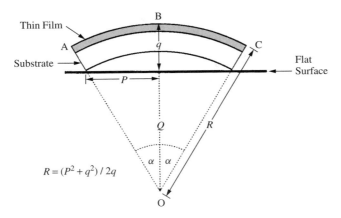

Figure 5.11   A schematic diagram showing measurement of curvature of a wafer by optical techniques.

be above it by a distance q. The radius of curvature R is related to q through:

$$R = (P^2 + q^2)/2q \quad \text{or} \quad R \cong P^2/2q \tag{5.17}$$

As a typical example, a wafer having diameter of 200 mm and with $q = 30\,\mu$m, the value of R will be 166.6 m. A decrease of 3 μm in q would mean an increase in R by $\sim$20 m. Similarly, for a site flatness of 0.17 μm over a site size of 26 $\times$ 26 mm^2 the value of R is $\sim$1 km.

In the high-resolution X-ray diffraction technique one utilizes the sharpness of diffraction curves for accurate measurement of bending or radius of curvature. In a bent crystal, the normal to lattice planes that are perpendicular to the crystal surface shows orientational change as one moves from one end of the crystal to the other, as shown in Figure 5.12. Let us suppose that the crystal is oriented for diffraction from one edge, say the right-hand edge on a multicrystal X-ray diffractometer. The specimen is moved across the X-ray beam in a stepwise manner and at each position the orientation of the diffraction vector **g** is determined from the rotation needed to bring the specimen back on the diffraction peak. If orientation of **g** is plotted as a function of linear position of the specimen across the X-ray beam, one would get a linear plot whose slope gives the value of the radius of curvature. If the crystal is absolutely flat, the slope would be zero and the curve would be a horizontal straight line. For crystals bent in a convex shape, the plot will be a straight line with positive slope whereas for concave-shaped crystals, the slope would be negative. The slope of the line gives the value of the radius of curvature. In this case, even when there is a 2 arcsec deviation in the orientation of diffraction vector **g** at the two extremes of the wafer, one can conveniently determine the radius of curvature. This essentially means that a radius of curvature of 20 km for a wafer of diameter of 200 mm can be measured.

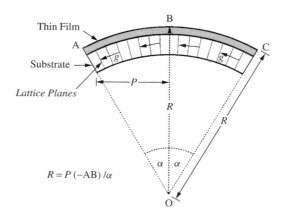

Figure 5.12   A schematic diagram showing measurement of curvature of a wafer by the high-resolution X-ray diffraction method.

Also, the resolution of measurement is high (less than 1 m for $R = 200$ m). For a site flatness of $0.17\,\mu$m over a site size of 26 mm $\times$ 26 mm the difference in orientation of **g** at two extremes will be $\sim$5 arcsec, which can be measured conveniently and with a low level of uncertainty. This technique is also being used for determination of biaxial stress in a variety of thin-film single-crystal substrate systems (Segmuller *et al.* 1989, Lal *et al.* 1990a, Goswami *et al.* 1994, Lal *et al.* 1996a).

5.6   CHARACTERIZATION OF PROCESS-INDUCED DEFECTS IN SEMICONDUCTORS: IMPLANTATION-INDUCED DAMAGE

Semiconductor crystals undergo various processing steps such as deposition of thin films of different types of materials under different temperature and pressure conditions, chemical and heat treatments, and introduction of impurities by diffusion or implantation. These processing steps produce changes in the real structure of the crystal. High-resolution X-ray diffraction techniques can be used for characterization of processing-induced defects. Here we shall consider as an example BF_2^+ implantation in silicon single crystals (Lal *et al.* 1991c, 1996c). BF_2^+ implantation is widely used to produce shallow submicrometer p^+-n junctions in silicon single crystals. BF_2^+ ion energies were 90 and 135 keV, with the fluence being kept constant at 1×10^{15} cm^{-2}. It was observed that implantation produced a very small broadening of the diffraction curves accompanied by a small but measurable decrease in the peak intensity. However, changes in DXS intensity distribution were significant.

Figure 5.13 shows a set of DXS intensity versus K^* plots for an unimplanted wafer and for specimens implanted with BF_2^+ ions of energy 90 keV and 135 keV.

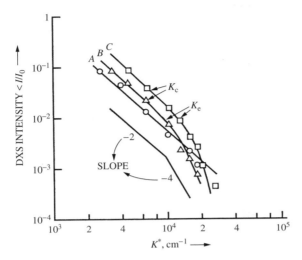

Figure 5.13 A set of DXS intensity versus K^* plots for unimplanted silicon wafer (curve A) and wafers implanted with BF_2^+ ions of energies 90 keV (curve B) and 135 keV (curve C). (Reprinted from Krishan Lal *et al.*, *J. Appl. Phys.* **69(12)** (1991) 8092–8095, copyright (1991) with permission from American Institute of Physics.)

It is seen that the DXS intensity for the unimplanted crystal varies as K^{*-2} at practically all values of K^* except in the highest-value range showing presence of isolated point defects (Huang scattering). Diffuse scattering from clustered point defects exhibits features that are distinct from the scattering by isolated point defects. At large values of K^* the scattered intensity varies as K^{*-4} or even faster than K^{*-4}. It is similar to the approximation given by Stokes and Wilson (1944) for scattering by dislocations. It is customary to refer to it as Stokes–Wilson scattering (see, e.g. Dederichs (1973)). We see from Figure 5.13 that at large values of K^* a Stokes–Wilson region is indicated. Anisotropy for this as well as the other plots (curves B and C) was positive, typical of Czochralski grown crystals (Figure 5.6). Due to implantation the overall DXS intensity increases indicating disordering of the crystal lattice. The Stokes–Wilson region (slope of -4) is quite prominent in implanted crystals showing clustering of point defects. The phenomenological model of Equation (5.14) has been employed to analyze the experimental data and the following values of R_{cl} and A_{cl} were obtained:

BF_2^+ energy   90 keV $R_{cl} = 1.47 \times 10^{-4}$ cm $A_{cl} = 2.92 \times 10^{-16}$ cm^3

BF_2^+ energy   135 keV $R_{cl} = 1.29 \times 10^{-4}$ cm $A_{cl} = 2.27 \times 10^{-16}$ cm^3

The values of N_{cl} (number of defects per cluster) have been determined as 1.46×10^7 (90 keV) and 1.37×10^7 (1.35 keV). This investigation has shown that all the implanted impurity has not been distributed atomically and a part of it has been segregated. Agglomeration of point defects had also been observed in BF_2^+-implanted silicon crystals by cross-sectional transmission electron microscopy (Queirolo *et al.* 1987). There is a good correlation between the two investigations.

We have also employed high-resolution X-ray diffractometry and absolute X-ray integrated intensity measurements to investigate top layers of BF_2^+-implanted silicon crystals. It is generally considered that at energies of $\sim$90 keV and fluence level of 2×10^{-15} cm^{-2} the top layer becomes amorphous (Wu and Chen 1985, Virdi *et al.* 1992). Precise diffractometry and absolute X-ray integrated intensity measurements have shown that the top layer is not amorphous (Lal *et al.* 1996c). It is disordered but crystalline in nature.

5.7 CONCLUSIONS

In this chapter we have described the basic principles of high-resolution X-ray diffraction and the essential features of a five-crystal X-ray diffractometer developed in the author's laboratory. A number of illustrative examples have been covered: characterization of defects in device-quality crystals; determination of crystallographic orientation; measurement of bending of blank wafers; and characterization of implantation-induced defects in silicon crystals. The range of techniques available is such that these can be suitably combined to meet any challenge concerned with characterization of crystal defects. Of course, there are limitations of spatial resolution in the case of X-rays. Nevertheless, the advantages of nondestructive characterization of large volumes and quantitative information with high sensitivity are remarkable.

5.7.1 ACKNOWLEDGEMENT

The author has benefited by collaborating with a large number of colleagues in his laboratory and other laboratories in India and abroad. Close collaboration with Drs Niranjana Goswami, G. Bhagavannarayana, (late) V. Kumar, S. K. Halder, R. V. A. Murthy, R. R. Ramanan, and A. Chaubey is acknowledged with pleasure. It has been a privilege and pleasure to have long in-depth discussions with Dr. A. R. Verma. Partial support received under an Indo-US project an Indo-Russian ILTP project and an Indo-German project is gratefully acknowledged.

REFERENCES

Alexander S., Hellemans L., Marti O., Schneir J., Ellings V., Longmire M. and Gurley G. (1989) *J. Appl. Phys.* **65**, 164.
American Society for Testing and Materials (1981a) *Annual Book of ASTM Standards Part 43* **F26-27**, 185.

American Society for Testing and Materials (1981b) *Annual Book of ASTM Standards Part 11* **E82-63**, 104.

American Society for Testing and Materials (1987) *Annual Book of ASTM Standards* **F847-87**, 744.

Authier A. (1977) *Topics in Applied Physics* edited by Queisser H. J. Springer-Verlag, New York, p. 145.

Barrett C. S. and Massalski T. B. (1968) *The Structure of Metals,* McGraw-Hill, New York, Asian Edition: Eurasia, New Delhi.

Batterman B. W. and Cole H. (1964) *Rev. Mod. Phys.* **36**, 681.

Binnig G., Rohrer H., Gerber C. and Weibel E. (1982) *Phys. Rev. Lett.* **49**, 57.

Binnig G., Quate C. F. and Gerber C. (1986) *Phys. Rev. Lett.* **12**, 930.

Bond W. L. (1976) *Crystal Technology,* Wiley, New York, p. 304.

Bonse U. and Hart M. (1965) *Appl. Phys. Lett.* **7**, 238.

Bullis W. M. and Huff H. R. (1993) *ULSI Science and Technology* edited by G. K. Geller, E. Middleworth and K. Hoh, The Electrochemical Soc, Pennington, USA, p. 103.

Claeys C. and Deferm, L. (1996) *Solid State Phenom.* **47−48**, 1.

Christiansen G., Gerward L. and Alstrup I. (1975) *Acta. Cryst.* **A31**, 142.

Cullity B. D. (1956) *Elements of X-ray Diffraction,* Addison-Wesley, Mass.

Dederichs P. H. (1973) *J. Phys. F.* **3**, 471.

Goswami S. N. N., Lal Krishan, Vogt A. and Hartnagel H. L. (1999), to be published.

Goswami S. N. N., Lal Krishan, Wurfl J. and Hartnagel H. L. (1994) *Physics of Semiconductor Devices,* edited by Lal Krishan, Narosa Publishers, New Delhi, 337.

Guinier A. (1952) *X-ray Crystallographic Technology,* Hilger and Watts, London.

Hansma P. K. and Tersoff J. (1987) *J. Appl. Phys.* **61**, R1.

Heydenreich J. (1982) *Synthesis, Crystal Growth and Characterization* edited by Lal Krishan, North-Holland, Amsterdam, p. 215.

Huff H. R. and Goodall R. K. (1996) *Solid State Phenom.* **47−48**, 65.

James R. W. (1950) *The Optical Principles of the Diffraction of X-rays,* G. Bell & Sons, London.

Kato N. (1975) *Crystal Growth and Characterization* edited by Ueda R. and Mullin J. B., North Holland, Amsterdam p. 279.

Kato N. (1992) *J. Acta. Cryst.* **A48**, 834.

Kholodnyi L. P., Terent'ev G. I. and Krol' I. M. (1984) *Meas. Technol.* **27**, 266.

Krivoglaz M. A. (1969) *Theory of X-ray and Thermal Neutron Scattering by Real Crystals,* Plenum Press New York.

Kumar V., Lal Krishan, Seetharaman and Gupta S. C. (1999) to be published.

Lal Krishan and Singh B. P. (1977) *Solid State Commun.* **22**, 71.

Lal Krishan, Singh B. P. and Verma A. R. (1979) *Acta. Crystallogr.* **A35**, 286.

Lal Krishan (1982) *Synthesis, Crystal Growth and Characterization* edited by Lal Krishan, North-Holland, Amsterdam p. 215.

Lal Krishan and Thoma P. (1983) *Phys. Stat. Sol. (a)* **80**, 491.

Lal Krishan (1985) *Advanced Techniques for Microstructural Characterization* edited by Krishnan R., Anantharaman T. R., Pande C. S. and Arora O. P., Trans Tech Pub., Aedermannsdorf, Switzerland, p. 143.

Lal Krishan and Goswami S. N. N. (1987) *J. Mater. Sci. Eng.* **85**, 147.

Lal Krishan and Goswami S. N. N. (1988) *Rev. Sci. Instrum.* **59**, 1409.

Lal Krishan and Bhagavannarayana G. (1989) *J. Appl. Cryst.* **22**, 209.

Lal Krishan (1989) *Prog. Cryst. Growth Charact.* **18**, 227.

Lal Krishan, Goswami S. N. N., Wurfl J. and Hartnagel H. L. (1990a) *J. Appl. Phys.* **67**, 4105

Lal Krishan, Goswami S. N. N. and Verma A. R. (1990b) *Pramana. J. Phys.* **34**, 506.

Lal Krishan, Bhagavannarayana G., Kumar Vijay and Halder S. K. (1990c) *Meas. Sci. Technol.* **1**, 793.

Lal Krishan (1991a) *Advances in Crystallography and Crystal Growth* edited by Lal Krishan, Indian National Science Academy, New Delhi p. 125.

Lal Krishan (1991b) *Crystalline Materials: Growth and Characterization*, edited by Rodriguez-Clemente R. and Paorici C., Trans Tech Pub, Zurich, p. 205.

Lal Krishan, Bhagavannarayana G and Virdi G. S. (1991c) *J. Appl. Phys.* **69**, 8092.

Lal Krishan (1993) *Bull. Mater. Sci.* **16**, 617

Lal Krishan, Mitra Reshmi, Srinivas G. and Vankar V. D. (1996a) *J. Appl. Crystallogr.* **29**, 2.

Lal Krishan, Goswami S. N. N. and Kuznetsov G. F. (1996b) *Semiconductor Devices* edited by Lal Krishan, Narosa Publishers, New Delhi, p. 113.

Lal Krishan, Bhagavannarayana G. and Virdi G. S. (1996c) *Solid State Phenom.* **47–48**, 377.

Lal Krishan (1998), *Proc. Indian National Sci. Acad.* **64**, 609.

Lal Krishan, Ramanan R. R., Bhagavannarayana G. (2000) *J. Appl. Crystallogr.* **33**, 20.

Lang A. R. (1958) *J. Appl. Phys.* **29**, 597 and (1970) *Modern Diffraction and Imaging Techniques in Materials Science* edited by Amelinckx S., Gevers R., Remant G. and Landyut L., North-Holland, Amsterdam 407.

Larson B. L. and Schmatz W. (1980) *Phys. Stat. Sol. (b)* **99**, 267.

Laudise R. A. (1972) *Analytical Chemistry: Key to Progress in National Problems*, edited by Meinke W. and Taylor J. K. NBS Special Publ 351, NBS, Washington p. 19.

Laue M. Von (1960) *Roentgenstrahlininterferennzen*, Acad Verlagsgescllschaft, Frankfurt.

Mathews J. W., Klokholm E., Sadagopan V., Plasket T. S. and Mendel E. (1973) *Acta. Met.* **21**, 203.

Murthy R. V A., Ravi Kumar M., Choubey A. and Lal Krishan., Kharchanko L., Shleguel , V. and Guerasimov, (1999) *J. Cryst. Growth* **197**, 865 .

Pinsker Z. G. (1978) *Dynamical Scattering of X-rays in Crystals*, Springer-Verlag, Berlin.

Queirolo G., Capraara P., Meda L., Guareschi C., Andrele M., Ottaviani G. and Armigliato A. (1987) *J. Electrochem. Soc. Solid State Sci. Technol.* **134**, 2905.

Ramanan R. R., Bhagavannarayana G. and Lal Krishan (1995) *J. Cryst. Growth* **156**, 377.

Sangwal K. S. (1987) *Etching of Crystals: Theory, Experiment and Application* (Series editors Amelinckx S. and Nihocal H.), North Holland, Amsterdam.

Schneider Jochen R. and Bouchard Roland, (1992) *Acta. Crystallogr* **A48**, 804.

Segmuller A., Noyan I. C. and Speriosu V. S. (1989) *Prog. Cryst. Growth Charact.* **18**, 21.

Spence J. (1984) *Experimental High Resolution Electron Microscopy* Clarendon Press, Oxford.

Stokes A. R. and Wilson A. J C. (1944) *Proc. Phys. Soc.* **56**, 174.

Trinkaus H. (1972) *Phys. Stat. Sol. (b)* **51**, 307.

Tanner B. K. (1976) *X-ray Diffraction Topography*, Pergamon Press, Oxford.

Virdi G. S., Rauthan C. M S., Pathak B. C., Khokle W. S., Gupta S. K. and Lal Krishan (1992) *Solid State Electron.* **35**, 535.

Wu I. W. and Chen (1985) *J. Appl. Phys.* **58**, 3032.

Zachariasen W. H. (1945) *Theory of X-ray Diffraction in Crystals*, Wiley, New York.

6   Computational Simulations of the Growth of Crystals from Liquids

ANDREW YECKEL and JEFFREY J. DERBY

Department of Chemical Engineering and Materials Science, Army HPC Research Center, and Minnesota Supercomputing Institute, University of Minnesota, Minneapolis, MN 55455-0132, USA

6.1   INTRODUCTION

In this chapter we review the current status and role of computer modeling of hydrodynamics and associated transport processes in bulk crystal growth. Hydrodynamics plays a key role in virtually all bulk crystal-growth systems, having a critical effect on both heat and mass transport. Indeed, hydrodynamics often is inextricably coupled to heat and mass transport, due to thermal and solutal buoyancy effects. Hydrodynamics strongly affects mixing in many systems, thereby affecting growth rates, interface shape, chemical composition of the crystal, and defect formation. Many physico-chemical phenomena, occurring on length scales ranging from atomistic to system scale, are involved in crystal growth. Nevertheless, it is fair to say that the future of crystal-growth modeling depends heavily on progress in hydrodynamics modeling. There already exists a large and rich body of literature on this topic, but much work remains to be done.

The scope of this chapter is limited to continuum representations of heat, mass, and momentum transport in melt and solution growth systems. Atomistic-scale phenomena are not considered, nor is the connection between macroscale phenomena and crystal properties. Nucleation phenomena, and other microscale phenomena, such as morphological instabilities and dendrite formation, also are excluded. All of these are active areas of research in crystal-growth modeling, however, space restrictions do not permit their inclusion. Nor is this chapter meant to serve as a 'how to' manual for hydrodynamic simulation, a subject that would require a lengthy book.

The next section reviews the current capabilities and limitations of mathematical methods used to model hydrodynamics in crystal-growth systems. A brief discussion is given on the subject of commercial software, both general-purpose software for hydrodynamic simulation and software specialized to the needs of crystal-growth researchers. In Section 6.4 we present examples from

Crystal Growth Technology, Edited by H. J. Scheel and T. Fukuda
© 2003 John Wiley & Sons, Ltd.   ISBN: 0-471-49059-8

our research that illustrate some of the capabilities and limitations of current methods. The first example is a one-dimensional model of multicomponent diffusion during diffusion-limited growth of cadmium zinc telluride (CZT), a ternary II-VI alloy. The model predicts the existence of a retrograde diffusion layer, in which zinc diffuses against its concentration gradient. The results illustrate the need for more research on fundamental model formulation. The second example is an axisymmetric model of the accelerated crucible rotation technique (ACRT), applied to vertical Bridgman growth of CZT. ACRT introduces a hydrodynamic time scale that is much shorter than other transport-related time scales inherent in vertical Bridgman growth, making this a two-dimensional problem that strains even today's largest computers. The last example is a three-dimensional bulk flow model of Czochralski (Cz) growth of bismuth silicon oxide, with crystal rotation. The simulations reveal that two distinct three-dimensional instabilities can occur simultaneously in this system. One instability leads to the formation of baroclinic waves deep within the melt, and the other causes the appearance of rotational spokes at the surface, a consequence of a Rayleigh thermal instability. The results demonstrate that the choice between using a two-dimensional or three-dimensional model can be a difficult one, with enormous consequences.

6.2   TRANSPORT MODELING IN BULK CRYSTAL GROWTH

6.2.1   GOVERNING EQUATIONS

Any attempt to analyze continuum transport phenomena in crystal growth begins with formulating a set of governing equations, which in general is based on well-known principles of conservation of mass, momentum, and energy. Here we provide a brief review of these equations. A general discussion of the equations as applied to crystal-growth systems can be found in Derby *et al.* (1998). A more detailed though somewhat dated discussion of crystal growth modeling is the excellent review article by Brown (1988). For a good general textbook on the subject of continuum transport and conservation principles, applied to physico-chemical systems, we refer the interested reader to Bird *et al.* (1960).

It is important to note that the equations presented here have been derived subject to a number of approximations, not all of which are discussed in the text. Furthermore, in certain special cases not all of these approximations will apply, in which case the equations may take on a somewhat different form. Nevertheless, the equations presented here are applicable to most bulk crystal-growth systems.

Conservation of mass, momentum, and energy, subject to the continuum approximation, are in general described by a set of partial differential equations. Conservation of momentum and continuity are given by the Navier–Stokes equations for an incompressible, Newtonian fluid:

$$\rho_0 \left(\frac{\partial \mathbf{v}}{\partial t} + \mathbf{v} \cdot \nabla \mathbf{v} \right) = -\nabla p + \mu \nabla^2 \mathbf{v} + \rho_0 \mathbf{g}[1 - \beta(T - T_0)$$

$$+ \beta_s(c - c_0)] + \mathbf{F}(\mathbf{v}, \mathbf{x}, t) \tag{6.1}$$

$$\nabla \cdot \mathbf{v} = 0 \tag{6.2}$$

where $\mathbf{v}$ is the velocity vector, p is pressure, T is temperature, c is molar concentration, t is time, μ is viscosity, $\mathbf{g}$ is the gravitational vector, and ρ_0 is the density at a reference temperature T_0 and reference concentration c_0. Density is treated as a constant, except in the gravitational body force term, where the effects of temperature and concentration variations are incorporated, using thermal and species expansivity coefficients β and β_s; this is the Boussinesq approximation. The term $\mathbf{F}(\mathbf{v}, \mathbf{x}, t)$ accounts for any additional body forces that may act on the fluid volume, e.g. the inductive force due to motion of a conducting liquid in a magnetic field or apparent forces that arise in a noninertial reference frame, such as the Coriolis force.

Conservation of energy is given by

$$\rho_0 C_p \left(\frac{\partial T}{\partial t} + \mathbf{v} \cdot \nabla T \right) = \kappa \nabla^2 T \tag{6.3}$$

where C_p is the heat capacity and κ is the thermal conductivity of the material. Whereas Equations (6.1) and (6.2) apply only to liquid or gas phases, Equation (6.3) applies to solid phases as well, within which $\mathbf{v} = \mathbf{V}_T$, the translation velocity relative to the reference frame.

Conservation of mass is given by

$$\frac{\partial c}{\partial t} + \mathbf{v} \cdot \nabla c = \mathcal{D}\nabla^2 c \tag{6.4}$$

where c is the mass density and $\mathcal{D}$ is the diffusion coefficient of the species. As with Equation (6.3), Equation (6.4) applies in all phases, though species transport is often neglected in solid phases in which the diffusivity is very small. For problems in solution growth it is usually convenient to substitute supersaturation, $\sigma \equiv \dfrac{c - c_s}{c_s}$, for concentration, since supersaturation is the driving force for crystallization (c_s is the saturation concentration). It is important to note that Equation (6.4) implies either that the species is dilute, or that the density of the liquid is nearly constant. In crystal growth from nondilute solutions it is possible for the density to vary considerably over the duration of growth, but the density change with time is typically so slow relative to mass transfer processes that Equation (6.4) remains accurate. Equation (6.4) may also be written for several species, provided each species is sufficiently dilute that diffusion occurs as a pseudo-binary process (the case of nondilute diffusion in a multicomponent system is discussed in the next section).

Except for the Boussinesq approximation, it has been assumed that all physical properties in Equations (6.1)–(6.4) are constant. It is straightforward to include the effect of temperature and concentration dependent properties, however, provided data are available.

6.2.2 BOUNDARY CONDITIONS

In order to apply Equations (6.1)–(6.4) to real systems, information must be provided about the velocity, temperature, and concentration at all boundaries. Because of the wide variety and complexity of crystal-growth systems, it is not possible to provide a comprehensive list of boundary conditions applicable to these systems, but we briefly mention here some of the most common ones.

Most often, velocity boundary conditions consist of no-slip at solid surfaces:

$$\mathbf{v} = \mathbf{U_s} \tag{6.5}$$

where $\mathbf{U_s}$ is the velocity of the solid surface (which often is zero). At liquid/gas interfaces it is usually assumed that the gas exerts no force on the liquid, in which case a no-stress condition is imposed:

$$\mathbf{n} \cdot (-p\mathbf{I} + \mu(\nabla\mathbf{v} + (\nabla\mathbf{v})^T)) = 0 \tag{6.6}$$

where $\mathbf{n}$ is the unit vector normal to the interface, and $\mathbf{I}$ is the identity tensor. However, Equation (6.6) may be inadequate to describe a system in which temperature or concentration varies along the interface. Such variations result in gradients of surface tension which, in turn, produce a net surface traction and drive fluid motion known as Marangoni flows.

Typically, energy boundary conditions include the specification that the temperature field is continuous, and heat fluxes balance, across all material interfaces. An exception occurs at interfaces at which a phase change occurs, where the energy balance must account for the latent heat of fusion:

$$(-\kappa_1 \nabla T|_1 + \kappa_s \nabla T|_s) \cdot \mathbf{n_{sl}} = \rho_s \Delta H_f \frac{\partial \mathbf{x}}{\partial t} \cdot \mathbf{n_{sl}} \tag{6.7}$$

where $\partial\mathbf{x}/\partial t$ is the velocity of the solid/liquid interface, ΔH_f is the latent heat of fusion, $\mathbf{n_{sl}}$ is a unit vector normal to the solid/liquid interface, oriented towards the liquid, and the subscripts l and s refer to the liquid and solid sides of the interface.

Perhaps the most difficult area to generalize regarding boundary conditions is the specification of thermal conditions connecting the domain of interest to the environment. For melt-growth models, these conditions are meant to represent the influence of the furnace, which can strongly affect the temperature field in the liquid and crystal. One approach is to account for furnace heating using simple approximations for convective and radiative transport in terms of a specified furnace temperature profile T_f:

$$-\kappa \mathbf{n}_b \cdot \nabla T = h(T - T_f) + \sigma \epsilon (T^4 - T_f^4) \tag{6.8}$$

where h is a heat transfer coefficient, ϵ is the emissivity, and σ is the Stefan–Boltzmann constant. There are several potential pitfalls to this approach, however. The heat transfer coefficient must account for all convective effects external to the domain, which may require reducing a gas flow through a furnace to a single empirical parameter. Likewise, Equation (6.8) implies that the radiation view factor equals unity everywhere, an unrealistic approximation in many high-temperature growth systems (Siegel and Howell, 1993; Kuppurao and Derby, 1993). Furthermore, transport of energy via internal radiation may be important in high-temperature crystal-growth systems that possess some transparency to infrared radiation (O'Hara et al., 1968; Brandon and Derby 1992), further complicating the situation. Lastly, information is needed regarding T_f, which may require costly experiments or exhaustive furnace heat transfer modeling effort. Nevertheless, the simple form of Equation (6.8) has proved adequate to elucidate key transport phenomena in many systems.

Concentration boundary conditions require that mass fluxes balance across all material interfaces, but unlike temperature, concentration need not be continuous across interfaces. Instead, equilibrium partitioning occurs, according to

$$c_s = k c_l \tag{6.9}$$

where k is the partition coefficient. The solute balance at the interface is given by

$$(-\mathcal{D}_l \nabla c|_l + \mathcal{D}_s \nabla c|_s) \cdot \mathbf{n}_{sl} = -(k - 1)c_l \frac{\partial \mathbf{x}}{\partial t} \cdot \mathbf{n}_{sl} \tag{6.10}$$

These conditions, coupled with diffusion and convection near the interface, have a profound effect on solute segregation in crystal-growth systems.

To complete the specification of governing equations requires a constraint that determines the location of the liquid/crystal interface. Here we distinguish between melt and solution growth systems. In melt-growth systems, the position of the interface is located at the melting-point isotherm:

$$T(r, z, t) = T_m \tag{6.11}$$

This condition neglects the effect of curvature on the free energy of the interface, which is negligible except when interface curvature occurs over sufficiently small scales, such as occurs during dendritic growth. This condition implies that the kinetics of crystallization are infinitely fast. In solution-growth systems, the location of the interface is controlled by kinetic factors and can be determined only from the initial condition and knowledge of the local interface velocity during growth. In the simplest case, growth from a single solute, the local interface velocity is related to the flux of solute to the interface:

$$\rho_s \frac{\partial \mathbf{x}}{\partial t} = -\mathcal{D}_l \mathbf{n} \cdot \nabla c|_l = \mathcal{R}(c_l, T) \qquad (6.12)$$

where ρ_s is the crystal density and $\mathcal{R}(c_l, T)$ represents surface growth kinetics.

Before we close this section, we note that theoreticians often rescale governing equations prior to analysis, yielding a set of nondimensional equations. In the course of rescaling, various dimensionless groups are introduced that incorporate the effects of geometric and physical parameters. There are several benefits to this rescaling. First, the introduction of dimensionless groups aids the theoretician to assess the magnitude, and therefore importance, of various driving forces that cause flow and transport to occur. Secondly, rescaling often eliminates or minimizes numerical problems that arise when using finite-precision computers to solve the governing equations. Thirdly, introduction of a proper set of dimensionless groups reduces the number of parameters to the minimum that characterize a problem, thereby simplifying parametric analysis of system behavior.

Some common dimensionless groups used in analysis of crystal growth include the Grashof and Prandtl numbers, defined by

$$\mathrm{Gr} \equiv \frac{\rho^2 g \beta \Delta T L^3}{\mu^2} \qquad (6.13)$$

$$\mathrm{Pr} \equiv \frac{\mu C_p}{\kappa} \qquad (6.14)$$

where L is the characteristic system length and ΔT is a characteristic temperature difference. The Grashof number represents the ratio of buoyancy force to viscous force and is characteristic of the intensity of buoyancy-driven convection. The Prandtl number, a physical property of the material, is the ratio of momentum diffusivity to thermal diffusivity. A low value of Pr, typical of liquid metals and most semiconductor melts, indicates that the thermal field is weakly affected by small changes in the velocity field. A high value of Pr, typical of many oxide melts, indicates the converse, namely that small changes in the velocity field strongly affect the temperature field. As a consequence, low Pr materials behave very differently from high-Pr materials in crystal-growth systems.

For a more detailed discussion of scaling analysis and dimensionless groups, in the context of crystal-growth modeling, the interested reader is referred to Derby *et al.* (1998) and Brown (1988).

6.3   COMPUTATIONAL ISSUES

6.3.1   NUMERICAL METHODS

The equations presented in Section 6.2 rarely can be solved using traditional methods of mathematical analysis, except under extremely restrictive simplifications. Numerical methods have become indispensable, therefore, to the analysis of crystal-growth systems. Here we comment briefly on these methods, including a discussion of commercial software for hydrodynamic analysis.

There exist several well-developed methods widely used to solve equations of continuum transport phenomena. These include finite-difference, finite-element, finite-volume, and spectral methods. Each has its enthusiasts who will hotly debate the superiority of their method of choice, but, in reality, the methods share several common features. Each method reduces the governing equations to a set of differential-algebraic equations (DAEs), which can be integrated in time using one of several standard methods. The result is a set of nonlinear algebraic equations, which must be solved iteratively at each time step. In all cases, the preponderance of computational effort is expended in obtaining a solution to this set of algebraic equations. In our work we have used both the Galerkin finite-element method and the Galerkin/least-squares method to discretize the governing equations (Hughes, 1987). Generally we use a second-order trapezoid rule to time integrate the DAEs (Gresho *et al.* 1980), and Newton's method is our method of choice to solve the set of nonlinear algebraic equations at each time step.

Within each Newton iteration it is necessary to solve a linearized set of algebraic equations, by far the most costly computational step. There are two general approaches to this step. The first approach is to use a direct method, i.e. some form of Gaussian elimination. The alternative is to use an indirect, or iterative method, of which there are many choices, the most popular being some sort of subspace projection method, e.g. the generalized minimal residual method of Saad and Schultz (GMRES) (1986). Direct methods are far more robust and are the method of choice for one- and two-dimensional problems, but are usually too costly for three-dimensional problems, even using the most powerful computers available today. Typically, the numerical analyst is forced to employ less robust iterative methods when tackling three-dimensional problems.

Many computing platforms are available to the numerical analyst, generally belonging to one of three broad classes: fast scalar computers, vector supercomputers, and parallel supercomputers. Vector supercomputers, once the workhorse of hydrodynamic simulation, have in recent years mostly been replaced by fast scalar machines. Indeed, desktop computers that exceed the computing power of a Cray C90 supercomputer, a state-of-the-art multimillion dollar machine 10 years ago, can be purchased today for about $1000. Such machines are adequate for nearly all two-dimensional hydrodynamic analyses in crystal growth (an exception will be presented in Section 6.4). Fully three-dimensional hydrodynamic calculations continue to be extremely costly and difficult to perform, however.

For many hydrodynamic problems, such calculations remain beyond the reach of fast scalar computers or vector supercomputers. In fact, many existing three-dimensional calculations that have been done using these types of computers have employed very coarse meshes. Accurate hydrodynamic simulation in crystal-growth systems often requires fine spatial discretizations in order to resolve thin boundary layers typical of these systems.

Recent advances in massively parallel supercomputers have dramatically affected the prospect of studying three-dimensional macroscopic transport effects in many systems. In Section 6.4, an example is presented that illustrates the application of massively parallel computing to analyze a complicated, three-dimensional flow in a crystal-growth system. We emphasize that such calculations are by no means routine, however. Parallel computers require special programming methods to exploit the architecture of these machines, and no commercial hydrodynamics codes known to us are yet available that take advantage of massive parallelism.

6.3.2   SOFTWARE: COMMERCIAL VERSUS RESEARCH, GENERAL VERSUS SPECIALTY

For the industrialist interested in modeling bulk crystal growth, but lacking in modeling experience, the question of how best to proceed is critical. Unfortunately there are no easy answers to this question. The choices are many. Some of the possibilities include: using a general-purpose commercial code; using a general-purpose commercial code that has been customized (either by the vendor or the user); using a specialized commercial code that has been tailored to the needs of the crystal-growth industry; contracting the work to a specialist; or developing software in-house.

There are numerous general-purpose codes on the market today that model fluid dynamics and transport phenomena, several of which have found use in the crystal-growth community. Each code has its strengths and weaknesses, and, as a result, each code is usually marketed more actively in some industries than others. We make no attempt to analyze these strengths and weaknesses nor to make any specific recommendations. Indeed, our experience in using such codes is limited; we have rather chosen to develop our own research codes, which are specialized to our needs. We do offer one opinion, however: we believe it is impossible for the commercial code developer to anticipate all possible physical problems, all possible boundary conditions, and so on. Since production crystal-growth systems usually are quite complex, and their behavior is affected by many physical phenomena, creating a general-purpose code that can accurately model such systems presents a special challenge to code developers. Indeed, for some problems there is no suitable choice among the general-purpose commercial codes.

An alternative is to use specialized software. General-purpose code vendors usually sell additional services, including custom code modification for specialized problems, and contract simulation on the request of the customer. There also have appeared in recent years several vendors of software products that are tailored to meet typical needs of the crystal-growth modeler. But even these specialized products are not suitable in every situation that might arise in crystal growth. So, like the general-purpose code developers, these vendors also usually offer customized products and contract simulation.

The last alternative we mention is developing and using software in-house. This could range from hiring a single specialist at smaller companies, to supporting a modeling department at larger companies. In-house code development in most cases is the most time- and capital-intensive approach, so it appears to be relatively rare at companies specializing in crystal growth. At this point it is worth mentioning that licensing or purchasing one or more commercial codes might require devoting one or more staff members to master the codes, which, due to the considerable art sometimes required of successful modeling, is by no means foolproof or trivial. Also, in all cases, whether using a commercial code, or software developed in-house, at least some knowledge of transport phenomena is required to both define the problem to be modeled and to interpret the output from the software.

Our principal advice is to re-emphasize that there is no single clear-cut approach that will be optimal in all situations. Furthermore, the industrialist who solicits opinions from commercial vendors and other experts in modeling is likely to hear several different opinions. In our view, the best approach is to first strive to carefully define the particular problems that are to be modeled, then to solicit opinions both from vendors that provide products and services and from other users of those products and services.

6.4   EXAMPLES OF ONE-, TWO-, AND THREE-DIMENSIONAL MODELS

6.4.1   CAN WE STILL LEARN FROM A 1D MODEL?

Prior to the widespread availability of digital computers, crystal-growth models were simple out of necessity, which usually meant they were restricted to describe one-dimensional behavior. Such models can be quite powerful when appropriately applied. An early example is the well-known Scheil equation (Scheil 1942), derived from a model of segregation during freezing from a completely mixed melt. Another example is the diffusion-only limit, studied by Tiller and coworkers (1953; 1955). A more general result was obtained by Burton *et al.* (BPS) (1953), based on the notion of a completely mixed bulk melt, separated from the crystal by a thin stagnant film in which transport occurs by diffusion only. The BPS model has been successfully fitted to experimental data obtained from many crystal-growth systems. Unfortunately, the model has little predictive capability, largely because the stagnant-film assumption is not a realistic representation of convection in the melt. Other variations on these one-dimensional models include effects of rotating the crystal and imposing an axial magnetic field; references and discussion can be found in Brown (1988).

Using modern computers, various researchers have compared these and other one-dimensional models to more rigorous two- and three-dimensional models, highlighting various shortcomings of simpler models. Since researchers no longer are confined to such simple models, it is relevant to ask whether anything new can be learned from them. We present an example here that suggests that one-dimensional models are still relevant to understanding crystal-growth systems, namely ternary segregation during diffusion-controlled growth of the II-VI alloy cadmium zinc telluride ($Cd_{1-x}Zn_xTe$, hereafter referred to as CZT).

CZT is employed in room-temperature gamma-ray and X-ray detectors, typically using material with 10% nominal zinc concentration ($x = 0.1$). Compositional inhomogeneity resulting from zinc and cadmium segregation (James et al., 1995) is an important issue in CZT-based detectors. An unresolved issue in modeling CZT growth, however, is whether pseudobinary diffusion (i.e. Equation (6.4)) is adequate to describe multicomponent mass transport and segregation phenomena in this system.

The pseudobinary diffusion model is based on the assumption that the flux of each species in a multicomponent system is simply proportional to the gradient of that species alone; this model is known as Fick's first law of diffusion (Bird et al., 1960). In true multicomponent diffusion, however, the flux of each species is represented by a more complicated, implicit relationship between gradients and fluxes of all species, known as the Stefan–Maxwell equations (Hirschfelder et al., 1954; Lightfoot et al., 1962). To better understand multicomponent mass transfer and segregation during diffusion-controlled growth of CZT, Ponde et al. have developed a one-dimensional model based on the Stefan–Maxwell equations (Ponde, 1999; Derby et al., 2000).

Figure 6.1 shows initial results obtained using this multicomponent diffusion model. Concentrations of zinc and cadmium are shown as a function of distance from the solid/liquid interface, under steady-state growth conditions. Zinc

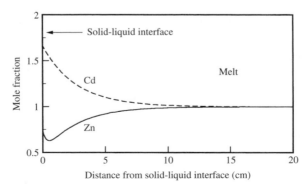

Figure 6.1 Compositional profiles in the melt ahead of the melt-crystal interface are shown for diffusion-controlled solidification of $Cd_{0.9}Zn_{0.1}Te$.

is preferentially incorporated into the growing crystal, causing its depletion at the interface. Conversely, cadmium is rejected at the interface, causing it to accumulate there. At steady-state conditions, zinc must diffuse towards the interface to balance its depletion there, whereas cadmium must diffuse away from the interface, to counter its accumulation. The figure shows that cadmium concentration falls monotonically with distance from the interface. Hence cadmium diffuses along its gradient everywhere, the same as predicted by the pseudobinary diffusion model. Zinc concentration, on the other hand, first decreases, then increases, with distance from the interface. Thus the multicomponent diffusion model predicts that zinc diffuses against its concentration gradient near the interface, a consequence of diffusion coupling between components. Ponde *et al.* described this feature as a retrograde diffusion layer during solidification. Such behavior may have significant consequences in setting compositional variations that occur during growth of II–VI ternary compounds.

6.4.2   IS 2D MODELING ROUTINE AND ACCURATE?

Two-dimensional modeling of crystal-growth systems is now widely regarded as routine and accurate, provided that realistic thermophysical properties are available. We generally agree with this conclusion, with a few caveats. One of these caveats is that multicomponent diffusion effects may have a significant effect on segregation in ternary systems, as discussed in Section 6.4.1. Incorporating the Stefan–Maxwell equations in any model of crystal growth adds considerable complexity to segregation calculations, however, and poses numerical difficulties not encountered when using pseudobinary transport equations. Another caveat is that two-dimensional systems, in which there are multiple, widely disparate, time scales, may require an onerous amount of computing time. One such system, the subject of this section, is vertical Bridgman growth of CZT, using the accelerated crucible rotation technique (ACRT). A final caveat is that two-dimensional models cannot account for three-dimensional behavior, the topic of the next section.

The purpose of ACRT is to promote mixing within the melt. Here, we assess the effectiveness of ACRT for that purpose and discuss the challenges associated with modeling this system. A schematic of a high-pressure vertical Bridgman system is shown in Figure 6.2 along with a sample finite element mesh. Previously we have reported simulations for this system using steady rotation; a detailed description of the model can be found in Yeckel *et al.* (1999). An in-depth analysis of the system using ACRT appeared elsewhere (Yeckel and Derby, 2000). Except for the use of accelerated rotation, values of all physical properties and operating parameters are the same as used in Yeckel *et al.* (1999). The effect of accelerated rotation is accounted for through the no-slip boundary condition at the ampoule walls and solid/liquid interface, given by

$$\mathbf{v} = \Omega(t)r\mathbf{e}_\theta \tag{6.15}$$

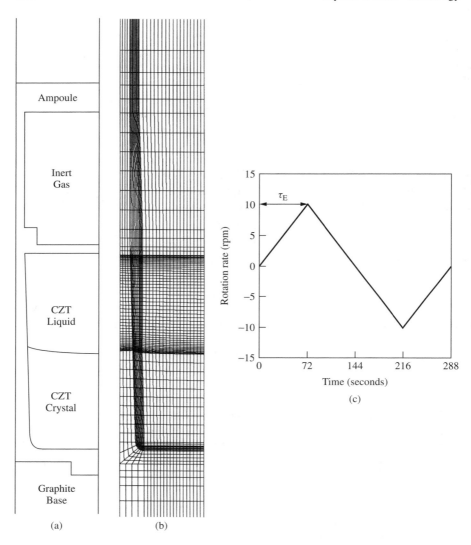

Figure 6.2 (a) Schematic of high-pressure vertical Bridgman system; (b) finite-element discretization; (c) rotation cycle used in this study.

where Ω is the crucible rotation rate, and $\mathbf{e}_\theta$ is a unit vector in the azimuthal direction. The periodic rotation cycle, Ω versus time, is shown in Figure 6.2.

Computing a complete segregation curve for this system presents a major computational challenge. The difficulty is brought on by the short time scale introduced when ACRT is used. Whereas a complete growth run typically lasts

100–150 h, the optimum rotation cycle is set by the Ekman time scale (Schulz-Dubois, 1972),

$$\tau_E = \left(\frac{R_o^2}{\Omega v} \right)^{1/2} \qquad (6.16)$$

equal to 72 s for the conditions studied here. The optimum rotation cycle is a few Ekman time units in length (Yeckel and Derby, 2000). Over the duration of the growth run, therefore, more than one thousand ACRT cycles are needed, each requiring hundreds of time steps for accurate time integration. Hence to compute a single segregation curve requires on the order of one million time steps, a task that would take several months using a modern engineering workstation or a traditional vector supercomputer. Such long computation times make it impractical to simulate a complete growth run for this system. Using a parallel supercomputer for this task holds more promise and is the subject of our ongoing work.

It is possible to learn much about this system without simulating a complete growth run, however. To do so, we proceed as follows. We begin by simulating approximately one-half of a growth run, without rotation. The simulation is then restarted subject to the application of ACRT, and continued for two more hours of growth (approximately 25 ACRT cycles). Figure 6.3 shows the evolution of zinc distribution that follows upon application of ACRT; the number of full

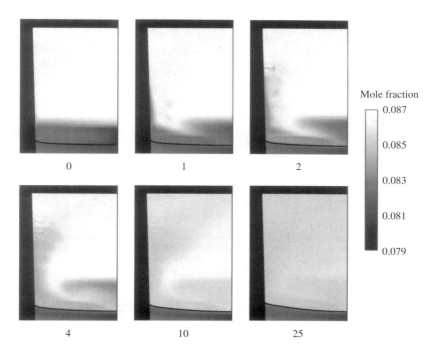

Figure 6.3   Mole fraction zinc versus ACRT cycle.

ACRT cycles undergone is shown below each visualization. At the start of ACRT (i.e. after zero cycles in Figure 6.3), the melt exhibits two distinct zones: a well-mixed bulk zone above, and a depletion layer below, separated by a diffusion layer characterized by a steep concentration gradient. These features, typical of CZT growth without rotation, are explained in Kuppurao *et al.* (1995).

After one ACRT cycle, a plume of zinc-poor liquid has been drawn from the depletion layer into the bulk, along the crucible wall. Likewise, some zinc-rich bulk fluid has been swept into the depletion layer. After two ACRT cycles, the interchange between the two regions continues. Mixing within the depletion zone has markedly increased; the region is much more uniform in concentration, particularly along the solid/liquid interface. Also, concentration variations begin to appear along the sidewall; these are the result of a Taylor–Görtler flow instability, as discussed in Yeckel and Derby (2000). After four ACRT cycles, there is substantial intermixing of the bulk and depletion regions, and, after ten ACRT cycles, substantial intermixing has occurred. The liquid concentration of zinc is nearly uniform throughout the melt after 25 ACRT cycles. The two-hour time span of ACRT growth is too short to meaningfully comment on the details of zinc segregation in the newly grown solid, but the results clearly illustrate the efficacy of ACRT at homogenizing the melt.

Whereas Figure 6.3 illustrates the pattern of mixing in the melt, Figure 6.4 shows a quantitative measure of mixing, the RMS mole fraction variation in the melt, defined by:

$$\Delta c_{\text{rms}} = \frac{1}{\bar{c}V} \int_V (c - \bar{c})^2 \, dV \qquad (6.17)$$

where $\bar{c}$ is the mean zinc mole fraction and V is the melt volume. Figure 6.4 shows that Δc_{rms} decreases steadily with ACRT cycle number for approximately 24 cycles, indicating an increasing degree of compositional homogeneity. After 24 cycles Δc_{rms} begins to slowly increase again. Note that Δc_{rms} cannot reach zero, a state of complete mixing, because zinc segregation ensures the existence

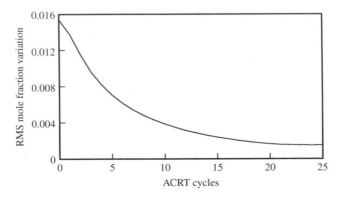

Figure 6.4   RMS mole fraction variation Equation (6.17) versus ACRT cycle.

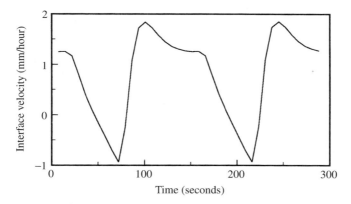

Figure 6.5 Average interface velocity (i.e. instantaneous growth rate) versus time.

of a nonvanishing concentration boundary layer at the solid/liquid interface, even though the bulk may be well mixed. The effect of this boundary layer on Δc_{rms} increases as the melt volume diminishes, causing the slight increase in Δc_{rms} after 24 cycles.

Figure 6.5 shows the average interface velocity (i.e. the instantaneous growth rate) as a function of time during the 25th ACRT cycle. Notably, the growth rate falls below zero during some parts of the cycle, which indicates that a portion of the crystal is melted during this phase, to be recrystallized later in the cycle. The rate of meltback is substantial, nearly reaching 1 mm/h at point D in the cycle. The large variation in growth rate, and the occurrence of meltback, raise the possibility that solute striations will occur. Kim *et al.* (1972) have linked striations with velocity and temperature field oscillations in Bridgman growth caused by an unstable, time-dependent flow. ACRT-induced striations were not observed in the Bridgman experiments of Capper *et al.* (1984; 1986); however, the striation spacing expected from the conditions of their experiments is an order of magnitude smaller than observed in Kim *et al.* (1972), making striations both more difficult to observe experimentally and faster to disappear by solid-state diffusion. On the other hand, the striation spacing expected under the conditions of our simulation, on the order of $2 - 4 \times 10^{-3}$ cm, is comparable to that observed in Kim *et al.* (1972), which indicates that stable striations in the grown solid may occur in the system modeled here. Due to the extreme computational challenges of accurately resolving these fine-scale striations, we do not further address these features here, but defer their analysis to ensuing work.

6.4.3   WHEN ARE 3D MODELS NECESSARY?

At first glance, many crystal-growth systems appear amenable to one- and two-dimensional modeling, and considerable knowledge has been gained over the years using such models. The assumption of axisymmetry for cylindrical geometries has been particularly useful and would seem to apply to vertical Bridgman, Float-Zone, Czochralski, and other melt-growth systems that produce unfaceted crystals. Furthermore, considerable effort is expended to ensure that these systems display two-dimensional, axisymmetric behavior. This can often be accomplished through careful design or through operation, for example by rotating the heater or crystal, to ensure that the crystal experiences an azimuthally uniform furnace environment, in a time-averaged sense.

Solution-growth systems are another story, since the faceted crystals grown in these systems usually are not well described as two-dimensional objects. Furthermore, solution-growth systems sometimes rely on a crystal support structure that cannot realistically be treated as two-dimensional, e.g. the platform used in rapid growth of KDP crystals (Zaitseva *et al.*, 1995), or the angled support rods used in KTP growth (Bordui and Motakef, 1989). Indeed, our simulations of these systems reveal flows that are strongly three-dimensional in character (Zhou and Derby, 1997; Yeckel *et al.*, 1998; Vartak *et al.*, 2000).

Though it may seem obvious whether a given system is predominantly two-dimensional or three-dimensional in character, several examples from our recent work demonstrate that many crystal-growth systems with axisymmetric, or nearly axisymmetric, geometries can produce strongly three-dimensional flow structures. For example, Xiao *et al.* (1996) demonstrated that flow in a slightly tilted vertical Bridgman system – with its axis tilted as little as one degree from vertical – exhibits a remarkable departure from axisymmetry. Even more confounding is that perfectly axisymmetric geometries sometimes exhibit three-dimensional flow instabilities; an example is the appearance of baroclinic waves in systems that have both rotation and buoyancy as driving forces (Xiao *et al.*, 1995; Xiao and Derby, 1995). Thus the choice between using a two-dimensional or three-dimensional model, of great importance given the drastic difference in computational cost, remains a difficult one. Two-dimensional geometries certainly do not guarantee two-dimensional behavior.

In this section, we present transient, three-dimensional simulations of an axisymmetric Czochralski growth system that simultaneously exhibits two different types of three-dimensional instabilities. The manifestation of these instabilities takes the form of annular wave patterns that appear both on the surface and within oxide melts during Cz growth. Both types of instabilities arise because of the nonlinear interaction of intense rotational and buoyant flows, to which the temperature field is strongly coupled in high Prandtl number oxide melts. But the instabilities are caused by fundamentally different mechanisms, meaning that each instability occurs independently of the other.

Figure 6.6 shows a schematic of the bulk flow model used to compute time-dependent flows during Cz growth. The bulk flow model is based on assumptions of fixed and flat crystal/melt interface and melt free surface; an extensive description of this model can be found elsewhere. Idealized temperature boundary conditions consist of constant temperature T_h at the wall and bottom of the

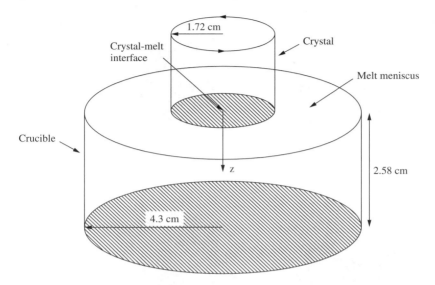

Figure 6.6 Schematic of Czochralski bulk flow model.

crucible, an insulated melt free surface, and a crystal/melt interface at the melt-ing point T_m. Physical properties used are those of bismuth silicon oxide (BSO), which has a Prandtl number of $Pr = 26$. Operating conditions correspond to $Gr = 1.3 \times 10^5$, equivalent to a system with a crucible radius of 4.3 cm, a cru-cible depth of 2.58 cm, a crystal radius of 1.72, and a temperature difference across the melt of 23 K. Four crystal rotation rates were used: 13.5, 15.2, 23.0, and 30.4 rpm. Here we briefly summarize the results of the simulations; more detailed analyses are found in Rojo and Derby (1999) and Derby et al. (1999).

Figure 6.7 shows the temperature distribution on the melt surface at various rotation rates. The surface distribution exhibits a low-temperature annulus cen-tered around the crystal. This inner region is separated by a sharp transition from a high-temperature outer region, which is nearly isothermal and very near the wall temperature. The size of the hot outer region decreases as rotation rate increases, until, at the highest rotation rate considered, the outer region is confined to a thin annular region next to the outer wall. What is most interesting, however, are the azimuthal temperature variations that appear near the crystal in the cooler region. These azimuthal variations, which we call rotational spokes, are convected by the azimuthal flow driven by the counter-clockwise rotation of the crystal. The radial extent of the spokes, between 12 and 16 in number, increases as the rotation rate is increased.

The rotational spokes in Figure 6.7 bear a strong resemblance to those observed in the experiments of Whiffin et al. (1976), who studied the effect of a rotating platinum disk in contact with molten BSO, under the same conditions mod-eled here. At issue is the mechanism by which rotational spokes occur; these

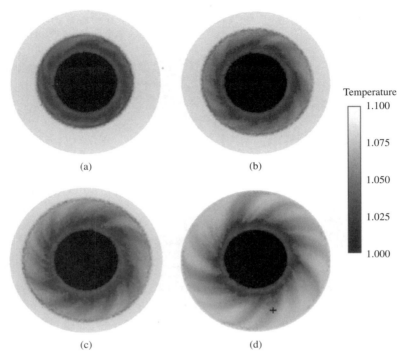

Figure 6.7 Temperature distribution on the melt surface at various rotation rates: (a) 13.5 rpm; (b) 15.2 rpm; (c) 23.0 rpm; (d) 30.4 rpm. Temperature is scaled by melting temperature of BSO.

have alternately been attributed to the baroclinic instability that arises in rotating stratified fluids (Whiffin *et al.*, 1976; Brandle, 1982), the Couette instability characteristic of centrifugal flows (Brice and Whiffin, 1977) or the Rayleigh instability caused by a destabilizing temperature gradient near the melt surface (Jones, 1983; Jones, 1985). The Rayleigh-instability mechanism was used by Jones to explain the appearance of radial spokes, observed in several oxide melts under conditions of no crystal rotation; results of our simulations indicate that a similar mechanism applies to rotational spokes, too.

The key to the Rayleigh instability is the presence of warmer, lighter fluid underlying cooler, denser fluid, which triggers the formation of roll cells. In the case of no crystal rotation studied by Jones, the cause of the destabilizing temperature gradient was assumed to be radiant cooling at the melt surface. A destabilizing temperature gradient also occurs in our simulations, as a consequence of centrifugal pumping of colder fluid, thrown outward along the melt surface by the rotating crystal. By performing a linear stability analysis applicable to high Prandtl number fluids, Jones (1985) calculated a critical value of the Rayleigh number, $Ra_c = 255$, above which the onset of roll cells would occur, where

$$\mathrm{Ra} = \mathrm{GrPr} \equiv \frac{\rho^2 g \beta C_p \Delta T L^3}{\mu \kappa} \tag{6.18}$$

In Jones' stability analysis, values of L and ΔT correspond to the depth and temperature difference across the thermal boundary layer at the melt surface. Figure 6.8 shows the vertical temperature profile beneath a location centered on one of the spokes in Figure 6.7(d). Using this temperature profile to measure values of L and ΔT yields a computed Rayleigh number of $\mathrm{Ra} = 278$, which compares favorably to the critical value predicted by Jones.

The simulations also reveal that azimuthal wave patterns occur deep within the melt, as shown in Figure 6.9 for the results obtained at 30.4 rpm. These patterns exhibit six- or seven-fold symmetry and occur far below the thermal boundary layer responsible for roll-cell formation near the melt surface. Similar patterns, with two- and four-fold symmetry, were also computed by Xiao and Derby (1995), for a different oxide melt. We attribute these wave patterns to the baroclinic instability (Greenspan, 1968), caused by the interaction of Coriolis and buoyancy forces. We recast our results in terms of dimensionless Rossby and Taylor numbers, to locate our result on the baroclinic stability diagram of Hide and Mason (1975) for a high Prandtl number fluid ($\mathrm{Pr} = 63$), as shown in Figure 6.10. The parameters of our case fall near the regular annular wave regime and are consistent with the six- or seven-fold wave structure that is expected.

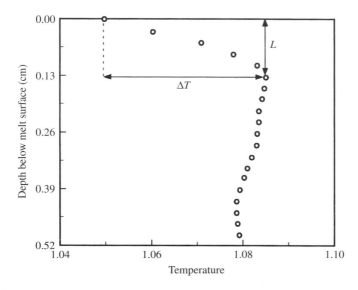

Figure 6.8 Axial temperature profile below the melt free surface at the point indicated in Figure 6.7b. Temperature is scaled by melting temperature of BSO.

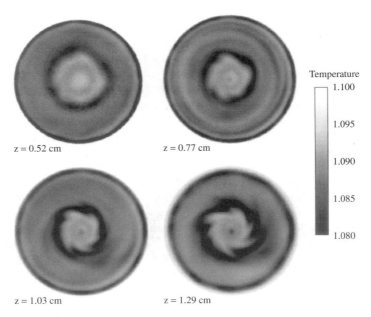

Figure 6.9 Temperature distribution on various horizontal planes within the melt; numbers refer to depth below the melt surface. Temperature is scaled by melting temperature of BSO.

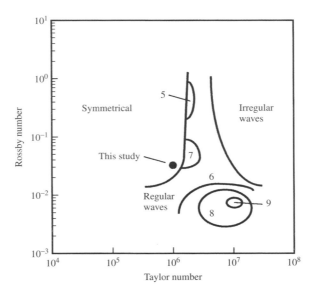

Figure 6.10 Stability diagram of Hide and Mason (1975) for flows in a rotating, differentially heated annulus filled with a high Prandtl number liquid. Asterisk shows parameter location corresponding to BSO calculations presented in Figure 6.9.

6.5   SUMMARY AND OUTLOOK

The examples presented in this chapter were chosen to demonstrate the power of simulation at elucidating answers to important questions in the study of crystal growth and to illustrate the current limitations of hydrodynamics modeling. The first example, the one-dimensional model of multicomponent diffusion in a ternary alloy, raises the question of how well we really understand basic model formulation in some crystal-growth systems. The second example, ACRT applied to vertical Bridgman growth, shows that not all two-dimensional problems are entirely routine, even with today's powerful computers. The third example highlights the risk of relying on a two-dimensional model, even when the geometry and boundary conditions would seem to indicate two-dimensional system behavior. Despite these caveats, however, hydrodynamic simulation has made enormous advances in the past decade, making two-dimensional analysis available at a low cost to a large number of researchers and making realistic three-dimensional analysis feasible at all.

The potential benefits of computer modeling in the analysis and design of bulk crystal-growth systems have been the subject of much discussion in the past decade. The realization of computer modeling as a comprehensive design tool still appears some way off, however. Rather, modeling at present often best serves to complement experimental studies, to provide a basic understanding of phenomena important to crystal-growth processing, and to identify qualitative, and in some cases quantitative, trends in system behavior, with respect to parametric variation. Such information, although failing to provide a complete picture of system behavior, often suggests beneficial modifications to both operation and design, and on occasion suggests entirely new designs. Nevertheless, modeling alone often is of limited use in the absence of good experimental data, human experience, and intuition, especially at the design stage. Furthermore, there yet remain some phenomena, and some physical systems, which are beyond our ability to model, except in the most rudimentary fashion.

The maturation of computational modeling of crystal-growth systems will rely on the increased availability of accurate thermophysical properties, the development of more capable supercomputer hardware and software, the advancement of experimental diagnostics in crystal-growth systems, and increased communication between crystal-growth experimentalists and theoreticians. Great advances in understanding and practice are to be expected in the coming years.

ACKNOWLEDGMENTS

The authors' research programs in crystal-growth modeling have been supported in part by Johnson Matthey Electronics (IRMP, Defense Advanced Research Projects Agency), Lawrence Livermore National Laboratory, the National Aeronautics and Space Administration (Microgravity Materials Science), the National

Science Foundation, Sandia National Laboratories, the University of Minnesota Supercomputer Institute, and the Army High Performance Computing Research Center under the auspices of the Department of the Army, Army Research Laboratory cooperative agreement DAAH04-95-2-0003/contract DAAH04-95-C-0008, the content of which does not necessarily reflect the position or policy of the government, and no official endorsement should be inferred. The authors also gratefully acknowledge significant input from J. C. Rojo and N. Ponde.

REFERENCES

Bird, R. B., Stewart, W. E., and Lightfoot, E. N. (1960) *Transport Phenomena*. John Wiley; New York.

Bordui, P. F., and Motakef, S. (1989) *J. Cryst. Growth*, **96**, 405.

Brandle, C. D. (1982) *J. Cryst. Growth*, **57**, 65.

Brandon, S., and Derby, J. J. (1992) *Int. J. Num. Meth. Heat Fluid Flow*, **2**, 299.

Brice, J. C., and Whiffin, P. A. C. (1977) *J. Cryst. Growth*, **38**, 245.

Brown, R. A. (1988) *AIChE J.*, **34**, 881.

Burton, J. A., Prim, R. C., and Schlicter, W. P. (1953) *J. Chem. Phys.*, **21**, 1987.

Capper, P., Gosney, J. J. G., and Jones, C. L. (1984) *J. Cryst. Growth*, **70**, 356.

Capper, P., Gosney, J. J. G., Jones, C. L., and Kenworthy, I. (1986) *J. Electron. Mater.*, **15**, 371.

Derby, J. J., Edwards, K., Kwon, Y.-I., Rojo, J. C., Vartak, B., and Yeckel, A. (1998) Large-scale numerical modeling of continuum phenomena in melt and solution crystal growth processes. *p. 119 of*: Fornari, R., and Paorici, C. (eds), *Theoretical and Technological Aspects of Crystal Growth*, vol. 276, Trans Tech Publications, Switzerland.

Derby, J. J., Kwon, Y.-I., Rojo, J. C., Vartak, B., and Yeckel, A. (1999) *Int. J. Comput. Fluid Dyn.*, **12**, 225.

Derby, J. J., Ponde, N., De Almeida, V. F., and Yeckel, A. (2000) *p. 93 of*: Roósz, A., Rettenmayr, M., and Watring, D. (eds), *Solidification and Gravity 2000*, vol. 329–330. Switzerland: Trans Tech Publ. Ltd.

Greenspan, H. P. (1968) *The Theory of Rotating Fluids*. Cambridge University Press, London.

Gresho, P. M., Lee, R. L., and Sani, R. L. (1980) *p. 27 of*: Taylor, C., and Morgan, K. (eds), *Recent Advances in Numerical Methods in Fluids, Vol. 1*, Pineridge, Swansea.

Hide, R., and Mason, P. J. (1975) *Adv. Phys.*, **24**, 47.

Hirschfelder, D. G., Curtiss, C. F., and Bird, R. B. (1954) *Molecular Theory of Gases and Liquids*. John Wiley, New York.

Hughes, T. J. R. (1987) *The Finite Element Method*. Prentice Hall, Englewood Cliffs, NJ.

James, R. B., Schlesinger, T. E., Lund, J., and Schieber, M. (1995) *p. 335 of*: Schlesinger, T. E., and James, R. B. (eds), *Semiconductors for Room Temperature Nuclear Detector Applications*, vol. 43. Academic Press, San Diego.

Jones, A. D. W. (1983) *J. Cryst. Growth*, **61**, 235.

Jones, A. D. W. (1985) *Phys. Fluids*, **28**, 31.

Kim, K. M., Witt, A. F., and Gatos, H. C. (1972) *J. Electrochem. Soc.*, **119**, 1218.

Kuppurao, S., and Derby, J. J. (1993) *Numer. Heat Transfer, Part B: Fundamentals*, **24**, 431.

Kuppurao, S., Brandon, S., and Derby, J. J. (1995) *J. Cryst. Growth*, **155**, 103.

Lightfoot, E. N., Cussler, E. L., and Rettig, R. L. (1962) *AIChE J.*, **8**, 708.

O'Hara, S., Tarshis, L. A., and Viskanta, R. (1968) *J. Cryst. Growth*, **3/4**, 583.

Ponde, N. (1999) M.S. thesis, University of Minnesota.

Rojo, J. C., and Derby, J. J. (1999) *J. Cryst. Growth*, **198**, 154.

Saad, Y., and Schultz, M. H. (1986) GMRES: A generalized minimal algorithm for solving nonsymmetric linear systems. *SIAM J. Sci. Stat. Comp.*, **7**, 856–869.

Scheil, E. (1942) *Z. Met. kd*, **34**, 70.

Schulz-Dubois, E. O. (1972) *J. Cryst. Growth*, **12**, 81.

Siegel, R., and Howell, J. R. (1993) *Thermal Radiation Heat Transfer, Third Edition.* McGraw-Hill, New York.

Smith, V. G., Tiller, W. A., and Rutter, J. W. (1955) *Can. J. Phys.*, **33**, 723.

Tiller, W. A., Jackson, K. A. Rutter, J. W., and Chalmers, B. (1953) *Acta Met.*, **1**, 428.

Vartak, B., Kwon, Y.-I., Yeckel, A., and Derby, J. J. (2000) *J. Cryst. Growth*, **210**, 704.

Whiffin, P. A. C., Bruton, M., and Brice, J. C. (1976) *J. Cryst. Growth*, **32**, 205.

Xiao, Q., and Derby, J. J. (1995) *J. Cryst. Growth*, **152**, 169.

Xiao, Q., Salinger, A. G., Zhou, Y., and Derby, J. J. (1995) *Int. J. Numer. Methods Fluids*, **21**, 1007.

Xiao, Q., Kuppurao, S., Yeckel, A., and Derby, J. J. (1996) *J. Cryst. Growth*, **167**, 292.

Yeckel, A., and Derby, J. J. (2000) *J. Cryst. Growth*, **209**, 734.

Yeckel, A., Zhou, Y. Dennis, M., and Derby, J. J. (1998) *J. Cryst. Growth*, **191**, 206.

Yeckel, A., Doty, F. P., and Derby, J. J. (1999) *J. Cryst. Growth*, **203**, 87.

Zaitseva, N. P., Rashkovich, L. N., and Bogatyreva, S. V. (1995) *J. Cryst. Growth*, **148**, 276.

Zhou, Y., and Derby, J. J. (1997) *J. Cryst. Growth*, **180**, 497.

7 Heat and Mass Transfer under Magnetic Fields

KOICHI KAKIMOTO

Institute of Advanced Material Study, Kyushu University, Kasuga 816–8580 JAPAN

ABSTRACT

The heat and mass transfer in the melts during Czochralski crystal growth of semiconductors significantly affect the quality of the single crystals. This chapter reviews the present understanding of heat and mass transfer under several kinds of magnetic fields from the results of flow visualization, and gives details of numerical calculation needed for quantitative modeling of melt convection under the magnetic fields. The characteristics of flow instabilities of melt convection under magnetic fields are also reviewed.

7.1 INTRODUCTION

The Czochralski (CZ) crystal-growth technique is widely accepted for fabricating high-quality substrates for silicon (Si) VLSIs, gallium arsenide (GaAs) monolithic and integrated devices, and indium phosphide (InP) optoelectronic devices.

The breakdown voltage of an oxide layer grown on Si substrates is well known to depend on the conditions under which the Si crystals were grown [1], such as the crystal pulling speed and/or the temperature distribution in the crystals. The origin of the degradation is the formation of voids in the substrate [2], which are formed during single-crystal growth. To avoid the excessive void formation is one of the key points to control characteristics of electronic silicon devices. This can be achieved by controlling the temperature distribution in the growing crystals. Simultaneously, the solid/liquid interface shape should also be controlled to obtain an appropriate temperature distribution near the solid/liquid interface in silicon.

For GaAs, the temperature distribution near the solid/liquid interface modifies the distribution of dislocations, which affects the threshold voltage of source–drain current in metal-semiconductor (MES) transistors [3]. Controlling heat and mass transfer in the melt during crystal growth can reduce this inhomogeneity.

Crystal Growth Technology, Edited by H. J. Scheel and T. Fukuda
© 2003 John Wiley & Sons, Ltd. ISBN: 0-471-49059-8

Although the melt convection should, therefore, be controlled, the actual flow in semiconductor melt has been difficult to monitor because these melts are opaque.

The distribution of impurities and point defects in crystals, which affects the degradation of the breakdown voltage for silicon crystals, is thought to depend on the amplitude of temperature fluctuation at a solid/liquid interface [4]. This is mainly caused by the instability of a flow containing laminar and/or turbulent components. Furthermore, temperature fluctuations in the melt cause local alternations between crystallization and remelting. This alternation produces a relatively high microdefect density in the crystals.

The origin of flow instability should therefore be clarified, so that this instability can be controlled and high-quality crystals obtained. Consequently, suppressing flow instabilities during crystal growth is important for fabricating crystals with homogeneous impurity distribution. Despite intensive research [5–8], there are still open questions such as flow structure and impurity transfer under magnetic fields [6, 7].

Application of stationary magnetic fields such as vertical, cusp-shaped and transverse magnetic fields is opening up a new field to controlling heat and mass transfer in electrically conducting melts such as silicon or GaAs. Witt et al. [5] first applied magnetic fields in crystal growth of semiconductors from the melt. Subsequently, Hoshi et al. [8] reported homogeneous oxygen distribution in silicon crystals by using transverse magnetic fields.

Three types of static-magnetic field such as vertical-magnetic-fields (VMF), cusp-shaped magnetic fields (CMF), and transverse magnetic fields (TMF) have been proposed. Suzuki et al. [9] applied TMF to silicon crystal growth, however, the thermal symmetry became asymmetric. Therefore, periodic rotational striations were observed [10].

To overcome the rotational striations, Hirata and Hoshikawa [11, 12] proposed VMF in which axially symmetric magnetic fields were applied perpendicularly to the growth interface. They reported that temperature fluctuations in molten silicon were reduced at a VMF larger than 0.1 T. Moreover, no striations were observed for a magnetic field of more than 0.05 T. However, oxygen concentration distributed inhomogeneously in the radial direction of the grown crystals. Hereafter, Hirata and Hoshikawa [12] developed a new type of magnetic field, which is named cusp-shaped magnetic field (CMF). They obtained crystals with homogeneously distributed oxygen and without rotational striations.

This chapter aims to introduce how silicon melt and oxygen are transferred by convection under the three types of magnetic fields.

7.2   MAGNETIC FIELDS APPLIED TO CZOCHRALSKI GROWTH

Figure 7.1 schematically shows a cross section of a VMF apparatus [13] that allows direct observation of molten silicon flow in the magnetic fields by X-ray radiography. This apparatus consists of four major parts: a set of magnets,

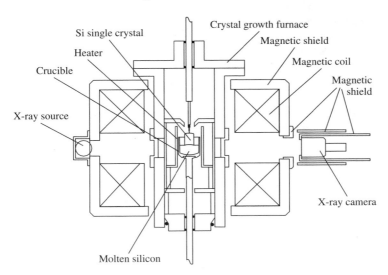

Figure 7.1 Schematic diagram of the magnetic field applied CZ crystal-growth furnace with X-ray radiography system.

an X-ray radiography system, a magnetic shield, and a crystal-growth furnace. Moreover, the system has two cylindrical-solenoid-coils to apply magnetic fields of VMF. Furthermore, CMF can be obtained by changing current direction of one of the two coils.

The X-ray radiography system consists of two sets of X-ray sources and cameras. X-ray sources and cameras are very sensitive to magnetic fields; therefore, shielding is required to observe melt flow under magnetic fields.

7.3 NUMERICAL MODELING

The control–volume method was used for discretizing the governing equations such as continuity, Navier–Stokes, energy, and impurity transfer equations in the present calculation [13]. The alternating directional implicit (ADI) method was used as a matrix solver to carry out a time-dependent calculation with three-dimensional geometry.

The governing equations of continuity, Navier–Stokes, energy, and impurity transfers are expressed in Equations (7.1)–(7.3).

$$\frac{\partial \rho}{\partial t} + \nabla \cdot (\rho u) = 0 \tag{7.1}$$

$$\frac{\partial \Phi}{\partial t} + u \cdot (\rho \nabla \Phi) = \nabla \cdot (\Gamma \nabla \Phi) + S_\phi (\Phi = u, T, c) \tag{7.2}$$

$$S = -\nabla p + F(= \rho g) + f \tag{7.3}$$

where ρ, T, and Γ are density, temperature, and diffusivity for the variables such as velocity, temperature, and impurity concentration. p, f, and g are the pressure, external forces such as Lorentz and viscous forces, and gravitational acceleration, respectively. u, T, and c are velocity, temperature, and impurity concentration, respectively. S is a source term of each variable of velocity, temperature, and impurity concentration.

When the effect of magnetic fields was taken into account, the Lorentz force (f) expressed by Equation (7.4) was included in Equation (7.3) as an external force,

$$f = J \times B \tag{7.4}$$

$$J = \sigma(-\nabla\Psi + u \times B) \tag{7.5}$$

where J, Ψ, σ, and B are the electric current, electric scalar potential, electric conductivity, and magnetic field, respectively.

When the effects of cusp-shaped magnetic fields (CMF) were calculated, the following equation based on Biot and Savart's law was used by taking into account the electric current, which was distributed in the solenoids with a finite volume,

$$dH = \frac{1}{4\pi\mu_0}\left(\frac{I\,dl \times r}{r^3}\right)dV \tag{7.6}$$

where H, μ_0, and I are the magnetic field strength, permeability, and current in solenoids, respectively. Here r and V are the distance between some specific point and a part of solenoids, and volume. Equation (7.6) was numerically integrated to obtain the magnetic field at specific points.

When oxygen transfer in the silicon melt was taken into account in the calculation, the following assumptions of equilibrium concentration at an interface between the melt and a crucible, and flux at the boundary melt-gas were imposed as expressed by Equations (7.7) [14] and Equation (7.8) [15],

$$O = 3.99 \times 10^{23}\exp(-2.9 \times 10^4/T)\ \text{atoms/cm}^3 \tag{7.7}$$

$$q = h(O(\text{melt}) - O(\text{gas})) \tag{7.8}$$

where h and O are the mass transfer coefficient at the interface between the melt and ambient gas, and the oxygen concentration in the melt, respectively. h and O(gas) are fixed as 2 and 0 atoms/cm^3, respectively [14]. Equation (7.7) was obtained from thermodynamical calculation under the equilibrium conditions [15]. Thermophysical properties used in the calculation are listed in Table 7.1.

A three-dimensional numerical simulation with a grid size of $50 \times 50 \times 50$ in the r, θ, and z directions was carried out. The geometry of the present calculation was based on the experimental one with 3-inch diameter crucible and

Table 7.1 Thermophysical properties for numerical simulation

Density (kg/m^3)	2520
Heat capacity (J/m^3 K)	2.39×10^6
Dynamic viscosity (kg/ms)	7×10^4
Thermal expansion coefficient (K^{-1})	1.4×10^{-4}
Melting temperature (K)	1685
Thermal conductivity (W/m K)	45
Emissivity	0.3
Electrical conductivity (S/m)	1.29×10^6
Crucible radius (m)	3.75×10^{-3}
Melt height (m)	3.40×10^{-3}

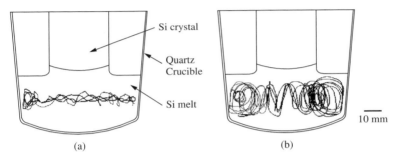

Figure 7.2 Particle paths of one specific tracer in silicon melt (a) with VMF and (b) without magnetic fields.

1.5-inch diameter crystal as shown in Figure 7.2 [16]. The temperature boundary conditions of the melt, which strongly affect the flow mode, were set to be axisymmetric to identify the origin of the nonaxisymmetric temperature, velocity, and oxygen concentration profiles in the melt.

7.4   VERTICAL MAGNETIC FIELD (VMF)

Figures 7.2(a) and (b) show particle paths observed from one direction in molten silicon under VMF with 0.035 T (a) and without magnetic field (b), respectively. The particle path in the figure shows an axisymmetric pattern originated by a buoyancy force induced by the temperature gradient in the melt. Since the flow pattern is axisymmetric, the particle path gave a torus-like pattern due to crucible rotation.

The velocity of molten silicon flow was obtained from tracking a specific tracer. The projected image of the tracer was observed from one direction in the experiment, therefore, the projected velocity of one specific tracer was obtained in the experiment. Hereafter, the projected velocity (V) of one specific tracer

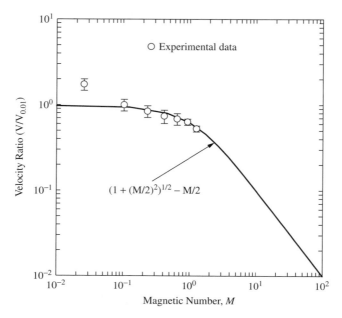

Figure 7.3 Measured and calculated velocity of one specific tracer as a function of magnetic number under VMF.

particle was defined as, $V = \sqrt{(\Delta x/\Delta t)^2 + (\Delta y/\Delta t)^2}$, where Δx and Δy are the horizontal and vertical displacements of the tracer image on a display, and Δt is the time interval between each frame of VTR film. Figure 7.3 shows the normalized velocity as a function of magnetic field strength [17]. Moreover, calculated results are also shown in the figure.

The velocity without VMF of 16.5 mm/s decreased to 5.6 mm/s when 0.035 T of VMF was applied to the melt. This clarifies that the flow velocity of molten silicon in the meridional plane was suppressed by VMF, which can be expressed by numerical simulation except in the case of nonmagnetic fields. The pattern of particle paths was axisymmetric even under VMF, which was kept during observation in the VMF range from 0 to 0.035 T. We estimated a general relationship between the flow velocity and VMF strength in order to speculate on the flow velocity in the range of magnetic field larger than 0.035 T, since it is difficult to observe a moving tracer under large magnetic field.

We proposed a nondimensional number based on Hartman number Ha and Reynold's number Re from analytical and numerical calculations as magnetic number M ($M = \sigma B_0^2 h/\rho v_0 = \text{Ha}^2/\text{Re}$) which was able to predict the numerical results [17], where σ is the electrical conductivity, B_0 is the applied filed, h is the melt height, ρ is the density and v_0 is the velocity of molten silicon without the magnetic field. Details of the derivation of this equation are shown in an original paper reported elsewhere [17]. A solid-line $(V/V_0 = \sqrt{1 + (M/2)^2} - M/2)$ represents a result obtained from calculation of magnetic number (M).

The above results show that the flow velocity decreased monotonically in the range above $M^{-2} = 1$. Moreover, the result indicates that flow velocity in the VMF with 0.1 T in the present configuration is about 0.1 times the flow velocity without the VMF.

The temperature boundary condition on the crucible wall determines the melt flow and the oxygen distribution in the melt under VMF. The temperature at the bottom of a crucible was set to two different values: 1412 °C (melting point), which is identical to type A in Figure 7.4(a), and 1430 °C, which is identical to type B in Figure 7.4(b), respectively. This study was carried out to clarify how Benard convection occurs under the two different temperature boundary conditions of types A and B [15].

An axisymmetric flow pattern resulted from the numerical simulation with the type-A heating system within the magnetic field from 0 to 0.3 T. However, a nonaxisymmetric flow pattern was observed in the simulation with the type-B heating system as shown in Figure 7.5(a) and (b), which indicate profiles of the velocity and temperature distribution at the top of the melt under a magnetic field of 0.1 T. With a magnetic field larger than 0.1 T, the flow velocity is reduced so that the temperature profile becomes almost axisymmetric. This means that the temperature distribution is mainly determined by heat conduction in view of the small velocity.

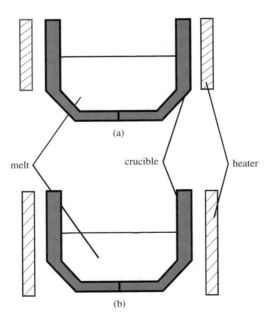

Figure 7.4 Schematic diagrams of two kinds of heating system. (a) type A, (b) type B.

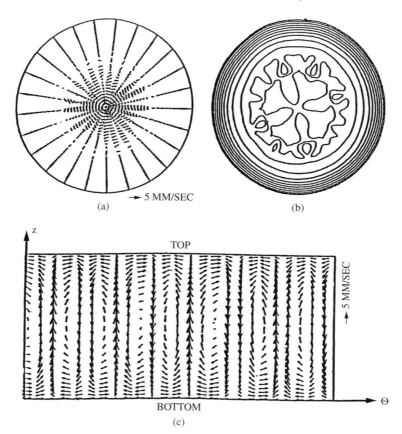

→ 5 MM/SEC

(a)   (b)

(c)

Figure 7.5 (a) Calculated velocity vectors, (b) temperature distribution and (c) velocity vectors in a $z-r$ plane.

To understand the nonaxisymmetric structure, the velocity profile in the $z-\theta$ plane is shown in Figure 7.5(c) at a position of 1 cm radius. A cell structure similar to Benard cells can be recognized in the $z-\theta$ plane [15]. This kind of cell structure can be observed only in the type-B heating system, while it was not observed in the type-A system. This suggests that the origin of the cell structure was Benard instability.

When vertical magnetic fields were applied to the melt, the radial flow was suppressed due to the Lorentz force [15]. Consequently, the temperature gradient in the radial and vertical directions in the melt increases; so that the system becomes hydrodynamically unstable. Since the formation of the Benard cells relaxes the unstable temperature distribution, the system becomes more stable.

Calculated oxygen concentrations as a function of applied magnetic fields in the center of the crystal grown under a condition of type B are indicated

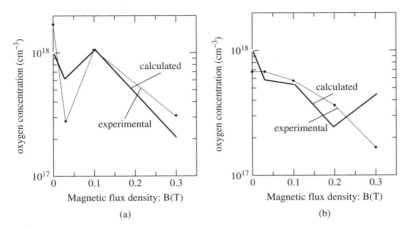

Figure 7.6 Calculated oxygen concentration as a function of magnetic fields for (a) type B and (b) type A.

in Figure 7.6(a) by the solid line. The broken line indicates the experimental results.

The absolute values of numerical results are slightly different from those of the experimental results. This discrepancy may be attributed to unreliable parameters such as the segregation coefficient, the evaporation rate at the free surface, and the dissolution rate at the melt/crucible interface used in the present numerical simulation. An anomaly in the oxygen concentration for both the experimental and numerical results was observed at 0.1 T in a case of type B, while the anomaly was not observed for the type-A heating system shown in Figure 7.6(b). The anomaly can be attributed to the formation of Benard cells, since the strength of the magnetic field in which the anomaly was observed was identical to that in the formation of Benard cells.

High oxygen concentration in crystals could be observed at the condition of the anomaly, since melt with high oxygen concentration transferred from the bottom to the crystal/liquid interface.

7.5   CUSP-SHAPED MAGNETIC FIELDS (CMF)

Two coils with a constant power (1120 A, 200 V) can produce three different types of CMF as follows. (1) The center of CMF was positioned 20 mm above the melt free surface, which was termed uppermost, (2) the distance was set to 10 mm, which was termed upper, (3) the center of CMF was positioned at the melt free surface (symmetric), (4) the center of CMF was positioned 10 mm below the surface (lower), (5) the distance was set to 20 mm below the surface, which was termed lowermost. Some of the configurations; uppermost, symmetric and lowermost are shown in Figure 7.7.

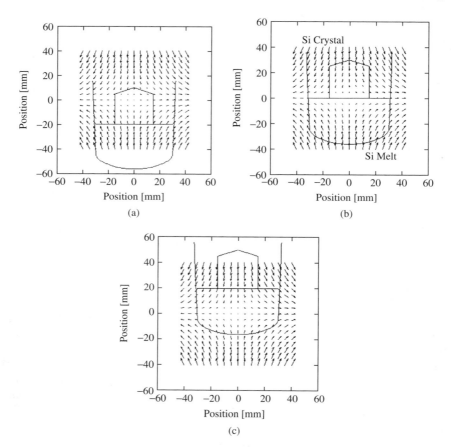

Figure 7.7 The configuration of CMF: (a) the center of CMF positioned 20 mm above the melt free surface (uppermost), (b) the center of CMF positioned at the melt free surface (symmetric) and (c) the center of CMF positioned 20 mm below the melt free surface (lowermost).

Figure 7.8 shows calculated profiles of velocity vectors in the $r-z$ plane for cases of lowermost (a), symmetric (b) and uppermost (c) [18]. The strength of the vertical magnetic fields at the center of the crucible bottom was 0.06 T for a symmetric case (a). The velocity profile characteristics for the surface are as follows: the velocity vectors near the free surface are almost parallel to the magnetic field with a cusp-shaped magnetic field since a flow parallel to a magnetic field does not induce a Lorentz force.

Figure 7.9 [19] shows calculated results of oxygen radial distributions at the top of the melt for cases of lowermost (a), lower (b), symmetric (c), upper (d) and uppermost (e). The crucible rotation rate was set to -3 rpm. The full and broken lines represent experimental and calculated results, respectively. The calculated

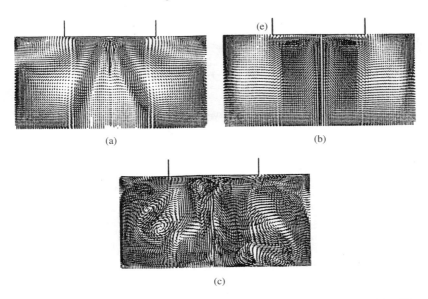

Figure 7.8 Velocity vectors under magnetic fields of the three types. (a), (b), and (c) correspond to cases symmetric and upper, and without magnetic fields, respectively.

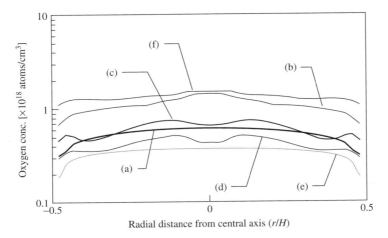

Figure 7.9 Oxygen concentration distribution at the interface along radial direction for cases of (a) lowermost, (b) lower, (c) symmetric, (d) upper and (e) uppermost.

results show that a homogeneous distribution with low oxygen concentration in the crystals can be achieved for upper, uppermost, and lowermost cases, while an inhomogeneous distribution and relatively high oxygen concentration was obtained in cases of lowermost and without the magnetic fields.

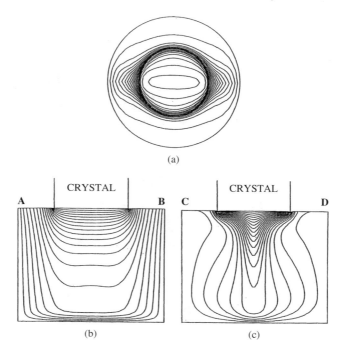

Figure 7.10 Temperature profiles in (a) planes horizontal, (b) planes parallel and (c) perpendicular to the magnetic fields.

7.6 TRANSVERSE MAGNETIC FIELDS (TMF)

Figures 7.10(a)–(c) [20] show temperature profiles of a horizontal plane, and of planes parallel and perpendicular to the magnetic fields in the melt under transverse magnetic fields, respectively. Figures 7.11(a)–(c), respectively, show velocity profiles in the melt in the same planes. Although crystal and crucible rotation rates were set to zero in this case, nonaxisymmetric temperature distribution can be recognized due to unidirectional magnetic fields, which induce nonaxisymmetric flow.

Therefore, the electric potential distribution in the melt becomes nonaxisymmetric as shown in Figure 7.12. These asymmetric temperature and velocity profiles might modify the oxygen concentration in the melt; therefore small oxygen concentration in the melt and crystal might be achieved. Further study should be carried out to clarify the mechanism of oxygen transfer in the melt.

7.7 SUMMARY

Molten silicon flow under magnetic fields of VMF, CMF, and TMF applied CZ crystal growth was discussed. Data obtained using an X-ray radiography

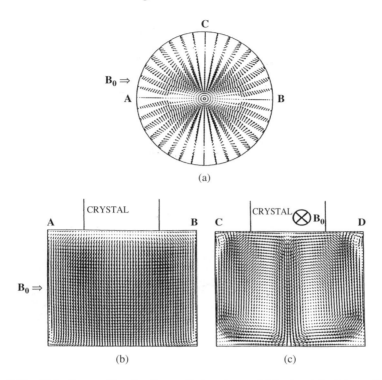

Figure 7.11 Velocity profiles in the melt in planes (a) horizontal, (b) parallel and (c) perpendicular to the magnetic fields.

technique were compared with the results of numerical simulation. In the VMF case, the flow velocity of axisymmetric flow decreased monotonically with increase of magnetic field strength in the range from 0 to 0.035 T. The analysis of flow-velocity reduction under VMF using a magnetic number M describes the reduction of flow velocity by the VMF.

For the case of CMF, molten silicon flow was discussed for three types of configuration between the center of CMF and the melt surface. The flow behavior in the CMF depended on the relative position between magnetic field and the melt.

Three-dimensional flow resulted in the case of TMF, since the distribution of the Lorentz force was nonaxisymmetric.

ACKNOWLEDGMENT

A part of this work was conducted as JSPS Research for the Future Program in the Area of Atomic-Scale Surface and Interface Dynamics.

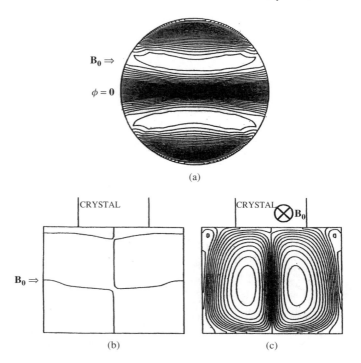

(a)

(b)   (c)

Figure 7.12   Electric potential distribution in the melt in the planes (a) at the top of the melt, (b) parallel and (c) perpendicular to the magnetic fields.

REFERENCES

[1]   H. Oya, Y. Horioka, Y. Furukawa and T. Shingyoji, *Extended Abstracts (The 37th Spring Meeting, 1990)*; The Japan Society of Applied Physics and Related Society.

[2]   M. Itsumi, H. Akiya and T. Ueki, *J. Appl. Phys.* **78** (1995) 5984.

[3]   S. Miyazawa, *Semi-Insulating III-V Materials* (1986) 3.

[4]   H. Yamagishi, I. Fusegawa, K. Takano, E. Iino, N. Fujimaki, T. Ohta and M. Sakurada, *Semiconductor Silicon* (1994) 124.

[5]   A. F. Witt, C. J. Herman and H. C. Gatos, *J. Mater. Sci.* **5** (1970) 882.

[6]   G. Meuller and R. Rupp, *Crystal Properties and Preparation*, Vol. 35 (Trans Tech. Switzerland, 1991) 138.

[7]   D. T. J. Hurle, *J. Cryst. Growth* **65** (1983) 124.

[8]   K. Hoshi, T. Suzuki, Y. Okubo and N. Isawa, *Extended Abstracts of E.C.S. Spring Meeting*, (The Electrochemical Soc., Pennington, 1980) 811.

[9]   T. Suzuki, N. Isawa, Y. Okubo and K. Hoshi, *Semiconductor Silicon 1981*, eds. H. R. Huff, R. J. Kriegler and Y. Takeishi (The Electrochem. Soc., Pennington, 1981) 90.

[10]   H. Hirata and N. Inoue, *Jpn. J. Appl. Phys.* **23** (1984) L527.

[11]   H. Hirata and K. Hoshikawa, *J. Cryst. Growth* **96** (1989) 747.

[12]   H. Hirata and K. Hoshikawa, *J. Cryst. Growth* **113** (1991) 164.

[13] M. Watanabe, K. Kakimoto, M. Eguchi and T. Hibiya, *The American Society of Mechanical Engineerings, FED* (1991) 255.
[14] K. Kakimoto, *Prog. Cryst. Growth Charact.* **30** (1995) 191.
[15] K. Kakimoto, Y. W. Yi and M. Eguchi, *J. Cryst. Growth* **163** (1996) 238.
[16] M. Watanabe, M. Eguchi, K. Kakimoto and T. Hibiya, *J. Cryst. Growth* **128** (1993) 288.
[17] K. W. Yi, M. Watanabe, M. Eguchi, K. Kakimoto and T. Hibiya, *Jpn. J. Appl. Phys.* **33** (1994) L487.
[18] K. Kakimoto, K. Eguchi and H. Ozoe, *J. Cryst. Growth* **180** (1997) 442.
[19] M. Watanabe, M. Eguchi, K. Kakimoto and T. Hibiya, *J. Cryst. Growth* **193** (1998) 402.
[20] Masato Akamatsu, Koichi Kakimoto and Hiroyuki Ozoe, *J. Mater. Process. Manuf. Sci.*, **5**, **No. 4** (1997) 329.

8 Modeling of Technologically Important Hydrodynamics and Heat/Mass Transfer Processes during Crystal Growth

V. I. POLEZHAEV

Institute for Problems in Mechanics, Russian Academy of Sciences 117526 Moscow, Prospect Vernadskogo 101

8.1 INTRODUCTION

Development of the mathematical and physical modeling of crystal growth in recent years gives a precise and adequate description of the fluid flow for the given type of crystal growth models. Czochralski growth as a main industrial method (see overview and references in Hurle, 1993) receives the most effort hour after early studies of flow visualization (Carruthers, 1977) and direct numerical simulation of the fluid flow on the basis of Navier–Stokes equations using idealized Czochralski method for axisymmetrical case (Kobayashi and Arizumi 1980, Langlois, 1977, see references Langlois, 1985).

Over 30 years a number of overviews were published by Carruthers, 1977, Pimputkar and Ostrach, 1981, Polezhaev, 1981, Scheel and Sielawa, 1985, Muller, 1988, Brown, 1988, Polezhaev, 1994, Dupret and Van Den Bogaert, 1994, Muller and Ostrogorsky, 1994 and many others. However, due to the nonlinear and multiparametrical nature of the problem of fluid flow in crucibles is not solved as yet. Development of direct numerical modeling on the basis of unsteady three-dimensional Navier–Stokes equations opens up new possibilities for analysis of most technologically important features of fluid flow and transport. However, a new problem of analysis of the complex fluid flows appears as well as development of special methodology – how to use most efficiently this 'store of knowledge' for improvement of technology and quality of crystals.

During the last decade analysis of axisymmetrical models was continued by Bottaro and Zebib 1988, Fontaine, *et al.* 1989, Sacudean, *et al.* 1989, Buckle and Schafer 1993, Kobayashi 1995, Mukherjee *et al.*, 1996, Polezhaev *et al.*, 1997 and many other studies. A new attack on this problem was attempted

Crystal Growth Technology, Edited by H. J. Scheel and T. Fukuda
© 2003 John Wiley & Sons, Ltd.   ISBN: 0-471-49059-8

in the last decade using direct 3D modeling (Bottaro and Zebib, 1989, Leister and Perec, 1992, Seidi *et al.*, 1994, Xiao and Derby, 1995). In more recent years a number of runs were done for parameters close to the instability (Yi *et al.*, 1995b, Nikitin and Polezhaev, 1999). Recently, direct simulation using 3D unsteady Navier–Stokes equations for the chaotic regime was done by Wagner and Friedrich, 1997. Nikitin and Polezhaev, 1999a, b presented results of transient and chaotic regimes for a benchmark configuration (Wheeler, 1991). Qualitative analysis using available knowledge of different types of instabilities (Carruthers, 1977, Polezhaev, 1984, Ristorcelli and Lumely, 1992) shows a comprehensive picture of the possible mechanisms of instabilities in Czochralski method. However, quantitative results of the values of critical governing parameters such as Grashof and Reynolds numbers for Czochralski model are very restricted.

Critical Grashof numbers for the onset of convective instability and temperature oscillations in an idealized Czochralski model for the 3D case were found by Polezhaev *et al.* 1997, Nikitin and Polezhaev, 1999a and b. Knowledge of the instabilities are not detailed in comparison with classical problems in fluid mechanics such as spherical Couette flow (Beliaev, 1997) or isothermal fluid flow in enclosures with lid rotation (Gelfgat *et al.*, 1996), however, progress is rapid in this direction. Because of the multiparametrical nature, multiscale, unsteady, spatial structure of the fluid flow as well as nonlinearity, instability in the real range of parameters and special technology demands control of the crystal quality and yields methods of the technological hydrodynamics as a separate discipline for quantitative analysis should be focused on the idealized Czochralski model.

Physical modeling using visualization techniques plays an important role in this concept. In the last decade a number of studies were carried out (Jones, 1984, Berdnikov *et al.*, 1990, and recently by Verezub *et al.*, 1995, Kosushkin, 1997, Krzyminski and Ostrogorsky, 1997). High Pr number transparent liquids are used in most cases and low Pr number melt in recent work for molten hot solution. For the lowest Pr number for semiconductor melts (real silicon melts) an X-ray technique was developed (Kakimoto *et al.*, 1989). For GaAs melt a special system was developed with a ceramic imitation crystal (Kosushkin, 1997). Control of the boundary conditions is one of the problems for quantitative comparison simulation and experimental data (Kakimoto *et al.*, 1989, 1993, Mukherjee *et al.*, 1996, Polezhaev, 1998). This problem needs special efforts and is not discussed here.

This chapter, following the strategy and methods of our previous works, presents a summary of the recent activity of the mathematical modeling by author and his colleagues (Polezhaev *et al.* 1997 and 1998, related to the benchmark problem (Wheeler, 1991) and extension of the benchmark problem to three-dimensional regimes using direct numerical modeling and linear-stability analysis. The hierarchy of the models is discussed and new results of multiparametrical research using parameters of the idealized industrial LEC GaAs configuration, and specialized computer video techniques are presented.

8.2 TECHNOLOGICALLY IMPORTANT HYDRODYNAMICS PROCESSES DURING CRYSTAL GROWTH

One of the well-known characteristics of the grown crystal is macroinhomogeneity, induced by impact of gravity-driven convection on the bulk segregation. It was found first for enclosures (cylinder, sphere, square, Polezhaev, 1974) and for the 2D Bridgman configuration: horizontal (Polezhaev *et al.* 1981 (see references Polezhaev, 1984), for 3D in a horizontal layer by Polezhaev *et al.* 1998c and axisymmetric vertical (Brown, 1988, Motakef, 1990). Spreading of the maximum of temperature/concentration macroinhomogeneity, induced by thermal gravity-driven convection in the cases of forced, buoyancy, and surface-tension driven convection under gravity/magnetic fields in uniform and conjugated cases are realized (see references in Polezhaev, 1992). The value of this maximum strongly depends on the crystallization rate and is of interest in low gravity (Alexander, 1990).

For the ground-based environment the problem of the 'optimal mixing' (how to provide transport of the species to the front and avoid striations due to instability) is of particular interest. The accelerated crystal/crucible rotation technique (ACRT) was proposed by Scheel, 1972 and efficiently used for high-temperature solution growth (see references in Scheel and Sielawa, 1985). However, as was discussed in the cited paper, the problem for semiconductor melts is more complicated. One of the possible domains on the amplitude-frequency diagram was found for real material and configurations in the cited paper. Note that for modern crystal growth applications the famous qualitative picture of the complex fluid flow in a crucible (Scheel and Sielawa, 1985) which includes forced types of flows like elementary Cochran, Taylor–Proudman-type, thermal convection including Rayleigh–Bernard and Marangoni flows should be realized in quantitative form, using realistic boundary conditions and taking into account nonlinear coupling phenomena, spatial effects, etc.

There are a number of new promising methods of control, using dynamical/thermal actions on the melt. Vibration of the seed is one of the possible control features (Verezub *et al.*, 1995). Acoustic control action (Kozkemiakin *et al.* 1992) is one of the so-called 'low energetic' possibilities for control of the transport phenomena during crystal growth. Temporal oscillations of the power heater (Zakharov *et al.*, 1998) is also one of them. The key problem of how to find 'optimal parameters' consists in using quantitative analysis of the fluid flow/transport fields in the general case of industrial regimes with coupling convection and rotation in the Czochralski model. This will be discussed below in Section 8.6.

Because of the trend to large-scale bulk-crystal production, inhomogeneities in transition and turbulence regimes as well as in the limiting case of the statistically average temperature on inhomogeneities in the Czochralski model as well as convective instabilities induced by the forced flows instabilities and nonlinear interaction of the forced and natural convection are of special interest now. Three-dimensional effects and asymmetry of the transport processes were studied by Polezhaev *et al.* 1998c. Order structure and spoke patterns in 3D convection were studied both experimentally and theoretically by Yi *et al.* 1995b). Characteristics of fluid flow in transient and turbulent regimes for the Czochralski model are one of the problems. There are only a few papers in this area (Berdnikov *et al.*, 1990, using physical experiments, Wagner and Friedrich, 1997 using mathematical modeling – with surface-tension driven effects on the melt surface and Nikitin and Polezhaev, 1999 – with gravity-driven convection). The above-mentioned problems correspond to the 'optimal mixing' as one of the key problem. For multiparametrical computer optimization adequate and operative computer tools are needed. Common and specialized models as well as techniques of multiparametrical analysis transition and turbulent fluid flows and, specifically, evaluation of 3D effects and comparison with axisymmetrical 2D one should be developed also for clear understanding of fluid-flow phenomena in a crucible.

We will focus below on new results concerning elementary critical values of the onset of the oscillations, induced by thermal gravity-driven convection as the leading mechanism, coupling with forced convection due to crystal/crucible rotation, including effects of 3D instability on the basis of the unsteady two- and three-dimensional models in the industrial technological range of parameters, which are unknown as yet.

8.3 BENCHMARK PROBLEM

The problem formulation and a range of problem parameters are described in the axisymmetrical case as suggested by Wheeler, 1991 and will be used below. The geometry of the problem is a vertical cylindrical crucible radius R_c, which is filled by the melt to the height H (Figure 8.1). The crucible can be rotated with constant angular velocity Ω_c and it is in a constant gravitational field with acceleration g. In the center of the upper free surface of the melt is placed the crystal of radius R_s. The crystal can be rotated with constant angular velocity Ω_s. It is supposed that the temperature T_s of the crystal surface is constant, the temperature of crucible side wall is T_c and the crucible bottom is adiabatic. The melt surface between crystal and crucible wall is free and plane and its temperature is a linear function of radius. It is supposed that the melt is a Boussinesque fluid and the flow is axisymmetrical.

Unsteady Navier–Stokes equations in velocity and pressure variables and the temperature equation using a cylindrical coordinate system (r,z) and nondimensional variables can be written in the following form:

$$(1/r)\partial(ru)/\partial r + \partial w/\partial z = 0 \tag{8.1}$$

$$\partial u/\partial t + u\partial u/\partial r + w\partial u/\partial z - v^2/r = -\partial p/\partial r + \nabla^2 u - u/r^2 \tag{8.2}$$

$$\partial v/\partial t + u\partial v/\partial r + w\partial v/\partial z + uv/r = \nabla^2 v - v/r^2 \tag{8.3}$$

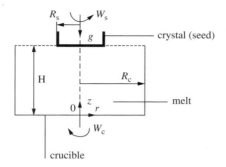

Figure 8.1 Scheme of the mathematical model of idealized Czochralski growth.

$$\partial w / \partial t + u \partial w / \partial r + w \partial w / \partial z = -\partial p / \partial z + \nabla^2 w + \mathrm{Gr} T \qquad (8.4)$$

$$\partial T / \partial t + u \partial T / \partial r + w \partial T / \partial z = (1/\mathrm{Pr}) \nabla^2 T \qquad (8.5)$$

Boundary conditions:

$$
\begin{aligned}
u = w = \partial T / \partial z = 0, \quad v = r \mathrm{Re}_c && \text{for } 0 \le r \le 1, \ z = 0 \\
u = w = 0, \quad v = \mathrm{Re}_c, \quad T = 1 && \text{for } r = 1, \ 0 \le z \le \alpha \\
\partial u / \partial z = \partial v / \partial z = w = 0, \quad T = (r - \gamma)/(1 - \gamma) && \text{for } \gamma \le r \le 1, z = \alpha \\
u = w = T = 0, \quad v = r \mathrm{Re}_s && \text{for } 0 \le r \le \gamma, \ z = \alpha \\
u = v = \partial w / \partial r = \partial T / \partial r = 0 && \text{for } r = 0, \ 0 \le z \le \alpha
\end{aligned}
$$

$$(8.6)$$

Here t-time, u, v, w-radial, azimuthal and axial components of velocity vector, p-pressure, T-temperature,

$$\nabla^2 = (1/r)(\partial / \partial r (r \partial / \partial r)) + \partial^2 / \partial^2 z$$

Nondimensional variables are introduced in the following way (dash over symbol denotes dimensional value)

$$(r', z') = R_c (r, z)$$

$$(u', v', w') = (v / R_c)(u, v, w)$$

$$p' = (\rho v^2 / R_c^2) p - \rho R_c g z$$

$$t' = (R_c^2 / v) t$$

$$T' = T_s + (T_c - T_s) T$$

There are the following nondimensional parameters in the equations and in the boundary conditions:

Two nondimensional geometrical parameters

$$\alpha = H/R_c, \quad \gamma = R_s/R_c$$

$$\Pr = \nu/\kappa - \text{the Prandtl number}$$

$$\text{Re}_s = \Omega_s R_c^2/\nu - \text{the crystal Reynolds number}$$

$$\text{Re}_c = \Omega_c R_c^2/\nu - \text{the crucible Reynolds number}$$

$$\text{Gr} = g\beta(T_c - T_s)R_c^3/\nu^2 - \text{the Grashof number.}$$

Here ν-kinematical viscosity, β-thermal expansion coefficient, ρ-melt density, κ-temperature conductivity.

For the benchmark configuration Wheeler, 1991 supposed for all test variants $H/R_c = 1.0$, $R_s/R_c = 0.4$, and $\Pr = 0.05$. The values of other parameters are given in (Wheeler, 1991, Buckle and Schafer, 1993). Benchmark cases are as follows: A1, A2, A3 – $\text{Re}_s = 10^2, 10^3, 10^4$ and C1, C2, C3 – $\text{Gr} = 10^5, 10^6, 10^7$. One can see other benchmark cases in Table 8.1.

The typical characteristic (for instance, amplitude of the temperature oscillations or value of macroinhomogeneity, etc.) of the fluid flow/transport in the melt

Table 8.1 Numerical results for benchmark configuration (axisymmetrical case)

Regime	Gr	Re$_s$	Re$_c$	Bessonov	Buckle and Schafer	S. Nikitin	Ermakov	N. Nikitin
A1	0	1.E2	0	−0.2198	−0.2345	−0.2198	−0.2172	
A2	0	1.E3	0	−5.0587	−5.3642	−4.9254	−5.0344	
A3	0	1.E4	0	−43.119	−40.443	−51.805	−43.206	−42.778
B1	0	1.E2	−2.5E1	−0.0454	−0.0502	−0.0475	−0.0452	
				0.1183	0.1180	0.1167	0.1186	
B2	0	1.E3	−2.5E2	−1.5061	−1.6835	−1.5301	−1.5097	
				1.1812	1.2414	1.1430	1.1832	
B3	0	1.E4	−2.5E3	−8.3106	−8.5415	−9.1298	−8.6607	−8.0881
				5.4813	5.2708	4.7729	5.5192	5.4294
D3	1.E5	1.E3	0	25.106	24.829	24.871	25.121	
C1	1.E5	0	0	28.440	28.437	28.404	28.420	
C2	1.E6	0	0	92.930	92.100	92.518	92.690	

Characteristics of the formulations and numerical schemes are shown below

	Bessonov	Buckle & Schafer	S. Nikitin	Ermakov	N. Nikitin
Mesh r, z	80 × 64	64 × 64	80 × 80	80 × 80	64 × 64
Numerical scheme	FV	FDM	FDM	CVM	FDM/S
Formulation	Velocity Pressure	Velocity Pressure	Vorticity Stream f.	Velocity Pressure	Vorticity Velocity

for the Czochralski model may be written as follows:

$$A = f(\mathrm{Re_c}, \mathrm{Re_s}, H/R_c, R_s/R_c, \mathrm{Gr}, \mathrm{Pr}, b\gamma, b\gamma_0) \qquad (8.7)$$

Here $b\gamma$, – type of boundary conditions and $b\gamma_0$ – type of initial values.

Motivation of the benchmark problem (besides being a problem of experimental study and difficulty for computer resolution of the fine boundary layer, secondary structures, etc.), presented by Wheeler, 1991 and confirmed now, is that the characterist function A of the fluid flow in the melt is multiparametrical even for the simplified (idealized) model of Czochralski growth for unicomponent melt flow without Marangoni convection, radiation, etc. Two geometrical parameters, H/R_c, R_s/R_c, two dynamical parameters for forced fluid flow – $\mathrm{Re_c}$, $\mathrm{Re_s}$ and one – Gr number – for gravity-driven natural convection, and Pr number as the parameter of physical properties, $b\gamma$ – type of boundary conditions, $b\gamma_0$ – type of initial values are very important for convection and heat transfer. It should be taken into account that during crystal growth some of parameters, for instance, H/R_c, as well as Gr and $b\gamma$ are time-dependent functions. Therefore the problem of fluid flow in the Czochralski model initiates the development of analysis of gravity-driven and rotational low Prandtl melt flow – nonlinear interaction, temperature oscillations, transition to chaos as a fundamental fluid dynamics problem in such a complicated situation.

It should be noted that in the early statement of the problem (Langlois, 1985) it was other additional factors such as radiation, Marangoni effect, magnetic fields, which are important for technological needs, but this makes the problem more complicated for analysis and control. This is why the so important thermal control parameter as temperature boundary conditions ($b\gamma$) was not discussed enough on the early stage of research. Because of the multiparametrical nature of the problem the parameters of crystal growth are different for most of the published works. A number of other peculiarities exist for real crystal technologies of the above-mentioned crystals: silicon (Kakimoto et al., 1993, bottom cooling, $\mathrm{Pr} = 0.01$), GaAs (Zakharov et al. 1998, bottom heating, adiabatic melt surface, $\mathrm{Pr} = 0.07$) with counterrotation for semiconductors and rotation of the crystal for oxides (Xiao and Derby, 1995, adiabatic bottom, $\mathrm{Pr} = 8$). There are different configurations when using special liquids for modeling: Berdnikov et al., 1990 (water, $\mathrm{Pr} = 7$, alcohol, $\mathrm{Pr} = 15$), Mukherjee et al., 1996 (silicone oil, $\mathrm{Pr} = 890$), Krzyminski and Ostrogorsky, 1997 (NaCl-CaCl$_2$ melt, $\mathrm{Pr} = 0.5$). Kobayashi, 1995 reported results of a multiparametrical analysis, using $H/R_s = 2.0$, $R_c/R_s = 2.5$, $\mathrm{Pr} = 1$, with the Marangoni effect and pulling velocity, however, without information related to the calculations, type of boundary conditions and initial values and different scales for nondimensional parameters. This situation does not help the analysis of a general picture of the fluid flow. Therefore the basic configuration for benchmark calculation and multiparametrical research is useful.

Certainly, the above-mentioned statement does not cover all important problems. In Polezhaev et al. 1998, and Nikitin and Polezhaev 1999b this benchmark problem was extended to the case of 3D Navier–Stokes equations. The general 3D case will be taken into account below (Section 8.4.6). Bottom heating for a GaAs crucible (Zakharov *et al.*, 1998) with an adiabatic melt surface and counter-rotation is another typical configuration that we propose for systematically research (Section 8.6).

8.4 HIERARCHY OF THE MODELS AND CODES AND SUMMARY OF BENCHMARK EXERCISES

Unsteady Navier–Stokes equations for the 3D case is a very delicate model and it is rational to develop a hierarchy of the models and codes on the basis of this system. New tools for modeling that were developed by the author and his colleagues during recent years in the IPM RAS are briefly described below with comments and a summary of the benchmarking results (references are made mainly to original methods and software). The sequence of these tools is as follows:

(1) Elementary two-dimensional nonlinear convective processes in enclosures for cartesian and cylindrical coordinates

A common PC-based system COMGA (computer laboratory) which includes a user-friendly interface and video visualization is an efficient tool for research and elucidation of unsteady convective processes on the basis of Navier–Stokes equations, specifically for the initial stage (Ermakov *et al.* 1992, Ermakov *et al.* 1997). This system includes most of the technologically important hydrodynamics, heat/mass transfer processes during crystal growth as mechanical (forced) convection (rotation, vibration and coupling) gravity-driven and surface-tension-driven. For nonuniform media each of these groups also consists of several elementary processes dependent on the direction of the heat/mass flux. For instance, the gravitational-type convection in a binary (double-diffusion) system consists of ten cases for different mutual directions of heat and mass flux and gravity vector. A similar situation exists for surface-tension-driven convection in a binary system with different mutual directions of heat and mass flux and free surface vector and the same for other groups of driven forces. Well-known international benchmark problems (De Vahl Davis and Jones, 1983, B. Roux, 1990) and different classical problems such as Rayleigh–Bernard, Marangoni, dynamical control actions with different angles of inclination, vibration, rotations have been analyzed during recent years (see classification and extended references Ermakov *et al.*, 1997). The computer laboratory also includes possibilities for analysis transient and chaotic fluid flow for all of the above-mentioned elementary processes.

(2) Axisymmetrical idealized Czochralski model.

The statement of this problem corresponds to Equations (8.1)–(8.7). Special versions of the above-mentioned system for the axisymmetric idealized Czochralski model, named 'Intex' (Ermakov *et al.*, 1997) and video techniques for visualization and saving of information, television and video recorder, as well as software for presentation are developed for this goal. A film for 200 s of a calculation run was presented by 2000 files that took up 37.7 Mb. Graphic information with 3400 Mb can be saved on a three-hour videocassette, which is efficient for multiparametrical computer modeling. Examples of the videofilm pictures are shown in Figures 8.2–8.4. The central difference approximation was used for the benchmark problem. Calculations were made on the uniform grids 21×21, 41×41, 81×81 for each variant to investigate the convergence of solutions. The results are given in Polezhaev *et al.* (1998d). Coincidence of results obtained with the data of the work of Buckle and Schafer, 1993 for variants C1 and C2 (thermal convection only) can be considered as good, but such coincidence is absent for variants A1–A3 (rotation of crystal). The discrepancy, for example, in the case of the slow flow regime A1 ($Re_s = 100$) is large for such fine grids. The calculations also show that the accuracy of results becomes worse

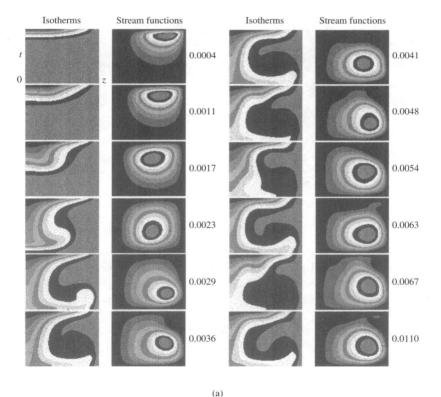

(a)

Figure 8.2 Temporal behavior of the temperature/isolines of stream function for benchmark problem ((Gr $= 6 \times 10^6$, $W_s = W_c = 0$). Values on the right show nondimensional time: (a) linear temperature profile on the melt surface, (b) adiabatic boundary conditions on the melt surface.

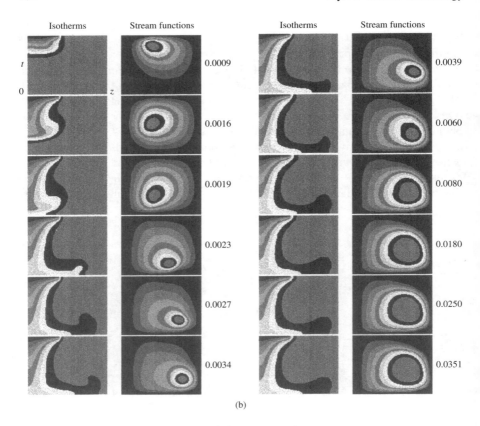

Figure 8.2 (*continued*).

when there is a counteraction of different mechanisms of flow (thermal convection, rotation of crystal and crucible). For variant C3 ($Gr = 10^7$) the oscillation regime of flow was obtained. A number of calculations were made to find the critical Grashof number that takes place for Grashof numbers between 2×10^6 and 3×10^6.

Besides this common type of system two other methods (control volume schemes for primitive – velocity-pressure formulation and spectral-difference) were employed. The first one is based on an implicit variant of the splitting method on a staggered mesh. Second-order derivatives (central differences for convective terms) are used. Calculations were made on the uniform meshes up to 160×160. For variant C3 ($Gr = 10^7$) in comparison with the same results in work Polezhaev *et al.*, 1998d. Coincidence nonperiodic oscillations were obtained instead of periodic ones in the mark of Buckle and Schafer, 1993. For the second method, based on the spectral-difference scheme (Nikitin, 1994) there is the best correlation for the $\psi_{\max}$ related to the C2 problem: asymptotic value

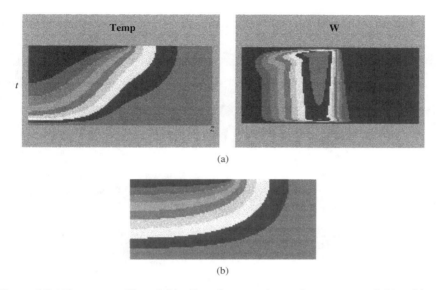

(a)

(b)

Figure 8.3 Temperature/flow fields (forced convection and temperature fields without convection, steady-state regime) for the model of Czochralski growth of GaAs (axisymmetrical case). (a) isotherms left/stream and functions (right) without thermal convection. Steady-state regime. (Gr = 0 W_s = −6 rpm, W_c = 16 rpm); (b) isotherms without convection and rotation. Steady-state regime. (Gr = W_s = W_c = 0).

ψ_{max} = 93.179 (1.2 % difference from the results of the work of (Buckle and Schafer, 1993) (64 × 64) and 0.7 % for the same result by FDM (81 × 81). However, for B3 the asymptotic value ψ_{max} = 55.833, which is 5.6 % different from the work of Buckle and Schafer, 1993 and 14.5 % from the same result of FDM. Nonperiodic behaviour of the temperature oscillations was also achieved by this method for the problem C3. Using this method, the critical value onset of oscillations was calculated. The most exact critical Grashof value for the onset of temperature oscillations for Re_s = Re_c = 0 using the axisymmetrical approach was found as Gr_c = 2.5 × 10^6, which is well correlated with the same result with the use of FDM.

Table 8.1 presents comparisons of results in the work (Polezhaev *et al.*, 1998a) for different regimes of the benchmark (H/R_c = 1, R_s/R_c = 0.4, Pr = 0.05, linear temperature profile on the surface melt), with the uniform meshes used – from 64 × 64 to 80 × 80. Values of the minimum (regimes A1, A2, A3, B1, B2, B3) and maximum (regimes B1, B2, B3, D3, C1, C2) of stream functions are shown. One can see that correlations of the results is adequate for steady-state regimes using such a uniform mesh. However, it is not so simple to provide accurate calculation results of the temperature oscillation even for the axisymmetrical model (Nikitin and Polezhaev, 1999a). Therefore stability analysis is a very important tool in this case.

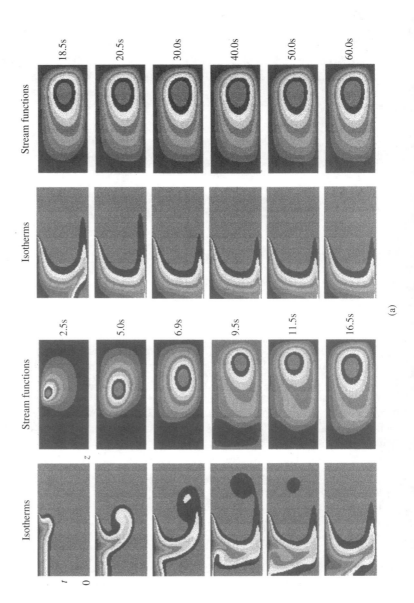

(a)

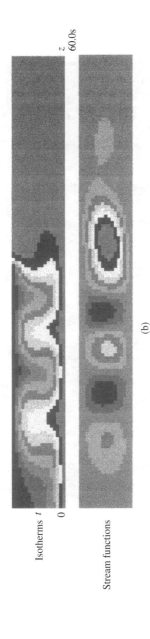

(b)

Figure 8.4 Interaction of gravity-driven and forced convection and instabilities for the model of Czochralski LEC growth of GaAs (axisymmetrical case) values on the right show dimensional time in seconds: (a) evolution of the thermal convection without rotation ($Gr_c = 7.8 \times 10^7$, $W_s = W_c = 0$); (b) convective cells of the Rayleigh–Bernard type of thermal convection for the case of reduced melt level ($H = 1$ cm); (c) evolution of the thermal convection and counter-rotation of crystal/crucible ($Gr = 7.8 \times 10^7$, $W_s = -6$ rpm, $W_c = 16$ rpm); (d) evolution of the thermal convection and crystal rotation ($Gr = 7.8 \times 10^7$, $W_s = -6$ rpm, $W_c = 0$); (e) evolution of the thermal convection and crucible rotation ($Gr = 7.8 \times 10^7$, $W_s = 0$, $W_c = 16$ rpm).

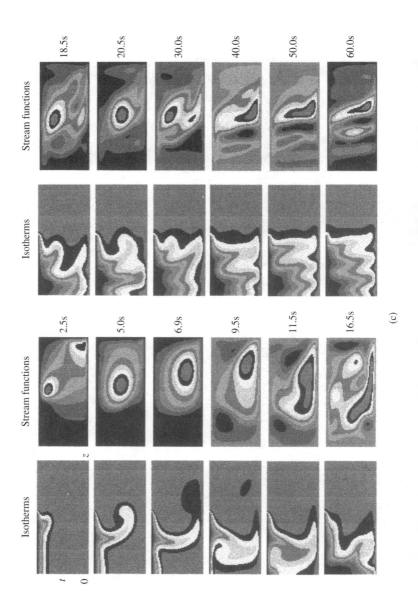

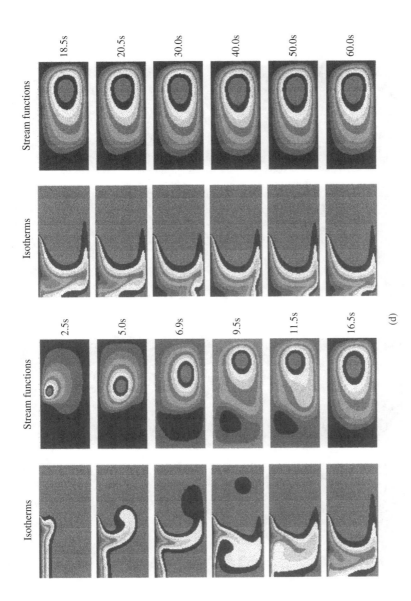

Figure 8.4 (continued)

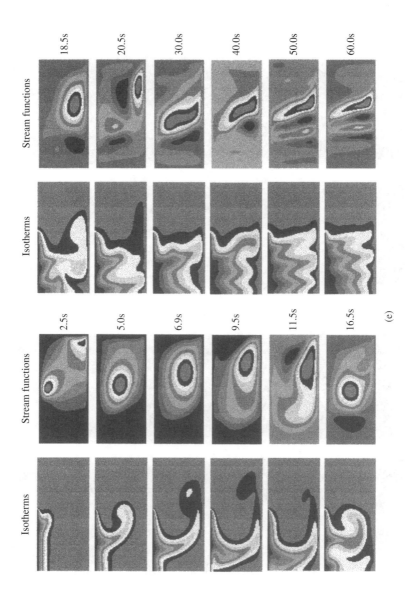

Figure 8.4 (*continued*)

(3) Direct numerical simulation of three-dimensional equations for the Czochralski model.

Using the spectral-difference method 2D and 3D problems as well as linear analysis were carried out (Nikitin and Polezhaev, 1999a, Polezhaev *et al.*, 1998b). The governing system of equations for the spectral-difference method is written for velocity-vorticity variables in the next vector form:

$$\partial V/\partial t = V \times \omega - \nabla\pi - \nabla \times \omega + \mathrm{Gr}T k_z$$

$$\nabla V = 0$$

$$\omega = \nabla \times V$$

$$\partial T/\partial t + \nabla(VT) = (1/\mathrm{Pr})\nabla^2 T$$

Here $V(t, r, \theta, z)$ – velocity, $\omega(t, r, \theta, z)$ – vorticity, $\pi = P/\rho + V^2/2$, $P(t, r, \theta, z)$ – pressure, ρ – density, k_z – unit vector along z-axis. Definitions of the nondimensional Gr and Pr numbers are the same as above, as well as boundary and initial values, α, γ, $\mathrm{Re_s}$ and $\mathrm{Re_c}$. For direct numerical simulation, spectral (Fourier) representations in the azimuthal direction and finite-difference approximations with staggered grids in radial and axial directions were used (Nikitin, 1994). The time advancement is performed by a semi-implicit Runge–Kutta scheme. Using direct simulation all the above-mentioned benchmark problems were carried out (Nikitin and Polezhaev, 1999a).

(4) Three-dimensional linear stability analysis code for Czochralski model.

For stability analysis small three-dimensional perturbations are superimposed on the steady axisymmetric solution and their behaviour is considered within linearized Navier–Stokes equations. The linear stability analysis code for Czochralski model developed by Nikitin (Nikitin and Polezhaev, 1999b) includes procedures for calculation of the axisymmetrical flow and temperature fields, input of the 3D disturbances and calculation of the disturbances evolution. It is possible, using this technique, to analyze types of instabilities (monotonic, oscillatory). Using stability analysis the critical Grashof (Reynolds) number for onset of flow/temperature oscillations in the three-dimensional case may be defined following the evaluation of the temperature disturbances T' in the fixed point and estimate the function $A^n(t)$ were

$$(A^n(t))^2 = (1/2\pi) \sum_0^{2\pi} (T'^n(t, r_o, z_o, \theta)^2 \, \mathrm{d}\theta$$

One can find examples of the growth of function $A(t)$ for different modes in paper Nikitin and Polezhaev, 1999a (see also Section 8.6.2a). Using the above-mentioned parameters of the benchmark configuration, the critical Grashof number for gravity-driven convection (oscillatory for linear temperature distribution on the surface melt and monotonic for adiabatic melt) were found for the 3D case (Nikitin and Polezhaev, 1999a). Validation of the stability analysis was checked, using direct numerical calculations of the 3D Navier–Stokes equations (Nikitin and Polezhaev, 1999, Polezhaev et al., 1998, see additional information and examples in Section 8.5 and in Section 8.6.2(a) and (b).

(5) Global thermal approach and thermostresses simulation for axisymmetrical Czochralski model.

This approach is widely used now (see Dupret and Van Den Bogaert, 1994, Muller, Ostrogorsky, 1994). The model developed by (Griaznov et al., 1992) includes the crystallization front, heat transfer in crystal, radiation and the field of thermostress. However, this approach is not used at this stage of research to make precise analysis of the fine structure of the fluid flow for analysis of the hydrodynamical reasons of the crystal quality and is very restricted in multiparametrical control possibility analysis. Baumgartl et al. (1993) carried out global simulation of heat transport including melt convection in a Czochralski crystal-growth process.

8.5 GRAVITY-DRIVEN CONVECTION INSTABILITY AND OSCILLATIONS IN BENCHMARK CONFIGURATION

Using the linear stability technique the critical Grashof number for onset of temperature oscillations in the three-dimensional case was defined following the evaluation of the temperature disturbances, as described above (Nikitin and Polezhaev, 1999a). The critical Grashof number for onset of temperature oscillations the due to convective instability in the benchmark example (linear temperature profile on the melt surface and adiabatic crucible bottom) is found to be 5.5×10^5 for the 3D case (2.5×10^6 in the axisymmetrical case). For $Gr_c > 6 \times 10^6$ nonregular oscillations are developed in the 2D case. Thermal isolation of the melt surface eliminates the temperature oscillations in the melt and the critical Grashof number is higher: $Gr_c = 1.2 \times 10^6$ for the 3D case ($Gr_c > 10^7$ for the axisymmetrical case).

Within these benchmark parameters special attention is paid to the structure of the temperature field during temperature oscillations and the possibility of damping using the temperature regime on the surface melt. Figure 8.2(a) shows isotherms and isolines of stream functions for $Gr = 6 \times 10^6$ without crystal/crucible rotation (benchmark case C2). Initial values were zero velocity and uniform temperature fields inside the melt with instantaneous cooling of the crystal (disc). There are three stages on this figure. First, development of the convective flow and temperature front propagation (until $t = 0.023$, when this front reaches the bottom and a boundary layer is developed on the heated side wall of the crucible. Second, penetration of the heated portion of the melt to the core of the melt (until $t = 0.048$), cooling and dissipating. After that (third stage) a new portion of the heated fluid is coming from the side wall to the top and so on. Videofilm shows this mechanism very clearly. Figure 8.2(b) shows the same run with only one change – an adiabatic melt surface. For this case stable stratification suppresses the above-mentioned mechanism of the periodic penetration and cooling of the heated layer from the side wall. This is why the critical Grashof number for onset of oscillations in this case for the axisymmetrical approach is more than 10^7.

Verification of these critical values in the 3D case was done by direct numerical calculation of full nonlinear equations. Direct numerical calculation on the basis of 3D equations (Polezhaev et al., 1998a) showed that there is no difference between the 2D and 3D results in stream function and isotherms fields for the cases Gr = 5.5×10^5 (linear temperature profile on the melt) and for Gr = 10^6 (adiabatic boundary condition). However, three-dimensional effects (oscillations) exist for Gr = 10^6 in the case of a linear profile on the surface melt, as was predicted by linear stability calculations. It is well correlated between the RMS of the oscillations, calculated by two different 3D codes – on the basis of finite difference and spectral-difference methods. In new calculations (Nikitin and Polezhaev, 1999) direct modeling of 3D the regime was done for the benchmark configuration (linear profile of the temperature on the melt surface) for Gr = 6×10^6 and 6×10^7. Nonregular oscillatory regimes with continuous spectra (turbulent) were shown for both cases.

Therefore, in the benchmark configuration two types of the onset of the oscillations are shown. The first one, by reason of the lost symmetry, with lower Grashof number. The second one, by reason of boundary layer-type convective flow instability, with higher Grashof number. Quantitative values are welcome to compare for other researchers. Both instability mechanisms are strongly dependent on the thermal boundary condition on the melt surface. Therefore it may be used for thermal control of the temperature oscillations.

8.6 CONVECTIVE INTERACTION AND INSTABILITIES IN CONFIGURATION OF INDUSTRIAL GaAs CZOCHRALSKI GROWTH

We will show in this section how hierarchy of the models/codes works for analysis of the actual industrial process. One of the possible configurations of the industrial regime for LEC semi-insulating GaAs growth is characterized (Kosushkin 1997, Zakharov et al., 1998) by the following values: H = 4.0 cm, R_s = 4.0 cm and R_c = 7.0 cm, therefore $H/R_c = R_s/R_c$ = 0.578. The dynamical regime is counter-rotation with W_s = 6 rpm, W_c = −16 rpm. Using physical properties for GaAs melt from the above-cited paper and the temperature difference between crystal and crucible as 30 °C, one can find nondimensional parameters Pr = 0.07, Gr = 7.8×10^7, Re_s = 6×10^3, Re_c = $−1.6 \times 10^4$. Dimensional rotation speed and time will be used below for easier analysis. There are some changes in thermal boundary conditions in comparison with the above-mentioned benchmark configuration: the bottom is not adiabatic – there is a bottom heat supply so that the temperature on the bottom is the same as on the side wall: $T_c = T_b$. Adiabatic melt surface is assumed to be as typical for the LEC configuration (Kakimoto et al., 1989).

The goal of this section is to analyze the nature of the oscillations in the melt for semi-insulating GaAs crystal growth with counterrotation and the damping possibilities. The oscillation mechanism for the case of zero rotation and for the coupling of nonlinear flows will be done in axisymmetrical and three-dimensional approaches. Different types of elementary (forced and gravity-driven) convective instabilities (due to cooling above the disc or bottom heating and side heating of crucible) as well as coupling will be discussed. For this part of the chapter related to the parametrical analysis of industrial ground-based regime a monotonic 'upwind' FDM scheme and nonuniform mesh (81 × 81 grids) for calculations on the basis of the 'Intex' system (Section 8.4) were used.

8.6.1 AXISYMMETRICAL APPROACH: NONLINEAR COUPLING FLUID FLOW AND CONTROL POSSIBILITIES

(a) Forced convection and diffusion.

For the analysis below it is important of clearly understand the impact on temperature field of the forced convection due to rotation. Figure 8.3(a) shows steady-state temperature and flow fields due to forced convection only, which corresponds to zero gravity ($W_s = 6$ rpm, $W_c = -16$ rpm, Gr $= 0$). Classical Taylor–Praadman flow may be recognized from this picture. The impact of the flow on the temperature field may be recognized by comparison with isotherms for the diffusion regime only with zero velocity (Figure 8.3(b)). The weak oscillations and relatively small impact on temperature field in comparison with gravity-driven convection below (Figure 8.4(a–e)) should be mentioned. Note, that for the 3D case this forced flow is definitely unstable due to a supercritical crucible rotation speed (see Section 8.6.2a).

(b) Thermal gravity-driven convection.

Using parametrical calculation on the basis of the axisymmetrical model with a change of Grashof number (similar to the benchmark configuration in Section 8.5), the critical Gr number for onset of the oscillations was found as Gr $= 1.5 \times 10^8$. Therefore the convective regime without any rotation in the axisymmetrical case should be steady state (note, however, that this critical value corresponds to the melt level $H = 4$ cm only). Ground-based convection without rotation (Gr $= 7.8 \times 10^7$, $W_s = W_c = 0$) was calculated, using for initial values of zero velocity field and uniform temperature in the melt, which is heated until the side wall temperature reaches T_c and the instantaneous cooling disc surface reaches T_s.

Figure 8.4(a) shows the temporal evolution of the thermal convection. One can recognize two kinds of thermal-convection mechanisms in the crucible: (a) local thermals due to the instantaneous crystal cooling from above and

(b) global circulation due to the side heating. However, the local mechanism in the form of thermals dominates in this case only on the initial stage (for about 10–15 s). Thermals were observed on the suddenly heated bottom in water by Sparrow *et al.* 1970 (see also realization of thermals by the computer system COMGA by Ermakov *et al.*, 1997). A similar structure of thermals penetrated from above the cooling water surface was reported by Bune *et al.*, 1985. After long time the distance global type of convection from the side heating dominates (Figure 8.4(a)). Note that this situation exists for low Pr number melts. For high Pr experiments Berdnikov *et al.*, 1990 reported regular oscillations induced by thermals near the above cooled front, which corresponds to recent calculations by Polezhaev *et al.*, 1998b). The long-term regime (Figure 8.4(a)) is steady state (in accordance with estimation of the critical Gr number for axisymmetrical case). It is similar to the benchmark configuration (Figure 8.2(b)) and helps to understand the stabilization.

However, it should be taken into account that the leading long-term convective mechanism is strongly dependent on the melt level. Figure 8.4(b) shows the calculation of convection for small height of the melt level $H = 1$ cm, where bottom heating (Rayleigh–Bernard mechanism) in the form of convective cells dominates. For this case the impact of dynamical control actions may be quite different. These comments show the importance of Pr number and the geometrical configuration for the flow field in the crucible and restricts the parametrical analysis here for the level of 4 cm in low Pr number melt.

(c) Coupling of the thermal convection and crystal/crucible rotation.

The initial stage of this industrial regime is similar to the previous one (Figure 8.4(a)). However the special oscillatory mechanism, which may be recognized as a W-type isotherm structure is realized for long-term duration as shown in Figure 8.4(c). An important feature of the flow field in this case in comparison with the long-term convection structure in Figure 8.4(a) is the strong disturbances due to interaction with the forced type of convection.

More detailed parametrical analysis was done to show the cause of this type of instability. Figure 8.4(d) shows the interaction of the thermal convection and crystal rotation only. One can see that convection is dominant again for a crystal rotation speed of 6 rpm ($Re_s = 6 \times 10^3$), because of the high value of the governing parameter $Gr/Re^2 > 5$. Only weak instability exists in the bottom region near the axis in this case in accordance with stability analysis (Nikitin, Polezhaev, 1999a). Therefore this type of interaction is not the reason for the flow-field structure in Figure 8.4(c). The next run with crucible rotation and zero velocity of the crystal rotation (Figure 8.4(e)) definitely shows that the reason for the strong instability in Figure 8.4(c) is coupling of thermal convection and crucible rotation because of the high crucible rotation speed (it is the small value of the governing parameter $Gr/Re^2 = 0.3$). There are a number of qualitative reasons for this type of instability, for instance, Kurpers and Lortz, 1956 (see Ristorcelli and Lumely 1992) reported a similar mechanism by reason of convection and rotation of the container. However, quantitative analysis is very important here because of the strong dependency of parameters. We will return to a discussion of this phenomena on the basis of the 3D model using stability analysis (Section 8.6.2(a)).

(d) Analysis of control possibility using axisymmetrical model.

There are two primitive possibilities for control of the above mentioned type of instability: (a) to reduce crucible speed rotation, (b) to reduce the gravity level. Using analysis on the basis of axisymmetrical model it was shown by Polezhaev *et al.* 1997, that for the last case temperature oscillations in GaAs melt may be avoided for $g/g_o = 0.01$. Temporal behaviour of this case in comparison with Figures 8.39(a) and 8.4(c) shows that W-type isotherms are completely eliminated. A similar effect of the elimination of the temperature oscillations may be achieved using reduction of crucible rotation speed, to 1.5 rpm. However, the impact of thermal convection on the temperature field in this case is quite different. Another more complicated possibility (time-dependent temperature on the side wall due to variation of the heater capacity) is discussed by Zakharov *et al.*, 1998. These control possibilities, termed 'low energetic possibilities alternative to microgravity', were studied in the framework of the project NASA-RSA (Polezhaev *et al.*, 1998b). However, the analysis on the basis of the axisymmetrical model shows a possibility to avoid only some (but probably the strongest) of mechanisms of oscillations and must be checked on the basis of 3D models.

8.6.2   THREE-DIMENSIONAL ANALYSIS

(a) Stability analysis of the elementary fluid flows.

Results of the linear stability analysis using the technique of Nikitin and Polezhaev, 1999a for this configuration (Figures 8.5–8.7) show structures of three-dimensional disturbances in two different (r, z and azimuthal, $z = $ const) cross sections and temporal behavior of 3D disturbances for elementary processes: thermal convection without any rotation Figure 8.5, crystal (disc) rotation without convection (Figure 8.6) and the same for only crucible rotation (Figure 8.7). The critical Grashof number for convection in the configuration discussed here was found to be 2.5×10^6 which is not far from $Gr_c = 1.2 \times 10^6$ for the benchmark configuration in the above-mentioned Section 8.5.2 and significantly smaller than 1.5×10^8 for the axisymmetrical case. Therefore it must be the oscillatory regime in the 3D case instead of steady state in Figure 8.4(a). There is a global structure of temperature disturbances in the $r-z$ plane (Figure 8.5(a)) and a periodically azimuthal temperature structure (Figure 8.5(b)) with half the period in the 3D benchmark configuration (Nikitin and Polezhaev, 1999a).

The critical rotation speed of the crystal under 3D disturbances was found to be $(W_s)_c = 7.7$ rpm. It corresponds to a critical $Re_s = 7.7 \times 10^3$, which is smaller than for the Cochran flow because of disturbances in the enclosure, as shown in the $r-z$ plane in Figure 8.6a. Azimuthal disturbances are nonsymmetrical in this case (Figure 8.6(b)). Anyway, the critical W_s value is higher

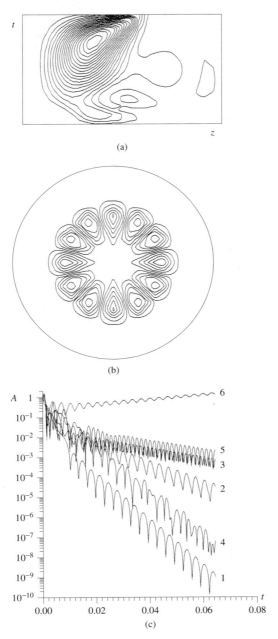

Figure 8.5 Three-dimensional instability for Czochralski LEC growth of GaAs configuration for thermal convection without rotation $Gr_{cc} = 2.5 \times 10^6$; (a) structure of the isotherms disturbances in $r-z$ plane; (b) structure of the isotherms disturbances in the azimuthal plane; (c) temporal behavior of the different temperature modes.

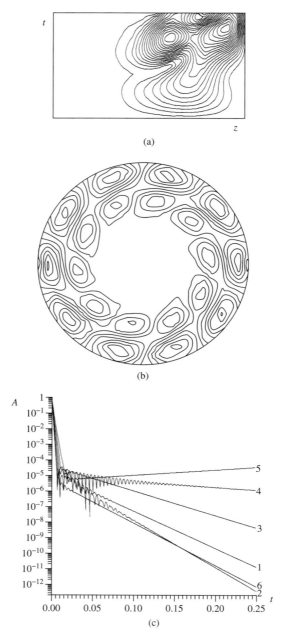

Figure 8.6 Three-dimensional instability for Czochralski LEC growth of GaAs configu-ration for disc rotation only $(W_s)_c = 7.7$ rpm: (a) structure of the isotherm disturbances in the $r-z$ plane; (b) structure of the isotherm disturbances in azimuthal plane; (c) temporal behavior of temperature modes.

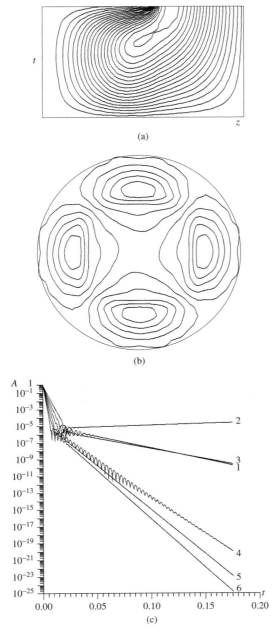

Figure 8.7   Three-dimensional instability for Czochralski LEC growth of GaAs configuration for crucible rotation only ($W_{cc} = -1.2$ rpm; (a) structure of the isotherm disturbances the in $r-z$ plane; (b) structure of the isotherm disturbances in azimuthal plane;
(c) temporal behavior of temperature modes.

than the industrial value 6.0 rpm. Therefore, it confirms, as was noted above (Section 8.6.1) that crystal rotation is not the reason for instability even in the 3D case. However, critical crucible rotation speed $(W_c)_c$, which was found to be about 1.2 rpm ($Re_c = 1.2 \times 10^3$), is more than 10 times lower than the industrial one 16 rpm. Azimuthal disturbances are symmetrical in this case (Figure 8.7(b)). Therefore temperature oscillations in this case will be present even in zero gravity. A monotonic instability in both cases of the elementary forced convection may be noted (Figures 8.6(c) and 8.7(c)). Therefore analysis of stability of the elementary mechanisms helps to recognize the possible impact of the elementary processes on the final structure of flow/temperature fields.

(b) Direct numerical simulation of the coupling regimes on the basis of unsteady 3D model.

Simulation of the three-dimensional industrial case (Gr $= 7.8 \times 10^7$, $Re_s = 6 \times 10^3$, $Re_c = -1.6 \times 10^4$), which was done using the spectral-finite difference technique (Nikitin and Polezhaev, 1999b and Polezhaev *et al.*, (1998b)) shows chaotic temporal behaviour of the flow/temperature fields. The results of the calculations are mainly qualitative, because a relatively coarse uniform mesh was used. There is not enough space for discussion of the computation technique and results, but in the cited paper the method of analysis chaotic (turbulent) regime

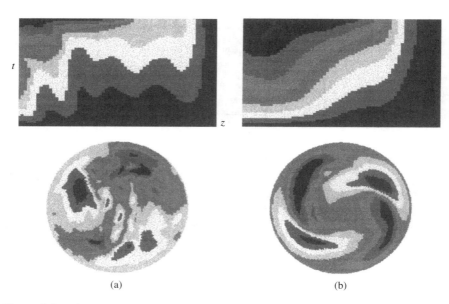

(a)   (b)

Figure 8.8 Direct simulation of thermal convection and counter-rotation in GaAs 'Czochralski method. Fragment of instanteneous isotherms of unsteady 3D modeling. Up: $r-z$ plane, down: azimuthal plane, $z = 0.95$: (a) ground-based industrial regime (Gr $= 7.8 \times 10^7$, $W_s = 6$ rpm, $W_c = -16$ rpm); (b) industrial regime under reduced gravity (Gr $= 7.8 \times 10^5$, $W_s = 6$ rpm, $W_c = -16$ rpm): laminar flow.

for the benchmark configuration was discussed in detail. Three-dimensional runs here show a typical turbulent temperature field for the industrial regime in the azimuthal plane (Figure 8.8(a)), but in the $r-z$ plane (Figure 8.8(b)) qualitatively similar to the axisymmetrical ones (Figure 8.4(c)). A special run was carried out with low gravity $Gr_c = 7.8 \times 10^5$ $(g/g_o = 0.01)$. A fragment of the instantaneous isotherms structure is shown in Figure 8.8(b) (Polezhaev et al., 1998b). In this case isotherms in the $r-z$ plane (Figure 8.8(b)) are similar to axisymmetrical ones for zero gravity (Figure 8.3(a)). Therefore conclusions related to the possibility to eliminate oscillations using thermal and dynamical control as alternatives to microgravity in industrial regimes of GaAs LEC, which was found using the axisymmetrical model, are confirmed qualitatively in the 3D case. The difference is that regular temperature oscillations exist in the 3D case for the industrial regime even in the low-gravity environment.

8.7   CONCLUSIONS

Instabilities induced by nonlinear interaction of the thermal convection and crystal/crucible rotation for the idealized Czochralski model for the industrial LEC GaAs crystal-growth regime, which characterize nondimensional parameters $H/R_c = 0.578$, $R_s/R_c = 0.578$, $Gr = 7.7 \times 10^8$, $Pr = 0.07$, $Re_s = 6.6 \times 10^3$, $Re_c = -1.6 \times 10^4$ (this corresponds to a GaAs melt $H = R_s = 4$ cm, $R_c = 7$ cm, $W_s = 6$ rpm, $W_c = -16$ rpm) are analyzed using an axisymmetrical approach, 3D stability and direct numerical simulation.

As a result of modeling of gravity-driven convection in a crucible in the Czochralski method that three different possible mechanisms of gravity-driven convection in the crucible are shown: side heating, above cooling near the crystal front and bottom heating. The critical Grashof number for onset of temperature oscillations due to convective instability in this configuration is found to be 1.5×10^8 in the axisymmetrical case and 2.5×10^6 for the 3D case.

The critical Reynolds number for disc rotation is found in the 3D case $(Re_s)_c = 7.7 \times 10^3$ and for crucible rotation $(Re_c)_c = 1.2 \times 10^3$ (this corresponds to the critical rotation speed of the crystal of 7.7 rpm and 1.2 rpm for crucible rotation). High stability of the elementary flows, induced by thermal convection, crystal/crucible rotation is found in the axisymmetrical case. Coupling of convection and crucible rotation is the main reason for instability, which is realized even in the 2D case. The results of the modeling encourage research into the possibility of symmetrization of the temperature and flow fields during Czochralski growth as an important control action.

Possibilities to reduce temperature oscillations in GaAs melt (thermal, dynamical, geometrical, etc., control actions including possibilities to reduce gravity) may be analyzed in quantitative form using an actual technological environment. However- it seems that industrial control actions are far from optimal. One of the possible reasons high crucible rotation speed (more than ten times higher than critical). It should be concluded also that for a given industrial ground-based technological process the microgravity level $g/g_o = 10^{-3}$ may be enough for damping the largest oscillations in LEC GaAs melt in the crucible for the growth of semi-insulating GaAs.

Tools for modeling using hierarchy of the models/codes for analysis of technologically important fluid dynamics, heat/mass transfer processes are developed and results using benchmark configurations for axisymmetrical and three-dimensional cases are summarized. For the benchmark configuration $Gr_c = 2.5 \times 10^6$ in the axisymmetrical case and 5.5×10^5 for the 3D case and $G_c = 3.5 \times 10^5$ for $Re_s = 10^3$, $Re_c = 0$ are presented for this benchmark as well as steady-state results for the axisymmetrical case. A strong impact of the boundary temperature conditions on the melt's surface (adiabatic: $Gr_c > 10^7$ in the axisymmetrical and $Gr_c = 1.2 \times 10^6$ for the 3D case) is shown. It is also shown that the two-dimensional axisymmetrical model is efficient for prediction of a number of technologically important fluid flows and heat/mass transfer in the $r - z$ plane and may be a powerful tool for parametrical research together with 3D stability and direct simulation.

ACKNOWLEDGMENTS

The author would like to make acknowledgment to V. A. Kosushkin for technological consulting, S. A. Nikitin, N. V. Nikitin and O. A. Bessonov for help in the developments of the concept and results, M. N. Myakshina for the help in calculations and graphics and O. A. Rudenko for the preparation of the manuscript.

This work was partly supported by the NASA-RSA program (TM-11 project).

REFERENCES

Alexander J. I. D. (1990), Low gravity experiment sensitivity of residual acceleration. A review. *Microgravity Sci. and Technol.*, **3**, 52.

Baumgartl J., Bune A., Koai K., *et al.* (1993) Global simulation of heat transport, including melt convection in a Czochralski crystal growth process – combined finite element/finite volume approach, *Mater. Sci. Eng.* A **173**, 9.

Beliaev Yu. N. (1997), Hydrodynamical instability and turbulence in spherical Couette flow, *Moscow State Univ.*, 348 (in Russian).

Berdnikov V. S., Borisov V. L., Markov V. A. and Panchenko V. I. (1990), Laboratory modeling of the macroscopic transport processes in the melt during crystal growth by the method of pulling In: Avduevsky V. S., Polezhaev V. I. (eds.) *Hydrodynamics, heat and mass transfer during material processing*, 68, Nauka, Moscow (in Russian).

Bottaro, A. and Zebib A., (1988) Bifurcation in axisymmetric Czochralski natural convection, *Phys. Fluids* **31**, 495.

Bottaro, A. and Zebib A., (1989) Three dimensional thermal convection in Czochralski melt, *J. Cryst. Growth* **97**, 50.

Brown R. A., (1988) Theory of transport processes in single crystal growth from the melt, *AIChE J.*, **34**, 881.

Buckle, U. and Schafer, M. (1993), Benchmark results for the numerical simulation of flow in Czochralski crystal growth, *J. Cryst. Growth* **126**, 682.

Bune A. V., Ginzburg A. T., Polezhaev V. I., and Fedorov K. N. (1985), Numerical and laboratory modeling of the development of convection in a water layer cooled from the surface Izvestiya, *Atmos. Oceanic Phys.* **21** (9), 736.

Carruthers J. R. (1977), Thermal convection instability relevant to crystal growth from liquids, In: *Preparation and Properties of Solid State Materials* (eds. W. R. Wilcox, R. A. Lefever), Vol. 3, 1, Marcel Dekker, Inc. New York and Basel.

De Vahl Davis G., Jones I. P. (1983) Natural convection in a square cavity: a comparison exercise, Int. *J. Numer Methods fluids*, (1983) **3**, 227–248.

Dupret F., Van Den Bogaert N. (1994), Modelling Bridgman and Czochralski Growth, In: *Handbook of Crystal Growth*, ed. D. T. J. Hurle, Vol. 2, 875.

Ermakov, M. K. Griaznov V. L., Nikitin S. A., Pavlosky D. S., Polezhaev V. I. (1992), A PC-based System for Modelling of Convection in Enclosures on the basis of Navier-Stokes Equations, *Int. J. Numer. Methods Fluid*, **15**, 975–984.

Ermakov M. K., Nikitin S. A. and Polezhaev V. I., (1997), A System and computer laboratory for modelling of the convective heat – and mass transfer processes, *Fluid Dynamics* **32** (3), 338.

Fontaine J.-P., Randriamampianina A., Extremet G. P., and Bontoux P. (1989), Simulation of steady and time-dependent rotation-driven regimes in a liquid-encapsulated Czochralski configuration, *J. Cryst.-Growth* **97**, 116.

Gelfgat Yu. M., Bar-Yoseph P. Z., and Solan A. (1996), Stability of confined swirling flow with and without vortex breakdown, *J. Fluid Mech.* **311**, 1.

Griaznov V. L., Ermakov M. K., Zakharov B. G. *et al.* (1992), Czochralski growth of gallium arsenide: technological experiments and numerical simulation. In *Fluid flow phenomena in crystal growth. Euromech colloquium* N. 284, Aussois, France, October 13–16, 1992, 15.

Hurle D. T. J. (1993) Crystal Pulling from the Melt, Springer, *J. Cryst. Growth*, **156**, 383.

Jones A. D. W. (1984), The temperature field of a model Czochralski Melt, *J. Cryst. Growth* **69**, 165.

Kakimoto K., Egushi M., Watanabe M. *et al.* (1989), Direct observation by X-ray radiography of convection of boric oxide in the GaAs liquid encapsulated Czochralski growth. *J. Cryst. Growth* **94**, 405.

Kakimoto K. Watanabe M., Egushi M., *et al.* (1993), Order structure in non-axisymmetric flow of silicon melt convection, *J. Cryst. Growth* **126**, 435.

Kirdyiashkin A. G. (1979), Structure of the thermal gravitational flows, In: Models in mechanics of continuous media, Proceedings of the Vth all-Union School on the models in mechanics of continuous media, 69, Novosibirsk 1979.

Kobayashi N. and Arizumi T. (1980), Computational studies on the convection caused by crystal rotation in a crucible, *J. Cryst. Growth* **47**, 419.

Kobayashi N. (1995), Steady state flow in a Czochralski crucible. *J. Cryst. Growth* **147**, 382.

Kosushkin V. G. (1997), Investigation of low frequency oscillation and crucible and crystal rotation control in GaAs crystal pulling technique. Proceedings of the Joint Xth

European and VIth Russian Symposium on Physical Sciences in Microgravity, St. Petersburg, Russia, June 15–21, 1997, 122.

Kozkemiakin G. N., Kosushkin V. G., Kurockin S.Yu. (1992), Growth of GaAs crystal pulled under the presence of ultrasonic vibrations, *J. Cryst. Growth* **121**, 240.

Krzyminski U., Ostrogorsky A. G. (1997), Visualization of convection in Czochralski melts using salts under realistic thermal boundary conditions, *J. Cryst. Growth* **174**, 19.

Langlois W. E. (1985), Buoyancy-driven flows in crystal-growth melts *Ann. Rev. Fluid Mech.* **17**, 191.

Leister H.-J., Peric'., M. (1992), Numerical simulation of a 3D Czochralski melt flow by a finite volume multigrid algorithm *J. Cryst. Growth* **123**, 567.

Mihelcic' M. and Wingerath K. (1984), Three dimensional simulations of the Czochralski bulk flow, *J. Cryst. Growth* **69**, 473.

Motakef S. (1990) Magnetic field elimination of convective interference with segregations during vertical-Bridgman growth of doped semiconductors *J. Cryst. Growth* **104**, 833.

Mukherjee D. K., Prasad V., Dutta P., Yuan T., (1996), Liquid crystal visualization of the effects of crucible and crystal rotation on Cz melt flows, *J. Cryst. Growth* **169**, 136.

Muller E. (1988) Convection and Inhomogeneities in Crystal Growth from the Melt Crystal, Vol. 12 (Springer, Berlin; (1988).

Muller, G. and Ostrogorsky A. (1994), Convection in melt growth, In: *Handbook of Crystal Growth* **2**, 711.

Nikitin N. V. (1994), A spectral finite-difference method of calculating turbulent flows of an incompressible fluid in pipes and channels, *Comput. Meth. Math. Phys.* **34**(6), 765.

Nikitin N. V., Polezhaev V. I. (1999a), Three dimensional convective instability and temperature oscillations in Czochralski model *Izv. RAS, MZG*, **6**, 26, (translated: *Fluid Dynamics* **26**, 3).

Nikitin N. V., Polezhaev V. I., Three dimensional features of transitional and turbulent convective flows in Czochralski model *Izv. RAS, MZG*, (accepted 1999b, translated: *Fluid Dynamics* **34**, 6).

Pimputkar M., Ostrach S. (1981), Convective effects in crystal growth from melt. *J. Cryst. Growth* **55**, 614.

Polezhaev V. I. (1974), Effect of the maximum temperature stratification and its application *Rep. Acad. Sci. USSR, ser. Hydrodynamics*, **218**, 783–786 (in Russian).

Polezhaev V. I. (1981), Hydrodynamics, heat and mass transfer processes during crystal growth In: *Crystals: Growth, Properties and Applications* **10**, 87.

Polezhaev V. I. (1992), Convective Processes in Microgravity. Proceedings of the First International Symposium on Hydrodynamics and Heat/Mass Transfer in Microgravity, **15**, Gordon and Breach

Polezhaev V. I., (1994), Modelling of hydrodynamics, heat and mass transfer processes on the basis of unsteady Navier-Stokes equations. Applications to the material sciences at earth and under microgravity, *Comput. Methods Appl. Mech. Eng.* **115**, 79.

Polezhaev V. I. (1998), Simulation aspects of technologically important hydrodynamics, heat/mass transfer processes during crystal growth, International School on Crystal Growth Technology, Beatenberg, Switzerland, 5–16 Sept. 1998, Book of lecture notes, (ed.) H. J. Scheel, 188.

Polezhaev V. I., Nikitin S. A., Nikitin N. V. (1997), Gravity-driven and rotational low Prandtl melt flow in enclosure: nonlinear interaction, temperature oscillations and gravitational sensitivity. Proceedings of the Joint Xth European and VIth Russian Symposium on Physical Sciences in Microgravity, St. Petersburg, Russia, June 15–21, 1997, 28.

Polezhaev V. I., Bessonov O. A., Nikitin N. V. and Nikitin S. A. (1998a), Three dimensional stability and direct simulation analysis of the thermal convection in low Prandtl melt of Czochralski model, The Twelfth International Conference on Crystal Growth, July 26–31, 1998, Jerusalem, Israel, Abstracts, 1998, 178.

Polezhaev V. I., Bessonov O. A., Nikitin N. V., Nikitin S. A., Kosushkin V. G., *et al.* (1998b), Low energetic possibility for control of crystal growth, The Inst. for Problems in Mech. RAS, Report on the Project TM-11.

Polezhaev V. I., Bessonov O. A., Nikitin S. A. (1998c), Dopant inhomogeneities due to convection in microgravity: spatial effects, *Adv. Space Res.* **22** (8), 1217.

Polezhaev V. I., Ermakov M. K., Nikitin N. V. and Nikitin S. A. (1998d), Nonlinear interactions and temperature oscillations in low Prandtl melt of Czochralski model: validation of computational solutions for gravity-driven and rotatory flows. Intern. Symp. on Advances in Computational Heat Transfer, 26–30 May, Cesme, Turkey, 1997. In.: Advances in Computational Heat Transfer, 492, Begell House.

Rappl H. O., Matteo Ferraz L. F., Scheel H. J., *et al.* (1984), *J. Cryst. Growth* **70**, 49.

Ristorcelli J. R. and Lumely J. L. (1992), Instabilities, transition and turbulence in the Czochralski crystal melt, *J. Cryst. Growth* **116**, 447.

Roux B., (ed.), (1990), Numerical simulation of oscillatory convection in low-Pr fluids. *Notes on Numerical Fluid Mechanics*, **27**, Vieweg.

Sacudean M. E., Sabhapathy P. and Weinberg F. (1989), Numerical study of free and forced convection in the LEC growth of GaAs, *J. of Crystal Growth* **94**, 522.

Scheel H. J., (1972) *J. Cryst. Growth* **13/14**, 560.

Scheel H. J., Sielawa J. S. (1985), Optimum convection conditions for Czochralski growth of semiconductors, 6th Internat. Symp. 'High Purity Materials in Science and Technology', Dresden, GDR, May 6–10, 1985, Proc. I. Plenary Paper, Akad. der Wiss. der DDR, 1985, 232.

Seidi A., McCord G., Muller G., Leister H.-J. (1994), Experimental observation and numerical simulation of wave patterns in a Czochralski silicon melt: *J. Cryst. Growth* **137**, 326.

Sparrow E. M., Husar R. B., Goldstein R. J. (1970), Observations and other characteristics of thermals, *J. Fluid Mech.*, **41**, 793.

Verezub N. A., Zharikov E. V., Mialdun A. Z. *et al.* (1995), Physical modeling of the low frequency vibration crystal actions on the flow and heat transfer in Czochralski method, The Inst. Probl. Mech. RAS, preprint N 543, 66, Moscow.

Wagner C., Friedrich R. (1997), Turbulent flow in idealized Czochralski crystal growth configuration, In: *New Results in Numerical and Experimental Fluid Mechanics*: Braunschweig, Germany, 367, ed. by Horst Korner and Reingard Hilbig.-Braunschweig: Vievig.

Wheeler A. A. (1991), Four test problems for numerical simulation of flow in Czochralski crystal growth. *J. Cryst. Growth* **102**, 691.

Xiao Q., Derby J. J. (1995), Three-dimensional melt flow in Czochralski oxide growth: high-resolution, massively parallel, finite element computations: *J. Cryst. Growth* **152**, 169.

Yi K.-W., Booker V. B., Egushi M., Shyo T., Kakimoto K. (1995a), Structure of temperature and velocity fields in the Si melt of a Czochralski crystal growth system, *J. Cryst. Growth* **156**, 383.

Yi K.-W., Booker V. B., Kakimoto K. Egushi M., *et al.* (1995b), Spoke patterns on molten silicon in Czochralski system, *J. Cryst. Growth* **144**, 20.

Zakharov B. G. Kosushkin V. G., Nikitin N. V., Polezhaev V. I. (1998), Technological experiments and mathematical modeling of the fluid dynamics and heat transfer in Gallium Arsenide single-crystal growth processes *Fluid Dynamics* **33** (1), 110.

Part 2

Silicon

9 Influence of Boron Addition on Oxygen Behavior in Silicon Melts

KAZUTAKA TERASHIMA

Silicon Melt Advanced Project, Shonan Institute of Technology Fujisawa, Kanagawa 251, Japan

ABSTRACT

The oxygen concentration in silicon crystals is markedly changed by adding impurities. The solubility of oxygen atoms in boron-doped silicon melt increases with increasing boron concentration. The dissolution rate of fused quartz also increases with increasing boron concentration. It has been found that the fraction of cristobalite area grown on a fused quartz rod surface changes the fused quartz dissolution rate. The evaporation rate of silicon monoxide has a tendency to increase with increasing boron concentration. The importance of understanding silicon melt will be discussed. The strategy to grow well-controlled silicon crystals and recent results are described.

9.1 INTRODUCTION

Large-diameter silicon crystals are of extreme importance for electronics. The diameter of silicon has increased with time and many silicon suppliers now have the capability to grow 12-inch diameter crystals. The formation of grown-in defects, detrimental to large-scale integrated circuits, changes with increase in crystal diameter. The origin of these defects is not fully understood because of the lack of data and of the understanding of the growing interface. In our opinion, the origin of defects is partly or mainly related to the melt conditions. Our strategy for research is shown in Figure 9.1. First, the dependence of melt properties on temperature and impurity concentration should be studied. The density of silicon melts is bound to vary significantly by impurity addition [1] resulting in changes in melt convection [2, 3], but the reason for the density change is not well understood. Secondly, we will study the evaporation of SiO and incorporation of ambient gases. Thirdly, the physics of silicon crystals at

Crystal Growth Technology, Edited by H. J. Scheel and T. Fukuda
© 2003 John Wiley & Sons, Ltd.   ISBN: 0-471-49059-8

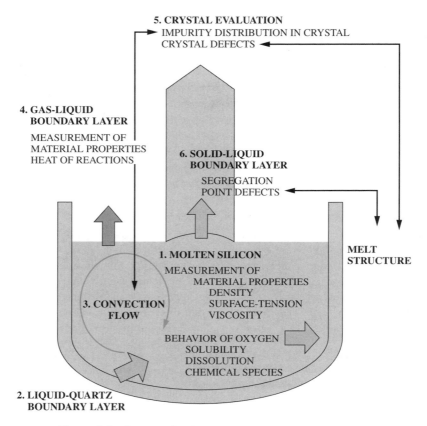

Figure 9.1   Strategy for the research of Si crystal growth.

high temperatures is important. Finally, the most important research area is the growing interface.

The melt properties should be studied more precisely, especially the variation with impurity species of fundamental physical properties such as density, surface tension and viscosity. The structure of the melt, its dependence on chemical bonding, and the dependence of the crystal/melt interface and induced defects on the melt structure are of essential interest. The oxygen concentration in Si crystals has been reported to change strongly with high boron doping [4, 5]. We report here the variation in oxygen solubility, fused quartz dissolution rate, and SiO evaporation from the melt surface with boron addition to silicon melts.

9.2   OXYGEN BEHAVIOR IN BORON-DOPED SILICON MELTS

Oxygen dissolves into silicon melts from the fused quartz crucible wall up to the solubility limit. The oxygen solubility changes with impurity addition and prediction of oxygen incorporation into the growing crystal requires data on oxygen dissolution rate, solubility and SiO evaporation rate from the free surface of the melt. Highly boron-doped silicon of large diameter has attracted much attention for p-type epitaxial wafers due to its iron-gettering ability. In this section, oxygen solubility, fused quartz dissolution rate and evaporation from the melt surface are reported as a function of boron concentration in the melt.

9.2.1 OXYGEN SOLUBILITY IN SILICON MELT

9.2.1.1 Experimental

Our experimental method is basically the same as Huang [6]. A silicon rod 13 mm in diameter and 33 mm in length was fixed in a 13-mm inner-diameter quartz ampoule with a flat bottom. The upper space in the ampoule was sealed with a quartz rod with a slightly smaller diameter and it was welded completely to the inside wall of the ampoule after evacuation to about 10^{-5} Torr. The quartz ampoule with a pure silicon rod or silicon rod with boron powder inside was covered with a graphite holder and heated to various temperatures in the range of $1430-1470\,°C$ for 120 min in an Ar atmosphere. The furnace used in this experiment was a horizontal three-zone type. The sample was quickly drawn from the furnace for quenching, after equilibration has presumably been attained in 120 min. The size of the specimen was $3 \times 3 \times 2.5\,\text{mm}^3$. The samples were mirror polished. The oxygen and boron concentrations were measured using a secondary in mass spectrometer (SIMS) (CAMECA, IMS-3f/4) instrument. The relative error of the SIMS measurements was less than 5 %.

9.2.1.2 Results and discussion

Oxygen concentration increases with increasing boron concentration as shown in Figure 9.2. When the boron concentration is 5×10^{20} atoms/cm^3, the oxygen concentration is 4×10^{18} atoms/cm^3. This oxygen concentration is about twice the value for undoped silicon. The results of undoped silicon in our laboratory showed good agreement with the reported value by Carlberg [7]. The dependence of the oxygen concentration on the silicon-melt temperature in boron-doped silicon melts is given in Figure 9.3. According to our experimental data, oxygen concentration increases with increasing boron concentration independent of the silicon melt temperature. The cause of this phenomenon is discussed below.

To discuss the relation between boron atoms and oxygen atoms in a silicon melt, first we assume the generation of boron oxide species. If X atoms of boron react with one oxygen atom, the following chemical reaction occurs:

$$X\text{B} + \text{O} \longleftrightarrow \text{B}_X\text{O} \qquad (9.1)$$

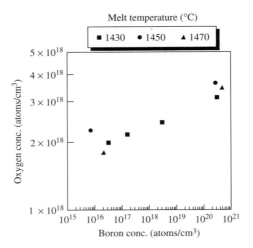

Figure 9.2 Dependence of the oxygen concentration on the boron concentration in a boron-doped silicon melt.

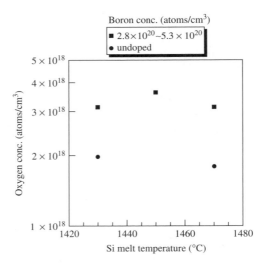

Figure 9.3 Dependence of the oxygen concentration on the Si melt temperature in a boron-doped silicon melt.

The equilibrium constant K is given by the following expression:

$$K = \frac{[B_X O]}{[B]^X [O]} \tag{9.2}$$

where [O], [B] and [B$_X$O] are the total oxygen concentration, the boron concentration, and B$_X$O concentration in the boron-doped silicon melt, respectively.

It is assumed that there are two kinds of oxygen in a boron-doped silicon melt: one is uncombined or isolated oxygen atoms, the other is combined with boron atoms in the form of B_XO. Then the following calculation can be carried out:

$$[O] = [O]_{Si} + [B_XO] = [O]_{Si} + K[B]^X[O] \tag{9.3}$$

$$\Delta[O]/[O] = K[B]^X \tag{9.4}$$

$\Delta[O] = [O] - [O]_{Si}$ is the increased oxygen concentration due to boron addition in the silicon melt. Equation (9.4) can be rewritten as follows:

$$\ln\{\Delta[O]/[O]\} = \ln K + X \ln\{[B]\} \tag{9.5}$$

The values of X and K are calculated by using our results. X and K values are obtained from the following process. The relationship between the increase of oxygen concentration and boron concentration in silicon melt is plotted in Figure 9.4. The gradient corresponds to X. In Figure 9.4, $\ln[B]$ being equal to O, the value of $\ln(\Delta[O]/[O])$ gives K. Table 9.1 shows X and K. According to

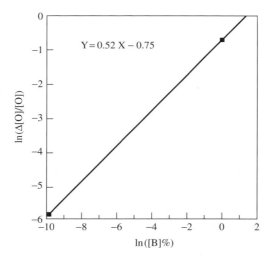

Figure 9.4 Relationship between the increase of oxygen concentration and boron concentration in a silicon melt: Silicon melt temperature = 1470 °C.

Table 9.1 Calculated X and the equilibrium constant K

Si melt temperature (°C)	X	Equilibrium constant: K
1430	0.16	0.51
1450	0.089	0.54
1470	0.52	0.47

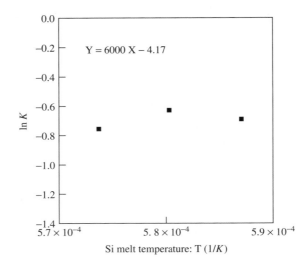

Figure 9.5 Relationship between equilibrium constant and Si melt temperature in a boron-doped silicon melt.

Table 9.1, the ratio of boron to oxygen varies with temperature. The dependence of K on the silicon melt temperature in boron-doped silicon melt is given in Figure 9.5. The value of K is smaller than unity and the dependence on the silicon-melt temperature has no definite tendency. The activation energy of the boron-oxygen reaction ΔE is given by the following expression:

$$\ln K = \frac{-\Delta E}{RT} + N \tag{9.6}$$

where R is the gas constant and N is a constant unrelated to the silicon-melt temperature. According to Equation (9.6) and Figure 9.4, ΔE is almost zero. So there is no activation energy in this reaction. This means the stable chemical species of $B_X O$ is not generated by boron addition in the silicon melt, even though boron seems to attract oxygen. The analysis indicates that about 52 % of all oxygen in a silicon melt relate to added boron. These results indicate that the influence of boron doping involves a shift in oxygen solubility above the values for an undoped melt.

Next, the diameter of boron and silicon ions should be considered as a measure to discuss the solubility variation. At room temperature the silicon ion diameter is 0.39 Å and the boron ion diameter is 0.2 Å. If boron ions substitute the silicon ion sites in the melt, the distance between boron ions and silicon ions as nearest neighbours should shrink. This tendency may attract oxygen ions. Oxygen ions relax the shrinkage of the melt structure. The structural variation of silicon melt with and without boron and oxygen atoms is now under investigation. We next consider equilibrium of silicon melt with silicon oxides. Carlberg [8] calculated

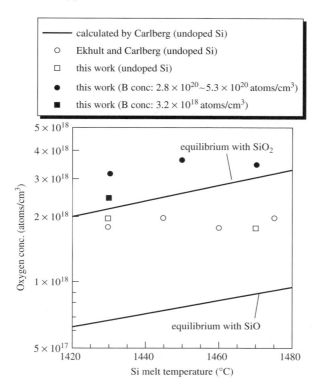

Figure 9.6 Relationship between oxygen concentration and temperature (This work: closed circles, closed squares and open squares).

oxygen solubilities in silicon melt in equilibrium with SiO_2 and SiO from thermodynamics. Our results are plotted in Figure 9.6 with the calculated equilibrium values given by solid lines. Our oxygen solubilities in boron-doped silicon melt are denoted by closed circles and a square, and those of undoped melt denoted by open circles and open squares, which all lie between the equilibrium values with SiO_2 and SiO.

For a boron-doped silicon melt the oxygen concentrations approach the equilibrium values with SiO_2. When the boron concentration is higher than about 1×10^{18} atoms/cm^3, oxygen concentrations increase above the values of equilibrium with SiO_2. In other words, the oxygen solubility shifts above the values for the undoped case. Also, in the case where the boron concentration is 1×10^{20} atoms/cm^3 the oxygen concentration values approach the equilibrium values that SiO_2 with increasing silicon melt temperature. When boron is doped in a silicon melt, then with increasing silicon melt temperature the oxygen concentration approaches a value that is consistent with an equilibrium with SiO_2.

A thin brown film was observed with increasing silicon melt temperature, but is not observed at 1430 °C. This brown-coloured layer was reported as an

$SiO_{2-\delta}$ layer by Carlberg [7]. The fraction of thin brown film area around the interface of the quenched silicon sample and the quartz varies depending on the boron concentration and temperature. The interface change strongly affects the oxygen solubility, because the equilibrium with the interface determines the oxygen concentration in the silicon melt. We believe the interface variation between Si melts and fused quartz is the most important phenomenon affecting oxygen solubilities. The holding-time dependence and the variation of chemical reaction rate by adding boron are now under investigation.

9.2.2 FUSED QUARTZ DISSOLUTION RATE IN SILICON MELTS

9.2.2.1 Introduction

Fused quartz is the material most commonly used as crucible material to hold the melt in silicon crystal-growth by the Czochralski method. During crystal-growth, the fused quartz crucible gradually dissolves in molten silicon, so that the grown crystal contains a high concentration of oxygen, on the order of 10^{17} to 10^{18} atoms/cm^3, which causes crystalline defects during the following heat treatments. As device dimensions shrink, the effective control of oxygen in silicon becomes increasingly important. Therefore it is desirable to control the dissolution of fused quartz in the molten silicon. But there is no data on the dependence of the fused-quartz dissolution rate on the boron concentration in silicon melts. It has been found that the dissolution rate in a silicon melt markedly varies in the presence of boron in silicon melt and due to its impact on oxygen concentration.

9.2.2.2 Experimental

Fused quartz rods 10 mm in diameter and 100 mm in length were prepared for this experiment. Silicon melts were contained in either a fused-quartz crucible (quartz crucible, hereafter) or a carbon crucible coated with SiC by chemical vapour deposition (SiC-coated crucible, hereafter). The size of the crucible was 50 mm in diameter and 84 mm in depth. The depth of the silicon melt in the crucible was about 50 mm. The starting material was polycrystalline silicon. The boron was put onto the crucible bottom, then polycrystalline silicon was put into the crucible. The rod was immersed to 20 mm in depth in the melts for 5 h and then withdrawn. The weight losses of the fused-quartz rods by dissolution were measured.

9.2.2.3 Results

By using SiC-coated crucibles there is initially no oxygen in the silicon melts. The fused-quartz dissolution rate increases with increasing boron concentration

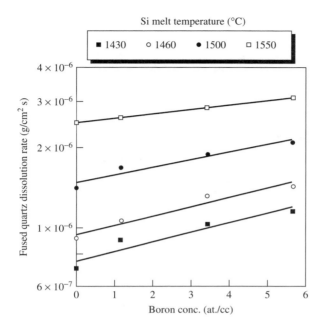

Figure 9.7 Dependence of the fused quartz-dissolution rate on the boron concentration in a boron-doped silicon melt by using a SiC-coated crucible.

as shown in Figure 9.7, and with increasing silicon melt temperature as shown in Figure 9.8. Also, the dependence of the dissolution rate on the boron concentration is small at 1550 °C. On the other hand, at the silicon melt temperature of 1430 °C the dissolution rate is strongly affected by boron as shown in Figure 9.8. Rods immersed in undoped silicon melts show an opaque surface that is brown at a 1430 °C melt temperature and white at 1550 °C. Rods immersed in boron-doped silicon melt ($>5.65 \times 10^{20}$ atoms/cm^3) at 1430 °C show white opaque parts on the surface. By observing the fused-quartz rod cross sections, there are no opaque parts at the rod centre, opaque parts are observed only on the rod surface. Rods immersed in boron-doped silicon melts ($>5.65 \times 10^{20}$ atoms/cm^3) at 1550 °C are markedly transparent. For the fused quartz samples immersed in undoped silicon melts at 1430 °C and 1550 °C and those immersed in boron-doped melts at 1430 °C the opaque surface fraction is crystallized and by X-ray diffraction shows the cristobalite pattern, whereas the transparent rods show the broad diffraction of amorphous silica.

9.2.2.4 Discussion

Our results show that the oxygen and boron concentrations in silicon melts affect the fused-quartz dissolution rate. The activation energy of the fused-quartz

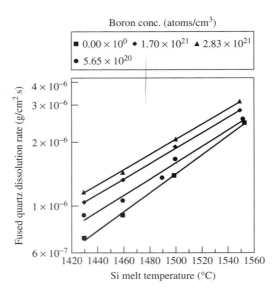

Figure 9.8 Dependence of the fused-quartz dissolution rate on the Si melt temperature in a boron-doped silicon melt by using a SiC-coated crucible.

dissolution is influenced by boron and by oxygen in silicon melts and is given by the following expression:

$$\text{Dissolution rate} = A_0 \exp(-\Delta E/RT) \tag{9.7}$$

where R is the gas constant and A is a pre-exponential factor. ΔE and A_0 are calculated according to Figures 9.7 and 9.8 and are shown in Table 9.2. A_0 expresses the frequencies and modes of molecular collisions. A_0 values are influenced by the boron concentration in silicon melts. The A_0 decreases with increasing boron

Table 9.2 Calculated activation energy: ΔE and pre-exponential factor: A_0

Boron conc. (atoms/cm^3)	Crucible	ΔE (kcal/mol)	A_0
0	SiC-coated carbon	151	172
6×10^{20}	SiC-coated carbon	128	11.6
2×10^{21}	SiC-coated carbon	121	5.54
3×10^{21}	SiC-coated carbon	119	4.73
0	fused quartz	122	2.82
6×10^{20}	fused quartz	145	46.4
3×10^{21}	fused quartz	157	180

concentration in the case of a SiC-coated crucible, and increases with increasing boron concentration when quartz crucibles are used. These phenomena are related to variations of the silicon melt structure. The structure of boron-doped silicon melt is now under investigation by using energy-dispersive X-ray diffractometry. If SiO_2 dissociates into silicon and oxygen, the combination energy is 145–153 kcal/mol [9]. This value is practically the same as our calculated ΔE. This result indicates that fused-quartz dissolution occurs with breaking of Si–O bonds.

Next we consider the chemical reactions on the fused-quartz rod surface. The following chemical reactions occur:

$$SiO_2 \longrightarrow Si + 2\ O \text{ (dissolution of amorphous } SiO_2) \qquad (9.8)$$

$$Si + 2\ O_{1-\delta} \longrightarrow SiO_{2(1-\delta)} \text{ (cristobalite growth)} \qquad (9.9)$$

$$SiO_{2(1-\delta)} \longrightarrow Si + 2\ O_{1-\delta} \text{ (dissolution of cristobalite)} \qquad (9.10)$$

And by using SiC-coated crucibles the following reaction occurs:

$$SiC + Si \longrightarrow 2Si + [C]_{Si,\ liquid} \qquad (9.11)$$

There is some carbon contamination in silicon melts when SiC-coated crucibles were used. Table 9.2 shows that the activation energy of the fused-quartz dissolution rates is not markedly affected by crucible materials. The carbon solubility in silicon melt is less than 0.1 at.% [10]. In this experiment, the boron concentration in silicon melts is ten times the carbon concentration. Whereas carbon slightly affects the fused-quartz dissolution rate, reaction (9.11) seems not to be dominant. Because oxygen evaporates as SiO from the silicon melt surface, the lack of oxygen is indicated as $O_{1-\delta}$ in reactions (9.9) and (9.10). The fused-quartz dissolution rate ΔV is expressed by

$$\Delta V = |V_i - V_{ii} + V_{iii}| \qquad (9.12)$$

where V_i, V_{ii} and V_{iii} are for Equations (9.8), (9.9) and (9.10), respectively. If oxygen transfer from the fused-quartz rod surface to silicon melt increases, reaction (9.8) is enhanced. So reaction (9.9) is retarded and $SiO_{2(1-\delta)}$ does not grow on the fused-quartz rod surface because this experiment is an open system. Because of $SiO_{2(1-\delta)}$ scarceness, reaction rate (9.10) is low. On the other hand, if reaction (9.9) is enhanced, $SiO_{2(1-\delta)}$ is apt to grow on a fused-quartz rod surface. The oxygen diffusion layer is related to reactions (9.8), (9.9) and (9.10). When the oxygen diffusion layer is formed, reactions (9.9)/(9.10) start and eventually all reactions (9.8), (9.9) and (9.10) reach steady state. The oxygen diffusion layer formation is affected by the oxygen diffusion rate in silicon melt. In the near future, we will publish the effect of boron concentration on oxygen diffusion. In the present work we assumed that oxygen diffusion in silicon melts was not affected by boron addition.

In the case of SiC-coated crucibles, there is initially no oxygen in silicon melts. Oxygen solubility in boron-doped silicon melts is about twice the value for undoped silicon melts [11]. Therefore the oxygen concentration on the fused-quartz rod surface in boron-doped silicon melt is about twice the value for undoped silicon melt. Because of SiO evaporation from the melt surface, oxygen concentrations are almost the same in undoped and boron-doped silicon melts. And these values are very low. So it is possible that the oxygen concentration gradient in the oxygen diffusion layer on the fused-quartz rod surface in boron-doped silicon melts is larger than the value for undoped silicon melt due to the higher oxygen solubility. Therefore when boron is doped in silicon melts, reaction rate (9.8) is enhanced. And because of oxygen scarceness in the bulk silicon melt, reaction (9.9) may be suppressed. So $SiO_{2(1-\delta)}$ is not apt to grow on the fused-quartz rod.

In the case of quartz crucibles there is much oxygen in silicon melts because of crucible dissolution. Oxygen concentration on the fused-quartz rod surface in boron-doped silicon melt is about twice as high as the value for undoped silicon melt in the case of SiC-coated crucibles. Just after withdrawing the fused-quartz rod from silicon melt we quenched the silicon melt in the quartz crucible and measured the oxygen concentration in this bulk silicon by SIMS. When the boron concentration is 5.65×10^{20} atom/cm^3, the oxygen concentration is 3.5×10^{18} atoms/cm^3. In the case of undoped melt the oxygen concentration is 8×10^{17} atoms/cm^3. So in the crucible used in our experiment, the oxygen concentration gradient in the oxygen diffusion layer on the fused-quartz rod surface in boron-doped silicon melts is smaller than the value of undoped silicon melts. In this case the dissolution rate of the quartz rod is decreased by doping boron in silicon melts.

9.2.3 EVAPORATION FROM FREE SURFACE OF BORON-DOPED SILICON MELTS IN FUSED-QUARTZ CRUCIBLE

9.2.3.1 Introduction

The influence of the addition of boron on the weight variation due to evaporation and the chemical species evaporated from boron-doped silicon melt were investigated. It was found that boron assisted the evaporation from the silicon melt and the dominant evaporated substance was found to be silicon monoxide.

9.2.3.2 Experimental

The influence of the addition of boron on the evaporation loss of the silicon melt was investigated by the analysis of its deposits. A thermogravimetric method, which is basically the same as that reported by Huang et al. [12–14], was used to measure the weight changes of the silicon melt due to evaporation. Quartz crucibles 20 mm in diameter and 60 mm deep, containing 20.00 g Si (purity 11N) and 0.15 g boron (purity 5N) which corresponded to 10^{21} atoms/cm^3, were hung on an electric balance with 0.1 mg sensitivity (METTLER AT200). The samples were heated to 1450 °C or 1550 °C and maintained within 3 °C of the set-point value for 12 h. A schematic of the experimental apparatus for collecting deposits is shown in Figure 9.9.

The power of the heater was controlled so that there were no temperature variations during the deposit sampling. The quartz crucible was supported by a high-purity carbon susceptor at the centre in the chamber. The crucible was covered by a special funnel-shaped deposit collector made of quartz (Figure 9.9). The evaporated species from the boron-doped silicon melt were deposited on the inner wall of the deposit collector. The temperature of the deposit regions was also measured by a B-type thermocouple set on the deposit collector side. The entire system was enclosed in a water-cooled stainless-steel chamber into which argon gas with purity 6N flowed at a rate of 2.0 l/min to protect the hot zone

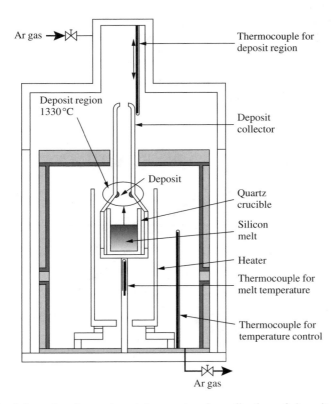

Figure 9.9 Schematic of experimental apparatus for collection of deposits evaporated from the boron-doped Si melt.

against oxidation. The quartz crucible had an internal diameter of 50 mm, a height of 84 mm and a thickness of 4 mm. The crucible contained 200.00 g of Si (purity 11N) and 0.15 g of boron (purity 5N) which corresponded to 10^{20} atoms/cm^3 in the silicon melt. The chamber was heated to 1550 °C at a heating rate of 750 °C/h to realize complete melting. The reference temperature of the silicon melt was measured using a B-type thermocouple fixed under the crucible. Then, the heater power was adjusted so that the reference temperature of the silicon melt was 1550 °C and was maintained within 3 °C for a period of 13 h.

The deposits on the inner wall of the collector were analysed using an electron-probe microanalyser (EPMA: JXA-8621MX) to identify the evaporated species. The EPMA was used with an acceleration voltage of 15 kV, a current of 50 mA and a beam diameter of 20 μm.

9.2.3.3 Results and Discussion

The weight variations due to the evaporation of the boron-doped silicon melt with different boron concentrations (undoped, 10^{21} atoms/cm^3 boron-doping) and at different temperatures (1450 °C, 1550 °C) under atmospheric pressure (770 Torr) in the chamber are shown in Figure 9.10. The vertical axis denotes the weight variations, while the horizontal axis denotes the holding time for the measurements. The weight variations were regulated by dividing the weight change ΔW by the initial weight W. The data denoted by dashed lines show the results of

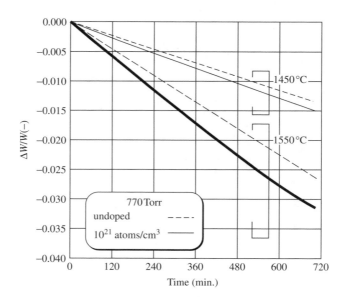

Figure 9.10 Weight variations of the boron-doped silicon melt caused by evaporation.

undoped silicon melt, while solid lines show those of 10^{21} atoms/cm^3 boron-doped silicon melt with respect to each temperature condition.

It is obvious that the weight variation of the boron-doped silicon melt is larger than that of the undoped one at 1450 °C. The increase of evaporation loss due to boron addition, however, is comparatively small at 1450 °C. The weight variations of both the boron-doped silicon melt and of the undoped one almost doubled with elevating the temperature by 100 °C from 1450 °C to 1550 °C. In particular, the influence of boron addition on the weight variation of the silicon melt is enhanced at 1550 °C compared to the result at 1450 °C. Boron assisted the evaporation of the silicon melt remarkably when the temperature of the silicon melt was high. Here it is interesting to note which chemical species is evaporating from the boron-doped silicon melt under such active evaporation conditions.

Silicon and oxygen atoms were clearly detected, while no boron was measured by EPMA. The evaporation species from the boron-doped silicon melt is thought to be silicon monoxide from these analyses. It can be concluded that boron in silicon melt enhances the evaporation of silicon monoxide. Details are given in [15].

9.3 CONCLUSION

The main results of this study are:

1. Oxygen concentration increases with increasing boron concentration in silicon melts.
2. Fused-quartz dissolution rate increases with increasing boron concentration.
3. Evaporation of silicon monoxide is enhanced with boron addition into silicon melts.

As mentioned above, the oxygen incorporation is associated with impurity doping. The impurity atoms will slightly change the structure of the oxygen-containing silicon melt. Our recent study of oxygen distribution on a wafer cut from a highly boron-doped crystal suggests that the oxygen transport is strongly affected by the presence of boron. The study of fundamental physical constants such as diffusion coefficients is now of interest. These parameters are believed to be very important to control the uniformity of large-diameter silicon crystals grown by the Cz method.

ACKNOWLEDGMENTS

The author deeply thanks Dr S. Kimura of the National Institute for Research of Inorganic Materials for his fruitful suggestion. The author is much indebted to Mitsubishi Materials Silicon Co. Ltd., Komatsu Electronic Metals Co. Ltd., and to Toshiba Ceramics Co. Ltd. for their positive support of this fundamental research of silicon melts. This work was conducted as JSPS Research for the Future Program in the Area of Atomic-Scale Surface and Interface Dynamics.

REFERENCES

1. J. Kawanishi, K. Sasaki, K. Terashima and S. Kimura: *Jpn. J. Appl. Phys.* **34** (1995) L1509.
2. S. Togawa, S. Chung, S. Kawasaki, K. Izunome, K. Terashima and S. Kimura: *J. Cryst. Growth* **160** (1996) 41.
3. S. Togawa, S. Chung, S. Kawasaki, K. Izunome, K. Terashima and S. Kimura: *J. Cryst. Growth* **160** (1996) 49.
4. K. Choe: *J. Cryst. Growth* **147** (1995) 55.
5. K. G. Barraclough, R. W. Series, D. S. Kemp and G. J. Rae: Vll Int. Conf. Crystal Growth, Stuttgart, Germany. ICCG-7, Sept. (1983) SY4/3.
6. X. Huang, K. Terashima, H. Sasaki, E. Tokizaki and S. Kimura: *Jpn. J. Appl. Phys.* **32** (1993) 3671.
7. T. Carlberg: *J. Electrochem. Soc.* **133** (1986) 1940.
8. T. Carlberg: *J. Electrochem. Soc.* **136** (1989) 551.
9. M. W. Chase, J. L. Curnutt, R. A. McDonald and A. N. Syverud: JANAF Thermo-chemical Tables, 1978 supplement, J. Phys. Chem. Ref, Darta, 7, 793 (1978).
10. A. R. Bean and R. C. Newman: *J. Phys. Chem. Solids* **32** (1971) 1211.
11. K. Abe T. Matsumoto, S. Maeda, H. Nakanishi, K. Hoshikawa and K. Terashima: *J. Cryst. Growth* **181** (1997) 41.
12. X. Huang, K. Terashima, H. Sasaki, E. Tokizaki, Y. Anzai and S. Kimura: *Jpn. J. Appl. Phys.* **33** (1994) 1717.
13. X. Huang, K. Terashima, H. Sasaki, E. Tokizaki, S. Kimura and E. Whitby: *Jpn. J. Appl. Phys.* **33** (1994) 3808.
14. X. Huang, K. Terashima, K. Izunome and S. Kimura: *Jpn. J. Appl. Phys.* **33** (1994) 902.
15. S. Maeda, M. Kato, K. Abe, H. Nakanishi, K. Hoshikawa and K. Terashima: *Jpn. J. Appl. Phys.* **36** (1997) L971.

10 Octahedral Void Defects in Czochralski Silicon

MANABU ITSUMI

NTT Lifestyle and Environmental Technology Laboratories 3-1, Morinosato Wakamiya, Atsugi-Shi, Kanagawa, 243-0198 Japan

10.1 BACKGROUND

During the past 20 years there have been several breakthroughs in the research on grown-in defects in Czochralski silicon (CZ-Si) for commercial use (Figure 10.1). The presence of oxide defects originating in CZ-Si (Figure 10.2) was reported from 1979 to 1982 (Nakajima *et al.* 1979, Itsumi *et al.* 1980, Itsumi and Kiyosumi 1982), and it was soon proposed that they could be eliminated by sacrificial oxidation (Itsumi and Kiyosumi 1982), nitrogen annealing (Kiyosumi *et al.* 1983), and hydrogen annealing (Matsushita *et al.* 1986). In those days, the origin of the defects was not clear. From 1990 onwards, when crystal-originated particles (COPs) were reported (Ryuta *et al.* 1990, Ryuta *et al.* 1992), the problem was studied in-depth (Yamagishi *et al.* 1992, Sadamitsu *et al.* 1993, Park *et al.* 1994, Itsumi 1994). Despite systematic examination of crystal-growth conditions and wafer-annealing conditions, the origin of the defects remained unclear until in 1995 octahedral void defects were found just under oxide defects (Itsumi *et al.* 1995, Miyazaki *et al.* 1995) (Figure 10.3) and in 1996 were also found in bulk CZ-Si (Ueki *et al.* 1996, Kato *et al.* 1996, Nishimura *et al.* 1996, Ueki *et al.* 1997) (Figure 10.4). Various observation methods for COP detection using particle counters (Ryuta *et al.* 1990, Ryuta *et al.* 1992), copper decoration (Itsumi and Kiyosumi 1982, Itsumi 1994, Itsumi *et al.* 1995), laser-scattering tomography (LST) (Ueki *et al.* 1996, Kato *et al.* 1996, Nishimura *et al.* 1996), sample thinning with a focused ion beam (FIB), observation with transmission electron microscopy (TEM), and other related techniques were refined and applied to many samples. It was then recognized that the octahedral void defects are a cause of oxide defects, of flow-pattern defects (FPDs), and of COPs (Yanase *et al.* 1996, Mera *et al.* 1996, Itsumi 1996, Tamatsuka *et al.* 1997, Inoue 1997a and b, Rozgonyi *et al.* 1997, Vanhellemont *et al.* 1997, Graf *et al.* 1998). These defects were then looked for in the megabit-level dynamic random access memories (RAMs) (Gonzalez *et al.* 1996, Furumura 1996, Yamamoto and Koyama

Crystal Growth Technology, Edited by H. J. Scheel and T. Fukuda
© 2003 John Wiley & Sons, Ltd.   ISBN: 0-471-49059-8

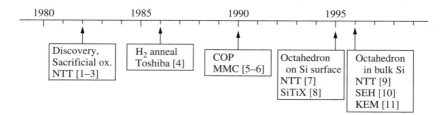

[1]   O. Nakajima, M. Itsumi, N. Shiono, and Y. Yoriume, Extended Abstract (The 26th Spring Meeting, Mar. 1979), Japan Society of Applied Physics 30p-R-1, 512 (1979) (in Japanese).
[2]   M. Itsumi, Y. Yoriume, O. Nakajima, and N. Shiono, Extended Abstract (The 27th Spring Meeting, Mar. 1980), Japan Society of Applied Physics Phys. 3p-E-1, 553 (1980) (in Japanese).
[3]   M. Itsumi and F. Kiyosumi, *Appl. Phys. Lett.* **40 (6)**, 496 (1982).
[4]   Y. Matsushita, M. Watatsuki, and Y. Saito, Extended Abstract, 18th Conf. Solid State Devices and Materials Tokyo, 529 (1986).
[5]   J. Ryuta, E. Morita, T. Tanaka, and Y. Shimanuki, *Jpn. J. Appl. Phys.* **29**, **No. 11**, L1947 (1990).
[6]   J. Ryuta, E. Morita, T. Tanaka, and Y. Shimanuki, *Jpn. J. Appl. Phys.* **31**, **Part 2**, **No. 3B**, L293 (1992).
[7]   M. Itsumi, H. Akiya, T. Ueki, M. Tomita, and M. Yamawaki, *J. Appl. Phys.* **78(10)**, 5984 (1995).
[8]   M. Miyazaki, S. Miyazaki, Y. Yanase, T. Ochiai, and T. Shigematsu, *Jpn. J. Appl. Phys.* **34**, 6303 (1995).
[9]   T. Ueki, M. Itsumi, and T. Takeda, (1996) Int. Conf. on Solid State Devices and Materials, Yokohama, LA-1, 862 (1996).
[10] M. Kato, T. Yoshida, Y. Ikeda, and Y. Kitagawara, *Jpn. J. Appl. Phys.* **35**, 5597 (1996).
[11] M. Nishimura, S. Yoshino, H. Motoura, S. Shimura, T. Mchedlidze, and T. Hikone, *J. Electrochem. Soc.* **143**, 243 (1996).

Figure 10.1   Representative research of grown-in defects in CZ-Si.

1996, Muranaka *et al.* 1997, Makabe *et al.* 1998). The octahedral void defects have since become of significant scientific interest.

10.2   OBSERVATION METHODS

The locations (2-dimensional coordinates) of COPs on a Si wafer treated with SC1 ($NH_4OH:H_2O_2:H_2O$ = 1:1:5, 75–85 °C) were detected as localized light scatterers by using a particle counter (or a surface-scanning inspection system). Atomic-force microscopy (AFM) was used to observe the shape of the COPs (Ryuta *et al.* 1990, Ryuta *et al.* 1992). Single pits or dual pits were observed. Viewed from the top, the etch pits on a (001) Si surface were square or rectangular and those on a (111) Si surface were triangular or hexagonal.

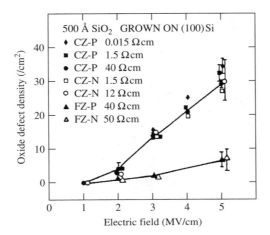

Figure 10.2 The dependence of oxide defect density on oxide electric field.

FPDs formed by vertical immersion of Si wafers in unstirred Secco etch solution can be easily observed using low-power optical microscopy (Yamagishi *et al.* 1992). We can observe a small etch-pit at the apex of a V-shaped etch feature, which is due to the vertical flow of etchant initiated by gas bubbles nucleating at the etch pit.

The locations of the oxide defects (weak spots) in thermal oxide grown on Si wafers were determined by using electrical copper decoration. The samples were then thinned to several hundred nanometers by using a FIB so that they could be observed by cross-sectional TEM (Itsumi *et al.* 1995). Copper decoration does not obliterate the shape of the oxide defects because it requires only a small electronic current if the decoration condition is optimized. The Si wafer with oxide film on its front side was immersed in methanol and the rear side of the wafer was brought downward into direct contact with a gold-coated cathode. A copper mesh anode was immersed in the liquid above the wafer and the required test voltage was applied. The voltage at the oxide surface was measured with a surface-voltage probe, and the oxide electric field strength was calculated as the applied voltage divided by the oxide thickness. Localized copper decorations at defect sites in the oxide were observed with a low-power optical microscope.

The locations (3-dimensional coordinates) of the grown-in defects in bulk Si were determined by laser-scattering tomography (LST) (Moriya *et al.* 1989, Ogawa *et al.* 1997) and then the thickness of the samples was reduced to several hundred nanometers by using FIB so that they could be observed by cross-sectional TEM (Ueki *et al.* 1996, Kato *et al.* 1996, Nishimura *et al.* 1996, Ueki *et al.* 1997). The sample-preparation procedure (Figure 10.5) is as follows: The wafers were cleaved to form specimens with the size of several millimeters by several centimeters. The locations (3-dimensional coordinates) of the grown-in defects were determined by LST. Then, four dots were made on one side

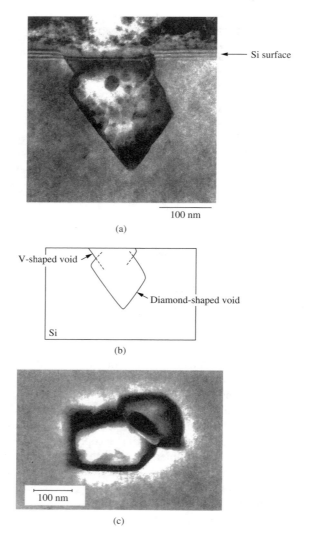

100 nm

(a)

V-shaped void

Diamond-shaped void

Si

(b)

100 nm

(c)

Figure 10.3 Octahedral-structured defects found on the surface of (001)-oriented CZ-Si: (a) cross-sectional view; (b) schematic illustration of (a); (c) plan view. (Part(a) – Reprinted from M. Itsumi *et al.*, *J. Appl. Phys.* **78/10** (1995) 5984–5988, copyright (1995) with permission from the American Institute of Physics.)

surface area surrounding the defect by using FIB (first marking). The arrangement of the four dots was then reconfirmed using LST, and a new set of four dots closer to the defect was made on the same surface (second marking). After that, the specimens were cut out using a diamond saw. Finally, the FIB thinning and TEM observations were repeated a few times so that the defects might be

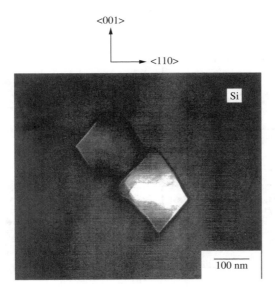

Figure 10.4 A dual-type octahedral void defect found in bulk CZ-Si. Cross-sectional view. (001)-oriented Si.

included in the specimens. The specimens were thinned by FIB to about 300 nm for TEM observation. The samples were also analyzed using energy-dispersive X-ray spectroscopy (EDS or EDX). Analysis by Auger electron spectroscopy (AES) with a more precise detection limit was also done using Ar sputtering.

It is not easy to make the specimen thin enough for TEM observation and at the same time be certain that it contains the defect. This is because the defect cannot be monitored in situ during the FIB specimen-thinning process. A stereo method was therefore developed to determine the precise location of the defects in specimens. First, FIB is used to form two dot markers on one side of the specimen and two dash markers on another side. Then the specimen is tilted by $-15°$, $0°$, $+15°$, and the defect and the two kinds of markers are observed by TEM at each of the tilt angles. The location of the defect can be calculated from the observed distances between the defect and the markers. This method enables us to obtain TEM specimens less than 200 nm thick.

10.3 CHARACTERIZATION

As shown in Figure 10.6, octahedral void defects are basically determined by eight (111) subplanes. But defects observed were often either an incomplete octahedral structure that is partially truncated by a (001) subplane (Figure 10.7), or two incomplete octahedral defects convoluting each other (dual-type void defects). This author observed twenty defects by TEM and found all of them to be

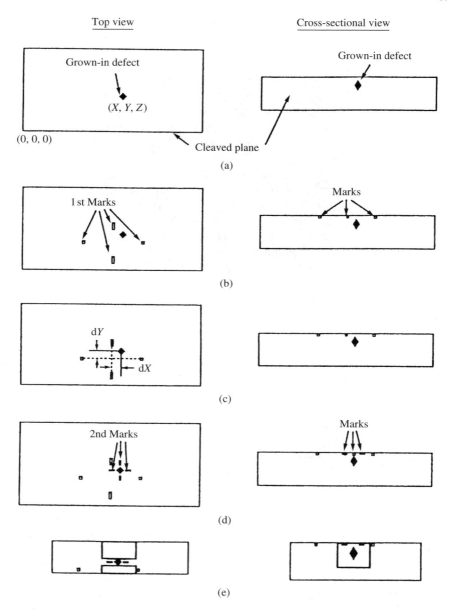

Figure 10.5 Sample-preparation procedure. (a) the first coordinate measurement with infrared tomography; (b) the first marking with a FIB; (c) the second coordinate measurement; (d) the second marking; (e) sample thinning with a FIB. (Reprinted from M. Itsumi *et al.*, *Jpn. J. Appl. Phys.* **37(1)** (1998) 1228–1235, copyright (1998) with permission from the Institute of Pure and Applied Physics.)

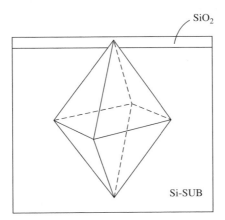

Figure 10.6 Schematic illustration of the octahedral structure.

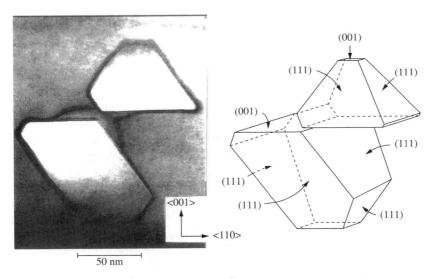

Figure 10.7 An incomplete dual-type octahedral void defect found in bulk CZ-Si.

dual-type void defects. The defects were about 0.1 μm across, the defect density per unit volume was 10^5–10^6 cm^{-3}, and the defect density per unit area was 1–10 cm^{-2}. The void defects were randomly distributed throughout the Si wafer. The defects are basically void defects, and the inner walls of the void defects are covered with oxide 2–4 nm thick. A top view of a defect – for example, Figure 10.3(c) – indicates that the shape of the defect is square-like and the sides are oriented along the [110] axes. A superimposed dual-type square structure is observed in many cases. These features are common to the crystal obtained from most prominent Si vendors because most of the vendors use similar growth conditions (for example, the growth rate for Si wafers 6–8 inches in diameter is about 1 mm/min).

The spatial relation of the two voids of the dual-type defect was examined and two types of configuration were found. In one type the two voids overlap each other (Figure 10.8) and in the other type the two voids are close to each other (Figure 10.9) (Ueki *et al.* 1998a). In the overlapping type, a hole is formed

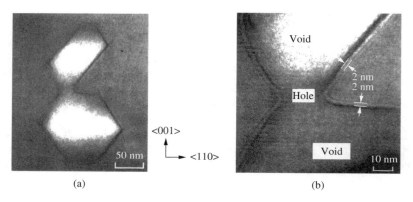

(a)   (b)

Figure 10.8 TEM micrographs of two overlapping voids: (a) cross-sectional view; (b) enlargement of the connection region.

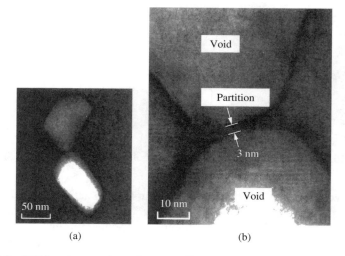

(a)   (b)

Figure 10.9 TEM micrographs of two adjacent voids: (a) cross-sectional view; (b) enlargement of the connection region. (Reprinted from M. Itsumi and T. Takeda, *Jpn. J. Appl. Phys.* **37**(1) (1998) 1667–1670, copyright (1998) with permission from the Institute of Pure and Applied Physics.)

in the connection part. It is also observed that the thickness (2–3 nm) of oxide on the side walls of one void is similar to that of oxide on the side walls of the other void. In the adjacent type, a partition layer of oxide 2–3 nm thick is located between the two voids. In this type, too, the side walls of both voids are covered with oxide 2–3 nm thick. How the two voids are generated is not clear and several models have been proposed (Kato *et al.* 1996, Tanahashi *et al.* 1997, Voronkov 1999).

Several researchers (Gonzalez *et al.* 1996, Furumura 1996, Yamamoto and Koyama 1996, Muranaka *et al.* 1997, Makabe *et al.* 1998) have found 0.1-μm dual-type or single-type square defects over the locally oxidized silicon (LOCOS) area in the memory cells of MOS RAMs. They reported that these defects cause excessive leakage current between neighboring cells and they asserted that this leakage current is more serious than that due to gate-oxide defects because the area of the LOCOS is much larger than that of the gate oxide. In any case, 0.1-μm defects will seriously affect the production yield and reliability of the next-generation Si LSIs with design rules of about 0.1 μm.

10.4   GENERATION MECHANISM

The formation of the octahedral void defects is closely related to the agglomeration of vacancies during the growth process. Two kinds of point defects in Si crystal are vacancies and self-interstitials, and their relative prevalence is determined by the growth conditions. It is widely believed that voids are vacancy-related defects, or that the agglomeration of (supersaturated) vacancies results in the formation of voids. It is also believed that the dislocation clusters are self-interstitial-related defects or the agglomeration of (supersaturated) self-interstitials results in the formation of dislocation clusters. The oxidation-induced stacking fault (OSF) ring is an important parameter: void defects as the vacancy-related defect are observed inside the OSF ring and dislocations as self-interstitials-related defects are observed outside the OSF ring. The radius of the OSF ring depends on the growth rate versus and the temperature gradient G at the solid/liquid interface of the growing crystal. When the growth rate decreases or the temperature gradient becomes larger (i.e., when v/G decreases), the radius of the OSF ring becomes small and finally the OSF ring shrinks closer to the center of Si crystal. When v/G further decreases sufficiently, the OSF ring reaches the center of the crystal and disappears there. In this case, a self-interstitial-rich region is formed throughout the wafer. When the growth rate increases or the temperature gradient becomes smaller (i.e., when v/G increases), on the other hand, the radius of the OSF ring becomes large and the OSF ring approaches the outer surface of the crystal. When v/G is sufficiently high, the OSF ring reaches the outer surface and disappears there. In this case, a vacancy-rich region is formed and void defects are observed throughout the wafer. Several researchers (deKock and Wijgert 1980, Voronkov 1982, Roksnoer 1984, Habu *et al.* 1994,

Brown *et al.* 1994, Dornberger and Ammon 1996, Nakamura *et al.* 1998, Abe 1999) have shown that there is a critical value of v/G (about 0.15 mm²/°C min in the growth-rate region from 0.3 to 5 mm/min) that separates the vacancy-rich region from the self-interstitial-rich region.

The formation process is shown schematically in Figure 10.10. First, vacancies and oxygen atoms are incorporated from the liquid phase into the solid Si crystal. As the crystal grows and cools, vacancy supersaturation is followed by vacancy agglomeration at some nucleation sites and voids begin to form when the temperature is about 1100 °C. From about 1100 °C to about 1070 °C, the size of the voids increases owing to vacancy agglomeration. Oxygen agglomeration is followed by the growth of oxide on the inner walls of the voids. As a result, 0.1-μm octahedral void defects are generated. The concentration and the diffusion coefficient of vacancies in the growing Si crystal are important factors in generating void defects.

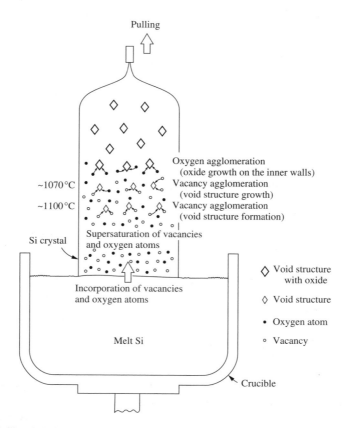

Figure 10.10 Schematic illustration of the formation of octahedral void defects during Si crystal growth.

The density and the size of the octahedral void defects strongly depend on the growth rate (i.e., cooling rate). When the cooling rate increases from 0.3 °C/min to 30 °C/min, their density increases and their size decreases (Saishoji et al. 1998).

10.5 ELIMINATION

Decreasing the Si growth rate was effective in decreasing the density of the octahedral void defects (Tachimori et al. 1990) but it increased their volume. Measures suggested to eliminate the void defects situated at the surface of Si wafers include sacrificial oxidation (Itsumi and Kiyosumi 1982), nitrogen annealing (Kiyosumi et al. 1983), hydrogen annealing (Matsushita et al. 1986), and epitaxial-layer growth. Measures suggested to eliminate void-related oxide defects after oxide growth include HCl-added oxidation (Itsumi and Kiyosumi 1982, Itsumi 1994, Itsumi et al. 1996), ion implantation through oxides (Itsumi 1994), and wafer rotation with deionized water (Itsumi 1994).

To examine the shrinkage process of the dual-type void defects, Ueki et al. prepared CZ-Si samples about 3 µm thick and annealed them in vacuum at about 1100 °C (Ueki et al. 1998b). They used ultrahigh-voltage electron microscopy (3 MeV) to observe these thick specimens and found that two voids did not shrink at the same time but that the smaller one shrunk first while the larger one remained unchanged until the smaller one had vanished (Figure 10.11). Then it too began to shrink. This selective extinction of the smaller void solely is very surprising. In addition, a detailed observation revealed that the smaller void begins to shrink from the adjacent region between the two voids. These phenomena suggest a selective emission of vacancies from the adjacent region of the smaller void into the Si bulk (or a selective injection of interstitial Si from the Si bulk into the adjacent region of the smaller void). This experimental finding might possibly imply the presence of the nucleus of the void defect. Before annealing, the thickness of the inner wall was about 2 nm for both the larger void and the smaller void. This thin wall was oxide. As the smaller void shrank, the thickness of its wall was zero while the thickness of the wall of the larger void was the same as the as-grown thickness (about 2 nm). These results suggest that the void shrinks after the side-wall oxide film disappears. The force driving void shrinkage may be related to minimizing the surface energy of the void walls during annealing. It is assumed that minimizing the surface energy first takes place selectively in the smaller void and, after the extinction of the smaller void, minimizing the surface energy takes place in the larger one. The mechanism of the preferential extinction of the adjacent region between two voids is not clear at present.

It was reported that the introduction of impurities (such as oxygen, nitrogen, or boron) into the CZ-Si significantly influences the generation of void defects (Graf et al. 1998, Ohashi et al. 1999, Kato et al. 1999). The size and density of the void defects tend to increase with increasing oxygen concentration. The generation of void defects was strongly suppressed by a boron concentration above a critical level (about 5×10^{18} atoms/cm^3). A small amount of nitrogen

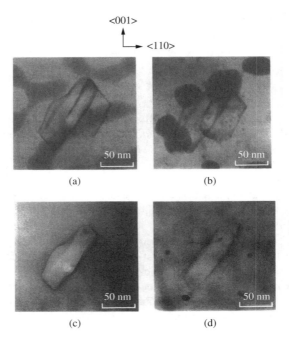

Figure 10.11 TEM micrographs of a void defect shrinking during annealing. (a) before annealing; (b) after annealing at 1080 °C for 30 min; (c) after annealing at 1090 °C for 30 min; (d) after annealing at 1110 °C for 30 min. (Reprinted from T. Ueki *et al.*, *Jpn. J. Appl. Phys.* **37**(2) (1998) L771–L773, copyright (1998) with permission from the Institute of Pure and Applied Physics.)

(about 1×10^{15} atoms/cm^3) results in the generation of smaller void defects and in the easier elimination of void defects during additional annealing. This nitrogen-doping method does not increase production cost and might deserve further study.

10.6   OXIDE DEFECT GENERATION

When octahedral void defects are situated on the Si surface (Figure 10.12), thermal oxidation is followed by the generation of oxide defects. The oxide defect density measured by using electrical copper decoration agrees well with that measured by using metal-oxide-silicon capacitor dielectric breakdown tests. The sensitivity of oxide defect detection depends strongly on the oxide thickness (Figure 10.13): oxide defects are more easily detected when the oxide thickness is from 20 to 100 nm. This strange oxide-thickness dependence, which was also reported by other researchers (Yamabe *et al.* 1983), can be explained in terms of the curvature radius of the corners of the octahedral voids and the size of the void defects themselves (Itsumi *et al.* 1998a).

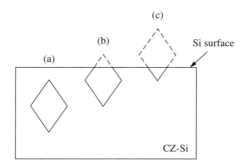

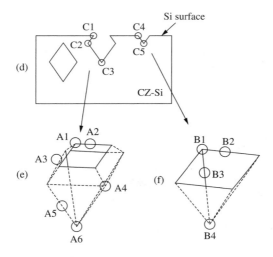

Location	Dimension	Shape
A1	3	Saddle (corner & edge)
A2	2	Edge
A3	2	Corner
A4	3	Concave
A5	2	Corner (=A3)
A6	3	Concave (=A4)
B1	3	Saddle (corner & edge)
B2	2	Edge
B3	2	Corner
B4	3	Concave

Figure 10.12 A Schematic illustration of an octahedral void defect and a Si surface. (a) when the void defect is in the bulk CZ-Si; (b) when the upper part of the void defect is truncated by the Si surface and a diamond-shaped structure appears; (c) when the bottom part of the void defect is truncated by the Si surface and a V-shaped structure appears; (d) cross section showing edges (C1 and C4) and corners (C2, C3, and C5); (e) bird's-eye view of diamond-shaped structure; (f) bird's-eye view of V-shaped structure. C1 corresponds to A1 or A2. C2 corresponds to A3 or A4. A1 is a corner and an edge. B1 is also a corner and an edge.

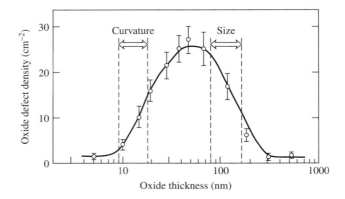

Figure 10.13 The dependence of oxide defect density on oxide thickness. Oxide electric field was 5×10^6 V/cm. This curve represents, in a sense, the detection sensitivity curve of oxide defect generation related to void defects. (Reprinted from M. Itsumi *et al.*, *J. Appl. Phys.* **84(3)** (1998) 1241–1245, copyright (1998) with permission from the American Institute of Physics.)

The curvature radius of the corners of the octahedral voids in bulk Si was about 5 nm (Figure 10.14(a)). When the octahedral voids were situated at the Si surface, the curvature radius of the corners of the octahedral voids was 10–15 nm (Figure 10.14(b)). When oxide with a thickness of less than 10 nm was grown, the effect of the corner curvature of the voids was really small. When oxide with a thickness of more than 10 nm was grown, the effect of the corner curvature of the voids became strong, inducing oxide thinning (defective spot) at the corners (Marcus and Sheng 1982, Hsueh and Evans 1983, Sunami *et al.* 1985). On the other hand, when the oxide thickness exceeded the size of the void defects (100–150 nm) (Figure 10.15), voids were filled with the oxide, eliminating the thin-oxide region (defective spot) (Figure 10.16).

In addition to the oxide thinning with the corner effect, contamination, which may be in the voids, may be related to the oxide-defect generation. HCl-added oxidation was effective in decreasing oxide defects related to octahedral void defects. This effect of HCl-added oxidation can be explained in terms of contamination gettering (or passivation) with HCl (Itsumi *et al.* 1996).

When a thick oxide was thermally grown on Si substrate, an oxide-surface pit was observed on the void defect (Itsumi *et al.* 1998, Itsumi *et al.* 1999b) (Figure 10.17). The diameter of the pit was 1 μm or less and this pit might affect the geometrical shape and electrical reliability of the aluminum interconnects on it. The number of pits (and also octahedral defects and COPs) detected by the particle counter increased with increasing oxide thickness. This shows that a thicker oxide leads to the enlargement of COPs, resulting in the number of COPs being larger than the detection limit of the particle counter. This experimental result resembles the famous findings that the number of COPs increases with

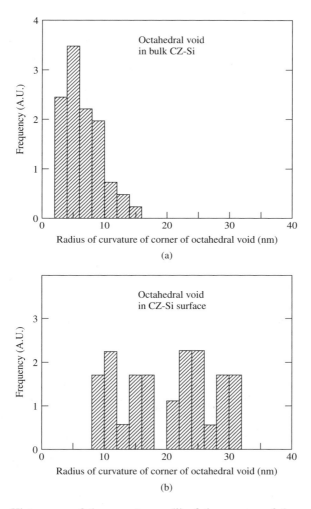

Figure 10.14 Histograms of the curvature radii of the corners of the octahedral void defects (a) in bulk CZ-Si; (b) on the surface of CZ-Si.

increasing SC1 cleaning cycles. In addition, the introduction of an epitaxial layer was effective for eliminating the oxide-surface pits with octahedral defects, which also resembles the result of SC1 cleaning.

10.7   CONCLUDING REMARKS

This review of the structural and chemical characteristics of octahedral void defects in CZ-Si outlined the mechanism of their generation, the way they

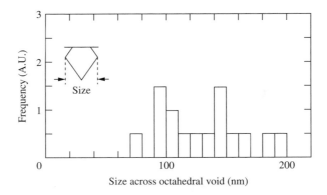

Figure 10.15 Histograms of the size of the octahedral void defects on the surface of CZ-Si.

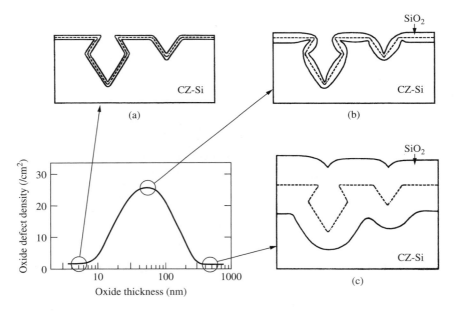

Figure 10.16 A Schematic illustration showing the dependence of oxide defect density on oxide thickness. The dotted lines represent the original Si surface before thermal oxidation. (a) when oxide thickness is less than 10 nm; (b) when oxide thickness is about 50 nm; (c) when oxide thickness is greater than 200 nm.

shrink, and how they can be eliminated. It also described a semiquantitative model explaining the noteworthy dependence of oxide defect density on oxide thickness. Many questions, however, remain. The most important concerns the dual voids. What is the mechanism of their formation? Do the two voids grow

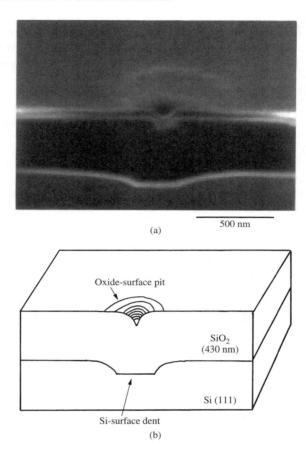

(a)

500 nm

Oxide-surface pit

SiO$_2$
(430 nm)

Si (111)

Si-surface dent

(b)

Figure 10.17   An oxide-surface pit and an underlying Si-surface dent: (a) SIM image; (b) schematic illustration [Itsumi 1999] (Reprinted from M. Itsumi *et al.*, *J. Appl. Phys.* **86**(3) (1999) 1322–1325, copyright (1999) with permission from the American Institute of Physics.)

simultaneously (from the same nucleus) or sequentially? If sequentially, what triggers the generation of the second void? Another important problem is the effect of impurities (oxygen, nitrogen, and carbon) on the generation of void defects, but we do not yet have enough information about this problem. It is necessary to examine the effect of these impurities more systematically and in more detail. This should lead to the development of a practical and economical way to grow Si with few void defects, partly because the introduction of small amounts of impurities during CZ-Si growth will not increase production cost much. And this scientific and practical study will lead to the production of the higher-quality and lower-priced large-diameter Si wafers needed for the rapidly expanding multimedia technologies.

REFERENCES

Abe T. (1999), The European Material Conference (E-MRS) Spring Meeting (Strasbourg, June), E-I.1, E-2.

Brown R. A., D. Maroudas, and T. Sinno (1994), *J. Cryst. Growth* **137**, 12.

Dornberger E. and W. von Ammon (1996), *J. Electrochem. Soc.* **143**, 1648.

Furumura Y. (1996), Proceedings of The 2nd International Symposium on Advanced Science and Technology of Silicon Materials (Nov. (1996), Hawaii), 418.

Gonzalez F., G. A. Rozgonyi, B. Gilgen, and R. Barbour (1996), *High Purity Silicon IV (ECS)*, C. L. Claeys, P. Rai-Choudhury, P. Stallhofer, J. E. Maurits, (eds), (The Electrochemical Society, Pennington, N.J., USA) 357.

Graf D., M. Suhren, U. Lambert, R. Schmolke, A. Ehlert, W. von Ammon, and P. Wagner (1998), *J. Electrochem. Soc.* **145**, No. 1, 275.

Habu R., T. Iwasaki, H. Harada, and A. Tomiura (1994), *Jpn. J. Appl. Phys.* **33**, 1234.

Hsueh C. H. and A. G. Evans (1983), *J. Appl. Phys.* **54(11)**, 6672.

Inoue N. (1997a), *Oyo-Buturi* (A Monthly Publication of The Japan Society of Applied Physics) **66**, No. 7, (ISSN 0369–8009), 715 (in Japanese).

Inoue N. (1997b), Proceedings of The Kazusa Akademia Park Forum on The Science and Technology of Silicon Materials Nov. 12–14, Chiba, edited by K. Sumino, 135.

Itsumi M., Y. Yoriume, O. Nakajima, and N. Shiono (1980), Extended Abstract (The 27th Spring Meeting, Mar. 1980), Japan Society of Applied Physics 3p-E-1, 553 (in Japanese).

Itsumi M. and F. Kiyosumi (1982), *Appl. Phys. Lett.* **40(6)**, 496.

Itsumi M. (1994), *J. Electrochem. Soc.* **141**, No. 9, 2460.

Itsumi M., H. Akiya, T. Ueki, M. Tomita, and M. Yamawaki (1995), *J. Appl. Phys.* **78(10)**, 5984.

Itsumi M. (1996), *Symposium Proceedings of Material Research Society* Volume 442, 'Defects in Electronic Materials II' (edited by J. Michel, Dec. 1996, Boston) 95.

Itsumi M., H. Akiya, T. Ueki, M. Tomita, and M. Yamawaki (1996), *J. Appl. Phys.* **80**, No. 12, 6661.

Itsumi M., M. Maeda, and T. Ueki (1998a), *J. Appl. Phys.* **84**, No. 3, 1241.

Itsumi M., Y. Okazaki, M. Watanabe, T. Ueki, and N. Yabumoto (1998b), *J. Electrochem. Soc.* **145**, No. 6, 2143.

Itsumi M., M. Maeda, T. Ueki, and S. Tazawa (1999), *J. Appl. Phys.* **86**, No. 3, 1322.

Kato M., T. Yoshida, Y. Ikeda, and Y. Kitagawara (1996), *Jpn. J. Appl. Phys.* **35**, 5597.

Kato M., M. Tamatsuka, M. Iida, H. Takeno, T. Otogawa, and T. Masui (1999), Extended Abstracts (The 46th Spring Meeting, March, 1999); The Japan Society of Applied Physics and Related Societies 29a-ZB-7, 470. (in Japanese)

Kiyosumi F., H. Abe, H. Sato, M. Ino, and K. Uchida (1983), Denshi-Tsushin Gakkai, Technical Report SSD 83–66, 1 (1983).

Kock A. J. R. de and W. M. van de Wijgert (1980), *J. Cryst. Growth* **49**, 718.

Makabe K., M. Miura, M. Kawamura, M. Muranaka, H. Kato, and S. Ide (1998), Extended Abstracts (The 45th Spring Meeting, 1998); The Japan Society of Applied Physics and Related Societies 30p-YA-5, 413 (in Japanese).

Marcus R. B. and T. T. Sheng (1982), *J. Electrochem. Soc.* **129**, No. 9, 1278.

Matsushita Y., M. Watatsuki, and Y. Saito (1986), Extended Abstract, 18th Conf. Solid State Devices and Materials, Tokyo, 529.

Mera T., J. Jablonski, M. Danbata, K. Nagai, and M. Watanabe (1996), Symposium proceedings of Material Research Society Volume 442, 'Defects in Electronic Materials II' (edited by J. Michel, Dec. 1996, Boston) 107.

Miyazaki M., S. Miyazaki, Y. Yanase, T. Ochiai, and T. Shigematsu (1995), *Jpn. J. Appl. Phys.* **34**, 6303.

Moriya K., K. Hirai, K. Kashima, and S. Takasu (1989), *J. Appl. Phys.* **66**, 5267.

Muranaka M., M. Miura, H. Iwai, M. Kawamura, Y. Tadaki, and T. Kaeriyama (1997), Extended Abstracts of the 1997 International Conference on Solid State Devices and Materials, Hamamatsu, 396.

Nakajima O., M. Itsumi, N. Shiono, and Y. Yoriume (1979), Extended Abstract (The 26th Spring Meeting, Mar. 1979), Japan Society of Applied Physics 30p-R-1, 512 (in Japanese).

Nakamura K., T. Saishoji, and J. Tomioka (1998), *Electrochem. Soc. Proc.* **98–13**, 41.

Nishimura M., S. Yoshino, H. Motoura, S. Shimura, T. Mchedlidze, and T. Hikone (1996), *J. Electrochem. Soc.* **143**, 243.

Ogawa T., G. Kissinger, and N. Nango (1997), *Electrochem. Soc. Proc.* **97–12**, 132.

Ohashi W., A. Ikari, Y. Ohta, A. Tachikawa, H. Deai, H. Yokota, and T. Hoshino (1999), Extended Abstracts (The 46th Spring Meeting, March, 1999); The Japan Society of Applied Physics and Related Societies 29a-ZB-1, 468 (in Japanese).

Park J. G., H. Kirk, K.-C. Cho, H.-K. Lee, C.-S. Lee, and G. A. Rozgonyi (1994), Semiconductor Silicon, edited by H. R. Huff, W. Bergholz, and K. Sumino (Electrochem. Soc., Pennington, 1994) p. 370.

Rozgonyi G., M. Tamatsuka, Ki-Man Bae, and F. Gonzalez (1997), Proceedings of The Kazusa Akademia Park Forum on The Science and Technology of Silicon Materials Nov. 12–14, Chiba, edited by K. Sumino, 215.

Roksnoer P. J. (1984), *J. Cryst. Growth* **68**, 596.

Ryuta J., E. Morita, T. Tanaka, and Y. Shimanuki (1990), *Jpn. J. Appl. Phys.* **29, No. 11**, L1947.

Ryuta J., E. Morita, T. Tanaka, and Y. Shimanuki (1992), *Jpn. J. Appl. Phys.* **31, Part 2, No. 3B**, L293.

Sadamitsu S., S. Umeno, Y. Koike, M. Hourai, S. Sumita, and T. Shigematsu (1993), *Jpn. J. Appl. Phys.* **32**, 3675.

Saishoji T., K. Nakamura, H. Nakajima, T. Yokoyama, F. Ishikawa, and J. Tomioka (1998), *Electrochem. Soc. Proc.* **98–13**, 28.

Sunami H., T. Kure, K. Yagi, Y. Wada, and K. Yamaguchi (1985), *IEEE J. Solid-State Circuits* **SC-2, No. 1**, 220.

Tachimori M., T. Sakon, and T. Kaneko (1990), The Japan Society of Applied Physics and Related Societies, The 7th Kessho Kougaku (Crystal Engineering) Symposium, 27. (in Japanese)

Tamatsuka M., Z. Radzimski, G. A. Rozgonyi, S. Oka, M. Kato, and Y. Kitagawara (1997), Extended Abstracts of the 1997 International Conference on Solid State Devices and Materials, Hamamatsu, 392.

Tanahashi K., N. Inoue, and Y. Mizokawa (1997), *Mater. Res. Soc. Symp. Proc.* **442**, 131.

Ueki T., M. Itsumi, and T. Takeda (1996), Int. Conf. on Solid State Devices and Materials, Yokohama, LA-1, 862.

Ueki T., M. Itsumi, and T. Takeda (1997), *Jpn. J. Appl. Phys.* **36, Part 1, No. 3B**, 1781.

Ueki T., M. Itsumi, and T. Takeda (1998a), *Jpn. J. Appl. Phys.* **37, Part 1, No. 4A**, 1667.

Ueki T., M. Itsumi, K. Yoshida, A. Takaoka, T. Takeda, and S. Nakajima (1998b), *Jpn. J. Appl. Phys.* **37, Part 2, No. 7A**, L771.

Vanhellemont J., E. Dornberger, D. Graf, J. Esfandyari, U. Lambert, R. Schmolke, W. von Ammon, and P. Wagner (1997), Proceedings of The Kazusa Akademia Park Forum on The Science and Technology of Silicon Materials Nov. 12–14, Chiba, edited by K. Sumino, 173.

Voronkov V. V. (1982), *J. Cryst. Growth* **59**, 625.

Voronkov V. V. (1999), The European Material Conference (E-MRS) Spring Meeting, (Strasbourg, June) E-IV.1, E-6.

Yamabe K., K. Taniguchi, and Y. Matsushita (1983), in Proceedings of the International Reliability in Physics Symposium, 184.

Yamagishi H., I. Fusegawa, N. Fujimaki, and M. Katayama (1992), *Semicond. Sci. Technol.* **7**, A135.

Yamamoto H. and H. Koyama (1996), Proceedings of the 2nd International Symposium on Advanced Science and Technology of Silicon Materials (Nov. 1996, Hawaii), 425.

Yanase Y., T. Kitamura, H. Horie, M. Miyazaki, and T. Ochiai (1996), Extended Abstracts (The 43rd Spring Meeting, 1996); The Japan Society of Applied Physics and Related Societies 26p-X-5, 184.

11 The Control and Engineering of Intrinsic Point Defects in Silicon Wafers and Crystals

R. FALSTER[1], V. V. VORONKOV[2] and P. MUTTI[2]

[1] MEMC Electronic Materials SpA, Viale Gherzi 31, 28100 Novara, Italy
[2] MEMC Electronic Materials SpA, via Nazionale 59, 39012 Merano, Italy

ABSTRACT

The control of intrinsic point defects in the growth of silicon crystals and the processing of silicon wafers is an increasingly important aspect of the manufacture of silicon materials for the microelectronic industry. This chapter summarizes many aspects of the problem of the incorporation of excess intrinsic point defects and reactions that produce harmful agglomerated defects and the enhanced precipitation of oxygen under certain conditions. Recent advances have led to the production of microdefect-free large-diameter silicon crystals. Point-defect engineering in thin silicon wafers has also led to an important advance in the control of oxygen precipitation for IC applications. Proper control of vacancy-concentration profiles installed during rapid thermal processing of silicon wafers effectively programs silicon wafers to produce robust oxygen-precipitate distributions ideal for internal gettering applications. The intrinsic point defect processes that underlie defect engineering in both crystal growth and wafer processing are the same and results from experiments in both can be joined to give useful information toward a coherent, unified picture of the intrinsic point-defect parameters and reactions.

11.1 INTRODUCTION

Increasing levels of integration and linewidth shrinking in integrated circuit design along with reduced thermal budgets IC processing are placing increasing demands on the silicon substrates used in their manufacture. If the polished Czochralski-grown silicon wafer – as opposed to epitaxial substrates – is to survive as a suitable electronic material for new generations of advanced IC devices, then it appears that new, cost-effective, approaches to the engineering of defects in CZ-grown silicon must be implemented. The problem is two-fold: (1) two

Crystal Growth Technology, Edited by H. J. Scheel and T. Fukuda
© 2003 John Wiley & Sons, Ltd.   ISBN: 0-471-49059-8

classes of defects *intrinsic* to melt-grown silicon, vacancy- and interstitial-type agglomerated defects and (2), the control of the precipitation of oxygen. Intrinsic point-defect-related agglomerates or *microdefects* are formed during the crystal growth process. The precipitation of oxygen occurs during processing of silicon wafers into ICs, but with a strong connection and coupling to details of crystal-growth processes and wafer heat treatments preceding device processing. These defect types are not new to silicon technology but the demands on their control have increased dramatically in recent years. Furthermore, as the silicon industry matures, there is a need to simplify material-selection processes. Uncertainty of performance must be eliminated and with it the need to highly *tailor* silicon products to specific applications. The complications that arise out of the complex interactions of defects from initial solidification of the silicon crystal ingot through device processing add undesirable costs and rigidity to the use of silicon as an electronic material. Native point defects play the key role in these problems. Control and engineering of these during both crystal-growth and wafer processing can produce solutions that offer significant simplifications in the use of CZ silicon polished wafers in advanced IC applications.

11.1.1 VACANCY-TYPE DEFECTS

The first challenge lies with a defect unique to silicon grown from the melt, an agglomeration of excess vacancies into a rather low-density (typically approximately $10^6 \, \mathrm{cm}^{-3}$ for the case of CZ silicon) void-type microdefect (Ueki *et al.* 1997). If present, this type of defect can result in the failure of gate oxides and they are of increasing importance in device geometries on the order of the void sizes (on the order of 100–150 nm), isolation failures and other topological faults (Park *et al.* 1999). Depending on the mode of detection, void-type defects are also referred to as D-defects, COPs, flow pattern defects, light-scattering tomography defects and GOI-defects.

11.1.2 SILICON SELF-INTERSTITIAL-TYPE DEFECTS

There exists another and, if anything, even more harmful defect type unique to silicon grown from a melt: agglomerates of excess silicon self-interstitials into large dislocation loops. These defects are sometimes referred as A-defects or large etch pit defects. Their density is generally far lower than that of the void microdefects but their size is usually orders of magnitude larger – hence their importance.

11.1.3 THE PRECIPITATION OF OXYGEN

The control of the behaviour of oxygen in silicon is undeniably one of the most important challenges in semiconductor materials engineering. In the 20 or so years since the discovery of the internal gettering effect in silicon wafers, many scientists and engineers have struggled with the problem of precisely and reliably controlling the precipitation of oxygen in silicon that occurs during the processing of wafers into integrated circuits (ICs). This has been met with only partial success in the sense that the 'defect engineering' of conventional silicon wafers is still, by and large an *empirical* exercise. It consists largely of careful, empirical *tailoring* of wafer type (oxygen concentration, crystal-growth method, and details of any additional pre-heat treatments, for example) to match the *specific* process details of the application to which they are submitted in order to achieve a good *and* reliable internal gettering (IG) performance. Reliable and efficient IG requires the robust formation of oxygen-precipitate-free surface regions (denuded zones) and a bulk defective layer consisting of a minimum density (at least about 10^8 cm^{-3} (Falster *et al.* 1991)) oxygen precipitates during the processing of the silicon wafer. Uncontrolled precipitation of oxygen in the near surface region of the wafer represents a risk to device yield.

11.2 THE CONTROL OF THE AGGLOMERATION OF INTRINSIC POINT DEFECTS DURING CRYSTAL GROWTH

11.2.1 THE v/G RULE FOR THE TYPE OF GROWN-IN MICRODEFECTS

Early studies of grown-in microdefects in float-zoned silicon (de Kock 1973) showed that swirl-microdefects (A- and B-defects) disappear if the growth rate v exceeds some threshold value. Crystals grown at larger v were found to contain a different kind of grown-in microdefect, D-defects (Veselovskaya *et al.* 1977, Roksnoer and van den Boom 1981). The threshold (or critical) growth rate for the change-over from A/B defects to D-defects was found to be proportional to the near-interface axial temperature gradient G (Voronkov 1982, Voronkov *et al.* 1984). In other words, the type of grown-in microdefects is controlled simply by the ratio v/G. Swirl (interstitial type) defects are formed if v/G is below some universal critical ratio ξ_t and D-defects are formed otherwise. This simple 'v/G rule' holds both for float-zoned and Czochralski-grown crystals (Voronkov and Falster 1998), in spite of a great difference in the oxygen content. The physical meaning of this rule is very simple (Voronkov 1982): the type of intrinsic point defects incorporated into a growing crystal is controlled by the parameter v/G, according to the defect transport equations for diffusion, convection and annihilation of point defects in the vicinity of the interface.

Growing a crystal at $v/G > \xi_t$ results in incorporation of vacancies, while the interstitial concentration is undersaturated and decays fast due to recombination with vacancies. The vacancies agglomerate into voids (D-defects) on further cooling – if the vacancy concentration is not too low (Voronkov and Falster 1998). The identification of D-defects with voids was recently demonstrated (Ueki *et al.* 1997, Nishimura *et al.* 1998). At low vacancy concentration (for instance, at v/G

only slightly larger than the critical ratio ξ_t) the dominant agglomeration path is production of oxide particles instead of voids (Voronkov and Falster 1998). Accordingly the main vacancy region of a crystal (that containing voids) is surrounded with a narrow band of oxide particles (P-band). The presence of P-band is manifested by the formation of ring-distributed stacking faults during high-temperature oxidation of wafers. Beside this OSF ring (P-band) there is a well-pronounced band of still lower vacancy concentration (the L-band) where vacancy agglomeration is suppressed, and appreciable vacancy concentration is frozen-in to enhance oxygen precipitation (Falster *et al.* 1998b).

Growing a crystal at $v/G < \xi_t$ results in incorporation of self-interstitials, while the vacancy concentration is undersaturated and decays rapidly. The self-interstitials agglomerate into A/B-swirl-defects. The A-defects were found to be extrinsic dislocation loops (Foell and Kolbesen 1975) while B-defects seem to be small globular clusters of interstitials (Petroff and de Kock 1975) (most likely, including carbon interstitials). Particularly, at low interstitial concentration only B-defects are formed. At still lower concentrations, no defects are formed. Accordingly, the main interstitial region of a crystal (that containing A-defects) is surrounded with a band of B-defects. This marginal B-band of interstitial region is analogous to the P-band of vacancy region.

Some crystals, due to axial and/or radial variation in v/G, consist of well-separated vacancy and interstitial regions. The boundary between these regions is approximately marked by the position of the OSF ring, and still better marked by the position of the L-band. An example of this defect-banding phenomenon is shown in Figure 11.1.

11.2.2 ALTERNATIVE VIEWS TO THE v/G RULE

There is an opinion (Abe 1999) that the type of microdefects, either A/B-defects or D-defects (that is, the type of incorporated point defects, either interstitial

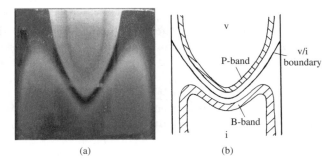

(a)   (b)

Figure 11.1 (a) An axial cross section of an etched cu-decorated CZ crystal section at the transition from vacancy-type defect growth conditions (the upper 'U-shaped' section) and interstitial-type defect conditions (the lower 'M-shaped' structure); (b) A schematic diagram of the structure.

or vacancy) is controlled simply by the value of the near-interface temperature gradient G: vacancies at lower G, interstitials at higher G. This notion is based on the fact that increasing the growth rate v is accompanied by some decrease in G. It was then thought (Abe 1999) that the actual reason for the change-over from interstitial to vacancy is not an increase in v (in v/G) but a decrease in G. This notion is certainly not true since the change-over in early float-zoned crystals occurs at very high values of $G(v)$, about 300 K/cm (de Kock 1973). A similar change-over in modern Czochralski crystals occurs at much smaller gradient, about 30 K/cm (Falster et $al.$ 1998b). In both cases the change-over occurs at about the same v/G ratio (around 0.15 mm^2/min K).

It should also be remarked that the measured (decreasing) dependence of G on v was considered (Abe 1999) as contradicting the heat balance equation at the interface ($Qv = \chi G - \chi'G'$ where Q is the heat of fusion per a unit volume of crystal and χ is the heat conductivity of crystal; χ' and G' refer to the melt). Actually there is no contradiction at all. The temperature field in the crystal, including the value of G, is completely decoupled from the above heat balance – if the interface shape is specified. This is so because the interface temperature is fixed: it is the melting temperature of 1412 °C. This boundary condition, together with the heat loss boundary condition at the crystal surface, defines the temperature field of the crystal. Increasing v means that

(a)   The convection heat flux from the interface into the crystal is increased, and all the crystal bulk becomes a little hotter; accordingly G is somewhat decreased. The computed function $G(v)$ for float-zoned crystals (Voronkov et $al.$ 1984) decreases on v only slightly, by less than 30 % within the relevant range of v. The measured gradient showed no resolvable dependence on v (Voronkov et $al.$ 1984).

(b)   The melt temperature gradient G' should be remarkably diminished, by a proper increment of the heater power, to keep the heat balance at the interface. The required decrease in $G' = (\chi G - Qv)/\chi'$ is caused both by an increase in v and a decrease in $G(v)$.

It is therefore clear that the decreasing dependence of G on v is a 'normal' (though not very significant) effect that casts no doubt on the v/G rule for the interstitial to vacancy change-over.

11.2.3   VOID REACTION CONTROL

One approach to an engineering improvement of the microdefect problem is to grow crystals in the vacancy mode but to engineer the reactions that produce voids. It is found that the resultant density of voids is proportional to the factor $q^{3/2}C_v^{-1/2}$ (Voronkov and Falster 1998), where q is the cooling rate at the temperature at which the reaction occurs and C_v the local concentration of vacancies. The v/G model (Voronkov 1982), although only a one-dimensional model, usually gives C_v accurately at distances greater than about a centimeter or so from the crystal surface. For a more complete and detailed picture, numerical simulation must be performed. The void reaction occurs over a very narrow temperature range (about 5 K) at a temperature that depends on C_v (Voronkov and Falster 1998). The range of typical reaction temperatures lies between about 1000 and 1100 °C (Falster *et al.* 1998b, Puzanov and Eidenzon 1992, Saishoji *et al.* 1998). The problem of reducing void defect density is thus a coupled engineering problem. A growing silicon ingot is a rigid object. One must engineer simultaneously the coupled problems of the incorporation of vacancies near the growth interface (v and G at about the melt temperature) and the cooling rate at the reaction temperature (v and G at a temperature that depends on C_v). This can be done, but for largely practical reasons, only about a factor of 10 reduction in void density is possible. For some applications, this is not enough improvement to insure device yields that are not limited by the starting substrate defectivity.

11.2.4 PERFECT SILICON

A more robust solution to the microdefect problem and one that circumvents the question and resultant complications of whether or not the material is sufficiently improved to meet the needs of an arbitrary IC process is not to control the microdefect reaction, but to suppress it. As described above, there exist two marginal bands straddling the transition from vacancy- to interstitial-type silicon in which the reactions that produce microdefects do not occur. In these regions, the concentrations of either vacancies or self-interstitials are too low to drive the reactions during the cooling of the crystal. They differ in character due to the differing effects of unreacted vacancies or interstitials on the subsequent nucleation of oxygen clusters at lower temperatures. Excess vacancies enhance the clustering, while excess interstitials suppress it. In terms of process control, the width in v/G space around the critical value that produces sufficiently low concentrations of both vacancies or interstitials is on the order of 10 %. Growth processes that can control v/G to within 10 % around the critical value both axially and radially are capable of producing microdefect-free silicon. This is not an easy task. Methods exist that result in the partial relaxation of this requirement such that the production of such 'perfect silicon' is practically possible along nearly the entire crystal length. In general, three types of perfect silicon are possible: (1) entirely vacancy-type, exhibiting a generally enhanced oxygen precipitation, (2) entirely interstitial-type perfect material in which oxygen precipitation is suppressed and (3) a mixed type. Figure 11.2 illustrates the difference between sections of vacancy defective, interstitial defective and interstitial-perfect silicon when decorated by copper.

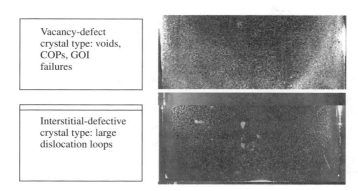

Figure 11.2   Images of silicon crystal cross sections after copper decoration and etching.

11.3   THE CONTROL OF OXYGEN PRECIPITATION THROUGH THE ENGINEERING OF VACANCY CONCENTRATION IN SILICON WAFERS: MAGIC DENUDED ZONE™ WAFERS

In the discussion above, it was noted that there exists a region of vacancy concentration accessible through crystal-growth in which no microdefects are formed but in which the clustering of oxygen is significantly enhanced. It is also possible to achieve such levels of vacancy concentration in thin wafers through the proper control of their heating and cooling. By doing so, it is possible to strongly affect the subsequent precipitation behaviour of the wafer. In fact, it is possible, through the judicious control of point-defect generation, injection, diffusion and recombination to install vacancy-concentration profiles into silicon wafers that result in the ideal precipitation performance for internal gettering (IG) purposes.

While high concentration of vacancies enhance oxygen clustering, it is found that there is a lower bound on vacancy concentration for which clustering is 'normal'. This is quite a sharp transition and lies around about $5 \times 10^{11}\,\mathrm{cm}^{-3}$. Installing a vacancy concentration profile that rises from the wafer surface into the bulk of the wafer crossing the critical concentration at some desired depth from the surface results in a wafer with a surface region of 'normal' silicon and a bulk of vacancy-enhanced precipitation. This is the basis underlying the 'magic denuded zone™' (or MDZ™) wafer (Falster *et al.* 1998a).

11.3.1   'TABULA RASA' SILICON AND THE SUPPRESSION OF OXYGEN PRECIPITATION IN LOW-VACANCY-CONCENTRATION MATERIAL

In modern low-vacancy silicon, oxygen precipitation is suppressed if grown-in pre-existing clusters are dissolved by simple high-temperature heat treatments. Such material is known as 'Tabula Rasa' silicon (Falster *et al.* 1997).

The dissolution of grown-in clusters is an integral part of the thermal treatments necessary to install the required vacancy concentration profile. From the 'Tabula Rasa' state, re-nucleation of stable oxygen clusters at low temperatures (approximately 450–700 °C) is inhibited in almost all practical cases by the requirement of relatively long incubation times, even though the oxygen concentration remains high. Thus 'normal' precipitating material actually means the suppression of precipitation in most practical situations. In the high vacancy concentration regions, the incubation times are reduced over a wide temperature range to very small values (Falster *et al.* 1998a). It is this very large difference in incubation times between high and low vacancy concentrations that is primarily responsible for the creation of the denuded zone effect.

Through control of process temperature, ambient and cooling rate a wide variety of precipitate profiles can be installed in thin silicon wafers. An example of the precipitate profiles achieved from two different installed vacancy-concentration profiles along with the resulting oxygen precipitate profiles is shown in Figure 11.3. Vacancy-concentration profiles were measured by a

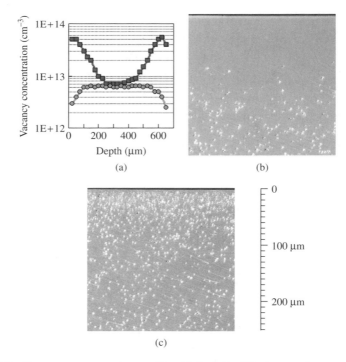

Figure 11.3 Vacancy concentration profiles (a) from Pt diffusion results on two samples heat treated at 1250 °C for 30 s in a nitrogen (squares) and argon (circles) ambient. Following this the wafers were given a precipitation test treatment (4 h, 800 °C + 16 h, 1000 °C); (b) argon treatment; (c) nitrogen treatment. The bulk precipitate densities in both cases are approximately $5 \times 10^{10}\,\mathrm{cm^{-3}}$.

Pt-diffusion technique (Jacob *et al.* 1997). In low vacancy concentration surface regions, the Pt technique is not very accurate due to competition with the kick-out mechanism for platinum incorporation.

11.3.2 MATERIAL 'SWITCHING' AND TRANSFER FUNCTIONS

The transition in vacancy concentration between 'normal' and 'enhanced' precipitation is very sharp indeed. Figure 11.4 is a plot of current estimates of the density of oxygen precipitates produced by a simple 'test' heat treatment (4 h at 800 °C followed by 16 h at 1000 °C) as a function of vacancy concentration at the outset of the treatment. The width of the bar represents some uncertainty in the determination of vacancy concentration. Such a sharp transition represents a kind of material-switching action and the curve a kind of material-transfer function that is nearly binary in nature. This switching action can be used to effectively 'program' silicon wafers to produce sharply defined zones of precipitation behaviour.

11.3.3 COMPARISON OF CONVENTIONAL AND VACANCY-
ENGINEERED CONTROL OF OXYGEN PRECIPITATION

Conventionally, denuded zones are prepared through the out-diffusion of oxygen by long, high-temperature heat treatments prior to the 'normal' (low vacancy concentration) very slow nucleation of oxygen precipitates. It relies on the very strong (typically 20-th order (Falster *et al.* 1998a)) oxygen concentration, dependence

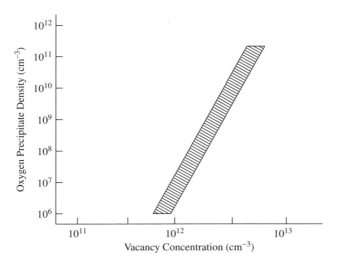

Figure 11.4 Oxygen precipitate density dependence on vacancy concentration.

of the nucleation processes. This very strong dependence on oxygen concentration while responsible for a denuded zone effect is, in fact, one of the major weaknesses of the conventional approach; it necessitates extremely tight controls on the concentration of oxygen in silicon crystal growth. A further difficulty is that with modern low thermal budget IC processing the times and temperatures required to produce sufficient out-diffusion of oxygen simply no longer exists. Another difficulty is the strong coupling of the 'normal' nucleation behaviour of the details of the crystal-growth process and the thermal cycles of the IC process to which it is submitted. The problem of the 'normal' nucleation of oxygen precipitation in silicon has proven to be an extraordinarily difficult one and has yet to be satisfactorily resolved. No models exist that can predict the outcome of this highly coupled problem. Thus the problem of precipitation control in 'normal' silicon remains a largely and unsatisfactorily empirical exercise in which silicon materials are 'tailored' to meet the needs of specific applications adding undesired costs and rigidity to the process.

In vacancy-engineered oxygen-precipitation control these problems are circumvented. The two-zoned precipitation required for internal gettering action is achieved by the installation of a vacancy concentration profile that rises from a low value at the wafer surface and crosses the vacancy-threshold concentration at some desired point below the surface defining the depth of the denuded zone. The two processes for the creation of denuded zones are schematically illustrated and compared in Figure 11.5.

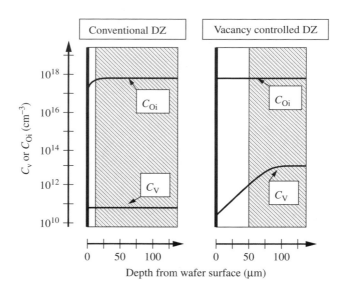

Figure 11.5 Schematic comparison of conventional and the vacancy-controlled denuded zone formation of the MDZ^{TM} wafer. The shaded areas represent the locations of the precipitated regions.

11.3.4   THE INSTALLATION OF VACANCY CONCENTRATION
PROFILES IN THIN SILICON WAFERS

The installation of appropriate vacancy-concentration profiles in a thin silicon wafers is a three-step process, all of which occur in a single rapid thermal processing (RTP) step (Falster *et al.* 1998a). (1) When silicon is raised to high temperatures, vacancies and interstitials are spontaneously produced in equal amounts through Frenkel pair generation, a very fast reaction. At distances far removed from crystal surfaces we thus have $C_I = C_V = \{C_{Ieq}(T)C_{Veq}(T)\}^{1/2}$, where T is the process temperature. Were the sample to be cooled at this point the vacancies and interstitials would merely mutually annihilate each other in the reverse process of their generation. (2) In thin wafers, however, the surfaces are not far away and this situation changes very rapidly. Assuming equilibrium boundary conditions (not oxidizing or nitriding) leads to coupled fluxes of interstitials to the surface and vacancies from the surface ($C_{Ieq}(T) < C_{Veq}(T)$ for the temperature ranges of interest (see below)) and the rapid establishment of equilibrium conditions throughout the thickness of the wafer. Experiments suggest that this occurs very rapidly – in a matter of seconds or even less. This equilibration is primarily controlled by the diffusivity of the fastest component, the self-interstitials, since the concentrations are roughly equal. (3) Upon cooling, two processes are important: direct recombination of vacancies and interstitials, and surface recombination and the resulting fluxes toward the surfaces. The slower vacancies are now the dominant species of the coupled diffusion ($C_V > C_I$) and hence the equilibration processes toward the equilibrium state at the surface is not as fast as the interstitial-dominated initial equilibration. It is thus possible to freeze-in excess vacancies at not unreasonable cooling rates. For samples cooled rapidly, the residual bulk concentration of vacancies following recombination with interstitials, C_V, is the initial difference $C_{Ve} - C_{Ie}$ (at the process temperature, T). Closer to the surfaces C_V is lower due to diffusion (again now primarily controlled by the dominating vacancies) toward the decreasing equilibrium values at the wafer surface. The relatively rapid cooling rates achievable (50–100 °C/s) in RTP systems insure that sufficiently high concentrations of vacancies can be frozen in the bulk of the wafer to achieve the desired effect. In general, the level of bulk precipitation is controlled by the value of C_V (determined by the process temperature) while the depth of the denuded zone is controlled by the diffusion of vacancies during cooling. Such information, coupled with numerical simulation of the coupled diffusion process, can be very important points of reference in analysing the parameters of the native point defects in silicon. For example, critical vacancy concentrations ($= C_{Ve}(T) - C_{Ie}(T)$) for precipitation enhancement (approximately $5–8 \times 10^{11}$ cm^{-3}), is achieved at about 1175 °C, while at about 1250 °C, much higher vacancy concentration is found, to be between about $2–5 \times 10^{12}$ cm^{-3}. The uncertainty in these numbers reflect our uncertainty in the platinum-diffusion technique. Further experiments with various cooling rates and subsequent relaxation rates of an installed profile during a second RTP process at a different temperature give insight into the diffusivity of vacancies at various temperatures. Such information gained from MDZ experiments is coupled with information gleaned from crystal growth experiments to compile a unified picture of the point-defect parameters. This is discussed below.

11.3.5   ADVANTAGES OF THE USE OF VACANCIES TO CONTROL OXYGEN PRECIPITATION IN WAFERS

The use of MDZ wafers is a very powerful tool for the resolution of the problem of oxygen precipitation control in silicon wafers. The installation of an appropriate vacancy-concentration profile effectively programs a silicon wafer to behave in a well-defined and ideal way. Essentially all the difficulties associated with the robust engineering of IG structures are removed at a stroke. At sufficiently high vacancy concentration levels the control of the nucleation reactions is decoupled from the oxygen concentration over the entire practical CZ range $(4-10 \times 10^{17} \, \text{cm}^{-3})$ thus eliminating the need for tight oxygen control in CZ crystal growth. The 'Tabula Rasa' character of the heat treatments that produce the MDZ wafer erase all details of oxygen clustering during crystal-growth. Finally, the extraordinarily rapid incubation of clusters at low temperatures insures that the installed vacancy 'template' is 'fixed' essentially as wafers are loaded into their first high-temperature heat treatment. The vacancies are consumed in this process and only the clusters in their well-defined distribution are left behind to grow. No further nucleation takes place, except in extraordinary circumstances. In short, an MDZ^{TM} wafer is a generic high-performance IG wafer decoupled from the need to tailor a specific material to a specific process. Our goal of simplicity is thus achieved.

11.3.6   THE MECHANISM OF THE VACANCY EFFECT ON OXYGEN PRECIPITATION

At low T, oxygen clusters are nucleated to high density even without vacancies but there is a long incubation (several hours at $650\,^{\circ}\text{C}$) before a subsequent high-temperature anneal results in an appreciable precipitate density, $10^9 \, \text{cm}^{-3}$ or higher. This incubation, most likely, is not related to a lag in the time-dependent nucleation rate but rather to the problem of subsequent survival of nucleated oxygen clusters. The clusters formed without volume accommodation (without assistance of vacancies) would inevitably dissolve at higher T since the strain energy per oxygen atom becomes larger than the oxygen chemical potential. The only way to survive is to relieve the strain energy by emitting enough self-interstitials, to acquire some space in the lattice. The emission rate is expected to be strongly size dependent. It becomes appreciable only if the cluster size is above some critical limit. To achieve this size, some incubation time is required. However, with vacancies present, the clusters will relieve the strain simply by absorbing vacancies. The vacancy species at low T exist predominantly as O_2V complexes (bound vacancies – see below). The O_2V species may be of high enough mobility, even at low T, to be absorbed by the clusters at the nucleation stage. Even if this is not the case, vacancies will be absorbed during the subsequent ramp-up of the temperature since the effective vacancy diffusivity (averaged over O_2V and V species) is a rapidly increasing function of T. Either way, at the beginning of the precipitation stage (for example at 1000 °C) each cluster will contain m vacancies ($m = C_v/N$ where N is the cluster density). The critical vacancy number m^* – for clusters in equilibrium with the equilibrium solution of point defects – is estimated to be about 100 at 1000 °C (it is slightly dependent on the oxygen concentration). This important estimate comes from the oxide/silicon interfacial energy fitted to account for the microdefect formation (Voronkov, to be published). If the actual value of m exceeds m^* then the cluster-controlled point-defect concentrations correspond to interstitial supersaturation (and vacancy undersaturation). Under such conditions, all the clusters can grow simultaneously since self-interstitials are removed to the sample surface thus providing more space for the clusters. This means that m will increase, and the number of oxygen atoms in each cluster (n) will follow the increase in m. However, if the starting value of m is below m^* (a much more likely situation), the point-defect solution in equilibrium with clusters corresponds to interstitial undersaturation; then the interstitials diffuse from the sample surface into the bulk decreasing m – and thus inducing the cluster dissolution. Along with this (presumably slow) process there will be fast coalescence with respect to the size m: the clusters of smaller m will lose vacancies, while the clusters of larger m will gain those vacancies. In this way the total cluster density N will decrease, while the average size m will increase, until it becomes larger than m^*. After that, all the remaining clusters can grow simultaneously. This consideration results in a simple rule for the final precipitate density N_p. It is on the order of C_v/m^* (about 10^{10} cm^{-3} for $C_v = 10^{12}$ cm^{-3}). It is remarkable that though the initial (nucleated) cluster density N may strongly depend on the oxygen content, the final precipitate density N_p is almost independent of the oxygen content: it is controlled by the vacancy concentration C_v. When the vacancy concentration is low, the starting size $m = C_v/N$ will be accordingly small. The clusters of small m would simultaneously lose the vacancies to the vacancy solution, and will be completely dissolved.

Another mechanism of vacancy-enhanced precipitation is direct agglomeration of the vacancy species (O_2V) during the nucleation stage. The production of this new cluster population – by agglomeration not of oxygen but of O_2V – goes in parallel with the nucleation of oxygen-only clusters considered above. Attachment of the O_2V species to a cluster may proceed either by direct migration of these species or by the dissociation with subsequent fast migration of a released free vacancy to a cluster followed by attachment of oxygen atoms. For both ways the nucleation rate is not very sensitive to the oxygen content but controlled primarily by the species concentration, C_v. If the vacancy size of clusters,

$m = C_v/N$, is below m^*, the precipitation anneal will result in a coalescence process (described above), and the cluster density N will be reduced to the final value on the order of C_v/m^*. At low C_v the nucleation rate of O_2V-clusters would be negligible.

The two mechanisms of vacancy-enhanced oxygen precipitation share some common features: (a) the basic effect of vacancies is to supply the oxygen clusters with space necessary for the cluster survival during the precipitation anneal; (b) the vacancy concentration must be over some limit to enhance the precipitation; (c) the produced precipitate density is controlled by the vacancy concentration rather than by oxygen content. It is likely that the first mechanism is important at low T (at $650\,^{\circ}$C or below). The second mechanism certainly operates at higher T (like $800\,^{\circ}$C) – when the effective vacancy mobility is enough for vacancy-assisted nucleation of oxygen clusters. The first mechanism is not operative at all at such T since the nucleation rate of oxygen-only clusters becomes negligible.

11.4 CONCLUSIONS DRAWN REGARDING THE INTRINSIC POINT-DEFECT PARAMETERS TAKEN FROM THE COMBINATION OF CRYSTAL GROWTH AND MDZ EXPERIMENTS

Both the problems of microdefect formation in crystal-growth and the principles behind the MDZ wafer involve the properties of point defects at high temperatures. At high temperatures in lightly doped material, the analysis of the problem should be free of the complications of defect complexes. This section discusses conclusions drawn from the analyses of both problems taken together to present a coherent view of the properties of free, neutral point defects in silicon and the nature of complexing at lower temperatures.

11.4.1 RECOMBINATION RATE

The v/G rule for grown-in microdefect type (the type of incorporated native defect) implies that the balance between defect recombination and pair generation is achieved very rapidly. The characteristic equilibration time τ_r must be much shorter than the time spent within the near-interface region where major defect annihilation occurs (for float-zoned crystals this dwell time may be as short as $20\,\mathrm{s}$). This conclusion is in line with the extremely short equilibration time observed in MDZ experiments in wafers at $1250\,^{\circ}$C (on the order of $1\,\mathrm{s}$). This result means that both equilibration processes (pair generation/recombination and diffusion) occur within this short time. The characteristic time τ_r is therefore on the order of $1\,\mathrm{s}$ (or less) at $1250\,^{\circ}$C. Since τ_r is a rapidly decreasing function of T, It will be still much lower (several milliseconds or less) at the melting point.

11.4.2   SELF-INTERSTITIAL DIFFUSIVITY

The short equilibration at 1250 °C also means that the characteristic diffusion time, $(d/\pi)^2/D_i$ – where d is the wafer thickness (0.07 cm) – is on the order of 1 s. Accordingly, the interstitial diffusivity, D_i, is on the order of 5×10^{-4} cm²/s. The estimate for D_i at the melting point (resulting from quenched-in microdefect patterns) is the same number (Voronkov 1982). This coincidence implies that (a) the interstitial diffusivity is very high, (b) the temperature dependence of D_i is not significant (at least over 1250 °C). The interstitial migration energy E_{mi} is thus rather low. We adopt $E_{mi} = 0.2\,\text{eV}$ following the data of Watkins 1999.

11.4.3   VACANCY DIFFUSIVITY

This value was deduced (Voronkov and Falster 1999) from the observed density and size of voids: D_v is about 2×10^{-5} cm²/s at the void formation temperature (around 1100 °C); a slightly higher number follows from the cavitation model of void production (Voronkov, to be published). Since the vacancy is mobile even at the room temperature (Voronkov and Falster 1999), the vacancy migration energy E_{mv} can be then deduced to be about 0.35 eV (Voronkov and Falster 1998). This is close to the number obtained by Watkins, 0.45 eV. On the other hand, the vacancy diffusivity is the parameter that largely controls the width of a denuded zone in the MDZ process: that zone is basically formed due to vacancy out-diffusion during fast cooling of wafers after the RTP anneal. Accordingly, a good estimate for D_v (assuming the above low migration energy) can be deduced from the observed quenched-in vacancy profiles under a variety of cooling-rate conditions. The deduced diffusivity at the reference temperature of 1100 °C is about 2×10^{-5} cm²/s, in excellent accord with the microdefect-based number for D_v.

11.4.4   THE DIFFERENCE OF EQUILIBRIUM VACANCY AND INTERSTITIAL CONCENTRATIONS

Both the world of grown-in microdefects and the world of MDZ are based on the inequality $C_{ve} > C_{ie}$. This inequality (at the melting point) insures the change-over from interstitial incorporation (at v/G below the critical ratio) to vacancy incorporation (at higher v/G). Incorporation of vacancies in the MDZ process is based on a similar inequality at the process temperature: the quenched-in vacancy concentration in the wafer bulk is equal to $C_{ve} - C_{ie}$ at the process temperature. This difference – for the two temperatures, 1250 and 1175 °C – was measured by the Pt technique with some uncertainty. The melting-point difference can be deduced from the amount of vacancies stored in voids (Saishoji et al. 1998) calculated from the LST data on the void density on size, in dependence of v/G. This difference lies within rather narrow limits, $(2–3) \times 10^{14}$ cm⁻³. The ranges

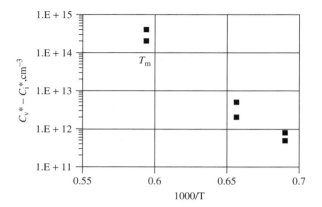

Figure 11.6 The range of the solubility difference for vacancy and self-interstitial. The melting point (T_m) range is based on the measured amount of vacancies in voids. The range for the two other temperatures (1250 and 1175 °C) comes from RTA-treated wafer measurement data.

for $C_{ve} - C_{ie}$ are thus available for the three temperatures; these are shown in Fig. 11.6.

11.4.5 FORMATION ENERGIES

The plot of Figure 11.6 imposes strong constraints on the values of formation energies, E_v for vacancy and E_i for self-interstitial. The melting-point value for the equilibrium interstitial concentration, C_{im}, is fixed by the estimated diffusivity D_i and the value of the $D_i C_{ie}$ product (2.3×10^{11} cm^2/s at the melting point) known from self-diffusion and zinc-diffusion data (Bracht 1999). Making allowance for some uncertainty in both D_i and $D_i C_{ie}$, a reasonable range for C_{im} is $(5-15) \times 10^{14}$ cm^{-3}. If one specifies the formation energies, then the temperature dependence of $C_{ve} - C_{ie}$ can be plotted for various C_{vm} and C_{im} within the fixed ranges for C_{im} and $C_{vm} - C_{im}$. If none of these curves falls within the ranges of Figure 11.6, the chosen energies E_v and E_i are rejected. Otherwise they are considered as consistent with the data of Figure 11.6. This procedure shows that acceptable energies are very close one to the other. The acceptable range for the average energy $E = (E_i + E_v)/2$ and for (very small and negative) energy difference $E_i - E_v$ is shown in Figure 11.7. Both formation energies are over 4.3 eV. To be consistent with self-diffusion data, E_i must be lower than the activation energy for the $D_i C_{ie}$ product, 4.9 eV (Bracht 1999). Since the interstitial migration energy is thought to be 0.2 eV (or perhaps somewhat larger), the actual estimate for the interstitial formation energy is $E_i \leq 4.7$ eV. The final range for E_i is then from 4.3 to 4.7 eV. The vacancy formation energy E_v is within the same range. An important implication of these estimates is that the

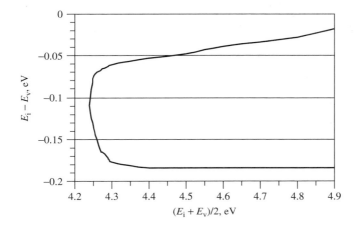

Figure 11.7   The values of average formation energy and the energy difference consistent with the data shown in Fig. 11.6. These values are in the area inside the curve.

activation energy for the vacancy contribution into the self-diffusivity, $E_v + E_{vm}$, lies within the range 4.7–5 eV. It is thus comparable to that for the interstitial contribution, $E_i + E_{im} = 4.9$ eV. The interstitial contribution to self-diffusivity prevails at high T (Bracht 1999). It will then prevail over the whole temperature range, with no cross-over to vacancy-dominated self-diffusivity at lower T. A presumed cross-over point is indeed refuted by recent data (Bracht 1999).

11.4.6   CRITICAL v/G RATIO

This is expressed (Voronkov 1982) through the average formation energy E, the $D_i C_{im}$ product and the concentration difference $C_{vm} - C_{im}$ – the same parameters that are relevant in the above analysis of the formation energies. On substituting the numerical values of these parameters one gets for $(v/G)_{cr}$ a rather narrow range consistent with the value found experimentally (Dornberger *et al.* 1996, Falster *et al.* 1998b) (about 2×10^{-5} cm^2/s or 0.12 mm^2/min K). This result shows the self-consistency of the present approach to the estimation of the point-defect parameters.

11.4.7   VACANCY BINDING BY OXYGEN

The reaction of vacancies with oxygen, to form a O_2V complex (a bound vacancy) must be invoked (Falster *et al.* 1998b) to account for nonzero residual concentration of vacancies in as-grown crystals. The residual vacancies are manifested in subsequent oxygen precipitation, which is strongly banded spatially, due to a banded distribution of residual vacancies (Falster *et al.* 1998b, Voronkov and Falster, to be published). The binding (complexing) reaction is strongly dependent on temperature: there is some characteristic binding temperature, T_b, separating the higher-temperature region (where the free vacancies dominate) from the lower-temperature region (where the vacancies exist predominantly as O_2V, with only a small fraction of free vacancies). The binding temperature was estimated to be somewhat above $1000\,°C$ ($1020\,°C$ or perhaps slightly higher). The basic implication of the binding reaction is that the effective vacancy mobility is rapidly reduced below T_b, and the vacancy loss to microdefects is thus strongly suppressed, giving an opportunity for some vacancies to survive – as O_2V complexes. The suppression of vacancy mobility below T_b is also of great importance in the processes that produce the MDZ wafer. If the quenched-in depth profile of vacancies is simulated without the binding reaction, the computed vacancy concentration is far lower than the measured value (for some cooling rates). Even if the computed concentration at the wafer centre is not low (at higher cooling rates) the profile shape is strongly different from that observed. When the vacancy binding is taken into account, the computed profiles are in accord with the experiment. The vacancy binding is also clearly manifested if the MDZ wafers are subjected to the second RTP process at a temperature T_2 lower than that of the first RTP anneal (at $T_1 = 1250\,°C$). At $T_2 = 1100\,°C$ (well above the binding temperature) the initially installed vacancy profile disappears within 20 s. At $T_2 = 1000\,°C$ a similar relaxation occurs but within a much longer time scale (at 60 s it is not yet fully relaxed) indicating a considerably reduced effective vacancy diffusivity. At $T_2 = 900\,°C$ the vacancy profile does not change at all within 15 min, indicating a strongly reduced effective vacancy mobility; therefore the temperature of $900\,°C$ is already well below T_b.

11.5 UNIFIED SCHEMATIC PICTURES OF VACANCY CONTROL FOR CRYSTAL GROWTH AND WAFER PROCESSING

The above discussion, which takes into account observations from both crystal-growth and wafer-processing experiments, can in part be summarized in a useful schematic graphical form illustrating the problem of vacancy-defect engineering. A similar set of schematics exists for the self-interstitial mode of growth and will be presented elsewhere. The vacancy schematics are illustrated in Figures 11.8–11.14.

Figure 11.8 illustrates the basic vacancy *reaction template*. Here is plotted the various important zones of vacancy reaction (void formation, O_2V binding and enhanced oxygen precipitation) in terms of temperature and vacancy concentration. As a main point of reference, the solubility of vacancies is also shown. Such a chart can be viewed as a kind of map through which the concentration of vacancies in a growing crystal or heat-treated wafer must be steered during the cooling of the crystal or wafer in order to achieve a desired result.

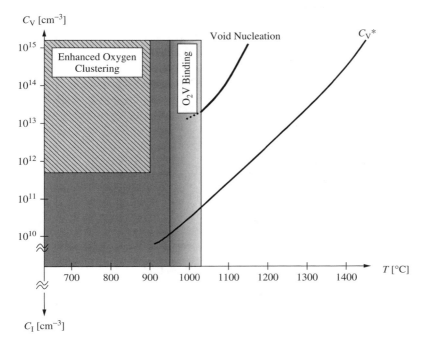

Figure 11.8 The basic vacancy reaction 'template'.

Figure 11.9 then uses this map to illustrate the first part of the crystal-growth problem: that of the incorporation of vacancies into a growing crystal. Starting at the equilibrium concentration at the melt temperature, the vacancy concentration is reduced as the crystal cools through a transient phase in which vacancy and self-interstitial fluxes compete with each other via mutual annihilation. Eventually this stops when the concentration of one or the other point-defect species becomes exhausted. At this point an essentially constant concentration of the surviving species is obtained. It is the residual vacancy (or self-interstitial) concentration, set at high temperature by the v/G rule, which is then launched into the possible reactions at lower temperatures. The reference value of the ratio of the actual to critical v/G value is given on the right-hand scale for several examples. The value for each example corresponds to the value at the vacancy-concentration saturation value. At the normalized V/V_t value of 0.8 the illustrated incorporated vacancy concentration drops below the equilibrium concentration. The crystal is at this point 'interstitial type'.

Figure 11.10 illustrates the most usual case in commercial CZ crystal-growth, that of void reaction. At a temperature of typically around 1100 °C the vacancies incorporated in the growing crystal become sufficiently supersaturated to drive the void reaction. Very shortly after this starts, a sufficiently large number of vacancies are consumed, thus shutting the reaction off. The resulting density of

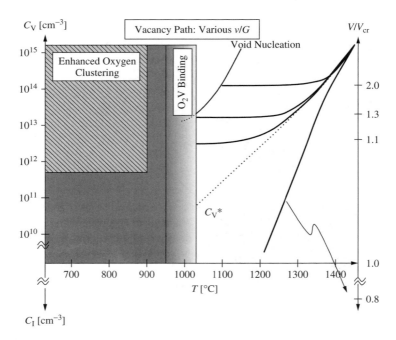

Figure 11.9 An illustration of the incorporation of vacancies into a growing crystal at various values of v/G relative to the critical value.

voids depends on the concentration of vacancies entering the reaction barrier and the cooling rate while the reaction proceeds. Upon further lowering of the crystal temperature no more voids are produced but vacancies continue to be consumed by the existing voids until the vacancies are fully bound in the relatively immobile O_2V complex. The residual concentration of vacancies (now in the form of O_2V complexes) is, however, usually below the critical concentration level for precipitation enhancement. If, however, the crystal is cooled rapidly through the growth phase, such as can happen when fast pull rates are used to produce crystal end cones, the residual vacancy concentration can be larger than the transition level to enhanced oxygen precipitation and regions of 'anomalously' high levels of precipitation can be observed. This is often the case in the tail-end regions of CZ crystals. Figure 11.11 illustrates this effect within the framework of the vacancy diagrams.

At lower values of v/G and in particular at values of v/G approaching the critical value, the concentration of incorporated vacancies is reduced. It is at these levels of vacancy concentration that strong defect banding effects are observed. Figure 11.12 illustrates the conditions for 'p-band' or OSF-ring formation. The consumption rate of vacancies is larger than the void case due to the typically 2 orders of magnitude higher density of the p-band particles (the vacancy 'sinks') than voids.

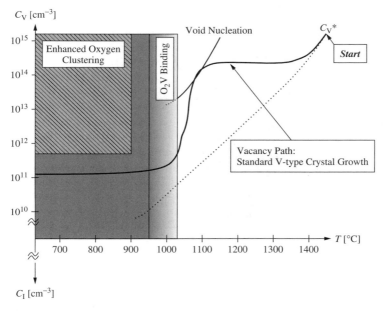

Figure 11.10 An illustration of the production of voids in a growing crystal in the most usual case.

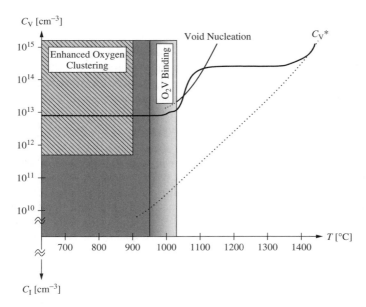

Figure 11.11 An illustration of the effect of more rapid cooling through the void growth regime resulting in 'anomalously' high values of oxygen precipitation.

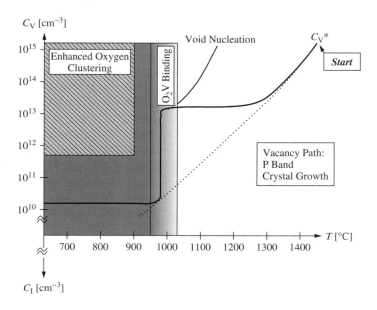

Figure 11.12   An illustration of the conditions under which the 'P-band' or OSF-ring is formed during crystal growth.

Figure 11.13 shows the case for 'L-band' formation. In this case the vacancy concentration goes underneath the void reaction and enters directly into the O_2V binding as illustrated in Figure 11.13. No voids can form as the process in now kinetically limited. The usual case is for the vacancies to enter the 'enhanced precipitation' condition at lower temperature giving rise to the precipitation peaks observed in the L-band region. Material grown under these conditions is 'vacancy perfect'-type material.

Finally, the same vacancy schematic used above for crystal-growth can be used to explain the incorporation of vacancies during the heat treatment of thin silicon wafers leading to the MDZ™ wafer. Figure 11.14 illustrates this. Here our starting point is not at the melt temperature of silicon but usually substantially lower. As a result, the maximum vacancy concentrations achievable are substantially lower than that of crystal-growth from the melt (unless point defects are additionally injected from the wafer surfaces). Initially, in the bulk of the wafer, the concentration of vacancies is somewhat less than equilibrium as a result of the fact that the only possible source of vacancies is Frenkel pair generation and mass action. Shortly after this, however, due to the close proximity of the wafer surfaces, the equilibrium vacancy concentration is rapidly reached throughout the bulk of the wafer primarily through interstitial in-diffusion coupled to IV recombination/generation. At the wafer surfaces, equilibrium conditions are always maintained. As the sample cools, the vacancy concentration

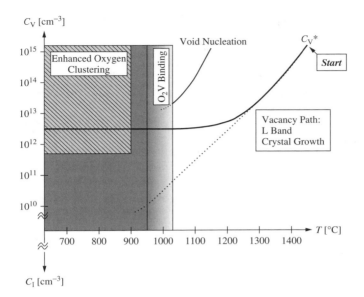

Figure 11.13 An illustration of the conditions under which the 'L-band' or 'vacancy-perfect' type material is formed during crystal growth.

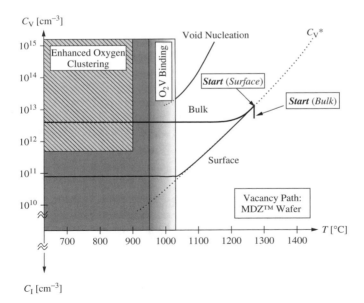

Figure 11.14 An illustration of the use of the vacancy template to illustrate the creation of dual-zoned precipitation effects in thin silicon wafers (the MDZ™ wafer) by means of high-temperature rapid thermal processing.

drops, following equilibrium at the surfaces, while the decrease in the bulk, far (relatively speaking) from the surfaces is limited to recombination with existing self-interstitials. These are rapidly exhausted, leaving behind a 'residual' vacancy concentration (which is equal to $C_{Ve}(T) - C_{Ie}(T)$ where T is the process temperature). If the cooling of the wafer is sufficiently rapid such that the expanding vacancy-diffusion profile does not penetrate into the wafer centre prior to reaching the O_2V binding temperatures, then this concentration is maintained. If this residual concentration is sufficiently large to be greater than the critical vacancy concentration for precipitation enhancement, then the desired two-zoned precipitation effect is created.

ACKNOWLEDGMENTS

Measurements of vacancy concentration by the platinum-diffusion technique were performed by Fabian Quast and Peter Pichler of the Fraunhofer Institut, Erlangen. Computer programs for the simulation of point-defect dynamics during MDZ treatments were developed by Marco Pagani. We also wish to acknowledge the contributions of Daniela Gambaro, Max Olmo, Marco Cornara, Harold Korb, Jeff Libbert, Joe Holzer, Bayard Johnson, Seamus McQuaid, Lucio Mule'Stagno, and Steve Markgraf of MEMC Electronic Materials to this work.

REFERENCES

Abe, T. 1999 in *ECS Proc.*, **99-1**, 414.
Bracht, H. 1999 in *ECS Proc.*, **99-1**, 357.
de Kock, A. J. R. 1973 *Philips Res. Repts Suppl.*, **1**, 1.
Dornberger, E. and W. von Ammon, W. 1996 *J. Electrochem. Soc.*, **143**, 1648.
Falster, R., Fisher, G. R. and Ferrero, G. 1991, *Appl. Phys. Lett.*, **59**, 809.
Falster, R., Cornara, M., Gambaro, D., Olmo, M. and Pagani, M. 1997, *Solid State Phenom.*, **57–58**, 123.
Falster, R., Gambaro, D., Olmo, M., Cornara, M. and Korb, H. 1998a *Mater. Res. Soc. Symp. Proc.*, **510**, 27.
Falster, R., Voronkov, V. V., Holzer, J. C., Markgraf, S., McQuaid, S. and MuleStagno, L. 1998b in *ECS Proc.*, **98-1**, 468.
Foell, H. and Kolbesen, B. O. 1975, *Appl. Phys.*, **8**, 319.
Jacob, M., Pichler, P., Ryssel, H. and Falster, R. 1997, *J. Appl. Phys.*, **82**, 182.
Nishimura, N.,Yamaguchi, Y., Nakamura, K., Jablonski, J. and Watanabe, M. 1998, in *ECS Proc.*, **98-13**, 188.
Park, J. G., Lee, G. S., Park, J. M., Chon, S. M. and Chung, H. K., 1999, *ECS Proc.*, **99-1**, 324.
Petroff, P. M and de Kock, A. J. R. 1975, *J. Appl. Phys.*, **30**, 117.
Puzanov, N. I and Eidenzon, A. M. 1992, *Semicond. Sci. Techol.*, **7**, 406.
Roksnoer, P. J. and van den Boom M. M. B. 1981, *J. Cryst. Growth*, **53**, 563.

Saishoji, T., Nakamura, K., Nakajima, H., Yokoyama, T., T. Ishikawa, T. and Tomioka, J. 1998, in *ECS Proc.*, **98-13**, 28.

Ueki, T., Itsumi, M. and Takeda, T. 1997, *Appl. Phys. Lett.*, **70**, 1248.

Veselovskaya, N. V., Sheyhet, E. G., Neymark, K. N. and Falkevich, E. S. 1977, in *Growth and Doping of Semiconductor Crystals and Films (in Russian)*, Part 2 (Nauka: Novosibirsk) 284.

Voronkov, V. V. 1982, *J. Cryst. Growth*, **59**, 625.

Voronkov, V. V. and Falster, R. 1998, *J. Cryst. Growth*, **194**, 76.

Voronkov, V. V. and Falster, R. 1999, *J. Cryst. Growth*, **198/199**, 399.

Voronkov, V. V., Voronkova, G. I., Veselovskaya, N. V., Milvidski, M. G. and Chervonyi, I. F. 1984, *Sov. Phys. Crystallogr.*, **29**, 688.

Watkins, G. D. 1999, in *ECS Proc.*, **99-1**, 38.

12 The Formation of Defects and Growth Interface Shapes in CZ Silicon*

TAKAO ABE

Shin-Etsu Handotai, Isobe R&D Center, Isobe 2-13-1, Annaka, Gunma 379-0196 Japan

ABSTRACT

The thermal distributions near the growth interface of 150 mm Czochralski (CZ) crystals were measured by three thermocouples installed at the centre, middle (half-radius) and edge (10 mm from surface) of the crystals. The results show that larger growth rates produced smaller thermal gradients (G) similar as reported for FZ crystals. The results seem not to contradict the balance equation because temperature distributions continue through the growth interface to the bulk of crystal. The growth rate (v) in Voronkov's theory does not effect directly the point-defect formation. Finally, it is proposed that the shape of the growth interface is also determined by the distribution of G across the interface. That is, small G and large G in the centre induce concave and convex interfaces to the melt, respectively.

12.1 INTRODUCTION

Investigations on native point defects in silicon crystals may be classified into three categories, depending on the temperature ranges at which they are formed.

---

*Note of the editor: The interesting measurements reported here are not yet fully understood and require further studies and interpretation. Generally one would assume that the thermal gradient in the crystal at the growth interface increases with increasing growth rate in order to remove the latent heat. Alternatively, in specific growth arrangements, the latent heat may be compensated by a decreasing thermal gradient in the liquid (near the interface) with increasing growth rate as postulated by Voronkov and in this work by T. Abe. In these latter cases the temperature gradient in the solid is either constant or even slightly decreasing with increasing growth rate. The effective temperature distribution at the growth interface and the thermal history of the crystal determine the interface shape, defect formation, and defect distribution HJS.

Crystal Growth Technology, Edited by H. J. Scheel and T. Fukuda
© 2003 John Wiley & Sons, Ltd. ISBN: 0-471-49059-8

The electron paramagnetic resonance (EPR) is carried out at cryogenic temperature in order to study the lattice vacancies and the silicon interstitials produced by electron irradiation. This investigation was extensively performed by Watkins *et al.* [1] in the 1960s. The second category of investigations is devoted to two phenomena: (i) the enhanced and retarded impurity diffusion induced by oxidation and nitridation, and (ii) the kick-out and dissociative diffusion of metals such as Au, Pt and Zn in the temperature range 44 to 1388 °C [2].

The third category concerns the secondary point defects produced during the crystal-growth process, from the melting point down to 1000 °C. These secondary defects form as a result of nonequilibrium phenomena, that is, the crystal-growth and subsequent cooling, which are characterized by the presence of thermal gradients. Up to now measuring procedures or parameters obtained with the investigations of the first and second type could not be applied in order to fully understand and then prevent the grown-in defect formation near the melting point.

When float-zone (FZ) silicon crystals were first grown without dislocations, the layer structure of defects perpendicular to the growth axis [3] and the so-called 'shallow pits' distributed in a swirl-like pattern [4] could be clearly observed. De Kock [5] first proposed that the shallow pits were agglomerates of vacancies. However, Foell and Kolbesen [6] later found by direct TEM observations that shallow pits are produced by excess interstitials that give rise to interstitial-type dislocation loops. Currently, such defects are commonly named A defects. Following this evidence, many researchers proposed their own models [7–13] as shown in Table 12.1. Foell *et al.* [7] and Petroff and de Kock *et al.* [8] assumed interstitial silicon to be in equilibrium and in nonequilibrium, respectively. Van Vechten [9] and Chikawa and Shirai [10] suggested that the vacancies are predominant at equilibrium. However, even when vacancies are assumed to be predominant, Chikawa's droplet model does not exclude the occurrence of A defects. The equilibrium model in which vacancy and interstitial concentrations are equal at the growth interface is accepted by many researchers. Sirtl [11] suggested that pair annihilation and creation is kept with the local equilibrium. Hu [12] and de Kock *et al.* [13] proposed that the A defects are generated due to the energy differences of recombination and agglomeration energies between interstitials and vacancies. Roksnoer *et al.* [14] reported another type of defects named D defects, consisting of excess vacancies created by growing FZ crystals with extremely large growth rate. Based on this observation, Voronkov [15] developed a theory according to which the species of intrinsic point defects created during growth depends upon the ratio or growth rate (v) and thermal gradient (G) near the growth interface. He defined a critical value of this ratio (C_{crit}), and suggested that when $v/G < C_{crit}$, interstitials are dominant, whilst if $v/G > C_{crit}$, vacancies are dominant. Tan and Goesele [16] tried to modify this theory introducing a new concept of 'local equilibrium', but this was not yet proven experimentally.

In 1990, Ryuta *et al.* [17] found the crystal-originated particles (COP) which are the result of the D defects in Czochralski (CZ) crystals, by using a particle

Table 12.1 Proposed models on the origin of point defects

	$C_I \gg C_v$	$C_I \ll C_v$	$C_I = C_v$
Equilibrium	1975 *Foell et al: Interstitials* [6] 1977 Foell, Goesele Kolbesen: In equilibrium [7]	1973 de Kock: Vacancies [5] 1978 Van Vechten: Cavity model [9]	1965 *Plaskett: Microdefects* [3] 1977 Sirtl: Local equilibrium [11] 1977 Hu: Recombination energy [12] 1981 *Roksnoer et al: D defect* [14] 1982 Voronkov: Thermodiffusion [15] 1985 Tan *et al.*: Local equilibrium [16] 1990 *Ryuta et al: COP* [17] 1995 *Itsumi et al: Voids* [18] 1994 Habu: Uphill diffusion [20] 1994 Brown: Thermodiffusion [21] 1995 Ammon: Voronkov's parameter [23] 1980 de Kock/van de Wijgert [13] 1984 Roksnoer: Carbon [29]
Nonequilibrium	1976 Petroff/de Kock: Impurity clustering [8]	1976 Chikawa/Shirai: *Interstitials by local remelting* [10] 1983 Abe, Harada/Chikawa: *Thermal gradient (stress)*	

counter based on light scattering techniques. Later, Itsumi *et al.* [18] reported that COPs are twinned structures of octahedral voids covered with thin oxide on inside (111) planes. After those two findings, many numerical simulations were started with the aim of explaining the generation of the A and D defects; see, for example, Wijaranakula [19], Habu *et al.* [20], and Brown *et al.* [21], who postulate almost equal concentrations of interstitials and vacancies at the growth interface. They calculated the flux of point defects using the gradient of chemical potentials. However, the parameters used for the computation were varied in a wide range: Habu *et al.* [20] postulated that both species of point defects at the growth interface are controlled by G and v but characterized by uphill diffusion [22] of vacancies. The conclusion resulted opposite to that of Voronkov who stated that if v is large enough, vacancies are predominant and uphill diffusion is negligible. However, the experimental results [29, 30] obtained by detaching the crystal from the melt, showed that the solid near the growth interface is filled with vacancies. On the other hand, it is experimentally known that the dominant species outside and inside of R-OSF (ring-likely distributed oxidation-induced stacking fault) are interstitial silicon and vacancy, respectively. Ammon *et al.* [23] an Hourai *et al.* [24] determined the value of v/G on the R-OSF to be 0.13 mm^2/min K [25] and 0.22 mm^2/min °C [24], respectively. The former used the commercially available finite element codes FEMAG [25] for the global heat transfer analysis, and the latter calculated the temperature distribution in the crystals by ABAQUS [24]. However, the present author and his coworkers measured the temperature distributions near the growth interface of growing FZ crystal surfaces by a two-colour thermometer [26]. The results clearly showed that larger v values induce smaller G values. Recently, we measured the temperature distributions [31] near the growth interface of growing CZ crystals using thermocouples buried in 150 mm crystals. From these results we have proposed that the generation of point defects only depends on G near the growth interface, but not explicitly on v, which just contributes to change G and to agglomeration of these defects during cooling.

In this chapter, we first discuss three points based on the G values actually measured on growing CZ crystals: first is the traditional balance equation used for the computer simulation, second is the meaning of Voronkov's theory, and third is the formation of growth-interface shape.

12.2 EXPERIMENTS

As schematically shown in Figure 12.1, two previously grown dislocation-free crystals of lengths of 250 mm and 350 mm with 150 mm diameter, having a conical shoulder but not a conical tail, were prepared for measuring temperature distributions during crystal-growth. Thermocouples were installed at the flat tail end. To avoid the radiation effects and to measure the temperatures in the crystal interior, three thermocouples at the centre, middle (half-radius) and edge

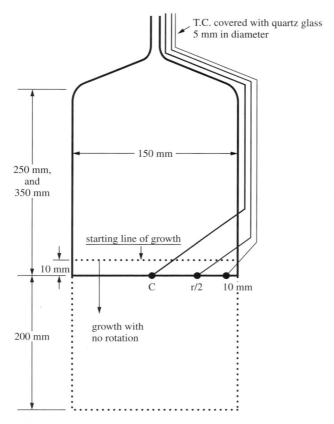

Figure 12.1   Schematic of test crystals installed with three thermocouples at the end surface.

(10 mm from edge) positions were inserted into the crystals. The diameter of the thermocouples is 1 mm and the outer diameter of their quartz cover is 5 mm. The test crystals were supported by multiple tungsten wires as if real crystals were growing. The weight of melted silicon was adjusted depending on the test crystal length to simulate an initial silicon melt weight of 40 kg. The test crystal was first melted to 10 mm from the tail end of the crystal by dipping a few mm into the melt at a time and waiting for melting to occur, achieving an average melting speed of 0.7 mm/min. Finally, the crystal-growth was started with a growth rate of 1.2 mm/min but no rotation. When the length of the newly grown crystal reached 210 mm from the re-melted position, the temperature measurement was halted and the re-melting with the above average melting speed was done again to the point 10 mm over the thermocouple positions. In the same manner, the temperature measurements were done for the other three cases of growth rates; 1.0 mm/min, 0.6 mm/min and 0.3 mm/min. The temperature distributions

of other test crystals of 350 mm length were also measured in the same manner. A standard CZ puller was used, but was equipped with a hot zone that cools crystals strongly, in order to observe the crystal length effects on G.

12.3   RESULTS

Figure 12.2 shows, for four different growth rates, the temperature distributions recorded at the centre, middle and edge thermocouple positions in the 250-mm crystal. However, all these profiles are intentionally adjusted to normalize the melting temperature to $1412\,°C$ because they showed lower temperatures than $1412\,°C$ at the point assumed as the growth interface. The actual position of the thermocouples may include a few mm errors around the 0-mm position in Figure 12.2. Since it is known from other experiments that the growth inter-face height with growth rate is postulated as concave to the melt as shown in Table 12.2, the thermocouple positions passing through the interface are also adjusted in Figure 12.2 according to Table 12.2. In the experiments on the stop-ping of growth of FZ crystal [26] and the decreasing growth rate of CZ crys-tals [27], secondary defects consisting of interstitials were observed in both crys-tals, and the effective region for defect generation as measured from the growth interface were several to several tens of millimeters depending on whether it is FZ or CZ growth. In this chapter, as we defined the effective length as 10 mm from the growth interface, the axial temperature gradient G is $G = (T_{int} - T_{10})\,°C/cm$. The G at the centre is the smallest and the G at the edge is the largest in the case of 1.2 mm/min growth rate as shown in Figure 12.2(a). On the other hand, in the case of the 0.3 mm/min growth rate the G values in all three positions are almost equal as shown in Figure 12.2(d). The temperature distributions shown in Figure 12.2 can be transferred to the isotherms as shown in Figure 12.3, if we postulate an axial symmetry of temperature distribution. When v is large, the temperature in the centre is higher than that at the edge. On the other hand, when v is small, a uniform temperature distribution in the cross section of the crystals is obtained. Concerning the interface height in Table 12.2, it can be said that our supposition is reasonable for Figure 12.2 and Figure 12.3 based on the follow-ing considerations. The smallest G in the centre of the crystal with 1.2 mm/min growth rate causes the deepest concave interface shape.

The axial temperature distributions for each thermocouple position, as it depends on the growth rate, are collected in Figure 12.4. A larger growth rate v produces a smaller G in the centre and middle positions, but at the edge the differences of G values are small for all growth rates. This means that the surface temperature distribution may be largely determined by the type of hot zone within 1.2 mm/min.

Previously reported papers concerning point-defect generation have not men-tioned the crystal-length effects, although some have discussed the thermal-history effects caused by the seed end of the crystal having a longer thermal

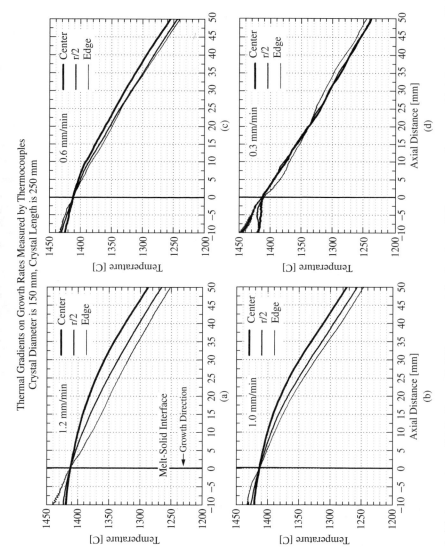

Figure 12.2 Axial temperature distributions of 250-mm crystals for four different growth rates: (a) 1.2 mm/min; (b) 1.0 mm/min; (c) 0.6 mm/min and (d) 0.3 mm/min.

Table 12.2 Postulated interface height (concave)

Growth rate (mm/min)	Center (mm)	Middle ($r/2$) (mm)	Edge surface (mm)
1.2	6	5	0
1.0	5	4	0
0.6	3	2	0
0.3	1.5	1	0

history than the tail of the crystal. In FZ crystal growth, both G and the thermal history do not vary with crystal length, due to rapid cooling effects. However, in CZ crystal growth, it is anticipated that the crystal length has a direct influence on G because of the large contribution of heat conduction through the crystal interior. So we also measured the temperature distributions of the 350-mm crystal in addition to the 250-mm crystal. The order of G values of the 350-mm crystal on growth rates is the same as that of the 250-mm crystal, although in the region far from the growth interface an anomaly in the temperature distributions is observed in the case of the 0.6-mm/min growth rate.

In order to consider the length effect, G values are collected as shown in Figure 12.5. It is first concluded that smaller v produces larger G in all positions in both Figure 12.5(a) and (b), independent of crystal length, with an exception in the case of a 1.2-mm/min growth rate for the 250-mm crystal. On the other hand, in actual crystal-growth when the melt level is descending, after-heating from the heater increases. As a result the G value at the edge becomes small. So to keep the same G, the v must be lowered. In a comparison of Figure 12.5(a) and (b), it is noted that the cross-sectional distribution of G changes with crystal length. The G at the edge is larger than that at the centre in Figure 12.5(a), but a flat distribution of G and a larger G at the centre are seen in Figure 12.5(b). This is a length effect.

12.4 DISCUSSION

12.4.1 BALANCE EQUATION

The balance equation is widely used as the boundary condition of heat transfer at a growth interface. If the thermal gradients in the solid and liquid at the interface are G_s and G_l then the continuity of heat flux requires

$$K_s G_s - K_l G_l = Lv \tag{12.1}$$

where K_s and K_l are the thermal conductivities of solid and liquid, L is the latent heat of solidification per unit volume, and v is the growth rate.

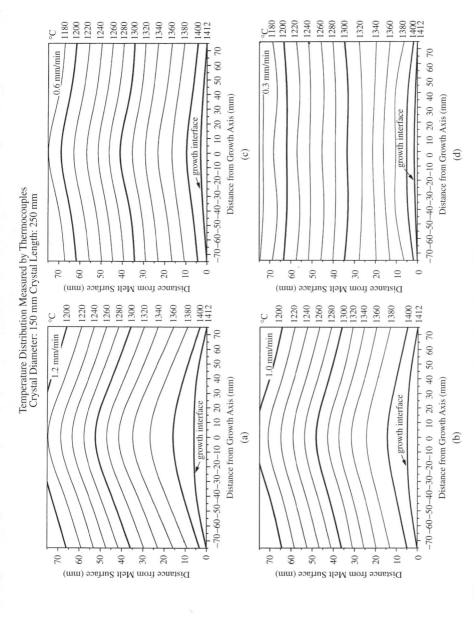

Figure 12.3 Isotherms near growth interfaces obtained from Figure 12.2. (a) 1.2 mm/min; (b) 1.0 mm/min; (c) 0.6 mm/min and (d) 0.3 mm/min.

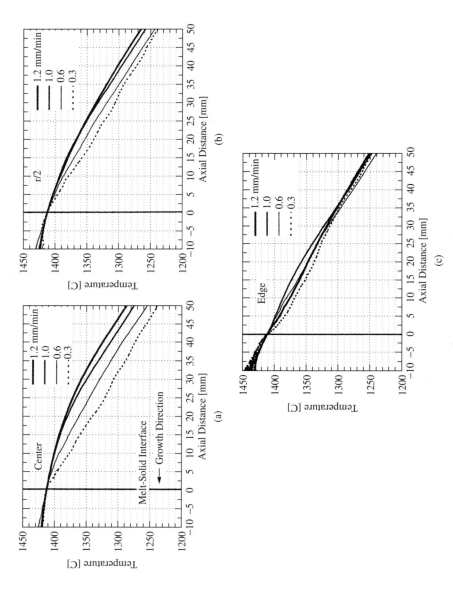

Figure 12.4 Axial temperature distributions of 250-mm crystal on (a) center; (b) middle and (c) edge positions obtained from Figure 12.2.

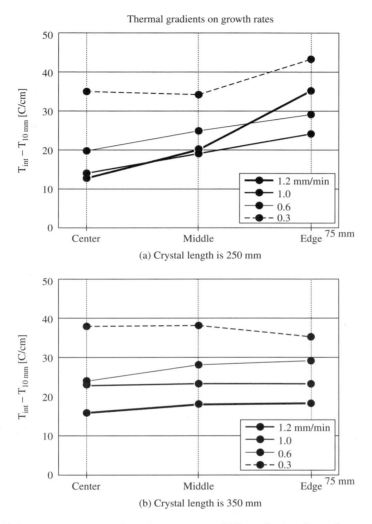

Figure 12.5 Temperature gradient G on center, middle and edge depending on growth rates. Crystal length (a) 250 mm and (b) 350 mm.

This equation satisfies the following four cases as seen in Figure 12.6. As there is no reason for G_l to increase with increasing v, Figure 12.6(a) may be excluded. Most researchers in their simulations suppose that the larger v produces the larger G_s with constant G_l as seen in Figure 12.6(b). However, both cases seem unrealistic although the latent heat term Lv becomes infinite with increasing v. In addition, Equation (12.1) is only valid in the vicinity of the interface and cannot be applied far from the interface. Voronkov proposes the case of Figure 12.6(c) based on the experimental results of Puzanov *et al.* in which G_s is independent

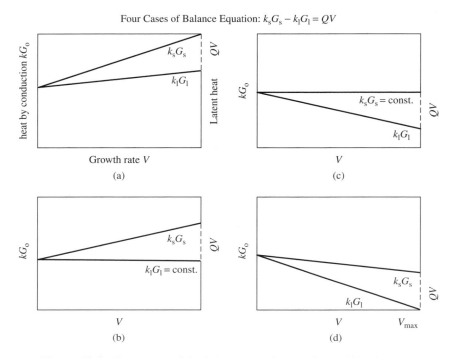

Figure 12.6 Four cases of the balance equation as discussed in the text.

of v. He speculates that the decreasing G_l is caused by the interface being lifted from the melt level. We propose the case of Figure 12.6(d), i.e. both G_s and G_l decrease with increasing v. Although this condition might only be valid in the case of transient phenomena during the re-solidification process in our experiment, we tend to believe the relationship shown in Figure 12.6(d).

12.4.2 DISCUSSION OF VORONKOV'S RELATION

G_s in Voronkov's relation is assumed as a constant value (independent of v) which is determined by the design of the hot zone. The relation is applied to the macroscopic range and thus is not limited to the interface. However, from our experimental results the thermal gradient G' in the crystal is slightly inverse of v. So the critical value C_{crit} for formation of R-OSF is written as

$$C_{crit} = v/G \sim a/G'G \tag{12.2}$$

where a is constant. When we use a hot zone with large G we have to achieve a small G' and a large v to compensate for the large G in order to obtain C_{crit}.

The occurrence of D-defects by Roksnoer may be interpreted by the very small G value in pedestal pulling of small-diameter (23 mm) crystals in combination with the very large v (>6 mm/min) and small G': vacancy-rich crystals are obtained with $C_{crit} > a/G'G$.

Von Ammon *et al.* maintain C_{crit} with the relationship between v and G, and the formation of R-OSF is explained by the above concept that large v induces small G'.

In growth experiments using a hot zone with very small G no R-OSF are formed even when C_{crit} is reached by small v. Thus it seems that G' is the important parameter and not v or G.

12.4.3 INTERFACE-SHAPE FORMATION

Macroscopic interface-shape formation has not been fully understood in silicon crystal-growth. In fact when the hot zone and the crystal diameter are constant, a small v creates a convex interface to the melt, and large v produces a concave interface. Our experimental results are shown in Figure 12.7(a) and (b). When v is small, as thermal conduction is dominant in the center, G' becomes large, whereas at large v a reduced G' is observed. The concave interface is formed

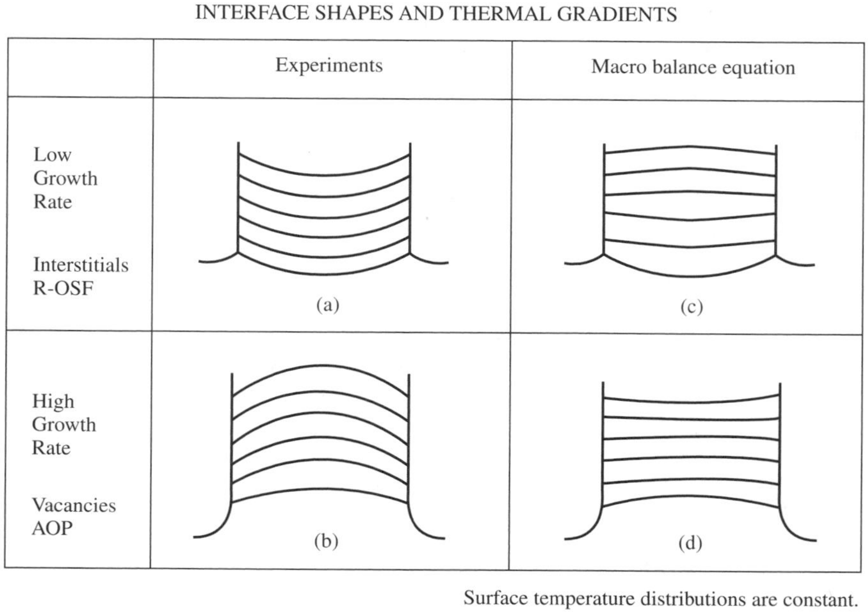

INTERFACE SHAPES AND THERMAL GRADIENTS

	Experiments	Macro balance equation
Low Growth Rate Interstitials R-OSF	(a)	(c)
High Growth Rate Vacancies AOP	(b)	(d)

Surface temperature distributions are constant.

Figure 12.7 Interface shapes and thermal gradients. (a) and (b) Experimental results, (c) and (d) traditional interpretation.

due to a reduced growth rate as explained by Figure 12.6(a). The growth rate v_{max} in the centre is limited by $G_1 = 0$, which is caused by increasing Lv. On the other hand, near the edge of the crystal heat is effectively removed from the surface so that large G values allow crystal-growth with large v.

Numerous researchers reported temperature distribution/growth rate dependences as shown in Figure 12.7(c) and (d), which differ from Figure 12.7(a) and (b). In the case of Figure 12.7(c) there is no reason for a convex interface when G' is small. It is concluded that the interface shape is determined by the distribution of G' across the interface. Small or large G' in the center of the crystal lead to concave or convex interfaces, respectively.

12.5 CONCLUSIONS

The thermal distributions near the growth interface of 150-mm CZ silicon crystals were measured by three thermocouples installed at the centre, middle (half-radius) and edge of the crystals. The results show that larger v produced smaller G' in analogy to a report for FZ crystals. The results do not contradict the heat balance equation because temperature distributions continue through the growth interface into the bulk crystal. Finally, it is proposed that the shape of the growth interface is also determined by the distribution of G' across the interface. It is evident that the knowledge of the actual G distributions across the interface is essential to understand the formation mechanism of grown-in defects and to predict the shape of the interface.

REFERENCES

[1]   G. Watkins, Defects and Diffusion in Silicon Processing, eds. T. Diaz, de la Rubia, S. Cofa, P. A. Stalk and C. S. Rafferty, *Mater. Res. Soc. Symp. Proc.* **469** (1997) 139.
[2]   H. Bracht, E. E. Haller, K. Eberl, M. Cardona, and R. Clark-Phelps, *Mater. Res. Soc. Symp. Proc.* **527** (1998) 335.
[3]   T. S. Plaskett, *Trans. AIME* **233** (1965) 809.
[4]   T. Abe, T. Samizo and S. Maruyama, *Jpn. J. Appl. Phys.* **5** (1966) 458.
[5]   A. J. R. de Kock, *Appl. Phys. Lett.* **16** (1970) 100.
[6]   H. Foell and B. O. Kolbesen, *Appl. Phys.* **8** (1975) 319.
[7]   H. Foell, U. Goesele and B. O. Kolbesen, *Semiconductor Silicon 1977*, eds. H. R. Huff and E. Sirtl (Electrochem. Soc., Princeton, 1977) p. 565.
[8]   P. M. Petroff and A. J. R. de Kock, *J. Cryst. Growth* **30** (1975) 117.
[9]   J. A. Van Vechten, *Phys. Rev. B* **17** (1978) 3179.
[10] J. Chikawa and S. Shirai, *Jpn. J. Appl. Phys.* **18, Suppl. 18-1** (1979) 153.
[11] E. Sirtl, *Semiconductor Silicon 1977*, eds. H. R. Huff and E. Sirtl (Electrochem. Soc., Princeton, 1977) p. 4.
[12] S. M. Hu, *J. Vac. Sci. Technol.* **14** (1977) 17.
[13] A. J. R. de Kock and W. M. Van de Wijgert, *J. Cryst. Growth* **49** (1980) 718.

[14] P. J. Roksnoer and M. M. B. Van Den Moom, *J. Cryst. Growth* **53** (1981) 563.
[15] V. V. Voronkov, *J. Cryst. Growth* **59** (1982) 625.
[16] T. Y. Tan and U. Goesele, *J. Appl. Phys. A* **37** (1985) 1.
[17] J. Ryuta, E. Morita, T. Tanaka and Y. Shimanuki, *Jpn. J. Appl. Phys.* **29** (1990) L1947.
[18] M. Itsumi, H. Akiya and T. Ueki, *J. Appl. Phys.* **78** (1995) 5984.
[19] E. Wijaranakula, *J. Electrochem. Soc.* **139** (1992) 604.
[20] R. Habu, I. Yunoke, T. Saito, and A. Tomiura, *Jpn. J. Appl. Phys.* **32** (1993) 1740.
[21] R. A. Brown, D. Maroudas, and T. Sinno, *J. Cryst. Growth* **137** (1994) 12.
[22] W. W. Webb, *J. Appl. Phys.* **33** (1962) 1961.
[23] W v. Ammon, E. Dornberger, H. Oelkrug, and H. Weidner, *J. Cryst. Growth* **151** No. **3/4** (1995) 273.
[24] M. Hourai, E. Kajita, T. Nagashima, H. Fujiwara, S. Ueno, S. Sadamitsu, S. Miki and T. Shigematsu, *Mater. Sci. Forum* **196–201** (1995) 1713.
[25] F. Dupret, P. Nicodeme, Y. Ryckmans, P. Wouters, and M. J. Crochet, *Int. J. Heat Mass Transfer* **33** (1990) 1849.
[26] T. Abe, H. Harada and J. Chikawa, *Physica* **116B** (1983) 139.
[27] T. Abe and K. Hagimoto, *Solid State Phenom.* **47–48** (1996) 107.
[28] N. Puzanov, A. Eidenson and D. Puzanov. Abstract of ICCG 12/ICVGE 10, July 1998, Jerusalem p. 342.
[29] P. J. Roksnoer, *J. Cryst. Growth* **68** (1984) 596.
[30] H. Harada, T. Abe and J. Chikawa, *Semiconductor Silicon 1986*, eds. H. R. Huff, T. Abe and B. O. Kolbesen (Electrochem. Soc. Penningston, 1986) p. 76.
[31] T. Abe, *J. Korean Assoc. Cryst. Growth* **9** (1999) 402.

13 Silicon Crystal Growth for Photovoltaics

T. F. CISZEK

National Renewable Energy Laboratory, Golden, CO 804 01-3393, USA

13.1 INTRODUCTION

Unlike silicon crystals used in the electronics industry, crystal perfection, purity, and uniformity are not necessarily highest on the list of desirable attributes for crystalline Si incorporated into commercial photovoltaic (PV) modules. Tradeoffs are routinely made, weighing these attributes against cost, throughput, energy consumption, and other economic factors. In fact, such tradeoffs for PV use have spawned far more alternative growth methods for silicon than has the many decades of semiconductor technology development. Semiconductor applications use the well-known Czochralski (CZ) technique almost exclusively, with a small contribution (on the order of 10 % to 15 %) from float-zone (FZ) growth.

Of course, some FZ material is used in the PV industry, and the highest recorded silicon solar cell efficiency (ratio of cell-output electrical power to solar power incident on the cell), 24 %, has been achieved for devices fabricated on FZ wafers (Zhao *et al.*, 1995). But the device-processing procedures needed to achieve the high efficiencies are expensive and time consuming. So, as in the semiconductor industry, more CZ wafers than FZ wafers are used for PV. What may be surprising though, is that more multicrystalline silicon than single-crystal material is currently used in PV modules. Some of this multicrystalline Si is not wafers from ingots, but rather ribbons or sheets of silicon solidified in a planar geometry.

Of the 152 peak megawatts (MW_p) of PV modules sold throughout the world in 1998, 132 MW_p, or 87 %, were crystalline Si. Comprising this 87 % was ~39 % fabricated from single-crystal Si ingots, ~44 % made from multicrystalline-Si ingots, and ~4 % based on multicrystalline-Si ribbons or sheets (Maycock, 1999). The other 13 % of modules sold are largely amorphous silicon or nonsilicon thin films, which will not be discussed in this chapter. There is an increasing PV research effort focused on thin-layer polycrystalline Si deposited on foreign substrates. These approaches have not yet reached a commercialization stage.

An issue common to all the Si PV growth approaches is the availability of low-cost polycrystalline Si feedstock. The PV industry has in the past relied

Crystal Growth Technology, Edited by H. J. Scheel and T. Fukuda
© 2003 John Wiley & Sons, Ltd. ISBN: 0-471-49059-8

on reject silicon from the electronics industry for use as feedstock. But the PV industry has been growing at an average rate of 20 % over the last five years, which is faster than the growth rate of the electronics industry, and the point has been reached where the supply of reject silicon is insufficient.

13.2 BASIC CONCEPTS

The criteria for crystalline silicon used in PV are somewhat different from those for silicon used in integrated circuits. In this section, the photovoltaic effect is described as a basis for understanding the importance of minority-carrier lifetime in PV device operation. Crystal-growth parameter effects on lifetime are discussed. The light absorption of silicon is relatively low because of the indirect bandgap. This limits how thin a crystalline-Si PV device can be and has some implications for thin-layer Si crystal-growth approaches.

13.2.1 THE PHOTOVOLTAIC EFFECT

A silicon solar cell (shown schematically in Figure 13.1) converts sunlight energy into direct-current (DC) electrical energy by the photovoltaic effect. As sunlight impinges on the top surface, some of the light is reflected off the cell's grid structure and some is reflected by the surface of the cell. The grid reflection loss can range from 3 % for advanced cell designs to as much as 20 % for some screen-printed cells. Antireflection coatings and texturing of the silicon surface can be used to minimize the surface reflection losses. Some light (long-wavelength

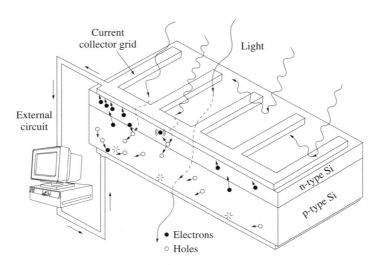

Figure 13.1   Operation of a Si solar cell.

infrared) does not have the threshold energy needed to free electrons from the Si atoms and passes through the cell without interacting. Some light (short-wavelength ultraviolet) has more than enough energy to create the electron–hole pairs. The excess energy transferred to the charge carriers is dissipated as heat. About 40 % of the incident light energy is effectively used in freeing electrons from silicon atoms so that they can wander in the crystal lattice. They leave behind holes, or the absence of electrons, which also wander. The solar cell has a p/n junction, like a large-area diode. The n-type portion (typically near the front of the device) has a high density of electrons and few holes, so generated electrons can travel easily in this region. The opposite is true in the p-type region. There, holes travel easily. The electric field near this junction of n- and p-type silicon causes generated electrons to wander toward the grid on the surface, while the holes wander toward the back contact. If the electrons survive their trip across the cell thickness without recombining at defects or impurities, they are collected at the grid and flow through an external circuit as current that can operate an electronic instrument or appliance. After that, they re-enter the solar cell at the back contact to recombine with holes, and the process repeats.

13.2.2 MINORITY-CARRIER LIFETIME, τ

If some of the generated carriers recombine at defects, impurities, surface damage, etc., before reaching the contacts, the current output of the solar cell is diminished. Because it is a quantitative measure of such phenomena, minority-carrier lifetime (τ) characterization (ASTM, 1993) is frequently used to qualify the crystalline-Si material before it is used in device processing. Quality in a silicon photovoltaic material is nearly synonymous with τ. The parameters used in crystal growth have a direct bearing on τ, because they determine impurity levels and defect structures that give rise to carrier-recombination sites. Impurity incorporation, segregation, and evaporation during the crystal-growth process can be altered via ambient choice, growth rate, number of solidification steps, choice of container, heat-source characteristics, selection of source material, and other factors that vary from one process to another.

Heavy doping imposes an upper limit on lifetime according to $\tau_A = 1/C_A N^2$, where τ_A is the Auger-limit lifetime. C_A is the Auger coefficient, and N is the doping concentration. At the lower limit, $\tau < 2\,\mu s$ is unlikely to be useful in most PV processes due to balance-of-systems costs. Poor-quality material cannot generate enough PV energy to justify the costs of the total PV system. Thus, the $\tau - N$ space available for photovoltaic applications is the nonhatched region in Figure 13.2, and the four labeled curves are 'quality' contours. In addition to the two limits, curves representing moderate τ (typical of Czochralski-grown, CZ, silicon) and high τ (typical of the best commercially available FZ silicon) are included. Note that there is a vast discrepancy between τ_A and the lifetime of the best available silicon. So there is potential for higher lifetimes and new

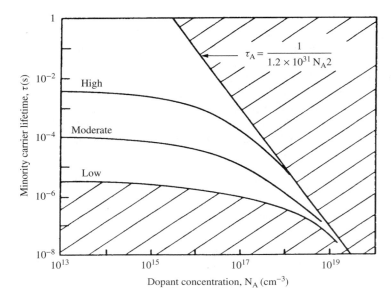

Figure 13.2 Material-quality contours of lifetime versus dopant concentration. The hatched region is not available or suitable for PV use.

device designs to take advantage of it. Transition-metal-impurity effects on τ and solar-cell efficiency as a function of their concentration levels are reasonably well understood from quantitative and detailed experimental studies (Davis *et al.*, 1980). Some metals such as titanium have a significant effect on τ even in concentrations as low as a few 100 ppta (parts per trillion, atomic). Others, such as copper, can be tolerated at a few ppma (parts per million, atomic). Fortunately, most of the detrimental impurities have small effective segregation coefficients, and their concentrations can be reduced during directional solidification (DS).

When no impurities are present in high enough concentration to affect τ, a myriad of structural defects can still act as recombination centers. Grain boundaries and their associated dislocation arrays usually constrain τ to $\leq 20\,\mu$s. The lifetime of Si decreases with decreasing grain area as reported by Ciszek *et al.* (1993) and illustrated in Figure 13.3. Even dispersed dislocations in a single crystal at a density $<5 \times 10^4\,\text{cm}^{-2}$ can reduce τ to $30\,\mu$s in material that, when grown in the same way except dislocation-free, yields $\tau = 450\,\mu$s. If grain boundaries, dispersed dislocations, and transition-metal impurities are present, as may be the case in ingots cast from low-grade silicon feedstock, it is not unusual to see $\tau < 10\,\mu$s.

Si crystals that are free of transition-metal impurities, dislocations, and grain boundaries unveil second-order structural effects on lifetime. These are most easily seen in FZ material because O and C effects somewhat obscure the issue in CZ crystals. Types A and B swirl microdefects (Si interstitial cluster defects)

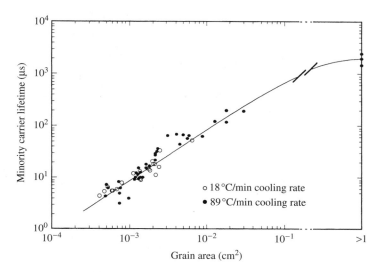

Figure 13.3 The effect of grain size on the minority-carrier lifetime of high-purity multi-crystalline silicon.

are present in dislocation-free FZ crystals that are grown at a speed v that is too slow or in a temperature gradient G that is too large. Eliminating these defects allows $\tau > 1$ ms. When A and B swirls are eliminated, a third-order effect is unveiled – τ varies inversely with cooling rate, the product of $v \times G$, in swirl-free crystals. Thus v should be chosen just fast enough to eliminate swirls, if very high lifetimes are required. The physical nature of this 'fast-cooling' defect is not understood at the present time. By appropriate choice of v and G, Ciszek *et al.* (1989) obtained $\tau > 20$ ms in lightly doped, p-type, high-purity silicon and were able to grow crystals on a quality contour an order of magnitude better than the one labeled 'high' in Figure 13.2.

13.2.3 LIGHT ABSORPTION

Besides variations in τ, another property of silicon that can impact solar-cell performance as a consequence of the growth method – in this case, because of the growth method's effect on geometry – is the absorption coefficient, α, defined as the inverse distance in cm required for the intensity of incident light to fall to 1/e of its initial value. The absorption coefficient is a function of wavelength. Because Si is an indirect-bandgap semiconductor, the absorption edge is not sharp, and for some of the thin-layer growth methods, useful light passes through the silicon without being absorbed. The absorbed useable fraction of photons as a function of Si layer thickness is shown in Figure 13.4 for the typical light spectrum incident on the Earth (Wang *et al.*, 1996). While 100 % of the light is

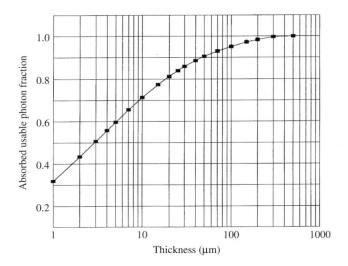

Figure 13.4 Fraction of solar photons absorbed, as a function of silicon thickness for air mass (AM) 1.5 illumination.

used in 300-μm wafers, this falls off to 90 % in 50-μm thin layers and 70 % in 10-μm thin layers. Thus if thin-layer Si solar cells are to be effective, it will be necessary to enhance the optical path length by appropriate surface coatings or texturing to cause multiple passes of the light in the thin structure. An advantage of thin Si layers is that shorter minority-carrier lifetimes can be tolerated because the generated carriers do not have as far to travel before reaching the contacts.

13.3 SILICON SOURCE MATERIALS

Silicon is the second most abundant element in the crust of the Earth (27 %), but it does not occur as a native element because SiO_2 is more stable. Many processing steps (Figure 13.5) are conducted to bring Si from its native ore, quartzite, to the crystalline substrates we use for solar cell fabrication or integrated circuit (IC) components. The starting silicon for both PV and IC applications is 99 % pure metallurgical-grade (MG) Si obtained via the carbon reduction of SiO_2 in an arc furnace. Although the overall reaction can be considered to be

$$SiO_2 + C \longrightarrow Si + CO_2 \tag{13.1}$$

there are a complex series of reactions that take place in different temperature regions of the arc furnace, with liquid Si finally forming from SiC. The Si liquid is periodically tapped from the furnace and typically allowed to solidify in shallow molds about 1.5×1 m in size. The major impurities are Fe, Al, and C. This

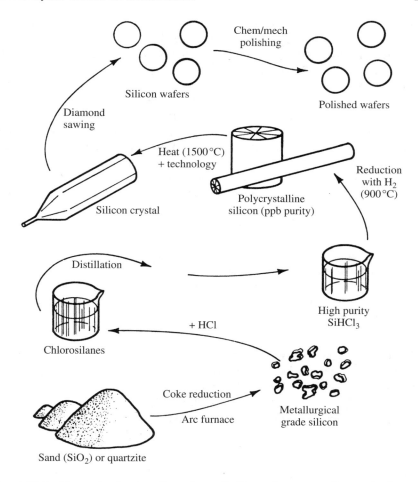

Figure 13.5 Stages in the transformation of silicon from quartzite to single-crystal wafers.

MG-Si material is inexpensive (~$1/kg), but the residual impurities degrade τ to unacceptably low values and some, like B and Al, electronically dope it too heavily for PV devices. B is particularly bothersome because it neither segregates nor evaporates significantly during melt processing.

IC industry chlorosilane purification and deposition steps increase the purity to more than adequate levels for PV use (99.999999 %), but also increase the cost to unacceptable levels for PV manufacturers. The current price of polysilicon from the chlorosilane IC process is about $50/kg, while PV users have a target price of about $20. So the Si PV industry has been using reject material from IC polysilicon and single-crystal production – material that is too impure for IC use but adequate for PV use. The IC industry now consumes about

20 000 metric tons of Si annually, and the PV industry uses about 3000 metric tons. But as production techniques improve and as the Si PV industry grows at a faster rate than the IC industry, the supply of reject material is becoming insufficient. New sources of polysilicon will be needed (Mauk *et al.*, 1997; Mitchell, 1998). Demand first exceeded supply of reject Si in 1996. The subsequent downturn in the IC industry temporarily relieved the PV feedstock shortage, but projections by one of the large polysilicon manufacturers indicate that the PV demand for reject Si will exceed the supply (8000 metric tons/yr) by a factor of 2 to 4 by the year 2010 (Maurits, 1998). This does not represent a fundamental material shortage problem, because the technology, quartzite, and coke needed to make feedstock are in abundant supply. The issue is to supply feedstock with necessary but only sufficient purity ($\sim$99.999 %) at an acceptable cost.

The trichlorosilane ($SiHCl_3$) distillation method is used to purify Si for more than 95 % of polysilicon production (even if converted to other chlorosilanes or silane for reduction), but is very energy intensive. It produces large amounts of waste, including much of the starting silicon and a mix of environmentally damaging chlorinated compounds. In addition, the feedstock produced by reduction following this distillation method exceeds the purity requirements of the PV industry, which are estimated to be:

- Electrical dopant impurities, iron, and titanium, each $<\sim$0.1 ppma (they will be reduced further in the crystal-growth process by impurity segregation)
- Oxygen and carbon concentrations below the saturation limits in the Si melt (i.e., no SiC- or SiO_2-precipitate formation)
- Total other nondopant impurities $<\sim$1 ppma.

Fresh approaches are needed to originate novel separation technologies that can extract B, Al, P, transition metal impurities, and other impurities from MG silicon, to meet the purity requirements listed above – but in a simpler process. Examples of new approaches that are in the early stages of investigation are: (i) use of electron-beam and plasma treatments with several DS steps to remove impurities from MG-Si (Nakamura *et al.*, 1998); (ii) directly purifying granular MG-Si using repeated porous-silicon etching, subsequent annealing, and surface impurity removal (Menna *et al.*, 1998); (iii) a method that uses MG-Si and absolute alcohol as the starting materials (Tsuo *et al.*, 1998); (iv) vacuum and gaseous treatments of MG-Si melts coupled with directional solidification (Khattak *et al.*, 1999) guided by thermochemical calculations (Gee *et al.*, 1998); and (v) use of impurity partitioning when silicon is recrystallized from MG-Si/metal eutectic systems (Wang and Ciszek, 1997). Approaches like these or other yet-to-be-determined innovative methods could have a major impact on the continued success and growth of the Si PV industry – especially if one is discovered that is intrinsically simpler than current technology, yet yields adequate Si purity.

13.4 INGOT GROWTH METHODS AND WAFERING

Basically, four methods are used to grow Si ingots for commercial PV use. These techniques are shown schematically in Figure 13.6. Two methods produce single-crystal material (FZ and CZ techniques shown in Figures 13.6(a) and 13.6(b),

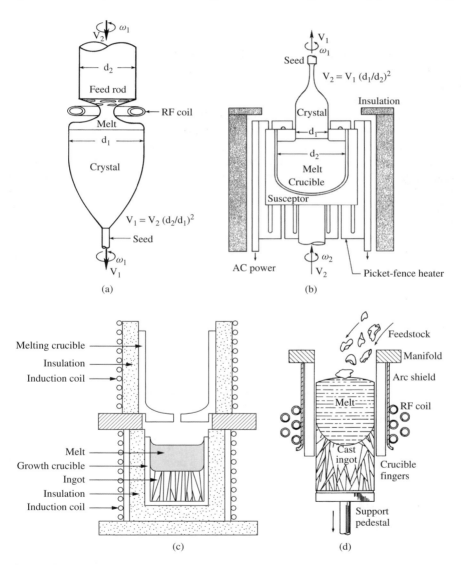

Figure 13.6 Schematic drawings of the four principal methods used to produce Si ingots for PV use: (a) float-zoning; (b) Czochralski growth; (c) directional solidification, and (d) electromagnetic semicontinuous casting.

respectively). Two others make large-grain (mm-to-cm grain size) multicrystalline silicon (directional solidification in square crucibles as shown in Figure 13.6(c), and electromagnetic semicontinuous casting shown in Figure 13.6(d)).

13.4.1 SINGLE-CRYSTAL GROWTH

The single-crystal growth methods, FZ and CZ, are relatively well known, and only some aspects pertinent to PV applications will be addressed here. Table 13.1 compares the characteristics of the FZ and CZ methods. There are two principal technological advantages of the FZ method for PV Si growth. The first is that large τ values are obtained due to higher purity and better microdefect control, resulting in 10 % to 20 % higher solar-cell efficiencies. The second is that faster growth rates and heat-up/cool-down times, along with absences of a crucible and consumable hot-zone parts, provide a substantial economic advantage. The main technological disadvantage of the FZ method is the requirement for a uniform, crack-free cylindrical feed rod. A cost premium (100 % or more) is associated with such poly rods. At the present time, FZ Si is used for premium high-efficiency cell applications and CZ Si is used for higher-volume, lower-cost applications.

Electrical power requirements for these two methods are on the order of 30 kWh/kg for FZ growth and 60 kWh/kg for CZ growth in the IC industry. The more cost-conscious PV industry has been achieving 35–40 kWh/kg for CZ growth, and some recent experiments indicate that levels on the order of 18 kWh/kg may be achieved for 150-mm diameter crystals by using improved insulation materials and lower argon gas-flow rates (Mihalik *et al.*, 1999). Not

Table 13.1 Comparison of the CZ and FZ growth methods

Characteristic	CZ	FZ
Growth speed (mm/min)	1 to 2	3 to 5
Dislocation-free?	Yes	Yes
Crucible?	Yes	No
Consumable Material Cost	High	Low
Heat-up/cool-down times	Long	Short
Axial resistivity uniformity	Poor	Good
Oxygen content (atoms/cm^3)	$>1 \times 10^{18}$	$<1 \times 10^{16}$
Carbon content (atoms/cm^3)	$>1 \times 10^{17}$	$<1 \times 10^{16}$
Metallic impurity content	Higher	Lower
Bulk minority charge carrier lifetime (μs)	5–100	100–20 000
Mechanical strengthening	10^{18} oxygen	10^{16} nitrogen
Production diameter (mm)	150–200	100–150
Operator skill	Less	More
Polycrystalline Si feed form	Any	Crack-free rod

only were energy requirements reduced, but also argon consumption was reduced from $3\,m^3/kg$ of Si to $1\,m^3/kg$ of Si. Also, oxygen content in the crystals was reduced by 20 %, crystal growth rate was increased from 1.28 kg/h to 1.56 kg/h, and relative solar-cell efficiency increased by 5 %.

In CZ Si PV technology, approximately 30 % of the costs are in the crystal ingot, 20 % in wafering, 20 % in cell fabrication, and 30 % in module fabrication. High-speed wire saws that can wafer one or more entire ingots in one operation have greatly improved the throughput of the wafering process. A wire saw can produce about 500 wafers/h compared to about 25 wafers/h for older inside-diameter (ID) saw technology. Furthermore, it creates shallower surface damage $(10\,\mu m)$ than the ID saws $(30\,\mu m)$, and allows thinner wafers to be cut, thus increasing the number of wafers per ingot. Currently about 20 wafers are obtained from 1 cm of ingot. Efforts are underway to obtain 35 wafers/cm. Problems with increased breakage are seen with the thinner wafers – especially in the sawing process. At 20 wafers/cm and a wafer thickness $>300\,\mu m$, breakage is on the order of 15 %. This can rise to on the order of 40 % when the wafer thickness is decreased to $200\,\mu m$. It is clear that wafer handling will be an important issue as wafers become thinner.

One clever way of dealing with the low fracture strength of thin (100) wafers from single-crystal CZ ingots is to deliberately introduce a controlled multicrystalline structure into the growing ingot. In particular, the tricrystalline structure described by Martinelli and Kibizov (1993) provides three grains propagating along the length of the ingot. Each has a $\langle 110 \rangle$ longitudinal direction. Two of the grain boundaries are first-order {111} twin planes, and the third is a second-order {221} twin plane. The three angles between boundaries are thus 125.27°, 125.27°, and 109.47°. The twins block any {111} planes from crossing across the entire ingot, and improve the resistance to cleavage or propagation of defects that takes place on {111} planes. Wafers from these tricrystals are observed to possess about 440 MPa fracture strength compared to about 270 MPa for (100) single-crystal wafers and 290 MPa for multicrystalline wafers (Endros et al., 1997). The measurements were made on wafers after etching in KOH at 100 °C to 310-μm thicknesses. Breakage during wire sawing of tricrystal ingots at $<200\,\mu m$ thickness is half of that for $\langle 100 \rangle$ ingots. The tricrystal ingots have been shown to maintain their structure for reasonably long lengths (150–400 mm at the present state of technology) with minimal degradation of minority-carrier-recombination properties (Wawer et al., 1997).

13.4.2 MULTICRYSTALLINE GROWTH

A multicrystalline structure with grains on the order of mm to cm in width and approximately columnar along the solidification direction is characteristic of DS or casting in crucibles and also of electromagnetic semicontinuous casting. Casting of multigrain silicon was reported more than 80 years ago (Allen, 1913).

In 1960, large multicrystalline-silicon domes and plates 330 mm in diameter by 50 mm thick were formed via ambit casting. Early applications were for infrared optics and other nonsemiconductor uses. A summary of early silicon casting approaches was compiled by Runyan (1965). Si casting into graphite molds as an alternative to CZ growth for PV applications was reported by Fischer and Pschunder (1976).

DS can be carried out in a separate crucible after (or as) silicon is poured into it from a melting crucible, as indicated in Figure 13.6(c). This process is usually referred to as silicon casting. Or the silicon can be melted and directionally solidified in the same crucible (e.g., the bottom crucible in Figure 13.6(c)). This technique is referred to as directional solidification. It is simpler than casting, because no melt pouring is involved, but there are longer reaction times at high temperature between the melt and the crucible, and longer turnaround times. Because melting and solidification are decoupled in casting, higher throughputs are possible. But the process and equipment are more complex.

DS of silicon in the same quartz (fused silica) crucible used to melt it would seem to be a logical approach for large-grained ingot production, but sticking and thermal expansion mismatch between the solidified Si and the crucible lead to significant cracking problems. This was alleviated to some extent by deliberately weakening the inner wall of the quartz crucible (Khattak and Schmid, 1978). High-density graphite crucibles were introduced for DS as a way of avoiding the cracking (Ciszek, 1979). Most crucible-based commercial methods of growing multicrystalline-Si ingots for PV consumption use either the casting method or DS in the melting crucible. The crucibles are either graphite or quartz (often coated with Si_3N_4 or other compounds to discourage sticking and cracking or enable reuse). The actual growth process is DS in either method.

Multicrystalline ingots as large as 690 mm × 690 mm in cross section and weighing as much as 240 kg are grown in total cycle times of 56 h. The resultant throughput is 4.3 kg/h. Thus the larger area, compared to CZ crystals, more than offsets the somewhat lower linear growth rates leading to higher throughputs for DS by a factor of $\sim$3. Both induction heating, as shown in Figure 13.6(c), and resistance heating can be used. The energy consumption for DS is in the range of 8–15 kWh/kg. Unlike CZ growth, the solid/liquid interface is submerged in DS, and precipitates or slag at the melt surface do not disrupt growth. DS is a simpler process requiring less skill, manpower, and equipment sophistication than CZ, which can make it a lower cost process. However, there are also drawbacks. There are numerous crystal defects (grain boundaries and dislocations) due to the multicrystalline structure. Impurity contents can be higher depending on the crucibles used, and portions of the bottom, sides, and top surface of the ingot are discarded. So the lower cost of DS is at the expense of solar-cell efficiency. DS solar cells are about 85 % as efficient as CZ cells. The best efficiency of small cells, with sophisticated processing, is 18.6 %. Typical large production cell efficiencies are 13–14 %, with good consistency.

Electromagnetic casting (EMC) has some similarities to the casting and DS methods described above, but also has several unique features that change the ingot properties and warrant a separate discussion. The method was first applied to semicontinuous silicon ingot casting by Ciszek (1985, 1986). EMC is based on induction-heated cold-crucible melt confinement, except that unlike the conventional cold crucible, there is no crucible bottom. A parallel, vertical array of close-spaced, but not touching, water-cooled, conducting fingers is attached at one end to a water-cooling manifold. The other end of each finger is closed. An internal distribution system carries cooling water to the tip and back again. The shape of the region enclosed by the close-spaced fingers determines the cross section of the cast ingot, and a wide variety of shapes are possible (circular, hexagonal, square, rectangular, etc.). Silicon is melted on a vertically moveable platform (typically graphite) located within the finger array. The melting is accomplished by induction heating after suitable preheating. The induction coil, placed outside the finger array (Figure 13.6(d)), induces a current to flow on the periphery of each finger, around the finger's vertical axis. Like a high-frequency transformer, each finger in turn induces a current to flow in the periphery of the silicon charge, about its vertical axis. The silicon is heated by its resistance to the current flow. There is a Biot–Savart-law repulsion between the current flowing in the periphery of the silicon melt and the currents flowing in the fingers, because they are induced to flow in opposite directions at any particular instant in the RF cycle. Thus, the melt is repulsed from the water-cooled fingers. The open-bottom arrangement allows the platform to be withdrawn downward, solidifying the molten silicon, while new melt is formed by introducing feed material from the top. In this way, a semicontinuous casting process can be carried out.

A variety of feed silicon geometries can be used (melts, rods, pellets, scrap, etc.). Because the interface is submerged, feed perturbances or slag at the melt surface do not affect the solidification front. Ingot lengths of nearly 3 m have been demonstrated. The cross section of the ingots has evolved over years of development and is currently about 350 mm × 350 mm. The cold fingers allow steep thermal gradients and fast growth speeds ($\sim$1.5–2 mm/min), even in ingots with large cross sections. But they also cause a steeply curved interface that is concave toward the melt. Thus, grains are neither as columnar nor as large as in conventional DS. The average grain size is on the order of 1.5 mm in large ingots. This decreases τ, but the relatively high purity and freedom from oxygen and carbon impurities (O $< 6 \times 10^{15}$; C $< 8 \times 10^{16}$) largely offset the grain-size effect, so that solar-cell efficiencies of about 14–15 % are obtained on 15 cm × 15 cm cells. The throughput of EMC is the highest of any ingot growth technique – up to approximately 30 kg/h. The power consumption can be as low as 12 kWh/kg.

13.5 RIBBON/SHEET GROWTH METHODS

More than a dozen techniques have been introduced over the years for growing silicon ribbons or sheets. Only the ones currently in use for commercial

PV substrates will be addressed here. For a description of the others, see Ciszek (1984). The four methods being developed commercially are dendritic web growth, shown in Figure 13.7(a), growth from a capillary shaping die (Figure 13.7(b)), growth with edge supports or 'strings' (Figure 13.7(c)), and growth on a substrate (Figure 13.7(d)). These methods can be placed in two categories: (i) those pulled perpendicular to a solid/liquid interface with the same shape as the ribbon cross section (web growth, capillary die growth, and edge-supported growth), and (ii) those pulled at a large angle to a solid/liquid interface that is much greater in area than the cross section of the sheet (growth on a substrate). There is a large difference in the limiting pulling rate v between type (i) and type (ii). For type (i) growth,

$$v_I = \frac{1}{L\rho_m} \left(\frac{\sigma\varepsilon(W+t)K_m T_m^5}{Wt} \right)^{1/2} \tag{13.2}$$

where L is latent heat of fusion, ρ_m is density at the melting temperature, σ is the Stefan–Boltzmann constant, ε is emissivity, K_m is the thermal conductivity at the melting temperature T_m, W is the ribbon width, and t is the ribbon thickness (Ciszek, 1976). For type (ii) growth,

$$v_{II} = \frac{4\alpha K_m b}{(2K_m - \alpha t)t L\rho_m} \Delta T \tag{13.3}$$

where α is the effective coefficient of heat transfer, b is the length of the solid/liquid interface (in the pulling direction), and ΔT is the temperature gradient between melt and substrate (Lange and Schwirtlich, 1990). For the case of a 250-μm thick ribbon, Equation (13.2) predicts a maximum type (i) growth rate of $\sim$8 cm/min. Experimentally, rates closer to 2 cm/min are realized. Equation (13.3) predicts a 6-m/min growth rate at $\Delta T = 160\,^\circ$C, and experimental pulling speeds near that value were realized. The indication is that type (ii) growth speeds can be hundreds of times faster than type (i) vertical pulling approaches, especially if b and ΔT are maximized.

Dendritic web growth, the oldest Si ribbon growth method, was introduced by Dermatis and Faust (1963). The technique arose from the observation that long, thin, flat dendrites with a (111) face and $\langle 2\bar{1}\bar{1} \rangle$ growth direction could be pulled from Ge and Si melts. As Figure 13.7(a) indicates, one such dendrite is used as a seed and a thermally defined 'button' is grown laterally from it. Then upward pulling is commenced with appropriate melt-temperature adjustments such that a dendrite of the same orientation propagates from each end of the button. A web of crystalline silicon solidifies between the dendrites. It is a single crystal except for an odd number (1, 3, 5, etc.) of twin planes in the central region. Web ribbons are currently grown at about 1.5 to 2 cm/min pulling rates, with a width of $\sim$5 cm, a thickness of 100 μm, and in lengths up to 100 m with continuous melt replenishment ($\sim$0.25 g/min). Furnace runs are typically one week in duration, and produce

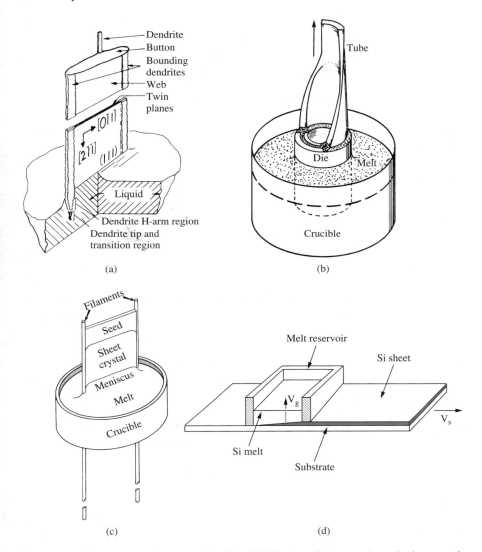

Figure 13.7 Schematic drawings of the four Si ribbon or sheet-growth methods currently under commercial development: (a) dendritic web growth; (b) growth from a capillary shaping die; (c) growth with edge support strings, and (d) growth on a substrate.

more than 1 m²/day. Material properties do not degrade over 100-m lengths. Dislocation etch pit densities are about $10^4/cm^2$, and τ is on the order of 100 μs or less. Growth is conducted from an 8-mm-deep melt contained in a shallow, rectangular quartz crucible. Thermal control is very important, not just for initiating the web, but also to maintain steady growth with proper dendrite propagation characteristics at the ribbon edges, low thermal stresses in the ribbon region, and continuous melt replenishment without disturbing the growing web ribbon. Edge dendrite thickness stability is an excellent indicator of melt-temperature stability. Both induction heating with molybdenum hot zones and resistance heating with graphite heaters and hot zones have been used. The electrical energy used for growth is about $200-300\,kWh/m^2$. The thin material is particularly well suited for PV applications that require some bending flexibility, or for bifacial solar-cell applications. Since the material is nearly single crystalline, relatively high cell efficiencies can be achieved. The best reported value is 17.3 % for a 4-cm^2 cell. Initial production cell efficiencies are expected to be $\sim$13 %. One growth furnace can produce web for about 50 kWp/y cell production.

Growth of crystals from the tips of capillary shaping dies was introduced for sapphire growth using molybdenum dies by LaBelle *et al.* (1971), and was first applied to silicon ribbons using graphite shaping dies by Ciszek (1972) and later to silicon tubes (Ciszek, 1975). As shown in Figure 13.7(b), liquid Si rises by capillarity up a narrow channel in the shaping die and spreads across the die's top surface, which defines the base of the meniscus from which the shaped crystal solidifies. The meniscus base is typically wider than the wall thickness of the crystal. Commercial development first concentrated on flat ribbons as wide as 100 mm, but edge-stability issues led to a preference for the tubular geometry (i.e., edges are eliminated). Octagonal tubes with 100-mm-wide flat faces are now used for production of PV substrates. Pulling rates are comparable to those used in web growth, but the 800-mm effective width increases the throughput to about $20\,m^2$/day. A graphite crucible and graphite shaping dies are used with induction heating. The electrical energy consumption for this method is approximately $20\,kWh/m^2$. After growth, rectangular 100-mm-wide 'wafers' are laser-cut from the tube faces. They provide 275-μm thick multicrystalline substrates with longitudinal grains that routinely make 14 % efficient solar cells. The best efficiency attained on a 10-cm^2 cell is 15.5 %. The capillary-die method is somewhat more susceptible to impurity effects from solar-grade feedstock than other methods, because the narrow channel impedes mixing of segregated impurities back into the melt and thus increases the effective segregation coefficient.

Edge-supported pulling of 'string ribbons' was introduced by Ciszek and Hurd (1980). This technique is similar to dendritic web growth with foreign filaments or strings replacing the edge-stabilizing role of the dendrites (Figure 13.7(c)). This greatly relaxes the temperature-control requirements and makes the technique easier to carry out than dendritic web growth. Simpler equipment can be used. A variety of carbon- and oxide-based materials were investigated for use as the filaments, with carbon-based filaments generating a higher density of grains at the edges of the ribbons than oxide-based filaments, but having a better thermal expansion match to silicon. The filaments are introduced through small holes in the bottom of either quartz or graphite crucibles. Ribbons as wide as 8 cm have been grown, with the standard commercial size now being 5.6 cm wide $\times$ 300 μm thick. The ribbons are grown at about $1-2$ cm/min pulling rates, giving a throughput of about $1\,m^2/day$, which is comparable to that obtained with web growth. Furnaces can be kept in continuous operation for weeks at a time by replenishing the melt. Ribbon sections of a desired length are removed by scribing while pulling is in progress. Continuous growth of more than 100 m of ribbon has been achieved, and lengths greater than 300 m have been obtained from a single furnace run (with successive seed starts). Dislocation densities are ~5 × $10^5/cm^2$ and τ is in the range $5-10\,\mu s$. The highest cell efficiency obtained is 16.3 %, although production efficiencies are <13 %. The steady-state grain structure contains longitudinal grains of about $1\,cm^2$ area, predominantly with coherent boundaries, in the central portion of the ribbons, and newly generated grains at the ribbon edges. The electrical energy used is about $55\,kWh/m^2$.

The first application of type (ii) sheet growth to a semiconductor material was by Bleil (1969), who pulled ice and germanium sheet crystals horizontally from the free surface of melts in a brim-full crucible. Many approaches have been considered for applying type (ii) growth to photovoltaic silicon, including horizontal growth from the melt surface. The ones currently under commercial development move a substrate through a hot zone tailored in such a way that a long region of molten silicon in contact with the upper surface of the substrate solidifies with a long wedge-shaped crystallization front. The front grades from 0 thickness at the tip to the sheet thickness t (where the sheet leaves the melt), over a distance b. An example of one such approach is shown in Figure 13.7(d). As indicated in Equation (13.3), the pulling speed is proportional to b/t and to ΔT. It is feasible to make b very large, on the order of tens of centimeters. Coupled with moderate ΔT values (160 °C), 250-μm thick sheets can then be grown with pulling speeds V_s as high as 6 m/min as mentioned earlier (Lange and Schwirtlich, 1990). If W is also tens of centimeters, extremely high throughputs can be achieved – in the vicinity of $1\,500\,m^2/day$. Heat removal is facilitated by the fact that the surface in which heat of crystallization is generated is nearly parallel to, and in close proximity to, the surface from which it is to be removed. The solid/liquid interface's growth direction V_g is essentially perpendicular to the pulling direction V_s. So, as grains nucleate at the substrate surface, their growth is columnar across the thickness of the sheet. This is in contrast to longitudinal grains aligned along the pulling direction obtained in the type (i) techniques in which V_g and V_s are 180° apart, pointing in opposite directions. The grains tend to be smaller in type (ii) growth methods, but are on the order of t. Production solar cell-efficiencies as high as 12 % are attainable at the present time, and the best small-cell efficiency is 16 %. The substrate does not have to remain with the grown sheet, and may be engineered for clean separation at some point after solidification.

13.6 THIN-LAYER GROWTH ON SUBSTRATES

Thin-layer Si is considered to be <50-μm thick silicon deposited on a foreign substrate. Potential advantages of thin-layer approaches include less Si usage, lower deposition temperatures relative to melt growth, monolithic module construction possibilities, and a tolerance for lower τ (the distance charge carriers have to travel is shorter). Disadvantages include incomplete light absorption (see Figure 13.3) and the probable need for light-trapping, a likelihood that grain sizes will be small, and difficulty in making rear contacts if the substrate is an insulating material.

The R&D challenge for successful thin-layer Si is to produce a 10–50-μm silicon layer of sufficient electronic quality with a diffusion length greater than the layer thickness and a grain size comparable to the thickness. A fast deposition rate of >1 μm/min on a low-cost substrate such as glass is needed. There is not yet any significant quantity of thin-layer crystalline Si in commercial production for PV because only partial successes have been achieved in meeting the challenge. What has been accomplished is:

- Fast epitaxy (1 μm/min) of high-quality Si layers at intermediate temperatures (700–900 °C), e.g., by liquid-phase epitaxy (LPE) but on Si substrates
- Low-T (<600 °C) epitaxial growth of high-quality Si layers (e.g., by chemical vapor deposition (CVD)), but at low growth rates (<0.05 μm/min) and on Si substrates.
- Low-T poly/microcrystalline 10 % cells, but at slow growth rates
- Low-T micro/amorphous direct-gap 13 % cells, but at slow growth rates
- Fast CVD at intermediate T on foreign substrates, but with submicrometer grain size
- Fast CVD of >1-μm grain-size layers on foreign substrates, but at high T (~1200 °C) and with contamination
- Smooth Si at intermediate T by solid-state crystallization, but at slow rates, from slowly grown a-Si layers, and highly stressed.

A new approach to iodine vapor transport growth (Wang and Ciszek, 1999) shows considerable promise, and has achieved 5–20-μm-thick Si layers with 5–10 μm grain size at 1–10-μm/min growth rate directly on hi-T glass at 850–950 °C. The layers have 5 μs effective minority-carrier lifetime, which implies diffusion length ≫ layer thickness and a low impurity content. A scanning electron microscopy (SEM) photomicrograph of a layer is shown in Figure 13.8.

Figure 13.8 SEM micrograph showing the grain size of a 20-μm-thick poly-Si layer grown directly on a glass substrate at 900 °C for 10 min (2 μm/min deposition rate).

13.7 COMPARISON OF GROWTH METHODS

Table 13.2 summarizes some of the technological characteristics of the methods used to grow silicon crystals for photovoltaic applications. A variety of approaches have viability for further development, which is a strong point for Si PV commercial growth. As mentioned in the introduction, cost-driven tradeoffs are made in Si PV crystal-growth technology. This is evident in both the ingot and sheet-growth approaches. For example, the highest throughput ingot method, electromagnetic casting, yields lower cell efficiencies because of smaller grain sizes. A similar situation is seen in ribbons and sheets, where substrate melt shaping has tremendous throughput potential. Again, the grain size and cell efficiency are smaller than for some of the lower-throughput ribbon methods. The diversity and redundancy in approaches is healthy for the industry and increases the probability that further reductions in PV module cost will be achieved.

13.8 FUTURE TRENDS

The PV industry is expected to continue to grow at an annual rate of about 20 %. The well-established technology base and ready availability, proven performance, and salubrity of silicon, coupled with economies of scale in larger factories, will likely allow Si to remain the dominant PV material for the foreseeable future. The demand for off-specification polycrystalline silicon feedstock for PV use is likely to exceed the available supply by a factor of at least 2 within the next 10 years, and this will probably be an impetus for development of alternative feedstock material with adequate but not excessive purity levels.

In ingot growth, the trend for single crystals will be away from the smaller 100- and 125-mm diameter sizes with more focus on 200-mm diameters. Despite the potential advantages of FZ material, it is unlikely that its role in PV will increase significantly because of higher costs for the crack-free, long cylindrical feedstock it requires and difficulty in producing the larger FZ diameters. In CZ growth, we are likely to see an increased effort to make hot zones more energy efficient, to grow larger diameters, and to achieve continuously melt-replenished long growth runs. An effort will continue to evaluate tricrystalline growth or other means of strengthening the ingots so as to improve breakage yields for thin wafers. There will be a continuing effort to achieve more wafers per length of ingot, and to take advantage of potentially higher cell efficiencies afforded by thinner wafers when back surface fields are used in the cell design. Multicrystalline casting, directional solidification, and electromagnetic casting are commanding an increasing share of the Si PV market (53 % of all ingot-based modules sold in 1998 were multicrystalline). This trend is likely to continue because the processes and equipment are simpler and the throughputs are higher (especially for electromagnetic casting) by a factor of 2.5 to 20.

In the ribbon- and sheet-growth technologies, a challenge for dendritic web growth and edge-supported pulling will be to increase areal throughput via wider

Table 13.2 Comparison of the silicon crystal-growth techniques used for photovoltaics

Method	Width (cm)	Weight (kg)	Growth rate (mm/min)	Growth rate (kg/h)	Throughput (m²/day)*	Energy use (kWh/kg)	Energy use (kWh/m²)**	Efficiency (typical %, best %)
Float-zone	15	50	2–4	4	80	30	36	<18,24
Czochralski	15	50	0.6–1.2	1.5	30	18–40	21–48	<15,20
Directional solidification	69	240	0.1–0.6	3.5	70	8–15	9–17	<14,18
Electromagnetic casting	35	400	1.5–2	30	600	12–20	14–24	<13,16
Dendritic web	5	–	12–20	–	1	–	200	<15,17
Capillary die growth	80	–	15–20	–	20	–	20	14,16
Edge-supported pulling	8	–	12–20	–	1.7	–	55	<13,16
Substrate melt shaping	20	–	1000–6000	–	>1000	–	–	<12,16
Thin-layer Si	2	–	$10^{-3}\perp$	–	–	–	–	– 13

*Areal throughput for ingots assumes 20 wafers/cm
– indicates data are not available or not appropriate
**Only the energy for growth is included
$\perp$ deposition rate perpendicular to the substrate

ribbons, multiple ribbons, or other approaches. Even though these methods have the advantage of minimal silicon consumption and elimination of wafering, it is unlikely that they can effectively compete with their current throughput of $1-2 \, m^2$/day. This is because the effective areal throughputs of ingot growth range between 30 and 600 m^2/day, and other sheet technologies produce 20 m^2/day to >1000 m^2/day. While capillary die growth of octagons produces about 20 m^2/day, experimentation is underway to grow large-diameter, thin-walled circular tubes (as depicted in Figure 13.7(b)) a meter in diameter and much thinner than current octagonal tubes. This would increase throughput to more than 75 m^2/day. So far, tubes with 0.5-m diameter have been made, effectively doubling the current octagon areal throughput (Roy *et al.*, 1999). We will probably see continued progress in horizontally pulled, large-area solid/liquid interface sheets by some variant of the method shown in Figure 13.7(d), because the throughput potential is enormous and one growth furnace could easily generate material for 35 MWp/year or more of solar-cell production.

The future is expected to bring continued exploration of thin-layer Si growth approaches, in search of ones that have significant economic advantages over the best ingot and sheet techniques. Successful ones will have fast deposition rates, large grain sizes, high efficiencies (at least 14 % production efficiency) compatibility with low-cost substrates, and amenability to low-cost cell-fabrication schemes. It is not likely that production of thin-layer Si PV modules will be a significant fraction of the mainstream PV market for at least 10 years, although they, like the ingot and sheet approaches, would have substantial advantages over many other thin-film PV approaches. These include the simple chemistry and relative abundance of the Si starting material. The Earth's crust contains 27.7 % Si, in contrast to 0.00002 % Cd, 0.00001 % In, 0.000009 % Se, and 0.0000002 % Te (commonly used thin-film elements). In addition, crystalline Si benefits from an extremely well-established technology base, compatibility with SiO_2 surface passivation, relative salubrity with respect to toxicity, and stability under light exposure.

REFERENCES

Allen, T. B. (1913) U.S. Patent 1,073,560.
ASTM (1993) F28-91 Standard. *1993 Annual Book of ASTM Standards*, Vol. **10.05**, Philadelphia, American Society for Testing and Materials, 30.
Bleil, C. E. (1969) *J. Cryst. Growth* **5**, 99.
Ciszek, T. F. (1972) *Mater Res. Bull.* **7**, 731.
Ciszek, T. F. (1975) *Phys. Stat. Sol. (a)* **32**, 521.
Ciszek, T. F. (1976) *J. Appl. Phys.* **47**, 440.
Ciszek, T. F. (1979) *J. Cryst. Growth* **46**, 527.
Ciszek, T. F. (1984) *J. Cryst. Growth* **66**, 655.
Ciszek, T. F. (1985) *J. Electrochem. Soc.* **132**, 963.
Ciszek, T. F. (1986) U.S. Patent 4,572,812.

Ciszek, T. F., and Hurd, J. L. (1980) in: *Proceedings of the Symposia on Electronic and Optical Properties of Polycrystalline or Impure Semiconductors and Novel Silicon Growth Methods* (K. V. Ravi and B. O'Mara, eds) p. 213. The Electrochemical Soc., *Proceedings Volume* 80-5, Pennington, NJ.

Ciszek, T. F., Wang, Tihu, Schuyler, T., and Rohatgi, A. (1989) *J. Electrochem. Soc.* **136**, 230.

Ciszek, T. F., Wang, T. H., Burrows, R. W., Wu, X., Alleman, J., Tsuo, Y. S., and Bekkedahl, T. (1993) *23rd IEEE Photovoltaic Specialists Conf. Record*, Louisville (IEEE, New York), 101.

Davis, J. R., Jr., Rohatgi, A., Hopkins, R. H., Blais, P. D., Rai-Choudhury, P., McCormick, J. R., and Mollenkopf, H. C. (1980) *IEEE Trans. Electron. Devices* **ED-27**, 677.

Dermatis, S. N., and Faust Jr., J. W. (1963) *IEEE Trans. Commun. Electron.* **82**, 94.

Endros, A. L., Einzinger, R., and Martinelli, G. (1997) *14th European Photovoltaic Solar Energy Conference Proceedings*, Barcelona, 112.

Fischer, H., and Pschunder, W. (1976) *IEEE 12th Photovoltaic Specialists Conf. Record*, IEEE, New York, 86.

Gee, J. M., Ho, P., Van Den Avyle, J., and Stepanek, J. (1998) in: *Proceedings of the 8th NREL Workshop on Crystalline Silicon Solar Cell Materials and Processes*, ed: B. L. Sopori (August 1998 NREL/CP-520-25232), 192.

Khattak, C. P., Joyce, D. B., and Schmid, F. (1999) in: *Proceedings of the 9th NREL Workshop on Crystalline Silicon Solar Cell Materials and Processes*, ed: B. L. Sopori (August 1999 NREL/BK-520-26941), 2.

Khattak, C. P., and Schmid, F. (1978) *IEEE 13th Photovoltaic Specialists Conf. Record*, IEEE, New York, 137.

LaBelle, H. E., Mlavsky, A. I., and Chalmers, B. (1971) *Mater. Res. Bull.* **6**, 571, 581, 681.

Lange, H., and Schwirtlich, I. A. (1990) *J. Cryst. Growth* **104**, 108–112.

Martinelli, G. and Kibizov, R. (1993) *Appl. Phys. Lett.* **62**, 3262.

Mauk, M. G., Sims, P. E., and Hall, R. B. (1997) *Am. Ins. of Phys. Conf. Proc.* **404**, 21.

Maurits, J. (1998) in: *Proceedings of the 8th NREL Workshop on Crystalline Silicon Solar Cell Materials and Processes*, ed: B. L. Sopori (August 1998 NREL/CP-520-25232), 10.

Maycock, P. D., ed. (1999) *PV News*, February.

Menna, P., Tsuo, Y. S., Al-Jassim, M. M., Asher, S. E., Matson, R., and Ciszek, T. F. (1998) *Proc. 2nd World Conf. on PV Solar Energy Conversion*, 1232.

Mihalik, G., Fickett, B., Stevenson, R., and Sabhapathy, P. (1999) Presentation at the 11th American Conference on Crystal Growth & Epitaxy, Tucson August 1–6. To be published, *J. Cryst. Growth*.

Mitchell, K. M., (1998) *Am. Ins. Phys. Conf. Proc.* **462**, 362.

Nakamura, N., Abe, M., Hanazawa, K., Baba, H., Yuge, N., and Kato, Y. (1998) *Proc. 2nd World Conf. on PV Solar Energy Conversion*, 1193.

Roy, A., Chen, Q. S., Zhang, H., Prasad, V., Mackintosh, B., and Kalejs, J. P. (1999) Presentation at the 11th American Conference on Crystal Growth & Epitaxy, Tucson August 1–6. To be published in, *J. Cryst. Growth*.

Runyan, W. R. (1965) *Silicon Semiconductor Technology* McGraw-Hill Book Company, New York.

Tsuo, Y. S., Gee, J. M., Menna, P., Strebkov, D. S., Pinov, A., and V. Zadde (1998) *Proc. 2nd World Conf. on PV Solar Energy Conversion*, 1199.

Wang, T. H., and Ciszek, T. F. (1997) *J. Cryst. Growth* **174**, 176.

Wang, T. H., and Ciszek, T. F. (1999) Presentation at the 11th American Conference on Crystal Growth & Epitaxy, Tucson, August 1–6. To be published.

Wang, T. H., Ciszek, T. F., Schwerdtfeger, C. R., Moutinho, H., and Matson, R. (1996) *Solar Energy Mater. Solar Cells* **41/42**, 19.

Wawer, P., Irmscher, S., and Wagemann, H. G., (1997) 14th *European Photovoltaic Solar Energy Conference Proceedings*, Barcelona, 38.

Zhao, J., Wang, A., Altermatt, P., and Green, M. A. (1995) *Appl. Phys. Lett.* **66**, 3636.

Part 3

Compound Semiconductors

14 Fundamental and Technological Aspects of Czochralski Growth of High-quality Semi-insulating GaAs Crystals

P. RUDOLPH[1] and M. JURISCH[2]

[1] *Institute of Crystal Growth, D-12489 Berlin, Germany*
[2] *Freiberger Compound Materials GmbH, D-09599 Freiberg, Germany*

14.1 INTRODUCTION

14.1.1 HISTORICAL BACKGROUND

Shortly after Teal and Little (1950) demonstrated the growth of germanium single crystals by the Czochralski technique it was also tried for the III-V compounds. Promptly, however, the first experiments of unprotected vertical pulling revealed the central problem of volatility of the group V elements (e.g. arsenic) and their rapid loss from the melt leading to an unacceptable deviation from stoichiometry (Gans *et al.* 1953, Welker 1953). As a consequence, considerable effort has been devoted to develop alternative technologies for Czochralski growth, especially of GaAs, InP and GaP.

First, Gremmelmaier (1956) described a GaAs hot-wall technology by using a gas-tight container made of fused silica, the inner walls of which were kept sufficiently hot ($>620\,°C$) in order to prevent the condensation of arsenic on them and to maintain a constant counter pressure against arsenic dissociation from the melt. A magnetic levitation system was applied for seed and crucible motion without any mechanical lead-throughs, which was somewhat later also adopted and perfected by Steinemann and Zimmerli (1967). Instead of such relatively complicated techniques various systems with liquid seals (molten metals, B_2O_3), allowing the application of mechanical lead-throughs, were proposed by Richards (1957) and Tanenbaum (1959). Furthermore, a syringe puller consisting of a well-fitting ceramic pull rod in a long bearing was used by Baldwin *et al.* (1965). The arsenic pressure was controlled by a separately heated source at the container bottom, and very low arsenic losses were observed.

Crystal Growth Technology, Edited by H. J. Scheel and T. Fukuda
© 2003 John Wiley & Sons, Ltd. ISBN: 0-471-49059-8

A breakthrough in GaAs growth was the liquid-encapsulation technique (Mullin *et al.* 1965), first introduced by Metz *et al.* (1962) for Czochralski growth of IV-VI compounds. The liquid-encapsulation Czochralski (LEC) method avoids the need for hot walls and permits the use of conventional pullrods. A liquid B_2O_3 layer, floating on the melt surface, acts as an impermeable liquid cover and prevents, in combination with an inert gas pressure greater than the partial arsenic pressure over the melt, the loss of dissociating arsenic. Additionally, it was observed that such liquid encapsulants can getter impurities from the melt during the growth process. This LEC technology allowed reproducible production of single crystals with circular cross section.

In 1999 by LEC approximately 90 % of all semi-insulating (SI) GaAs is produced for the use in high-speed microelectronics, i.e. HBTs, HEMTs and MMICs (the remaining 10 % were produced by the vertical-Bridgman (VB) or vertical-gradient-freeze (VGF) methods). Nowadays, the share of VGF material is rapidly increasing and was about 50 % in 2000.

Of course, the LEC technology is not restricted to the growth of SI single crystals but is also used for crystal growth of doped (mainly Te and Zn) semiconducting material for optoelectronic applications. But due to the high dislocation density of LEC-grown crystals and its detrimental effect on lifetime and brightness of light-emitting diodes (LED), a significant amount of semi-conducting GaAs is grown by the horizontal-Bridgman (HB) technique introduced by Folberth and Weiss (1955). This also applies to Si-doped crystals for laser applications, which have to be grown at low thermal gradients. Furthermore, in LEC growth from heavily Si-doped GaAs melt a scum is formed in the B_2O_3 layer often inducing twins or even polycrystallinity (Cockayne *et al.* 1988).

There are, however, considerable economical drawbacks of the HB technique reviewed by Rudolph and Kiesling (1988). Mainly the noncircular (D-shaped) cross section makes an adaptation to device manufacturing difficult and leads to significant waste when circular wafers are produced. Furthermore, the maximum width of the shoulder of about 75 mm limits the wafer size to 2.5 inch diameter in regular production (such diameter is insufficient for the large-scale production of laser arrays, for example).

Thus, considerable efforts were made to overcome this disadvantage by vertical growth methods. On the one hand-modified LEC techniques with markedly reduced temperature gradients like vapour-pressure-controlled Czochralski (VCz), fully encapsulated Czochralski (FEC) or the renewed hot-wall Czochralski (HWC) methods are under investigation as described in detail in Section 14.3. On the other hand the VB technique, first applied by Beljackaja and Grishina (1967), was rediscovered for growth of low-dislocation GaAs single crystals (Gault *et al.* 1986 and Liu, 1995). Nowadays, diameters even up to 150 mm have been achieved by the fully computer-controlled VGF method (see reviews of Monberg 1994 and of Tatsumi and Fujita 1998). It is not the aim of this chapter to discuss this powerful competing growth technique. Here we will concentrate on the features and developments of the Czochralski growth for large SI GaAs crystals. In the following the

SI material requirements (Section 14.1.2), LEC growth principles (Section 14.2.1), and growth-related phenomena will be discussed in Sections 14.2.2 to 2.5. A brief overview on unsolved problems like minimization of the as-grown thermoelastic stresses in large-diameter crystals as well as improvement of the controllability and uniformity of their structural and electrical parameters will be given. Finally, in Section 14.3 alternative modifications of GaAs Czochralski technology under investigation will be presented. A comparison of the LEC method with other growth methods like HB and VGF was recently reviewed by Mullin (1998), Müller (1998), Tatsumi and Fujita (1998), and Rudolph and Jurisch (1999).

14.1.2   THE IMPORTANCE OF SI GaAs AND ITS PERFORMANCE

At present, there is a rising need of low-noise high-frequency devices for micro-electronics like HBTs, HEMTs and MMICs (Mills 1998) which require high-purity semi-insulating (SI) GaAs wafers and lower specific costs per die in device manufacturing. This forces the crystal grower to optimize the crystal and wafer production technologies as well as the transition to 150 mm (6-inch) SI GaAs crystals with new equipment and technologies.

It is well known that the SI properties of GaAs are fundamentally connected with the intrinsic mid gap donor EL2°, which is most probably a simple As antisite (As_{Ga}). This causes pinning the Fermi level in the midgap position with the condition

$$N_{EL2°} > [C] - (N_{SD} - N_{SA}^*) > 0, \tag{14.1}$$

where $[C]$ is the carbon concentration, N_{SA}^* is the concentration of shallow acceptors other than carbon, N_{SD} is the concentration of shallow donors, and $N_{EL2°}$ is the concentration of deep donor EL2° (further named EL2). For the first time Holmes *et al.* (1982) showed that the EL2 concentration in as-grown crystals is related to the deviation from stoichiometry and increases with increasing As atom fraction starting at 0.487. Taking into account the highest available material purity, i.e. $(N_{SD} - N_{SA}^*) \leq 10^{14} \, cm^{-3}$, the EL2 concentration is usually fixed at a value near to $10^{16} \, cm^{-3}$ in order to control the resistivity (i.e. semi-insulating behaviour) by the carbon content in the range $10^{14} - 10^{16} \, cm^{-3}$ (see Section 14.2.4).

Presently about 66 % of the total semi-insulating wafer area for high-speed electronics is processed using ion implantation for MESFET fabrication, the remainder mainly by MBE and MOCVD for heterostructure devices like HBTs. Fluctuations of the turn-on threshold voltage (V_{th}) of the MESFETs across a wafer must be minimal for ensuring high yield of devices. V_{th} is proportional to the channel carrier concentration given by the difference between the concentration of the implanted electrically active donor and that of the net shallow acceptor, which effects the electrical resistivity of the SI substrate. The implantation efficiency is determined by the concentration of boron and EL2. Also, arsenic

Table 14.1 SI-GaAs specifications required for microwave devices.

Microwave device	Production technology	Demanded SI GaAs wafer properties				
		Diameter (cm)	Impurity content (cm^{-3})	EPD (cm^{-2}) (uniformity %)	EL2 (cm^{-3}) (uniformity %)	Electrical Resistivity (Ω cm) (uniformity %)
MESFET	Ion-implantation	$100 \to 150$	$<10^{14}$ (low boron!)	$\leq 10^5$ (<20)	1.2×10^{16} (<3 %)	$5 \times 10^6 - 1 \times 10^8$ (<10 %)
HBT	Epitaxy MBE, MOCVD	$100 \to 150$	$<10^{14}$	$\leq 10^4$(?) (<10 %?)	$\approx 10^{16}$ (10 %)	$\approx 1 \times 10^8$ (10 %)

precipitates seem to affect the threshold voltage. Essential for the applications and the device yield is a high uniformity within the wafer and along the crystal: uniformity of EL2, carbon, background impurities, dislocations, and electrical resistivity. Homogeneously distributed dislocations up to densities $<10^5\,\mathrm{cm}^{-2}$ are presently accepted for this class of devices.

The influence of dislocation density on V_{th} is still not clear. Miyazawa et al. (1982) found that dislocations may affect the threshold voltage of GaAs MES-FETs, which was doubted by Winston et al. (1984). Recently, the detrimental effect of dislocations was confirmed by Fantini et al. (1998). Additionally, there is an indication that the intrinsic point-defect concentration (i.e. EL2) and distribution are correlated with dislocation density (Tajima 1982). Therefore, the current efforts on dislocation reduction and on their more homogeneous distribution in SI GaAs crystals are justified.

Compared with ion implantation, the requirements on wafer homogeneity for epitaxy-based devices are believed to be lower due to the role of buffer layers to reduce the influence of wafer defects. Nevertheless, threading dislocations penetrate through the epilayers and are responsible for the degradation of devices. Therefore, a low and homogeneously distributed dislocation density is important for GaAs microelectronics. Important specification requirements for semi-insulating GaAs crystals for high-frequency microelectronics are summarized in Table 14.1.

14.2   FEATURES AND FUNDAMENTAL ASPECTS OF LEC GROWTH OF SI GaAs CRYSTALS

14.2.1   THE PRINCIPLE OF MODERN LEC TECHNIQUE

As discussed in Section 14.1.1 the liquid-encapsulation Czochralski growth has matured to a leading production method for large semi-insulating GaAs crystals (even the first 200 mm LEC crystal was reported by Seidl et al. 2000). Figure 14.1 shows the principle of a modern LEC arrangement consisting of a high-pressure vessel, two pressure-tight through-shafts for translation and rotation of the pulling and crucible rods, an electronic weighing cell coupled with the pulling rod, and a gas-flow system for supply and control of the CO (CO_2) content in the inert gas atmosphere important for controlling the carbon and oxygen potential in the melt (Section 14.2.5) and, thus, in the crystal (see for example Higgins et al. 1997).

In the LEC process a pre-synthesized polycrystalline GaAs charge or an in situ synthesis from elemental Ga and As in the growth equipment can be used. The latter needs a high-pressure equipment because about 6 MPa of As pressure are needed to prevent As evaporation through the B_2O_3 encapsulation layer during heating up and before the GaAs formation reaction starts. Typically, the pulling process pressure is between 0.2 and 2.0 MPa independently of both in-situ or presynthesized GaAs. Axial temperature gradients and pulling rates are in the

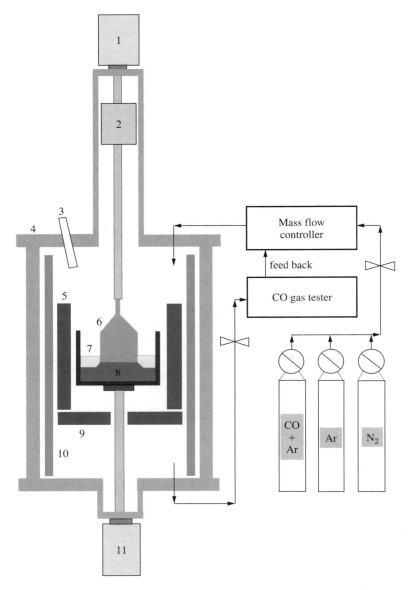

Figure 14.1 Principle of a modern LEC arrangement for growth of SI GaAs crystals with controlled carbon content. 1 – crystal drive assembly (pulling system), 2 – electronic weighing cell, 3 – optics, 4 – high-pressure vessel, 5 – main heater, 6 – growing crystal, 7 – liquid boron oxide, 8 – melt, 9 – bottom heater, 10 – heat shield, 11 – crucible shaft drive assembly.

region of $100-150\,\mathrm{K\,cm^{-1}}$ and $7-14\,\mathrm{mm\,h^{-1}}$ (dependent on the crystal diameter), respectively. For more details see the reviews of Hurle and Cockayne (1994) and Müller (1998).

At present, the LEC method has matured to grow single crystals up to 6 inch in diameter from melts up to about $28-40\,\mathrm{kg}$ (Kuma and Otoki 1993, Ware *et al.* 1996, Flade *et al.* 1999, Inada *et al.* 1999) using multiheater growth furnaces including an afterheater for controlled cooling down of the crystals. Crucibles up to 16-inch (Inada *et al.* 1999) in diameter from high-purity pyrolytic boron nitride (pBN) are applied without exception. As will be detailed later the crystals are grown under carbon- and (limited) EL2-control. This implies the establishment of reproducible initial conditions for growth by optimized evacuation and purging procedures of the growth chamber, heating up, melting and homogenization of the charge. The carbon content is controlled by the (initial) water content of the boron oxide and a proper CO fugacity of the growth atmosphere in the course of the growth process (Korb *et al.* 1999, see also Section 14.2.5). The EL2 is established by post-growth ingot annealing.

There are numerous advantages of the LEC method, namely, (i) the free growth without container contact, (ii) circular cross section, (iii) use of the gettering ability of boron oxide to influence background impurities, (iv) 'conditioning' of the melt by its defined overheating before the seed is dipped, (v) precise carbon control at all growth stages, (vi) visual control of the whole growth process offering the possibility of re-melting if crystallization-defects appear like twins, for example, and (vii) technological variability allowing uncomplicated changes of crystal diameter, boule length, heater and insulator configurations. Problems arising are (i) stoichiometry control due to uncontrolled gallium and arsenic losses through the boron oxide encapsulant, (ii) selective As evaporation of the crystal surface, emerging from the encapsulant, resulting in Ga droplets or trails that reduce yield, (iii) high-temperature nonlinearities in the growing crystal near the solid–liquid phase boundary and at the emerging region resulting in a rather high and inhomogeneously distributed dislocation density in the range of $(0.5-1) \times 10^5\,\mathrm{cm^{-2}}$, (iv) unsteady convection in the melt and turbulence in the gas phase causing dopant inhomogeneities and fluctuating temperature and stress fields in the growing crystal, (v) high investment and process costs, and (vi) necessity of a post-growth multistep heat treatment in order to reduce the residual stress level and to improve homogeneity of the electrical properties. It will be shown in Section 14.3 that proper technological modifications are under development (e.g. VCz) to overcome these drawbacks.

Important topics of R&D in the industrial LEC-growth of 6-inch crystals are (i) reduction of thermal stresses and dislocation density (including their distribution inhomogeneity) toward values of the well-established 4-inch technology at least, (ii) the control of intrinsic point defect situation (i.e. defined off-stoichiometry) and uniformity of electrical properties, (iii) enhancement of crystal weight (large crystals) and, (iv) further improvement of axial and radial homogeneity. Of course, these problems are also of principal importance for all other growth methods introduced in Section 14.1.1. Subjects (i) and (ii) will be discussed in more detail in the following sections.

14.2.2   CORRELATION BETWEEN HEAT TRANSFER, THERMOMECHANICAL STRESS AND DISLOCATION DENSITY

It is experimentally (e.g. Motakef *et al.* 1991) and theoretically (e.g. Jordan *et al.* 1984) well established that, independent of the growth method, the density and distribution of dislocations in melt-grown crystals are due to a thermoplastic relaxation of thermally and, to a much lower extent, constitutionally induced stress during growth. To a first approximation thermally induced stresses arise from temperature nonlinearity, i.e. divergence of the isotherm curvature from an idealized linear course (Indenbom 1979). Theoretically, this implies the simplified but quite useful formula

$$\sigma = \alpha_T E L^2 (\partial^2 T / \partial z^2) \approx \alpha_T E \delta T^{\text{max}} \tag{14.2}$$

with σ the thermal stress, α_T the coefficient of thermal expansion, E Young's modulus, L characteristic length (about the crystal diameter), T the temperature, z the given coordinate (pulling axis), and δT^{max} the maximum deviation of the isotherm from a linear course. The extremely critical situation in GaAs will be obvious by using the material constants near to the melting point ($\alpha_T = 8 \times 10^{-6}\,\text{K}^{-1}$, $E = 7.5 \times 10^4\,\text{MPa}$). As can be seen, only very small isotherm deviations from linearity δT^{max} of $1-2$ K are enough to reach the critical-resolved shear stress (CRSS) for dislocation multiplication (0.7 MPa). This is one order of magnitude lower than in silicon where much more isotherm curvatures are tolerated to still ensure dislocation-free growth.

Even during pulling of a cylindrical crystal from the melt steep temperature curvatures can occur due to different temperature gradients in the inner and outer regions of the crystal. As a result hoop stress can be created, which acts on the {111} planes to produce slip and dislocations. Note that already from the simplified Equation (14.2) follows a direct correlation between stress level and crystal diameter (characteristic length L) which has been observed in reality. The larger the crystal diameter the higher the mean dislocation density. In undoped 3-inch, 4-inch and 6-inch LEC GaAs crystals typical mean dislocation densities are 2×10^4, 5×10^4 and $10^5\,\text{cm}^{-2}$, respectively. Therefore, the content of dislocations is determined by the (time- and space-dependent) stress level during growth and the thermophysical and mechanical properties of the solid. The stress level (and its local and temporal fluctuations) is unambiguously related to the temperature field (including its fluctuations) in the crystal during growth and cooling-down procedure. Hence, the knowledge and control of the temperature field at all process stages are of essential significance. Due to the difficulties of measurements, numerical simulation is of increasing importance for heat flow analysis and 'tailoring'. Two approaches have been used so far:

1. Calculation of thermoelastic stress field (linear theory, isotropic and anisotropic analysis) of a crystal for a given temperature field using available computer packages and comparison of the resolved shear stress RSS in the glide systems or the von Mises invariant with the critical-resolved shear stress CRSS taking into account its temperature dependence known from high-temperature creep experiments (σ_{CRSS} kPa $= 22.72 \exp(4334.0/T$, Meduoye $et\ al.$ 1991)). The (local) dislocation density can then be concluded from a dislocation density parameter σ_{ex} (Jordan $et\ al.$ 1981) given by $\sigma_{ex} = \sum_1^{12} \sigma_i^e$ where

$$
\sigma_i^e = \begin{cases} |\sigma_{RSS,i}| - \sigma_{CRSS}, & \text{for } |\sigma_{RSS,i}| > \sigma_{CRSS} \\ 0, & \text{for } |\sigma_{RSS,i}| \leq \sigma_{CRSS} \end{cases} \tag{14.3}
$$

i.e. the maximum stress at any time of growth determines the local dislocation density. Examples of this approach for LEC growth of GaAs can be found at Jordan $et\ al.$ (1980) and Miyazaki $et\ al.$ (1991).

2. Estimation of the local dislocation density from the constitutive law linking the plastic shear rate and dislocation density with the applied stress in the course of the cooling-down procedure of the crystal. The constitutive law for SI GaAs reads (Motakef 1991)

$$
d\varepsilon/dt = \rho b V_0 (\tau_{app} - D\sqrt{\rho})^m \cdot \exp\left(-\frac{Q}{kT}\right) \tag{14.4}
$$

where ρ is the density of moving dislocations, b the Burgers vector (0.4 nm), V_0 a pre-exponential factor (1.8×10^{-8} m^{2m+1}N^{-m}s^{-1}), m is a material constant (1.7), Q the Peierls potential (1.5 eV), D a parameter relating ρ and τ (3.13 N/m), and τ_{app} is the applied stress. Details of this approach can be found in Maroudas and Brown (1991) and Tsai $et\ al.$ (1993), for example. For a profound review see Völkl (1994).

Figures 14.2(a) and (b) show calculated stress and dislocation distributions within the cross section of [100]-grown LEC GaAs crystals using approaches 1 (Jordan $et\ al.$ 1980, see also Thomas $et\ al.$ 1993) and 2 (Tsai $et\ al.$ 1993), respectively. A good agreement between EPD mapping of LEC wafers and calculated stress fields has been observed.

Typically, stresses are minimal in a circular region around $R/2$, ($R =$ radius) of the crystal and slightly increase in the centre, which results in a W-shaped distribution. Dupret $et\ al.$ (1989) showed that the stress maximum in the centre, of the crystal increases relative to the near surface maximum with increasing length of the crystal. If, as usual, the solid/liquid interface is convex curved indicating radial temperature gradients, a second maximum of stresses is observed at the interface near the periphery (Motakef and Witt 1987). A concave part of the interface was shown to enhance considerably the stress level (Lambropoulos and Delameter 1988, Shibata $et\ al.$ 1993). As concavity of the interface increases with growth rate there is an influence of growth rate on stress level at the interface, too. Shibata

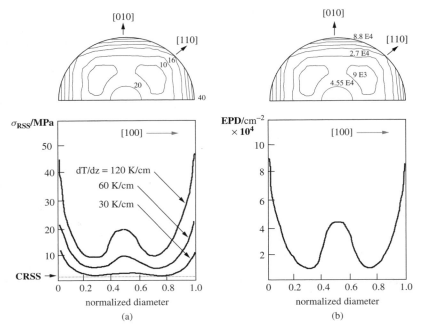

Figure 14.2 (a) Sketched thermomechanical stress distribution along cross section and diameter in a 3-inch GaAs crystal growing by LEC technique in different axial temperature gradients calculated by use of linear theory 1 (see text; after Jordan, 1980 and Thomas *et al.* 1993); (b) Sketched dislocation density distribution along cross section and diameter of a hypothetic GaAs ingot having 2 cm diameter and 10 cm length (without B_2O_3) with assumed initial dislocation density of $1\,cm^{-2}$ using the constitutive law of dislocation dynamics 2 (see text; after Tsai *et al.* 1993).

et al. (1993) presented a successful LEC concept to grow large single crystals with lengths up to 500 (3-inch), 350 (4-inch) and 170 mm (6-inch) without concave interface sections: they used large melts in crucibles with diameters 2–3 times larger than the crystal diameter. Recently, it was demonstrated by Inada *et al.* (1999) that a decrease of EPD for 6-inch crystals to $(6-7) \times 10^4\,cm^{-2}$ is possible by further optimization of such a growth technique.

Theoretically, a more uniform stress distribution and related reduction of dislocation density could be achieved with decreasing radial temperature gradient, which helps also to flatten the growth interface. However, in the case of Czochralski growth, a so-called unconstrained meniscus-controlled method, radial gradients are inevitably necessary to realize diameter control and, therefore, there exists some kind of lower stress level and a corresponding dislocation density that cannot be reduced further. Besides, a promising low axial temperature gradient would promote the arsenic dissociation from the very hot as-grown crystal surface as it emerges from the B_2O_3. These are quite serious restrictions of the conventional LEC process, especially for SI GaAs, where the approach of lattice hardening by doping is not permitted due to the required high doping concentrations (typically above $10^{18}\,\mathrm{cm^{-3}}$).

But first, one has to find out the actual parameters of this limitation for a given LEC arrangement, crystal diameter and length very accurately. Of course, this requires a well-coordinated interplay between modelling, constructive measures, and growth experiments. For this, a global computer modelling of the whole temperature field in the crystal, melt and surrounding gas phase is very helpful (see review of Dupret and van den Bogaert 1994). Especially in LEC growth of GaAs at medium and high gas pressures, convection in the gas phase in addition to conduction and radiation is known to influence the heat balance significantly and must be included in the model. But due to the large geometrical dimensions and high temperature gradients in an industrial LEC system resulting in very large Rayleigh numbers ($>10^9$ for the gas phase) this convection is turbulent and therefore difficult to model. Turbulent convection in the gas phase has been included only recently (Zhang et al. 1996, Fainberg et al. 1997). Convection in the melt is mostly assumed to be laminar. In fact, it has been demonstrated that models neglecting gas convection cannot predict accurately the thermal field in the growth system. First, the increasing gas flow rates at higher pressures promote the heat-transfer coefficient and, therefore, the curvature of radial isotherms, i.e. deflection of the growing interface (Fainberg et al. 1997). Secondly, one has to consider that turbulent gas pressure pulses can enhance and rearrange dislocations within the growing crystal (but this effect has not yet been studied in detail and requires a careful analysis in future). In fact, it was observed experimentally that the dislocation density in LEC crystals depends on the inert gas pressure and increases sensitively even in the region of about 1 MPa (Seifert et al. 1996), where turbulence becomes important.

Furthermore, from a thermoelastic point of view the presence of a B_2O_3 layer is not favorable. Due to the low thermal conductivity an (absolute) maximum of stress is found in the near-surface region of the crystal where it emerges from the boron oxide encapsulant (e.g. Jordan et al. 1984). The stress peak at the boron oxide level decreases during the growth process due to the decreasing melt height and a reduced radiative heat transfer to the surroundings. For LEC growth it has been found that thicker boron oxide layers reduce the stress level at the emerging crystal interface.

The VCz, FEC and HWC methods discussed in Sections 14.3.1–14.3.3 are alternative Czochralski modifications enabling growth in low axial temperature gradients resulting in markedly reduced stresses. Besides, the HWC technique refrains from the B_2O_3 encapsulant completely.

14.2.3   DISLOCATION PATTERNS

Elementary mechanisms of thermoplasticity are conservative glide- and stress-induced diffusional climb motion of dislocations, multiplication of dislocations by interaction and their rearrangement. Dislocations are introduced from the dislocated seed crystal, initiated at outer surfaces or internal phase boundaries, and generated at different kinds of dislocation sources. Homogeneous nucleation of dislocations is insignificant. Generally, the as-grown density of dislocations and their distribution cannot be significantly reduced or changed by post-growth heat treatment. Hence, measures to reduce dislocation density and to influence their distribution have to be taken in the growth process.

In GaAs dislocations form a typical cellular structure that is most evident for LEC-grown crystals. Cells have a globular shape except for regions between the centre and periphery of the crystal where elongated cells in ⟨110⟩ directions are observed. The cell size decreases with increasing crystal diameter due to an increased average dislocation density. With decreasing average dislocation density, however, the cell walls disintegrate into fragments and the cell size increases. Typical mean cell diameters <500 μm for 4-inch SI GaAs LEC crystals have been observed (Figure 14.3(a)). The cellular structure is due to static and dynamical, i.e. stress-assisted polygonization of dislocations (van Bueren 1961) and hence influenced by the temperature field and cooling process in the crystal region behind the interface. Constitutional supercooling and the associated loss of stability and cellular breakdown of the interface shape causing the cellular structure can be excluded as shown by Wenzl *et al.* (1993). Hence, the most powerful measure to control or even prevent the dislocation cell structure seems to be the application of low temperature gradients, i.e. the achievement of a nearly stress-free situation in the crystal region behind the growing interface. The complexity of such a concept for LEC growth was discussed already in Section 14.2.1. But the validity of such an approach was demonstrated by low-gradient Czochralski methods (Section 14.3) and also by low-gradient vertical

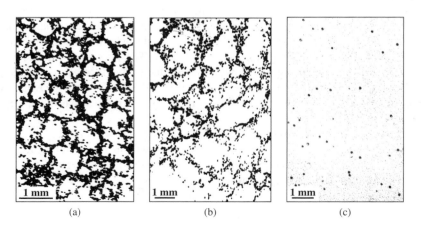

(a)   (b)   (c)

Figure 14.3 Typical cell structure patterns decorated by dislocation etch pits in (a) LEC crystals (grown in usual high axial temperature gradients), and (b) and (c) VCz crystals (grown in low temperature gradients).

Bridgman growth (Tatsumi and Fujita 1998). Markedly enlarged cell diameters (>1–2 mm) and extended regions with a homogeneous dislocation arrangement without substructure have been observed in VCz wafers (Figures 14.3(b) and (c)).

Simultaneously with isolated dislocations occasionally slip line bundles, localized dislocation clusters, subgrain boundaries and lineages have been observed that deteriorate yield. Slip lines are due to crystallographic glide and therefore indicate still significant thermal stress at lower temperatures where diffusional creep is unimportant. As the back-stress is proportional to the actual dislocation density (see constitutive law given by Equation (14.3)) slip lines are more evident where the dislocation density is low and have to be avoided by a proper adjustment of the temperature field even during the cool-down process.

According to Tower *et al.* (1991) and Shibata *et al.* (1993) the appearance of dislocation clusters is related to concavely curved regions of the solid/liquid interface exhibiting local maxima of thermal stress. As growth proceeds the local dislocation density increases, subgrain boundaries are formed, and finally transition to polycrystalline growth takes place. Consequently, the localized density of dislocation clusters can be suppressed by establishing a slightly convex shape over the whole interface during the growth process (see Section 14.2.1).

Lineages and subgrain boundaries, typically propagating along the crystal axis for rather long distances, seem to be related to the solid/liquid interface. But details of their formation and the role of experimental parameters are still insufficient.

Figure 14.4 schematically summarizes the most typical dislocation patterns in LEC GaAs crystals and the chronology of their generation. More details are given by Ono (1988), Schlossmacher and Urban (1992), Prieur *et al.* (1993), Möck and Smith (2000), and Naumann *et al.* (2001).

In general, single dislocations as well as dislocation cell walls are known to influence the local thermodynamic potential of the different components including intrinsic defects due to a mechanical (stress field) or electrical interaction. This results in equilibrium segregation to or from dislocations/cell walls, i.e. an enhancement or depletion of components/native defects in the surroundings of dislocations/cell walls, a phenomenon commonly called gettering of dislocations. It follows that the distribution of defects across cells can be influenced by heat treatment and a corresponding freezing-in procedure cooling down resulting in a homogenization of physical properties, as will be discussed in the following section.

14.2.4 PRINCIPLES OF NATIVE-DEFECT CONTROL

According to the phase diagram of Ga-As (Wenzl *et al.* 1993) the compound GaAs is characterized by a congruent melting point (CMP) slightly deviating from stoichiometric composition towards the As-rich side (at about 10^{16} cm^{-3} according to Terashima *et al.* 1986 and Inada *et al.* 1990; about 10^{19} cm^{-3} after Hurle 1999).

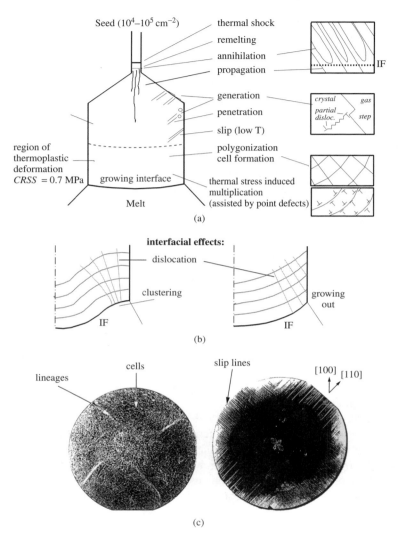

Figure 14.4   The dislocation dynamics in GaAs Czochralski crystals. (a) possible generation and multiplication mechanisms; (b) interfacial effects leading to dislocation bunching,
i.e. lineages at convex-concave curvature of the growing phase boundary; (c) LEC GaAs
wafers with typical dislocation arrangements.

This, and a temperature-dependent homogeneity range including the stoichiometric
composition and retrograde solid solubilities at both sides, are schematically represented in Figure 14.5. But details of the phase diagram like the position, temperature
and coexisting As partial pressure of the CMP as well as shape and temperature
dependence of the homogeneity range are still controversial (Wenzl *et al.* 1993,

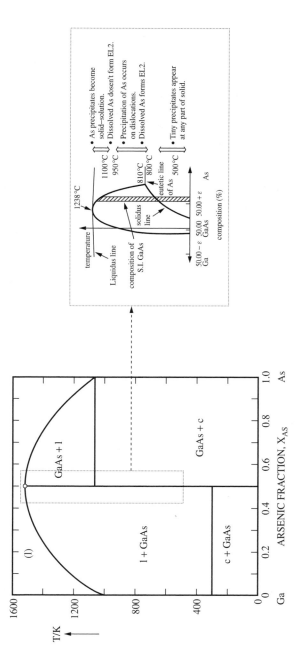

Figure 14.5 The $T-x$-phase projection of GaAs showing the sketched region of homogeneity and generation mechanism of As precipitates. (Reprinted from S. Kuma *et al.* in Gallium Arsenide and Related Compounds (1993) eds. H. S. Rupprecht and G. Weimann, copyright (1993) with permission from the Institute of Physics.)

Hurle 1999). Consequently, a GaAs single crystal of stoichiometric composition has to be grown from a Ga-rich melt. This is of special interest because the dislocation density of GaAs single crystals exhibits a slight minimum at stoichiometric composition, which was ascertained by Parsey and Thiel (1985) and is explained by Nishizawa *et al.* (1986) as a result of the lowest intrinsic point-defect concentration taking part in the propagation of dislocations. In particular, the HWC growth technology was used to control the stoichiometry in-situ (Section 14.3.3). But this finding is not used in commercial LEC crystal growth. One of the reasons is the risk of constitutional supercooling and loss of morphological stability if a markedly Ga-rich diffusion boundary layer is piled up in front of the propagating interface due to considerable Ga rejection. Furthermore, as was already shown in Section 14.1.2 the precondition for the semi-insulating behaviour of GaAs is an As excess favouring the EL2-related As-antisite defect, which requires growth of GaAs crystals from a melt above a critical $x_{As} > 0.483$ (Hurle 1999). Practically, SI GaAs is almost exclusively obtained from an As-rich melt resulting in crystals with off-stoichiometric arsenic-rich composition. Moreover, arsenic excess helps to avoid twin formation at crystal growth (Hurle 1995) and is a prerequisite for better homogeneity of electrical properties (Terashima *et al.* 1986).

At the CMP the As-equilibrium pressure amounts to about 0.2 MPa (Wenzl *et al.* 1991) which defines the minimum inert gas counter pressure in LEC GaAs growth from an As-rich melt. With increasing As concentration above the CMP level, the working gas pressure required to avoid As bubble formation increases significantly.

It follows from Figure 14.5 that the nonstoichiometry of melt-grown GaAs should be controllable by defining the As content of the melt. In reality, however, this method is restricted to an As concentration range rather close to the congruently melting concentration (approximately corresponding to the maximum solid solubility). This is due to macrosegregation, i.e. continuous increase of As in the melt with increasing solidified melt fraction. This limits crystal growth by As bubble formation, constitutional supercooling and even eutectic crystallization when reaching the eutectic concentration. Furthermore, a precise adjustment of the As/Ga ratio of the melt is difficult due to the fact that Ga and As can be oxidized and thus incorporated into the liquid boron oxide encapsulant depending on the O-potential in the system (Korb *et al.* 1999). Because of the higher affinity of Ga to O and the greater vapour pressure of Ga oxides over liquid boron oxide compared to As oxides, a higher loss of Ga from the GaAs melt occurs leading to a continuous increase of the As excess and a corresponding increased boron concentration in the melt. So far there is no analytical method for determining the deviation from stoichiometry with a detection limit lower than about 2×10^{17} cm^{-3} (Holthenrich 1997) necessary to control nonstoichiometry in solid GaAs. Therefore, instead of an in-situ stoichiometry control via melt composition (or As-partial pressure) the amount of As used in addition to the stoichiometric concentration in (ex-situ or in-situ) synthesis has been empirically optimized and is kept constant by a reproducible and reliable technology.

Initially, at crystallization temperature the As-excess-related native point defects (mainly As_i, V_{As} and V_{Ga} with $[As_i] \gg [V_{As}] > [V_{Ga}]$ according to Hurle 1999) are dissolved completely within the range of homogeneity following thermodynamic regularities (Rudolph, Chapter 2 of this book). During cooling down as soon as the concentration-dependent retrograde solubility line (solidus) is reached, nucleation and growth of As precipitates occur. Their size, distribution and density follow in principle the relationships of nonstationary nucleation theory and Ostwald ripening, i.e. they are linked to the available nucleation sites (homogeneous and heterogeneous nucleation), the actual supersaturation (supercooling), the resting time and the temperature below the solubility limit determining the average diffusion length, but details of the atomistic mechanism of precipitation are not yet understood.

Due to the lower nucleation energy, precipitation starts at dislocations, as has been concluded from IR-laser-scattering tomography (Moriya and Ogawa 1983, Otoki et al. 1995, for example). As these decoration precipitates are present in as-grown crystals it follows that formation and rearrangement of dislocations are already finished when precipitation starts. In contrast, dislocations belonging to slip lines formed at lower temperatures are not or only partly decorated by arsenic precipitates (Kuma and Otoki 1993, and Naumann et al. 2001). The excess arsenic incorporated in the decoration precipitates is lower by 2–3 orders of magnitude as compared to $[As_i]$, which mainly remains dissolved. This has been concluded from TEM studies (Wurzinger et al. 1991).

Quenching experiments of GaAs samples, dissolution-annealed at temperatures $>1100\,°C$, have demonstrated that $As_{Ga} \equiv EL2$ is generated cooling the crystals to around $900\,°C$ according to the reaction $As_i + V_{Ga} = As_{Ga}$. This removes the supersaturation of V_{Ga} at this temperature and, therefore, the As_i content is controlled by the V_{Ga} concentration. Correspondingly, the EL2-concentration is enhanced near the dislocation cell structure as a result of dislocation climb by the Petroff–Kimerling mechanism (Petroff and Kimerling 1976) supplying V_{Ga} occupied by As_i to form EL2.

Usually, as-grown LEC crystals contain a mean EL2 concentration in the range of $(1-2) \times 10^{16}\,cm^{-3}$ showing the above-discussed nonuniform distribution with maxima in the cell walls and minima in the cell interior. As a result, an electrical nonuniformity, a so-called mesoscopic parameter inhomogeneity, is observed, the degree of which depends on the heat treatment.

Typically, different mapping analyses reflect the above-mentioned correlation to the cellular structure. For instance, the point-contact current is high in the cell-wall region and decreases to the centres of the cells indicating an inhomogeneous electrical resistivity across the cells exceeding possibly more than one order of magnitude. Recently it has been proved (Wickert et al. 1998) that the enhancement of EL2 concentration at the cell walls is not sufficiently high to explain the drop of the electrical resistivity. Therefore, a simultaneous segregation of a donor shallower than the EL2 has to be assumed. Furthermore, an increased band-to-band (BB) photoluminescence intensity at the cell walls is observed in

LEC-grown crystals indicating a reduced concentration of nonradiative recombination centres with regard to the cell interior, which is in agreement with an enhanced carrier lifetime at the cell walls. From the correlation of BB-PL intensity and EL2 concentration it follows that EL2 is not the carrier-lifetime-determining recombination centre. It is supposed that the native EL6 centre, which is probably a complex containing an arsenic vacancy V_{As}, determines carrier lifetime in GaAs (Müllenborn et al. 1991). EL2 and EL6 were demonstrated to be anticorrelated.

Mesoscopic homogeneity can be influenced by a heat treatment of the crystal boules (Jurisch 1998). Usually, a two-step process is applied consisting of a dissolution period at the highest possible temperature and a holding time long enough for complete homogenization of the excess arsenic followed by an accelerated cooling to a temperature of about 800 °C and a resting at that temperature. Except for regions around the cell walls, which are, furthermore, exhausted by the decoration precipitates, matrix precipitates as haze-like scattering centres are found.

Generally, the native-point-defect homogenization by two-step bulk annealing improves the global properties of the crystals. To give an example, an average standard deviation of the radial EL2-concentration of 6-inch wafers of 2.5 % has been realized by this heat treatment. The average Hall mobility in low-carbon material is $8000 \, \text{cm}^2 \, \text{V}^{-1} \, \text{s}^{-1}$ at an electrical resistivity of $10^7 \, \Omega \, \text{cm}$ (Flade et al., 1999). At present the bulk annealing procedure is a well-established step within the matured technology of SI GaAs large-scale production. On the other hand, wafer annealing as proposed by Oda et al. (1992) allowing to influence the concentration of excess As by an annealing in a controlled arsenic gas ambient is a further effective post-growth possibility.

14.2.5   CARBON CONTROL

Improvements in raw material purity, state-of-the-art cleanliness in material handling, and progress in the understanding of the thermochemistry of the gettering ability of boron oxide and of the influence of gas-phase composition on the GaAs-melt in synthesis and growth have led to a significant reduction of extrinsic defects (impurities) and to the transition from LEC-growth of nominally undoped to carbon-controlled semi-insulating GaAs-crystals on a production scale in the last few years (see also Figure 14.1). But there is a remaining interest concerning the incorporation behaviour of those species belonging to the ingredients of the synthesis and growth systems like boron, nitrogen, hydrogen and oxygen in addition to carbon, among which oxygen is electrically active and boron influences the activation efficiency of dopants incorporated in the Ga sublattice. The concentrations of these components are basically controlled by the chemical potentials of oxygen and carbon in the growth systems. These potentials can be influenced by the water content of the boron oxide and/or the nitrogen

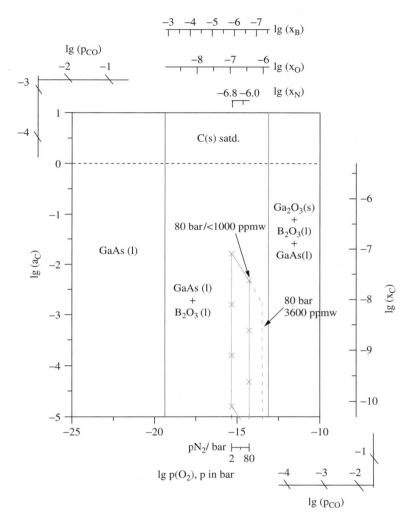

Figure 14.6 Species concentration in liquid GaAs in dependence on the O and C potential. a_C – carbon activity, x – mole fraction, p – pressure, 1000 and 3600 ppmw – water contents in B_2O_3.

fugacity as well as the carbon oxides in the working atmosphere. To get a more detailed insight, thermochemical modelling, instead of analysing the numerous interrelated chemical reactions, was performed using the concept of minimizing the free energy of the complex reaction system (Oates and Wenzl 1998, Korb *et al.* 1999). As an example, a calculated carbon potential versus oxygen potential diagram with the domain of stability of liquid GaAs in contact to boron oxide is given in Figure 14.6. Liquid phases were described as subregular solutions according to the Redlich–Kister–Muggianu model (Redlich and Kister 1948). The operating region for crystal growth is marked for the case that the oxygen and carbon potentials are determined by the nitrogen partial pressure, the water content of the boron oxide, and the CO partial pressure in the system. Having fixed the oxygen and carbon potential there are no degrees of freedom left in the system. So, by using the corresponding activity coefficients the concentrations of C, B, O, N and H dissolved in liquid GaAs can be estimated. The resulting B, O and N concentrations in liquid GaAs are indicated on the oxygen potential axis in Figure 14.6. They depend only on the oxygen potential and are independent of the C potential in the system. As can be seen from Figure 14.6 the boron content of liquid GaAs can be reduced by adding nitrogen to the growth atmosphere. In the presence of nitrogen water determines the oxygen potential only at high enough contents in the system. On the other hand, it is obvious from Figure 14.6 that fixing of the CO partial pressure is not sufficient to define the C potential and thus the C concentration in liquid GaAs. The oxygen potential must be fixed too. This may be why controversial results about a C control by means of CO control have been published in the literature (Oates and Wenzl, 1998). By controlling the CO partial pressure and the O potential in the growth system, the C content in GaAs single crystals can be controlled over an extended range from 1×10^{14} to $<2 \times 10^{16} \, \text{cm}^{-3}$. This is demonstrated in Figure 14.7.

But it must be kept in mind that extraction and addition of carbon from and to the GaAs melt requires the transport of the reactants through the different phases (mainly boron oxide acting as a diffusion barrier) to the phase boundaries, where thermodynamic equilibrium exists. Therefore, the control of transport phenomena is also of great importance in controlling the C concentration in the melt.

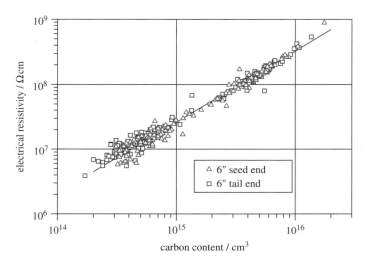

Figure 14.7 Electrical resistivity of 6-inch crystals in dependence on the C concentration.

14.3 MODIFIED CZOCHRALSKI TECHNOLOGIES

14.3.1 VAPOUR-PRESSURE-CONTROLLED CZOCHRALSKI (VCz) METHOD

A powerful method to realize low temperature gradients in the range 15–35 K/cm in a Czochralski growth system for 4–6-inch diameters is the VCz method. The main constructive feature is the presence of an inner chamber leading to the shielding of the growing crystal and hot gas from the water-cooled walls of the outer high-pressure vessel and, therefore to the promising precondition of markedly reduced temperature gradients in axial and radial directions (Figure 14.8). In order to avoid the decomposition of the very hot crystal surface by arsenic dissociation and formation of Ga trails (as the crystal emerges from the top of the B_2O_3 layer) a temperature-controlled As pressure within the inner chamber is maintained during the whole growth process. The inert gas pressure and boron oxide encapsulant are still employed allowing a quite accurate stoichiometry control by use of slightly As-rich pre-synthesized GaAs starting charge. A typical mirror-like surface of VCz-grown crystals is the evidence for its thermodynamic equilibrium (Tatsumi *et al.* 1994 and 1998, Neubert *et al.* 1996 and 1998, Frank *et al.* 2000).

For the first time the VCz principle was presented in Japanese patent descriptions in the mid-1980s (Azuma 1983, Tada and Tatsumi 1984). The first paper on GaAs VCz growth, using a transmission X-ray shadow-imaging system for in-situ crystal-shape control, was published by Ozawa *et al.* (1985). Low dislocation densities in order of $10^2 \, cm^{-2}$ were reported for undoped 2-inch GaAs crystals. Today, much larger crystals of high-quality with diameters of 100–150 mm can

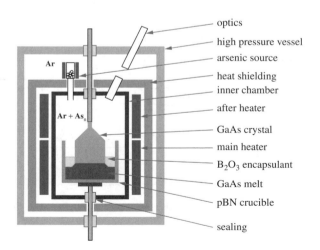

Figure 14.8 Principle of the VCz method for growth of GaAs crystals in low temperature gradients in Ar gas, for example.

be reproducibly grown (see reviews of Rudolph *et al.* 1997, Tatsumi and Fujita 1998, and Neubert and Rudolph 2001).

There are some VCz-specific technological problems that have been solved during the recent years. First- the complicated diameter control in low temperature gradients (see Section 14.2.2) was achieved by well-programmed pulling and rotation rates especially during the shoulder growth. Furthermore, the difficulties of the visual process control in the inner chamber as well as the As outflow have been overcome by proper construction measures so that the VCz process became a matter of routine quite similar to the conventional LEC process. A detailed overview is given by Neubert and Rudolph (2001).

Typically, a very low residual thermomechanical stress below 1 MPa (equal to relative residual strain $S_r - S_t \approx 2 \times 10^{-6}$) has been found over whole VCz wafers without annealing. As a result the dislocation density is reduced by about one order of magnitude compared to LEC crystals as demonstrated in Figure 14.9. EPDs markedly below 10^4 cm^{-2} with minima in the central region of 2×10^3 cm^{-2} to 5×10^3 cm^{-2} has been ascertained in 4-inch and 6-inch crystals, respectively. From Figure 14.9 it follows that the dislocation distribution of VCz crystals along the radial direction is more homogeneous and resembles a U-shaped curvature rather than the usual W-type in LEC crystals. A further consequence is a modified substructure (Figure 14.3(b)). Cell dimensions of more than

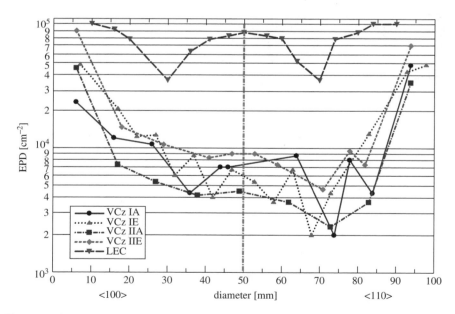

Figure 14.9 Radial EPD distribution in 4-inch VCz SI-GaAs crystals grown at IKZ Berlin compared with an LEC wafer grown at the same laboratory (upper curve). A- and E-marked wafers from starting and end regions of the same VCz boule (samples I and II), respectively.

1–2 mm have been reported (Tatsumi *et al.* 1994, Neubert *et al.* 1996). Moreover, the cell walls are often not completely closed because of the low absolute dislocation density and, obviously, due to incomplete polygonization under conditions of low temperature gradients. No evident differences of the electrical parameters have been observed. After two-step bulk annealing a more homogeneous distribution of the electrical resistivity with standard deviations <10 % has been found in as-grown VCz crystals (Neubert and Rudolph 2001).

The EL2 concentration in as-grown VCz crystals is usually significantly lower than in as-grown LEC crystals amounting to $(5-9) \times 10^{15}$ instead of $>10^{16}$ cm^{-3}. At first sight such behaviour can be explained by the lower dislocation density in VCz crystals (see also Section 14.2.3) but more detailed investigations are necessary. Note that the EL2 content of as-grown VCz GaAs can be increased to values typical for LEC GaAs by post-growth annealing at about 850 °C (Neubert and Rudolph 1999 and 2001).

A higher carbon content of $>10^{15}$ cm^{-3} as compared to LEC crystals has been reported (Tatsumi *et al.*, 1994, Tatsumi and Fujita, 1998). However, the results of a recently completed program of carbon control in VCz crystals at IKZ Berlin demonstrated that carbon concentrations below this value (down to $<10^{14}$ cm^{-3}) may be realized (Jacob *et al.* 2001, Neubert and Rudolph 2001).

In conclusion, the main advantages of VCz are: (i) the markedly reduced thermal stresses lead to a reduced density of dislocations ($\leq 10^4$ cm^{-2}) which are distributed more homogeneously and arranged randomly or in cells with considerable larger diameter, (ii) there are no gas turbulences inside the VCz chamber by shielding the flow from the outer vessel and, (iii) the mirror-like surface of the crystal indicates thermodynamic-equilibrium post-growth conditions. At present there are still some disadvantages, like (i) relatively high complexity of the equipment and therefore increased investment costs, (ii) enlarged process time (low pulling rates of about 5 mm h^{-1} are typical due to the low temperature gradient) and, (iii) somewhat higher operational costs.

Nevertheless, VCz is a promising technique for constraint-free crystal growth of GaAs with large diameters in a 'tailored' temperature field to ensure minimal thermally induced stresses. This improved GaAs is of interest for HBT production. Recently, the VCz growth of nearly dislocation-free Si-doped 3-inch semiconducting GaAs crystals have been reported (Hashio *et al.* 1997). Also in the case of InP the VCz growth of crystals with diameter of 75 mm and lengths of 240 mm has been introduced into production (Oda *et al.* 1998).

14.3.2   FULLY-ENCAPSULATED CZOCHRALSKI (FEC) GROWTH

Aimed at a reduction of stress level during growth, especially at the contact boundary between crystal and B$_2$O$_3$ surfaces, and avoidance of selective As evaporation from the free crystal surface the full encapsulation of the growing crystal by boron oxide was used to grow SI GaAs single crystals by Nakanishi *et al.* for the first time in 1984. Newer results are described by Elliot *et al.*

(1992). A significantly lower dislocation density could be demonstrated. Even dislocation-free 50-mm diameter SI GaAs crystals have been reported (Kohda *et al.* 1985). Encouraging electrical properties were observed. The main problems, however, concern limitations in length and diameter of the crystals as well as the control of the carbon content. Therefore, this method did not prevail in large-scale production.

14.3.3 HOT WALL CZOCHRALSKI (HWC) TECHNIQUE

As already mentioned in Section 14.1.1 HWC systems were first used for the growth of GaAs (Gremmelmaier 1956, Steinemann and Zimmerli 1967). Because of the absence of a liquid encapsulant this method is capable of an in-situ control of stoichiometry via the total As pressure in the growth chamber neglecting the low Ga pressure at melting temperature. A gas-tight growth chamber with hot walls requiring additional heating above the As source temperature to about 620 °C, and feedthrough mechanisms for translation and rotation of seed and crucible are the most important ingredients of a HWC system. Figure 14.10 shows

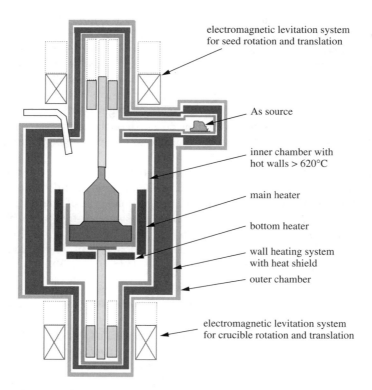

Figure 14.10 Principle of hot wall Czochralski (HWC) growth.

the sketch of the arrangement of a modern HWC system investigated by Wenzl (1998). Due to much lower temperature gradients inherent to the hot wall system, low-dislocation GaAs crystals have been grown as demonstrated by Tomizawa *et al.* (1984) and Nishizawa (1990). Even dislocation-free samples with diameter <20 mm were reported (Steinemann and Zimmerli 1967). After Nishizawa (1990) there is a direct correlation between deviation from stoichiometry, dislocation density and electrical parameters. An optimum arsenic vapour pressure corresponding to a source temperature of 612–615 °C reduced the deep-level EL2 concentration to values of 10^{15} cm^{-3}.

As a consequence there is a certain interest in this method, especially from the thermodynamic point of view. But there are serious technical and scientific disadvantages of the method preventing its application in industrial production until now: (i) complex system requirements, (ii) pressure control inside and outside the growth chamber, (iii) difficulties to control stoichiometry due to temperature gradients across the free GaAs melt surface, (iv) control of carbon content and suppression of background impurities and, (v) high investment and operating costs.

14.4 CONCLUSIONS AND OUTLOOK

GaAs is of rising importance for opto- and microelectronics, especially for high-frequency devices like HBTs, HEMTs and MMICs. To meet the rapidly increasing demands the efforts are directed towards growth of high-quality semi-insulating crystals with enlarged diameter of 150 mm. The well-established and further improved LEC technique will keep its pronounced position for mass production of SI wafers in the future (recently, the growth of the first 200-mm LEC crystals has been demonstrated). A certain share of GaAs wafers with markedly reduced residual stress and dislocation density for special devices could be produced by VCz.

Problems in LEC growth to be solved in future are reduction of thermomechanical stress in the growing crystal by 'tailoring' of the temperature field at all process stages. For this, powerful numerical-simulation programs including gas convection and based on improved physical models will be helpful. The optimization of the temperature field and the cooling program is important to control the density and rearrangement of dislocations. At the same time the correlation between unsteady thermal stress and dislocation generation and clustering has to be studied more carefully.

Furthermore, due to the correlation between the dislocation structure and physical properties, including their mesoscopic and macroscopic homogeneity, a more detailed knowledge about the interaction of dislocations with extrinsic and intrinsic defects and their relation to stoichiometry and to the physical properties of SI GaAs is required.

ACKNOWLEDGEMENT

The authors wish to thank Dr B. Weinert from Freiberger Compound Materials GmbH for critically reading the manuscript and helpful discussion.

REFERENCES

Azuma, K. (1983), Japanese Patent 60-11299.

Baldwin, E. M. N., Brice, J. C., Millet E. J. (1965), *J. Sci. Instrum.* **42**, 883.

Beljackaja, N. S., Grishina, S. P. (1967), *Izvest. AN Neorg. Mater.* **3**, 1347.

Cockayne, B., MacEwan, W. R., Hope, D. A. O., Harris, I. R., Smith, N. A. (1988), *J. Cryst. Growth* **87**, 6.

Dupret, F., Nicodeme, P., Ryckmans, Y. (1989), *J. Cryst. Growth* **97**, 162.

Dupret, F., van den Bogaert, N. (1994) in: *Handbook of Crystal Growth*, Hurle, D. T. J. (ed.). Amsterdam: North Holland, Vol. 2b, p. 875.

Elliot, A. G., Flat, A., Vanderwater, D. A. (1992), *J. Cryst. Growth* **121**, 349.

Fainberg, J., Leister, H.-J., Müller, G. (1997), *J. Cryst. Growth* **180**, 517.

Fantini, F., Salviati, G., Borgarino, M., Cattani, L., Cova, P., Lazzarini, L., Fregonara, C. Z. (1998), *Inst. Phys. Conf. Ser.No.* **160**, 503.

Flade, T., Jurisch, M., Kleinwechter, A., Köhler, A., Kretzer, U., Prause, J., Reinhold, T. H., Weinert, B. (1999), *J. Cryst. Growth* **198/199**, 336.

Folberth, O. G., Weiss, H. (1955), *Z. Naturforsch.* **10a**, 615.

Frank, Ch., Jacob, K., Neubert, M., Rudolph, P., Fainberg, J., Müller, G. (2000), *J. Cryst. Growth* **213**, 10.

Gans, F., Lagrenaudie, J., Seguin, P. (1953), *C. R. Acad. Sci. Paris C* **237**, 310.

Gault, W. A., Monberg, E. M., Clemans, J. E. (1986), *J. Cryst. Growth* **74**, 491.

Gremmelmaier, R. (1956), *Z. Naturforsch.* **11a**, 511.

Hashio, K., Sawada, S., Tatsumi, M., Fujita, K., Akai, S. (1997), *J. Cryst. Growth* **173**, 33.

Higgins, W. M., Ware, R. M., Tiernan, M. S., O'Hearn, K. J., Carlson, D. J. (1997), *J. Cryst. Growth* **174**, 208.

Holmes, D. E., Chen, R. T., Elliott, K. R., Kirkpatrick, C. G. (1982), *Appl. Phys. Lett.* **40**, 46.

Holthenrich, A. (1997), *Berichte des Forschungszentrums Jülich*, Jül 3370.

Hurle, D. T. J. (1995), *J. Cryst. Growth* **147**, 239.

Hurle, D. T. J., Cockayne, B. (1994) in: *Handbook of Crystal Growth*, Hurle, D. T. J. (ed.). Amsterdam: North-Holland, Vol. 2a, p. 99.

Hurle, D. T. J. (1999), *Appl. Phys. Rev.* **85**, 6957.

Inada, T., Otoki, Y., Kuma, S. (1990), *J. Cryst. Growth* **102**, 915.

Inada, T., Komata, S., Ohnishi, M., Wachi, M., Akiyama, H., Otoki, Y. (1999), *Int. Conf. on GaAs Manufacturing and Technology (MANTECH) in Vancouver, Canada*, St. Louis. p. 205.

Indenbom, V. L. (1979), *Krist. Tech.* **14**, 493.

Jacob, K., Frank, Ch., Neubert, M., Rudolph, P., Ulrici, W., Jurisch, M., Korb, J. (2001), *Cryst. Res. Technol.* **35**, 1163.

Jordan, A. S., Caruso, R., von Neida, A. R. (1980), *Bell. Syst. Tech. J.* **59**, 593.

Jordan, A. S., Caruso, R., von Neida, A. R., Nielsen, J. W. (1981), *J. Appl. Phys.* **52**, 3331.

Jordan, A. S., von Neida, A. R., Caruso, R. (1984), *J. Cryst. Growth* **70**, 555.

Jurisch, M. (1998) in: *Proceedings Symp. on Non-Stoichiometric III-V Compounds*, Kiesel, P., Malzer, S., Marek, T. (eds), *Physik mikrostrukturierter Halbleiter Bd.6*, Erlangen-Nürnberg: Friedrich-Alexander-Universität, p. 135.

Kohda, H., Yamada, K., Nakanishi, H., Kobayashi, T., Osaka, J., Tomizawa, K. (1985), *J. Cryst. Growth* **71**, 813.

Korb, J., Flade, T., Jurisch, M., Köhler, A., Reinhold, Th., Weinert, B. (1999), *J. Cryst. Growth* **198/199**, 343.

Kuma, S., Otoki, Y. (1993), *Inst. Phys. Conf. Ser.* **No. 135**, 117.

Kuma, S., Shibata, M., Inada, T. (1994) in: *Gallium Arsenide and Related Compounds 1993*, Rupprecht, H. S., Weimann, G. (eds) Bristol: IOP Publishing 136, p. 497.

Lambropoulos, J. C., Delameter, C. N. (1988), *J. Cryst. Growth* **92**, 390.

Liu, X. (1995), *III-V Rev.* **8**, 14.

Maroudas, D., Brown, R. A. (1991), *J. Cryst. Growth* **108**, 399.

Meduoye, G. O., Bacon, D. J., Evans, K. E. (1991), *J. Cryst. Growth* **108**, 627.

Metz, E. A. P., Miller, R. C., Mazelsky, R. (1962), *J. Appl. Phys.* **33**, 2016.

Mills, A. (1998), *III-Vs Review* Vol. 11, No. 4, 46.

Miyazaki, N., Uchida, H., Hagihara, S., Munakata, T., Fukuda, T. (1991), *J. Cryst. Growth* **113**, 227.

Miyazawa, S., Mizutani, T., Yamazaki, H. (1982), *Jpn. J. Appl. Phys.* **21**, L542.

Möck, P., Smith, G. W. (2000), *Cryst. Res. Technol.* **35**, 541.

Monberg, E. (1994) in: *Handbook of Crystal Growth*, Hurle, D. T. J. (ed.). Amsterdam: North-Holland, Vol. 2a, p. 51.

Moriya, K., Ogawa, T. (1983), *Jpn. J. Appl. Phys.* **22**, L207.

Motakef, S. (1991), *J. Cryst. Growth* **108**, 33.

Motakef, S., Kelly, K. W., Koai, K. (1991), *J. Cryst. Growth* **113**, 279.

Motakef, S., Witt, A. F. (1987), *J. Cryst. Growth* **80**, 37.

Müllenborn, M., Alt, H. Ch., Heberle, A. (1991), *J. Appl. Phys.* **69**, 4310.

Müller, G. (1998) in: *Theoretical and Technological Aspects of Crystal Growth*, Fornari, R., Paorici, C. (eds) Zurich: Trans Tech Publ. p. 87.

Mullin, J. B. (1998) in: *Compound Semiconductor Devices*, Jackson, K. A. (ed.) Weinheim: Wiley-VCH. p. 1.

Mullin, J. B., Straughan, B. W., Brickell, W. S. J. (1965), *Phys. Chem. Solids* **26**, 782.

Nakanishi, H., Kohda, H., Yamada, K., Hoshikawa, K., (1984) in: *Ext. Abstracts 16th Conference on Solid-State Devices and Materials*, Kobe, p. 63.

Naumann, M., Rudolph, P., Neubert, M., Donecker, J. (2001), *J. Cryst. Growth* **231**, 22.

Neubert, M., Seifert, M., Rudolph, P., Trompa, K., Pietsch, M. (1996), in: *1996 IEEE Semiconducting and Semi-Insulating Materials Conference*, *IEEE SIMC-9, Toulouse 1996*, Fontaine, C. (ed.) Piscataway: IEEE, p. 17.

Neubert, M., Rudolph, P., Seifert, M. (1998) in: *1997 IEEE International Symposium on Compound Semiconductors*, Melloch, M., Reed, M. A. (Eds) Bristol: IEEE. p. 53.

Neubert, M., Rudolph, P. (1999), *Mitteilungsblatt der DGKK (Deutsche Gesellschaft für Kristallwachstum und Kristallzüchtung e.V.)* **Nr. 69**, p. 11.

Neubert, M., Rudolph, P. (2001), *Prog. Cryst. Growth Charac. Mater.* **43/2-3**, 119.

Nishizawa, J. (1990), *J. Cryst. Growth* **99**, 1.

Nishizawa, J., Okuno, Y., Suto, K. (1986), *JARECT Semiconductor Technologies* **19**, 17.

Oda, O., Kohiro, K., Kainosho, K., Uchida, M., Hirano, R. (1998) in: *Recent Developments of Bulk Crystal Growth*, Isshiki, M. (Ed.). Trivandrum: Research Signpost. p. 97.

Oda, O., Yamamoto, H., Seiwa, M., Kano, G., Inoue, T., Mori, M., Shimakura, H., Oyake, M. (1992,) *Semicond. Sci. Technol.* **7**, A215.

Ono, H. (1988), *J. Crys. Growth* **89**, 209.

Otoki, Y., Sahara, M., Shibata, M., Kuma, S., Kashiwa, M. (1995) *Materials Science Forum* **196–201**, p. 1431.

Ozawa, S., Miyairi, H., Nakajima, M., Fukuda, T. (1985), *Inst. Phys. Conf. Ser.* **79**, ch. 1, p. 25.

Oates, W. A., Wenzl, H. (1998), *J. Cryst. Growth* **191**, 303.

Parsey, J. M., Thiel, F. A. (1985), *J. Cryst. Growth* **73**, 211.

Petroff, P., Kimerling, L. (1976), *Appl. Phys. Lett.* **29**, 461.

Prieur, E., Tuomi, T., Partanen, J., Yli-Juuti, E., Till, M. (1993), *J. Cryst. Growth* **132**, 599.

Redlich, O., Kister, A. T. (1948), *Ind. Eng. Chem.* **40**, 345.

Richards, J. L. (1957), *J. Sci. Instrum.* **34**, 289.

Rudolph, P., Jurisch, M. (1999), *J. Cryst. Growth* **198/199**, 325.

Rudolph, P., Kiesling, F. (1988), *Cryst. Res. Technol.* **23**, 1207.

Rudolph, P., Neubert, M., Arulkumaran, S., Seifert, M. (1997), *Cryst. Res. Technol.* **32**, 35.

Schlossmacher, P., Urban, K. (1992), *J. Appl. Phys.* **71**, 620.

Shibata, M., Suzuki, T., Kuma, S., Inada, T. (1993), *J. Cryst. Growth* **128**, 439.

Seidl, A., Eichler, S., Flade, T., Jurisch, M., Koehler, A., Kretzer, U., Weinert, B. (2000), *J. Cryst. Growth* **225**, 561.

Seifert, M., Neubert, M., Ulrici, W., Wiedemann, B., Donecker, J., Kluge, J., Wolf, E., Klinger, D., Rudolph, P., (1996), *J. Cryst. Growth* **158**, 409.

Steinemann, A., Zimmerli, U. (1967), in: *Crystal Growth*, Peiser H. S. (ed.) Oxford: Pegamon Press, p. 81.

Tada, K., Tatsumi, M. (1984), Japanese Patent 60-264390; see also: US Patent 5256389 (1993).

Tajima, M. (1982,) *Jpn. J. Appl. Phys.* **21**, L227

Tanenbaum, M. (1959), in: *Semiconductors*, Hannay, N. B. (ed.) New York: Reinhold.

Tatsumi, M., Kawase, T., Iguchi, Y., Fujita, K., Yamada, M. (1994), in: *Semi-insulating III-V Materials*, Godlewski, M. (ed.). Singapore: World Scientific, p. 11.

Tatsumi, M., Fujita, K. (1998) in: *Recent Developments of Bulk Crystal Growth*, Isshiki, M. (ed.) Trivandrum: Research Signpost. p. 47.

Teal, G. K., Little, J. B. (1950), *Phys. Rev* **25**, 16.

Terashima, K., Washizuka, S., Nishio, J., Okada, A., Yasuami, S., Watanabe, M. (1986), *Inst. Phys. Conf. Ser.* **79**, 37.

Thomas, R. N., Hobgood, H. M., Ravishankar, P. S., Braggins, T. T. (1993), *Prog. Cryst. Growth Charact.* **26**, 219.

Tomizawa, K., Sassa, K., Shimanuki, Y., Nishizawa, J. (1984), *J. Electrochem. Soc.: Solid-State Technol.* **131**, 2294.

Tower, J. P., Tobin, R., Pearah P. J., Ware, R. M. (1991), *J. Cryst. Growth* **114**, 665.

Tsai, C. T., Gulluoglu, A. N., Hartley, C. S. (1993), *J. Appl. Phys.* **73**, 1650.

van Bueren, H. G. (1961), *Imperfections in Crystals*. Amsterdam: North-Holland, p. 221.

Völkl, J. (1994) in: *Handbook of Crystal Growth*, Hurle, D. T. J. (ed.) Amsterdam: North Holland, Vol. 2b, p. 821.

Ware, F. M., Higgins, W., O'Hearn, K., Tiernan, M. (1996), *GaAs IC Symposium – IEEE Gallium Arsenide Integrated Circuit Symposium*, Piscataway: IEEE. P. 54.

Welker, H. (1953), *Z. Naturforsch.* **8a**, 248.

Wenzl, H. (1998), private communication.

Wenzl, H., Dahlen, A., Fattah, A., Petersen, S., Mika, K., Henkel, D. (1991), *J. Cryst. Growth* **109**, 191.

Wenzl, H., Oates, W. A., Mika, K. (1993) in: *Handbook of Crystal Growth*, Hurle, D. T. J. (ed.) Amsterdam: North-Holland, p. 103.

Wickert, M., Stibal, R., Hiesinger, P., Jantz, W., Wagner, J., Jurisch, M., Kretzer, U., Weinert, B. (1998) in: *Proc. of the 10th Conf. on Semi-Conducting and -Isolating Materials (SIMC-X)*, Miner, C. (ed.), IEEE Publisher, p. 21.

Winston, H. V., Hunter, A. T., Olsen, H. M., Bryan, R. P., Lee, R. E. (1984), *Appl. Phys. Lett.* **45**, 447.

Wurzinger, P., Oppolzer, H., Pongratz, P., Skalinsky, P. (1991), *J. Cryst. Growth* **110**, 769.

Zhang, H., Prasad, V., Bliss, D. F. (1996), *J. Cryst. Growth* **169**, 250.

15 Growth of III-V and II-VI Single Crystals by the Vertical-gradient-Freeze Method

T. ASAHI, K. KAINOSHO, K. KOHIRO, A. NODA, K. SATO and O. ODA

Central R & D Laboratory, Japan Energy Corporation, Toda, Saitama 335-8502, Japan

15.1 INTRODUCTION

II-VI and III-V compound semiconductors with their advantageous physical and optical properties are important for optical and electronic devices. Semi-insulating GaAs is applied for high-frequency devices such as metal-semiconductor field-effect transistors (MESFETs), high-electron-mobility transistors (HEMTs), and HBTs (heterojunction-bipolar transistors). In fact, their main applications are for cellular phones and broadcasting-satellite (BS) systems. Conductive GaAs substrates are used for light-emitting diodes (LEDs), and for laser diodes (LDs) used for displays and CD players. Conductive GaP is a key material for yellow-green display LEDs. Conductive InP substrates are needed for LEDs, LDs and photodiodes (PDs) for optical-fiber communications. Semi-insulating InP is required for HEMTs and HBTs, exceeding GaAs-based devices, to be used for anticollision systems and millimeter-wave communications (Hagimoto *et al.* 1997; Wada and Hasegawa 1999; Sano 1999). CdTe and CdZnTe substrates are used for HgCdTe epitaxial films for far-infrared detectors. Semi-insulating CdTe is applied for X-ray or gamma-ray detectors. Even ZnTe single crystals are now required for pure-green LEDs based on intrinsic p-n junctions (Sato 1999).

GaAs, InP and GaP have >80 % of the compound semiconductor market and are mainly grown by the liquid-encapsulated-Czochralski (LEC) method except for conductive GaAs, which is grown by the horizontal Bridgman (HB) method. However, the LEC method has several disadvantages: Crystals are grown at a large temperature gradient exceeding several tens of degrees/cm, so that dislocation densities are normally higher than $10\,000\,cm^{-2}$. In the case of LEC-GaAs and LEC-InP applied for high-frequency devices, the wafer breakage in the device-production process due to the large thermal stress in wafers is a significant problem. Therefore it is indispensable to grow crystals at low thermal stress. For

Crystal Growth Technology, Edited by H. J. Scheel and T. Fukuda
© 2003 John Wiley & Sons, Ltd.   ISBN: 0-471-49059-8

LEC growth of large-diameter crystals the cost of apparatus with large-diameter crucibles makes a large fraction of the final wafer cost.

Compared with the LEC method the vertical bridgman (VB)/vertical gradient freezing (VGF) method has several advantages: Since crystals are grown in the crucible itself, the temperature gradient is very small ($1-10\,°C/cm$) so that dislocation densities are small compared with crystals grown by the LEC method. The apparatus cost is lower than that of the LEC apparatus, because mechanics for both rotating and moving the crystal and the crucible are not necessary. Since the crystal diameter is nearly the same as the crucible diameter, the dimension of the apparatus becomes smaller compared with the LEC method. Another advantage of VB/VGF is that the process requires little operator attention. All these advantages explain the increasing importance of the VB/VGF methods for growing large-diameter crystals.

The VB/VGF methods were first used in research for growing many oxide and semiconductor crystals (Kikuma et al. 1986; Singh et al. 1988; Brandon and Derby 1986; Kennedy et al. 1988), but they were not widely employed then in commercial production because of low crystal productivity compared with the LEC method. Only recently have the VB/VGF methods been used for producing large single crystals of $75-150\,mm$ diameter GaAs and 75 mm diameter $Li_2B_4O_7$. The VGF method is also used to grow single crystals of the II-VI compound semiconductors CdTe and ZnTe that could not be grown by the LEC method. CdTe single crystals with diameter larger than 75 mm can only be grown by VGF (Asahi et al. 1995).

In the next chapter we will show our experimental results of seeded growth of GaAs and InP, and of CdTe and ZnTe growth without seed crystals by the VGF method.

15.2   InP CRYSTAL GROWTH BY THE VGF METHOD

Conductive InP for optoelectronic devices is mainly grown by the LEC method. In the conventional LEC method the temperature gradient is larger than $100\,°C/cm$, so that many dislocations are generated in the crystal by the large thermal stress. It is difficult to grow single crystals under low temperature gradients in the conventional LEC method, because phosphorus evaporates from the growing crystals due to the high temperature above the B_2O_3 surface. In order to grow crystals with low dislocation densities without any phosphorus loss, improved LEC methods such as thermal-baffled LEC (TB-LEC) (Hirano and Uchida 1996; Hirano 1998), phosphorus-vapor-pressure-control LEC (PC-LEC) (Kohiro et al. 1991; Kohiro et al. 1996), and vapor-pressure-controlled czochralski (VCZ) methods have been developed (Morioka et al. 1992; Hosokawa et al. 1998). 50-mm diameter crystals with very low dislocation densities can be grown by the TB-LEC method. 75-mm diameter InP crystals with low dislocation densities are industrially grown by PC-LEC and by VCZ. Even 100-mm diameter InP crystals with low dislocation densities have recently been grown by the VCZ method. However, these modified LEC methods seem to be limited for growing crystals with larger than 100 mm diameter, because the dislocation density is increased with the crystal diameter.

The first VB/VGF growth of InP <111> oriented single crystals was reported by Gault *et al.* 1986. Since dislocation densities of grown crystals were one order of magnitude lower than those of LEC-grown crystals, VB/VGF became attractive for growing high-quality single crystals. Furthermore, the dislocation densities of 50-mm diameter crystals (Gault *et al.* 1986) and 100 mm diameter crystals (Asahi *et al.* 1999) do not differ much as shown in Figure 15.1. Even for larger-diameter crystals, lower dislocation densities are expected. However, in the early phase of VGF development, it was very difficult to grow <100> oriented InP single crystals due to twinning (Monberg *et al.* 1987; Matsumoto *et al.* 1986; Müller *et al.* 1992). Twins generate easily due to the low stacking-fault energy of InP (Thomas *et al.* 1990). There are several origins for twinning. The most probable one seems to be the variation of the solid/liquid interface position due to temperature fluctuations. In VB/VGF, the solid/liquid interface varies easily due to the low axial temperature gradient thus causing twinning. Therefore it is important to reduce temperature fluctuations for growing twin-free crystals.

Since the vapor pressure of phosphorus at the melting point of InP is about 30 atm, crystals have to be grown under high pressure to prevent phosphorus

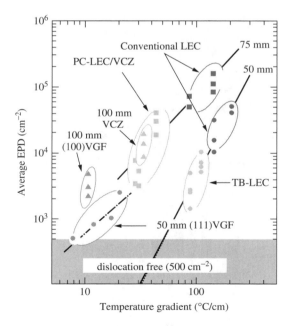

Figure 15.1 Dislocation densities as a function of temperature gradient in several growth methods.

evaporation. The high-pressure VB/VGF methods have some difficulties in grow-ing single crystals compared with conventional VB/VGF methods. Strong gas convection due to the high pressure causes large temperature fluctuations. It is very difficult to control the heater temperature precisely because of this large gas convection. The growth rate is varied easily by the temperature fluctuation at the solid/liquid interface because of the low temperature gradient in the VB/VGF methods. In particular, the temperature near the seed crystal with a small ther-mal capacitance due to a small diameter is strongly affected by the temperature fluctuation. It seems that this temperature fluctuation causes twin generation. It is therefore most important to reduce the temperature fluctuations by controlling the gas flow during crystal growth in order to grow twin-free InP single crystals.

We applied at first a precise temperature-measurement system to measure the temperature fluctuation during crystal growth. In the system, ice mixed with water in a thermos of about five liters was used to obtain the reference temperature. A Keithley digital voltmeter 2010 and metal-sheathed K-type thermocouples were used in the system. This temperature measurement system was able to measure the temperature with an accuracy of 0.01 °C. The sheathed thermocouples were inserted into alumina tubes to prevent their reaction with phosphorus.

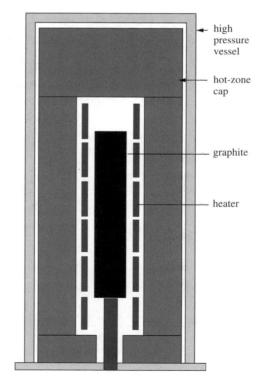

high pressure vessel

hot-zone cap

graphite

heater

Figure 15.2 Simulation model of the high-pressure VGF furnace.

Since the temperature fluctuation is caused by the gas flow, we investigated the gas flow in the furnace by computer simulation in which the simulation program STREAM and an IBM workstation RS6000 were used. Figure 15.2 shows the schematic diagram of the simulation model. To make the calculation easier, most material properties in the growth system are substituted to carbon properties except for the thermal insulating materials.

Figure 15.3 shows a schematic diagram of our high-pressure VGF furnace. It has six heater zones whose temperatures were independently controlled. The inner diameter of the graphite heaters was about 150 mm and the heights of each heater were 100 and 200 mm. The pBN crucible had an 8-mm inner-diameter seed pocket with 70 mm in length. A <100>-oriented 8-mm diameter InP seed crystal that was prepared from a LEC crystal and placed in this seed pocket of the crucible, and then 5 kg of InP polycrystal, prepared by the HB method or the LEC method, was charged in a pBN crucible with 104–110 mm inner diameter. The crucible was covered with a semiopen quartz ampoule and the melt was encapsulated by B_2O_3. The crucible was not perfectly sealed in the quartz ampoule in order to prevent bursting of the ampoule due to the gas pressure difference between the inside and outside of the ampoule. Therefore, B_2O_3 was used for preventing the phosphorus evaporation from the melt. In the VB/VGF methods, B_2O_3 is used to encapsulate the melt, to prevent the contact between melt and crucible, and thus prevent polycrystallization from the crucible wall

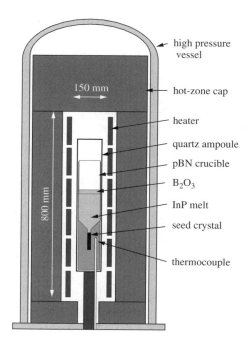

Figure 15.3 Schematic diagram of the high-pressure VGF furnace for InP crystal growth.

(Hoshikawa *et al.* 1989; Rudolph *et al.* 1996). About 150 g of B_2O_3 with low water content of 90 ppm H_2O was charged in the pBN crucible. After the ampoule was set in the furnace, the high-pressure vessel was evacuated to 10^{-4} torr and Ar gas was charged in the vessel. The ampoule was heated to the melting point of InP under the Ar pressure of 45 atm and held for about 30 h. The seed crystal was melted at a distance of about 20 mm down from the top of the seed pocket. The axial temperature gradient was less than 10 °C/cm and the growth rate was higher than 0.4 mm/h. After solidification, the grown crystal was cooled at a rate of about 100 °C/h.

(100) wafers were obtained by slicing the grown crystal perpendicular to the <100> growth axis. After wafers were polished and etched by a Br_2-CH_3OH solution, they were etched by a Hüber etchant to measure dislocation densities.

The computer-simulated gas flow explains the temperature fluctuations as shown in Figure 15.4. The gas flows from the bottom to the top along the inside of the heaters, and then from the top to the bottom along the outside of heaters. The outside gas flows between the heaters and collides with the gas inside the heaters near the ampoule as seen in Figure 15.4. These collisions seem to cause the temperature fluctuations. In the case of conventional hot-zones, the collisions between gases disrupt the flow near the ampoule because both gas-flow velocities inside and outside of the heaters are similarly small. The instability of gas flow thus seems to cause the temperature fluctuation. On the other hand, in the case of improved hot zones, the flow near the ampoule seems not to be influenced by the collisions, because it is much faster than the other gas flows.

The temperature fluctuation seemed to be reduced in the improved hot-zone arrangement that we have designed from the simulation results. Figure 15.5 shows

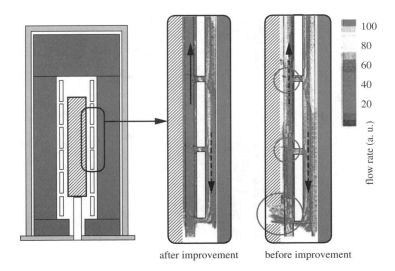

Figure 15.4 Results of the gas-flow computer simulation.

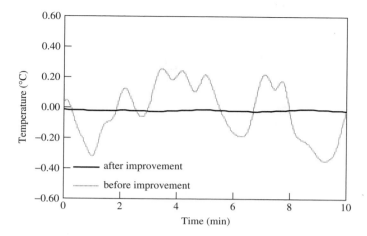

Figure 15.5 Temperature fluctuations near the seed crystal.

the temperature fluctuation near the seed crystal in the conventional and improved hot-zones. The temperature rapidly fluctuated in the range of $\pm 0.3\,°C$ in the conventional hot zone. If the temperature at the solid/liquid interface also fluctuates, the temperature fluctuations cause twin formation because the growth rate is significantly influenced by the temperature fluctuation. On the other hand, the fluctuation was reduced to less than $\pm 0.03\,°C$ and thus the growth-rate variation decreased after improving the hot-zone arrangement, as seen in Figure 15.5.

As mentioned above, twins are easily generated during crystal growth in InP because its stacking-fault energy is lower than that of GaAs. In fact, there were hardly any reports on twin-free large-diameter <100> InP single-crystal growth, even though the VB/VGF methods are commercially used for GaAs production. In the VB/VGF methods, the growth rate is largely varied by the temperature fluctuation due to the low temperature gradient. The temperature fluctuation in the LEC method is larger than that of the VB/VGF methods, but the variation in the growth rate is smaller because the temperature gradient is large. Twinning therefore takes place very easily due to the large variation in the growth rate during crystal growth by the VGF method. In this work, after the temperature fluctuations were reduced to less than $\pm 0.03\,°C$ and some suitable growth conditions were ascertained, twin-free <100> InP single crystals of 100 mm diameter and 80 mm length could be obtained as shown in Figure 15.6.

Figure 15.7 shows the etch pit density (EPD) distribution measured after Hüber etching. The EPD was measured across the <110> direction on the 100-mm diameter InP wafers. The EPD ranged from about $300\,cm^{-2}$ to about $3000\,cm^{-2}$ and the average EPD was $2000\,cm^{-2}$. This average EPD is quite low compared with that of conventional 75-mm diameter InP LEC crystals. This low average EPD is due to the low temperature gradient during crystal growth in the VGF method. The temperature gradient of our VGF method was less than $10\,°C/cm$,

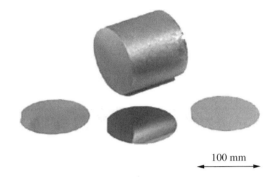

Figure 15.6   100 mm diameter InP single crystal and wafers.

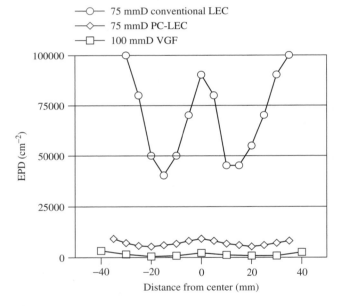

Figure 15.7   EPD distribution of InP wafers obtained by various growth methods.

so that it lowers the thermal stress in crystals during growth. In the conventional LEC method, because the temperature gradient is larger than 100 °C/cm, many dislocations are generated in the crystal by the large thermal stress. It is very difficult to grow single crystals under low temperature gradients using the conventional LEC method, because the phosphorus evaporates from the growing crystals due to the high temperature on the B$_2$O$_3$ surface. In order to prevent the phosphorus evaporation, the phosphorus-pressure-controlled LEC method, such as the PC-LEC method and the VCZ method, was developed. Since it is possible to grow crystals at the lower temperature gradient in the phosphorus atmosphere, low average EPD crystals can be obtained. However, the EPD of PC-LEC and VCZ crystals is still higher than that of VGF crystals.

Large-diameter semi-insulating InP substrates are needed for high-frequency devices such as HEMTs and HBTs. As in the case of LEC-GaAs, the wafer breakage of large-diameter, more fragile InP substrates is a significant problem in their wafer processing for high-frequency devices. Therefore, VGF crystals grown under the lowest thermal stress will be preferable for the industrial production of large-diameter substrates for high-frequency devices.

15.3   GaAs CRYSTAL GROWTH BY THE VGF METHOD

GaAs crystal growth by the VB/VGF methods in several modifications has been studied by many groups. For example, some groups (Gault *et al.* 1986; Hoshikawa *et al.* 1989; Amon *et al.* 1998; Kremer *et al.* 1990) studied growth in high-pressure furnaces, other groups (Okada *et al.* 1990; Buhring *et al.* 1994) studied growth under arsenic vapor pressure in a quartz ampoule near to ambient pressure. In this case there is the problem of ampoule-shape deformation due to the difference between inner and outer gas pressures during growth at temperatures of around 1200–1300 °C. However, the method has the advantage of melt-stoichiometry control through the arsenic pressure being adjusted by the temperature of the arsenic reservoir in the ampoule. An encapsulant was recently employed for VB/VGF growth of GaAs by many groups, because it prevents defect generation and polycrystallinity from the crucible wall. We have examined VGF growth with and without encapsulant and found liquid-encapsulated VGF (LE-VGF) very effective to grow high-quality crystals.

15.3.1   GROWTH OF UNDOPED GaAs

Our VGF furnace with six molybdenum silicide heaters of 150 mm inner diameter and 800 mm height for crystal growth is shown in Figure 15.8. A three-zone heater was used to control the arsenic reservoir temperature. The quartz ampoules used as growth vessel bulge because the arsenic pressure for the stoichiometric melt composition during growth is slightly larger than the atmospheric pressure (Arthur 1967). The bulged ampoule damaged the furnace. In order to solve this problem, we covered the ampoule with a ceramic tube and prevented the contact of the quartz ampoule with the furnace.

In this VGF growth, the temperature gradient was larger than 4 °C/cm as shown in Figure 15.9. Because the ampoule was deformed at the growth temperature, it was very difficult to measure the temperature distribution in the real melt. Therefore the melt was replaced by an alumina block Although the thermal conductivity of Al_2O_3 is lower than that of GaAs, the temperature distribution

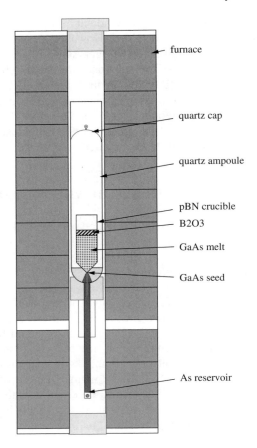

Figure 15.8   Schematic diagram of the VGF furnace for GaAs crystal growth.

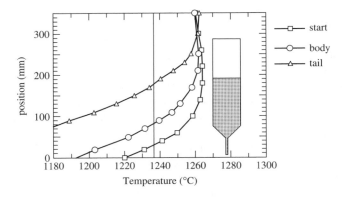

Figure 15.9   Temperature distributions measured by using Al_2O_3 blocks.

may not differ largely from that of the real material. This is because the starting position of the real crystal growth was consistent with the measured temperature distribution using the Al_2O_3 block.

Polycrystals of about 5 kg and a <100> seed crystal with about 8 mm diameter were charged in an 85-mm inner diameter pBN crucible with an 8 mm inner diameter and 70 mm length seed pocket. Around 30 g of B_2O_3 encapsulant material was charged in the crucible. Arsenic was charged in the reservoir of the ampoule, and the crucible with source materials was charged in the ampoule, which was sealed under 10^{-6} torr. The ampoule was set in the furnace and heated to above the melting point. Crystal growth was started after the seed crystal was melted down to 20–30 mm and held for several tens of hours. The temperature of each heater was controlled for the growth rate of 1 mm/h. The grown crystal was cooled at around 100 °C/h to room temperature.

After grown crystals were cut into wafers with the growth direction of <100>, the dislocation densities and electrical properties were characterized. The dislocation density was measured by counting the number of pits on the wafer after the wafer was etched by KOH melt at 400 °C for 10 min.

Figure 15.10 shows a 75 mm diameter VGF-grown crystal of 200 mm length. When B_2O_3 was not used in the crystal growth, grown crystals often had grain boundaries. On the other hand, most crystals grown by the VGF method using B_2O_3 as the encapsulant did not have grain boundaries.

Figure 15.11 shows the EPD distribution of a VGF-grown undoped crystal and an LEC-grown undoped crystal. The average EPD of 3700 cm^{-2} of the VGF crystal is much lower than that of the LEC crystals. Figure 15.12 shows wafers cut from near the seed crystal after EPD etching. Although etch pits distributed uniformly, a line of etch pits like a lineage was found in the center of the wafer as shown in Figure 15.12(a). It was found that the line has propagated from

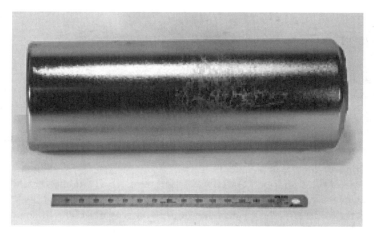

Figure 15.10   75-mm diameter undoped GaAs single crystal of 200-mm length.

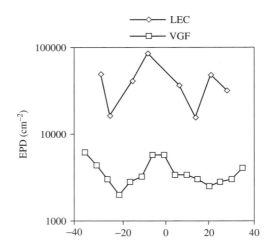

Figure 15.11   EPD distribution in undoped GaAs.

(a) using a LEC-crystal seed   (b) using a VGF-crystal seed

Figure 15.12   GaAs wafers near seed crystals after molten KOH etching.

the seed, because the seed crystal was prepared from a LEC crystal with high dislocation densities. In the VGF method, low dislocation density crystals, such as VGF, crystals should be used as seed crystal to prevent the propagation of dislocations. When VGF-grown crystals with around 3000 cm^{-2} EPD were used as seed crystals, although the average EPD of the wafer was not different from that of LEC-grown seed crystals, there were no lineages propagating from the seed crystals as shown in Figure 15.12(b).

The resistivity of undoped GaAs was larger than 10^8 Ω cm and the carbon concentration in the crystals was around 10^{16} cm^{-3} measured by the FT-IR method. Although the source of the carbon contamination was not clear, it seemed to be introduced from the graphite susceptor and the residual carbon impurity in the source materials. It is well known that the control of the carbon concentration is very important to obtain semi-insulating GaAs in the LEC method. In GaAs crystal growth using B_2O_3 as encapsulation, the carbon concentration in the grown crystal depends on the concentrations of impurities such as H_2O, CO, C and Ga_2O in the B_2O_3 and the melt (Nishio and Terashima 1989). In the case of the LEC method, the carbon monoxide gas pressure is regulated for controlling the carbon concentration in the GaAs melt (Nosovsky *et al.* 1991; Inada *et al.* 1994). On the other hand, it is difficult to control the carbon concentration in the VB/VGF methods using quartz ampoules, because the carbon monoxide gas can not be charged in the ampoule during crystal growth. VB/VGF-grown crystals with uniform resistivities were recently reported (Kawase *et al.* 1996). In the case of VB/VGF, the carbon monoxide pressure seems to be related with the oxygen gas pressure ($2C + O_2 = 2CO$) in the ampoule. If the oxygen pressure is controlled by any oxygen source, the carbon oxide pressure can be controlled. For example, when solid As_2O_3 is charged as the oxygen source in the reservoir of the ampoule and the temperature is controlled, the oxygen gas seems to be controlled according to the equation ($As_4 + 3O_2 = 2As_2O_3$) (Miyata and Oda 1997).

In analogy to InP discussed above the breakage of LEC substrates is a significant problem in the device processing. Therefore, one can expect that VB/VGF-grown semi-insulating substrates with low dislocation densities will be favored for electronic devices, when the precise control of the carbon concentration is achieved.

15.3.2 GROWTH OF SI-DOPED GaAs CRYSTALS

Conductive GaAs with n-type Si-doping is used for LEDs, photodiodes, LDs and solar cells and had been industrially grown mainly by the HB method. However, since the VB/VGF methods have the advantages of circular wafers and very low dislocation density (compared to HB-grown crystals), VB/VGF wafers are increasingly used for these optoelectronic devices.

Figure 15.13 shows the relationship between Si concentration and carrier concentration. The carrier concentration does not correspond to the Si concentration and hardly increases when the Si concentration exceeds 5×10^{18} cm^{-3}. Group IV element Si can occupy both Ga and As sites. When the Si concentration is lower than 5×10^{18} cm^{-3}, most Si atoms are located on Ga sites and become n-type impurities. On the other hand, when the Si concentration exceeds 5×10^{18} cm^{-3}, Si atoms are also located on the As site, so that the carrier concentration is not increased with the Si concentration (Greiner and Gibbons 1984).

The grown Si-doped crystal contained several weight-ppm boron from the encapsulant. Si in the GaAs melt seems to react with B_2O_3, and boron is released from B_2O_3, because the Gibbs free energy of B_2O_3 is larger than that of SiO_2 (Barin *et al.* 1989). Although boron does not affect electrical properties largely, we could reduce the boron contamination by using B_2O_3 with 5 mol % Si in the crystal growth.

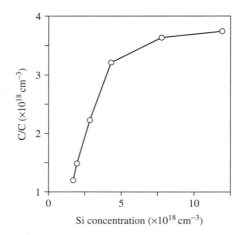

Figure 15.13   Carrier concentration of Si-doped GaAs.

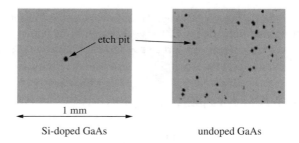

Figure 15.14   Etch pits for Si-doped and undoped GaAs crystals after molten KOH etching.

It is well known that dislocation-free (EPD $< 500\,\text{cm}^{-3}$) crystals can be obtained by Si-doping. Figure 15.14 shows photographs of the surfaces of Si-doped and undoped GaAs wafers after molten KOH etching. There are far fewer etch pits in the Si-doped GaAs wafer, than those of the undoped wafer, corresponding to dislocation densities of less than $100\,\text{cm}^{-3}$ in the whole of crystals.

15.3.3   GROWTH OF Zn-DOPED CRYSTALS

Conductive p-type GaAs substrates are sometimes used for high-power LDs such as V-grooved inner-stripe LDs. The p-type GaAs is prepared by Zn doping in epitaxial growth. GaAs crystals with a very high carrier concentration, in excess of $10^{19}\,\text{cm}^{-3}$, can be grown by Zn doping, which is about 10 times higher than

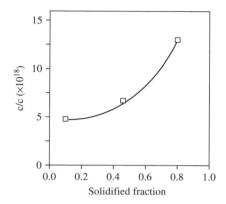

Figure 15.15   Carrier-concentration distribution in Zn-doped GaAs.

that of Si doping. Boron contamination is not a problem in the growth of Zn-doped crystals.

In order to achieve Zn doping effectively, it is preferable to use Zn-doped GaAs polycrystals as source material. When Zn or Zn_2As_3 are charged with GaAs polycrystals and heated to the GaAs melting point, most Zn has evaporated before the GaAs polycrystals are melted. For preparing Zn-doped polycrystals, Zn and GaAs are charged in a quartz ampoule and heated over the melting point of GaAs for several days. Zn-doped GaAs polycrystals and undoped GaAs of around 3 kg were charged in a pBN crucible. Crystals with <100> direction were grown by the VGF method under similar conditions mentioned above for undoped GaAs growth. The dislocation density of the Zn-doped crystals was similar to that of undoped GaAs.

The carrier concentration was higher than $10^{19}\,cm^{-3}$, thus higher than that achieved by Si-doping, and ranged from 0.5 to $1.5 \times 10^{19}\,cm^{-3}$ as shown in Figure 15.15. The carrier concentration of the crystal was consistent with the Zn concentration.

15.4   CdTe CRYSTAL GROWTH BY THE VGF METHOD WITHOUT SEED CRYSTALS

CdTe has low thermal conductivity, low stacking-fault energy, and a low critical-resolved shear stress. These physical properties render the growth of large-diameter high-quality CdTe crystals more difficult than the growth of GaAs and InP. However, there have been many attempts and progress in CdTe crystal growth (Thomas *et al.* 1990; Oda *et al.* 1986; Konkel *et al.* 1988; Casagramde *et al.* 1993; Kyle 1971; Mühlberg *et al.* 1993; Neugebauer *et al.* 1994; Asahi *et al.* 1996; Iwanov 1998). We have succeeded in growing 100-mm diameter single crystals without any twins by the VGF method (Asahi *et al.* 1996).

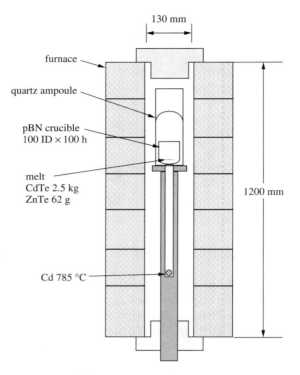

Figure 15.16   Schematic diagram of the VGF furnace for CdZnTe crystal growth. (Reprinted from T. Asahi *et al.*, *J. Cryst. Growth* **149** (1994) 23–29, copyright (1994) with permission from Elsevier Science.)

Figure 15.16 shows a schematic diagram of the VGF furnace. It has seven heater zones whose inner diameter was about 130 mm and the height was about 1200 mm in total. The upper four zones were used for crystal growth, and the lower three zones for controlling the Cd reservoir temperature. Temperature fluctuations of all zones were less than 0.1 °C.

The high-purity polycrystalline CdTe source had been synthesized from 6N Cd and 6N Te in a quartz ampoule using a high-pressure furnace at 5 atm. 2.5–3 kg CdTe and 60–70 g polycrystalline 6N ZnTe were put in a pBN crucible of 100 mm inner diameter and 100 mm height. The crucible with source materials and about 10 g 6N-Cd was placed in the quartz ampoule that was then sealed under 2×10^{-6} torr. After positioning the ampoule in the furnace, the source materials were melted at about 10 °C above the melting point. The Cd reservoir was controlled to 785 °C. The axial temperature gradient near the crucible was controlled within 10 °C/cm and the cooling rate was about 0.1 °C/h during crystal growth. The ampoules were cooled at about 100 °C/h after the crystal growth. Most of the 100 mm diameter $Cd_{0.97}Zn_{0.03}Te$ single crystals were grown in the <111> direction although we used the self-seeding method without any seed crystals. It seems that the growth direction <111> is preferred because the atomic planes have the highest packing density.

The grown crystal was sliced to wafers perpendicular to the growth axis. Figure 15.17 shows the grown crystal and sliced wafers. There were neither grain boundaries nor twins from top to tail except at the periphery and the tail of the ingot. The polycrystallinity of the tail part was caused by the higher growth rate. The growth rate increased from top to tail, because the temperature gradient became gradually small from top to tail and the cooling rate was constant during growth. The growth rate at the tail, calculated from the relationship between temperature gradient and cooling rate, was higher than 1 mm/h, the typical growth rate in VGF growth of InP and GaAs of higher thermal conductivity. This explains the polycrystallinity of CdTe/CdZnTe crystals at the tail.

The grown crystals were characterized by etch-pit density (EPD), X-ray rocking curve, GD-MS purity measurement, low-temperature photoluminescence (PL) spectra, electrical properties and Te precipitate observation. Several etchants have been studied for the characterization of dislocations in CdTe and CdZnTe crystals (Inoue *et al.* 1963; Nakagawa *et al.* 1979; Hahnet and Schenk 1990). The etch pits formed on the CdTe (111)A planes by the Nakagawa etchant (30 ml HF + 20 ml H_2O_2 + 20 ml H_2O) clearly correspond to dislocations (Nakagawa *et al.*, 1979). Figure 15.18 shows EPD distributions of top, middle and tail wafers measured in the <011> and <112> directions across the wafer. The EPD uniformly distributed in the wafer, and the difference of EPD among top, middle and tail wafers was very small. The average EPD was $(4-6) \times 10^4 \, cm^{-2}$ over the whole crystal.

The full-width-at-half-maximum (FWHM) of X-ray rocking curves is often used to characterize the crystalline perfection. High-resolution X-ray diffractometry with four Ge (440) front crystals as monochromator was used for measuring the $Cd_{1-x}Zn_xTe$ (333) reflection with $CuK\alpha_1$ radiation and $1 \times 5 \, mm^2$ radiation area. Figure 15.19 shows FWHMs of the diffraction peaks ranging from 8 to 13 arcsec except at the edge of the wafers. The large peak widths at the edge of the wafers are due to lineages and microtwins at the periphery.

100 mm

Figure 15.17   100 mm diameter CdZnTe.

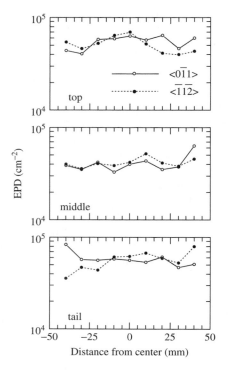

Figure 15.18 EPD distributions in CdZnTe. (Reprinted from T. Asahi *et al.*, *J. Cryst. Growth* **149** (1994) 23–29, copyright (1996) with permission from Elsevier Science.)

The Zn concentration decreases along the CdZnTe crystal due to segregation as shown in Figure 15.20. The measured Zn compositions are in good agreement with the calculated ones for the segregation coefficient $k = 1.35$.

The impurities of crystals are analyzed by GD-MS. Many elements can be simultaneously detected, and the detection limit of most elements is lower than 0.05 ppm. Only 0.1 ppm Na and 0.05 ppm Al were detected in the CdZnTe crystals, and Cu, Ag, Ca, Al and other impurities were below the detection limit (0.001–0.05) ppm. This purity level was satisfactory for the applications.

For the electric characterization the electrode was prepared by electroless Au plating on (110) planes perpendicular to the (111) wafers. The $I-V$ curve showed a good ohmic contact within a 0.01-V bias voltage. Resistivities, carrier concentrations, and Hall mobilities along the crystal were $(30-37)\,\Omega\,cm$, $(2.1-2.2) \times 10^{15}\,cm^{-3}$, and $(80-100)\,cm^2\,V^{-1}\,s^{-1}$, respectively. These electrical properties seem to be caused by Cd vacancies or Te interstitials.

PL measurement at low temperatures are sensitive to detect impurities at very low concentrations that cannot be detected by a chemical analysis method such as GD-MS. Bound exciton (BE) or free-exciton (FE) peaks are very strong in

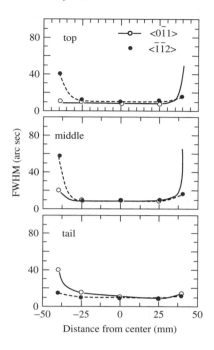

Figure 15.19 X-ray FWHM distributions in CdZnTe. (Reprinted from T. Asahi *et al.*, *J. Cryst. Growth* **149** (1994) 23–29, copyright (1994) with permission from Elsevier Science.)

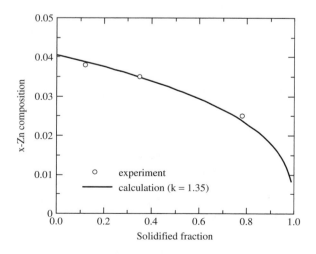

Figure 15.20 Zn distribution in CdZnTe. (Reprinted from T. Asahi *et al.*, *J. Cryst. Growth* **161** (1996) 20–27, copyright (1996) with permission from Elsevier Science.)

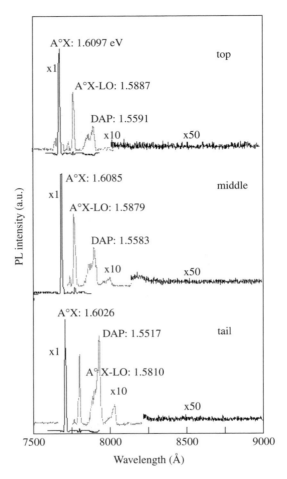

Figure 15.21 Photoluminescence spectrum of CdZnTe at 4.2 K. (Reprinted from T. Asahi *et al.*, *J. Cryst. Growth* **161** (1996) 20–27, copyright (1996) with permission from Elsevier Science.)

high-purity CdTe, while a broad luminescence peak at around 1.4 eV is shown in contaminated CdTe (Allred *et al.* 1987; Feng *et al.* 1988). The PL spectra were measured at 4.2 K in liquid He using a 5145 Å Ar laser. Figure 15.21 shows a PL spectrum of the grown crystal. The PL spectrum at the top was shifted about 0.02 eV higher than that of CdTe due to the larger band-gap energy of the $Cd_{1-x}Zn_xTe$ ($x = 0.04$). The strongest peak at 1.609 eV is assigned to the exciton bound to a neutral acceptor (A^0X) transition because of p-type conductivity caused by the Te-rich crystal composition. The donor–acceptor pair emission at the tail is stronger than that of the top of the crystal, so that it seems that the donor impurities are increased from top to tail by segregation. Compared with A^0X, other peaks were weak, especially the broad peak at 1.4 eV was not found, so that it is concluded that the present $Cd_{1-x}Zn_xTe$ crystals are of high purity.

There are many large precipitates in CdTe because of the large deviation from stoichiometry at the congruent point of about 10^{16} cm^{-3} excess to the Te side. On the other hand, the deviation of composition of the solidus line at room temperature is only 10^{15} cm^{-3} excess to the Te side. Therefore, the difference between the crystal compositions at the congruent point and at room temperature leads to Te precipitates (Zanio 1978). The Cd pressure must be controlled to prevent this large deviation from stoichiometry during crystal growth. In our experiment, the precipitate diameter was smaller than 5 μm at the Cd reservoir temperature from 780 °C (about 1.2 atm) to 813 °C (about 1.7 atm) as shown in Figure 15.22. There were many precipitates larger than 10 μm when the Cd pressure was lower than 1.2 atm. On the other hand, precipitates larger than 30 μm were found in the crystal under a Cd pressure higher than 1.7 atm. In the present work, we controlled the Cd reservoir temperature to 785 °C, which corresponds to a Cd pressure of 1.2 atm, in order to minimize the size of Te precipitates.

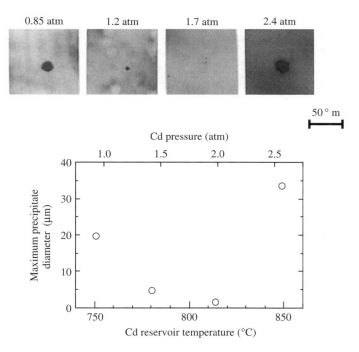

Figure 15.22 Te precipitates in CdZnTe. (Reprinted from T. Asahi *et al.*, *J. Cryst. Growth* **161** (1996) 20–27, copyright (1994) with permission from Elsevier Science.)

15.5   ZnTe CRYSTAL GROWTH BY VGF WITHOUT SEED CRYSTALS USING THE HIGH-PRESSURE FURNACE

ZnTe crystals of limited size have been grown from high-temperature solutions (Crowder and Hammer 1966; Steininger and England 1968; Triboulet and Didier 1975; Sato *et al.* 1997) and by vapor transport (Su *et al.* 1993; Taguchi *et al.* 1977). Only by hetero-seeding with sapphire substrate, crystals up to 20 mm diameter and 10 mm length could be grown from Te solution in a VB arrangement (Seki *et al.* 1997). However, solution growth has the disadvantage for industrial production of the low growth rate compared with growth from melts. The VGF method with ZnTe melt using a sapphire seed and B_2O_3 encapsulation was attempted to increase the productivity (Sato *et al.* 1997). By this hetero-seeding method, single crystals could not be obtained because B_2O_3 partially disturbed the ZnTe melt contacting with the sapphire seed. In view of the successful unseeded VGF growth of CdTe/CdZnTe discussed in Section 15.4, this approach has been examined for growing ZnTe single crystals.

Figure 15.23 shows the high-pressure furnace in which two graphite heaters were used for crystal growth. The pBN crucible of 70 mm inner diameter and 150 mm height was covered by a graphite plate for preventing contamination from the hot-zone. About 400 g 6N-Zn, 780 g 6N-Te, and 80 g B_2O_3 were charged into the pBN crucible, which had a lid and was positioned in the graphite holder. The high-pressure vessel was evacuated to 10^{-3} torr and N_2 gas was charged to about 15 atm. The source materials were heated to the ZnTe melting point at the rate of 500 °C/h. Zn reacted with Te at about 500 to 800 °C, and the ZnTe polycrystal was synthesized at about 30 atm N_2 pressure. The synthesized ZnTe polycrystal was melted and the N_2 gas pressure was reduced to 10 atm. The melt was held at about 5 °C above the melting point (1295 °C), and then the heaters were controlled for the growth rate of 1 to 4 mm/h. The temperature was measured by W-WRe; 5–26 % Re thermocouples. The temperature gradient in the melt was lower than 15 °C/cm and the temperature fluctuation was ±0.3 °C. The self-seeded crystals were cut perpendicularly to the growth orientation (nearly <111>), and polished and etched by NaOH solution (Fuke *et al.* 1971) for 10 min for investigating the dislocation density. Impurities in the crystal were measured by GD-MS.

Figure 15.24 shows the grown crystal with 70 mm diameter and 45 mm length. There was no grain boundary at the top of the ingot. However, polycrystallization occurred near the tail of the ingot, because the solid/liquid interface shape was not suitable due to the high growth rate and the very low temperature gradient at the end of the growth process.

In the case of ZnTe the correlation of etch pits revealed by NaOH solution with dislocations was not proven. However, the etch pits are in good agreement with dislocations measured from CL micrographs in our work (Sato *et al.* 1997) and were around 2000 cm^{-2}. For VGF this is very low and comparable with the EPD of solution-grown crystals (Seki *et al.* 1997). The high structural perfection of these VGF-grown ZnTe crystals is the result of careful control of the solid/liquid interface shape in the low temperature gradient during the growth process.

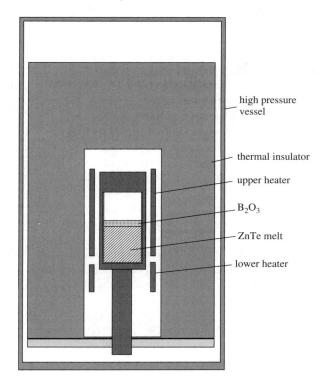

high pressure
vessel

thermal insulator

upper heater

B$_2$O$_3$

ZnTe melt

lower heater

Figure 15.23   Schematic diagram of the high-pressure VGF furnace for ZnTe crystal growth.

cleavage
plane

<111> growth

75 mm

Figure 15.24   70-mm diameter ZnTe.

15.6 SUMMARY

The VGF growth of InP single crystals of 100 mm diameter, low-dislocation GaAs crystals of 75 mm diameter, and large-diameter CdZnTe and ZnTe crystals with diameters of 75–100 mm has been presented. These results could be achieved by optimizing the growth conditions, especially the temperature distributions and minimized temperature fluctuations.

Even though most III-V crystals are industrially grown by the LEC method, our work showed that the VGF method is also attractive for the production of low-dislocation crystals. The twinning problem of VGF, compared with LEC, was overcome mainly by reducing the temperature fluctuations and optimizing the temperature distributions.

In the case of II-VI compounds, VGF is the main industrial method for production of CdZnTe crystals, also high-quality ZnTe crystals for pure-green LEDs can be grown by VGF.

The present work shows that VGF production of various III-V and II-VI crystals is promising, as the LEC method approaches its limitations for large diameters, low dislocation densities, and production furnace costs.

REFERENCES

Allred, W. P., Khan, A. A., Johnson, C. J., Giles, N. C., and Schetzina, J. F. (1987) *Mater. Res. Soc. Symp. Proc.* **90**, 103.

Amon, J., Dumke, F., and Müller, G. (1998) *J. Cryst. Growth* **187**, 1.

Arthur, J. R. (1967) *J. Phys. Chem. Solids* **28**, 2257.

Asahi, T., Oda, O., Taniguchi, Y., and Koyama, A. (1995) *J. Cryst. Growth* **149**, 23.

Asahi, T., Oda, O., Taniguchi, Y., and Koyama, A. (1996) *J. Cryst. Growth* **161**, 20.

Asahi, T., Kainosho, K., Kamiya, T., Nozaki, T., Matsuda, Y., and Oda, O. (1999) *Jpn. J. Appl. Phys.* **38** 977.

Barin, I., Sauert, F., Rhonhof, E. S., and Sheng, W. S. (1989) Thermochemical Data of Pure Substances, eds. H. F. Ebel and C. D. Brenzinger, VCH Verlagsgesellschaft, Weinheim.

Brandon, S., and Derby, J. J. (1986) *J. Cryst. Growth* **121**, 473.

Buhring, E., Frank, C., Gärtner, G., Hein, K., Klemm, V., Kühnel, G., and Voland, U. (1994) *Mater. Sci. Eng. B* **28**, 87.

Casagramde. L. G., Marzio, D. D., Lee, M. B., Larson Jr., D. J., Dudley, M., and Fanning, T. (1993) *J. Cryst. Growth* **128**, 576.

Crowder, B. L., and Hammer, W. N. (1966) *Phys. Rev.* **150**, 541.

Feng, Z. C., Bevan, M. J., Choyke, W. J., and Krishnaswamy, S. V. (1988) *J. Appl. Phys.* **64**, 2595.

Fuke, S., Washiyama, M., Otsuka, K., and Aoki, M. (1971) *J. Appl. Phys.* **10**, 687.

Gault, W. A., Monberg, E. M., and Clemans, J. E. (1986) *J. Cryst. Growth* **74**, 491.

Greiner, M. E., and Gibbons, J. F. (1984) *Appl. Phys. Lett.* **44**, 750.

Hagimoto, K., Kataoka, T., Yonenaga, M., and Kobayashi, I. (1997) *Proc. 9th Int. Conf. on InP and Related Mater.*, Hyannis, 233.

Hahnet, I., and Schenk, M. (1990) *J. Cryst. Growth* **101**, 251.

Hirano, R. (1998) *Proc. 10th Int. Conf. on InP and Related Materials*, Tsukuba, 80.

Hirano, R., and Uchida, M. (1996) *J. Eletron. Mater.* **25**, 347.

Hoshikawa, K., Nakanishi, H., Kohda, H., and M. Sasaura (1989) *J. Cryst. Growth* **94**, 643.

Hosokawa, Y., Yabuhara, Y., Nakai, R., and Fujita, K. (1998) *Proc. 10th Int. Conf. on InP and Related Materials*, Tsukuba, 34.

Inada, T., Wachi, M., Suzuki, T., Shibata, M., and Kashiwa, M. (1994) *Hitachi Cable Review* **13**, 53.

Inoue, M., Teramoto, I., and Takayanagi, S. (1963) *J. Appl. Phys.* **34**, 404.

Iwanov, Yu. M. (1998) *J. Cryst. Growth* **194**, 309.

Kawase, T., Hagi, Y., Tatsumi, M., Fujita, K., and Nakai, R. (1996) *9th IEEE Semiconducting and Semi-Insulating Materials Conference*, Toulouse, 275.

Kennedy, J. J., Amirtharaj, P. M., Boyd, P. R., Qadri, S. B., Dobbyn, R. C., and Long, G. G. (1998) *J. Cryst. Growth* **86**, 93.

Kikuma, I. M., Sekine, M., and Furukoshi, M. (1986) *J. Cryst. Growth* **75**, 609.

Kohiro, K., Kainosho, K., and Oda, O. (1991) *J. Electro. Mater.* **20**, 1015.

Kohiro, K., Ohta, M., and Oda, O. (1996) *J. Cryst. Growth* **158**, 197.

Konkel, W. H., Tighe, S. J., Bland, L. G., Sharma, S. R., and Taylor, R. E. (1998) *J. Cryst. Growth* **86**, 111.

Kremer, R. E., Francomano, D., Beckhart, G. H., and Burke, K. M. (1990) *J. Mater. Res.* **5**, 1468.

Kyle, N. R. (1971) *J. Electochem. Soc.* **118**, 1790.

Matsumoto, F., Y. Okano, I. Yonenaga, K. Hoshikawa and T. Fukuda (1986) *J. Cryst. Growth* **132**, 348.

Miyata, A., and Oda, O. (1997) Japanese Patent No. P09-100860.

Monberg, E. M., Gault, W. A., Simchock, F., and Dominguez, F. (1987) *J. Cryst. Growth* **83**, 174.

Morioka, M., Yokogawa, M., Fujita, K., and Akai, S. (1992) *Proc. 4th Int. Conf. on InP and Related Materials*, Newport, 258.

Mühlberg, M., Rudolph, P., and Laasch, M. (1993) *J. Cryst. Growth* **128**, 571.

Müller, G., G. Hirt and D. Hofmann (1992) 7th Conf. on Semi-Insulating III-V Materials Ixtapa, Mexico.

Nakagawa, K., Maeda, K., and Takeuchi, S. (1979) *Appl. Phys. Lett.* **34**, 574.

Neugebauer, G. T., Shetty, R., Ard, C. K., Lancaster, R. A., and Norton, P. (1994) *Proc. SPIE Int. Soc. Opt. Eng.* **2228**, 2.

Nishio, J., and Terashima, K. (1989) *J. Cryst. Growth* **96**, 605.

Nosovsky, A. M., Bolsheva, Yu. N., Ilyin, M. A., Markov, A. V., Mikhailova, N. G., and Osvensky, V. B. (1991) *Acta Phys.* **70**, 211.

Oda, O., Hirata, K., Imura, K., and Tsujino, F. (1986) *Jpn. Soc. Appl. Phys (Oyo Buturi)* **55**, 1084 (in Japanese).

Okada, Y., Sakuragi, S., and Hashimoto, S. (1990) *Jpn. J. Appl. Phys.* **29**, 1954.

Rudolph, P., Matsumoto, F., and Fukuda, T. (1996) *J. Cryst. Growth* **158**, 43.

Sano, E. (1999) *Proc. 11th Int. Conf. on InP and Related Mater.*, Davos, 299.

Sato, K. (1999) *9th Int. Conf. on II-VI Compounds*, Kyoto.

Sato, K., Seki, Y., Matsuda, Y., Oda, O. (1997) *J. Cryst. Growth* **197**, 413.

Seki, Y., Sato, K., and Oda, O. (1997) *J. Cryst. Growth* **171**, 32.

Singh, N. B., Gottlieb, M., Conroy, J. J., Hopkins, R. H., and Mazelsky, R. (1988) *J. Cryst. Growth* **87**, 113.

Steininger, J., and England, R. E. (1968) *Trans. Met. Soc. AIME* **242**, 444.

Su, C. H., Voltz, M. P., Gillies, D. C., Szofran, F. R., Lehoczky, S. L., Dudley, M., Yao, G.-D., and Zhou, W. (1993) *J. Cryst. Growth* **128**, 627.

Taguchi, T., Fujita, S., and Inuishi, Y. (1977) *J. Cryst. Growth* **45**, 204.

Thomas, R. N., Hobgood, H. M., Ravishankar, P. S., and Braggins, T. T. (1990) *J. Cryst. Growth* **99**, 643.

Triboulet, R., and Didier, G. (1975) *J. Cryst. Growth* **28**, 29.

Wada, O., and Hasegawa, H. (1999) in: *InP-Based Materials and Devices – Physics and Technology*, eds Wada, O., and Hasegawa, H. (Wiley-Interscience New York), 1.

Zanio, K. (1978) in: *Semicond. Semimet.* **13**.

16 Growth Technology of III-V Single Crystals for Production

TOMOHIRO KAWASE, MASAMI TATSUMI
and YASUHIRO NISHIDA

Sumitomo Electric Industries, Ltd. Itami, Hyogo, 664-0016 Japan

16.1 INTRODUCTION

III-V compound semiconductors were invented by Welker in 1952, three years after the invention of the silicon transistor by Shockley (1949).

The III-V compounds GaAs and GaP are used in light-emitting diodes for electrical appliances and displays, and in laser diodes, one of the key devices in optical communication. GaAs is also important for electronic devices for portable telephones. The properties of these crystals strongly depend on the process of both crystal growth and surface treatment. Basic properties, the crystal-growth technique and applications of GaAs and InP, the most promising III-V compound semiconductors, are reported in this chapter.

16.2 PROPERTIES OF III-V MATERIALS

Table 16.1 summarizes the characteristic parameters of GaAs and InP in comparison with Si. One of the effective features of GaAs and InP is the direct electron transition in band structure, which gives high efficiency of light emission and detection for light-emitting diodes, laser diodes and photo-detectors. Another important characteristic is the higher electron mobility in comparison with Si, by which these materials are advantageous for high-speed high-frequency electrical devices with very small power consumption, suitable for mobile communication.

Despite the superior properties of GaAs- and InP-based electronic devices their applications are limited due to the higher price of GaAs (factor of 10) and InP (factor of 50) compared to Si. These high prices of compound semiconductors are caused by high prices of raw materials (such as gallium, indium, arsenic and phosphorous) and by the lower productivity of single crystals.

GaAs and especially InP have high dissociation pressures at the melting point. Special equipment and growth techniques are necessary to suppress dissociation.

Crystal Growth Technology, Edited by H. J. Scheel and T. Fukuda
© 2003 John Wiley & Sons, Ltd.   ISBN: 0-471-49059-8

Table 16.1 Characteristic parameters of GaAs and InP at room temperature (except dissociation pressure and critical resolved shear stress) in comparison with Si

	Si	GaAs	InP
Crystal structure	diamond	zinc blend	zinc blend
Band structure	indirect	direct	direct
Band gap energy (eV)	1.11	1.40	1.35
Electron mobility (cm^2/V/s)	1900	8800	4600
Lattice constant (nm)	0.543	0.565	0.587
Density (g/cm^3)	2.33	5.32	4.79
Melting Point (°C)	1410	1238	1062
Dissociation pressure at m.p. (atm)	10^{-7}	0.9	25
Thermal expansion coefficient (K^{-1})	2.33×10^{-6}	5.93×10^{-6}	4.5×10^{-6}
Thermal conductivity (W/cm/K)	1.41	0.46	0.7
Critical shear stress at 40 °C below the m.p. (dyn/cm^2)	1.850×10^7	0.587×10^7	0.6×10^6

Their low thermal conductivity limits the growth speed and thus the productivity. Additionally, the lower critical shear stress induces high dislocation density under lower thermal stress and thus degrades the crystal quality. These disadvantages explain the high prices and limited applications of compound semiconductors so that their future depends on innovations for cost reduction.

16.3   GROWTH TECHNOLOGY OF III-V MATERIALS

Typical growth techniques of GaAs and InP single crystals are summarized in Table 16.2. Development of 2-inch diameter GaAs single crystals for device applications was advanced by the horizontal Bridgman (HB) (Suzuki *et al.* 1978, Akai *et al.* 1983) and liquid encapsulated Czochralski (LEC) techniques, by which mass production of conductive and semi-insulating GaAs substrates have been achieved.

InP single crystal was developed mainly using the LEC technique, which was modified in order to decrease dislocation density: Vapor-pressure-controlled Czochralski (VCZ) technique developed by Tada (Tada *et al.* 1987; Tatsumi *et al.* 1989; Kawase *et al.* 1990, 1992a, 1992b; Hashio *et al.* 1997) became a typical growth method of InP single crystals with low dislocation density.

Gault *et al.* (1986) showed that the vertical gradient freeze (VGF) method was very promising to grow III-V compound semiconductor single crystals with very low dislocation density. Hoshikawa *et al.* (1989) reported liquid encapsulated vertical Bridgman (LE-VB) techniques to make 3-inch GaAs single crystals with good reproducibility, which shed light on VB and VGF as a production method. Mass production of 4-inch and 6-inch semi-insulating GaAs single crystals has

Table 16.2 Typical growth techniques of GaAs and InP single crystals

	HB/HGF	LEC	VCZ	VB
	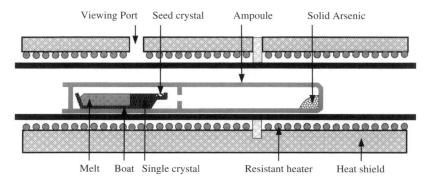			
Growth chamber	Fused silica	Stainless steel	Stainless steel	Stainless steel Fused silica
Wafer shape	Rectangular Half circular	Circular	Circular	Circular
Large diameter	Poor	Very good	Very good	Very good
Long crystal	Good	Very good	Medium	Good
Low dislocation density	Good	Poor	Good	Very good

been achieved (Kawase *et al.* 1996, 1999a, 1999b; Nakai *et al.* 1998; Hagi *et al.* 2001). Now, over 50 % of semi-insulating GaAs substrates in the world are manufactured by these techniques.

16.3.1 HB AND HGF TECHNIQUES

HB and horizontal gradient freeze (HGF) techniques are typical growth techniques of GaAs single crystals. Figure 16.1 shows the schematic drawing of the growth system for the HB technique. A seed crystal and presynthesized GaAs polycrystals, placed in a fused silica boat, and solid arsenic are sealed in a fused

Figure 16.1 Schematic drawing of the growth system by the HB technique.

silica ampoule. The ampoule is moved into a furnace and heated to over the melting point of GaAs. A part of the seed crystal is immersed in the molten raw material, and subsequently crystal growth is started. Instead of using presynthesized polycrystals, GaAs melt can be prepared in situ by the synthesis with liquid gallium filled in a fused-silica boat and arsenic vapor evaporated from solid arsenic. The yield of single crystal is lower in the latter one-step process because of the complexity. Therefore the former two-step process is adopted in mass production.

In the HB method the fused-silica boat or the resistance heater are moved horizontally to grow a single crystal under a substantially fixed temperature profile. In the HGF method the temperature profile of the resistance heater is continuously changed under a fixed position of the components. For manufacturing, a long crystal is advantageous for cost reduction. To grow a single crystal without defects, the temperature profile around the solid/liquid interface must be kept constant. The HB technique has the advantage of growing a long crystal in comparison with HGF, because the temperature profile around the solid/liquid interface can easily be kept constant.

Since Si-doped n-type and Zn-doped p-type GaAs crystals with low dislocation density can be obtained by these techniques, HB and HGF became popular in mass production of GaAs substrates for photonic devices such as light emitting diodes (LED) and laser diodes (LD). On the other hand, only a few % of Cr-doped crystals are used for semi-insulating substrates for electronic devices, because these growth techniques are not appropriate to produce large-diameter circular wafers. Most of the semi-insulating GaAs substrates are manufactured by the LEC, VB and VGF techniques mentioned below.

16.3.2 LEC TECHNIQUE

Figure 16.2 shows a schematic drawing of a LEC puller. Presynthesized GaAs or InP polycrystals and liquid encapsulant boric oxide are placed in a fused silica or pyrolytic boron nitride (PBN) crucible. In the case of GaAs having lower dissociation pressure than InP, direct synthesis was also adopted in mass production, where GaAs is synthesized from gallium and solid arsenic placed in a crucible. A resistant-type graphite heater surrounded by a graphite heat shield is equipped in a water-cooled stainless steel puller. The puller is evacuated and pressurized by nitrogen or argon, and subsequently the crucible is heated. At about 450 °C, the boric oxide encapsulant begins to soften. At the melting point, the raw material is melted and covered by liquid boric oxide. The pressure in the puller is adjusted from 1 to 2 MPa for GaAs and from 4 to 5 MPa for InP, whereby evaporation from the melt is suppressed.

The crucible is rotated typically from 3 to 20 rpm to improve uniformity of temperature and composition of the melt. After the temperature is adjusted, the seed crystal attached on the lowest part of the pulling rod is immersed into the

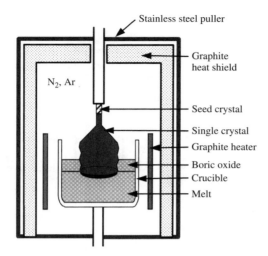

Figure 16.2   Schematic drawing of the LEC technique.

melt. The seed is pulled up slowly typically from 5 to 10 mm/h, whereby a single crystal is grown. The crystal is pulled through the boric oxide layer and exposed to argon or nitrogen atmosphere. When the temperature of the surface of the crystal is higher than a critical value, arsenic or phosphorous is dissociated from the surface of the crystal. Residual gallium or indium aggregates and drops down along the surface of the crystal thereby eroding the surface. If the droplet of gallium or indium liquid reaches the growth interface, nucleation of polycrystalline material is unavoidable. Therefore, the surface temperature of the crystal must be kept low enough to suppress dissociation by growing the crystal under a large temperature gradient, which, however, results in high thermal stress and high dislocation density in the crystal, the drawback of the LEC technique.

The LEC technique is advantageous to grow high-purity crystals. Si contamination can be suppressed using a PBN crucible, and additionally boric oxide purifies the melt by gettering the impurities. The LEC technique also has the advantage of growing long and large-diameter crystals, because there is no wetting problem between liquid GaAs and the crucible, which is the most serious problem in horizontal boat and vertical crucible techniques, such as HB, HGF, VB and VGF. Because of these advantages, the LEC method became the preferred growth method of semi-insulating GaAs. Large-diameter and long semi-insulating GaAs single crystals have been manufactured (Shibata *et al.* 1993).

16.3.3   VAPOR-PRESSURE-CONTROLLED CZOCHRALSKI (VCZ) TECHNIQUE

VCZ was developed to overcome the above problem of the LEC technique. A schematic drawing of a VCZ puller is shown in Figure 16.3. Basically we

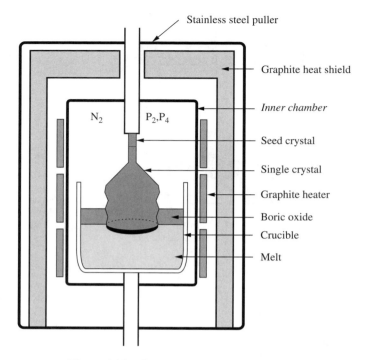

Figure 16.3 Concept of the VCZ technique.

can employ conventional LEC pullers for the VCZ technique but we should place a gas-impermeable inner chamber inside the heater. Since dissociation on the surface of the crystal can be suppressed by controlling the partial pressure of arsenic or phosphorous in the inner chamber, the crystal can be grown under a low temperature gradient. Figure 16.4 shows a temperature profile in a VCZ puller measured at the typical growth condition of GaAs in comparison with LEC. GaAs crystals with low dislocation density were successfully developed by the VCZ technique (Kawase *et al.* 1990, 1992a, 1992b; Hashio *et al.* 1997). Though VCZ is advantageous to grow the crystals with large diameter and low dislocation density, mass production of GaAs by VCZ has not been achieved, because of the rapid progress of vertical crucible techniques, such as VB and VGF.

On the other hand, mass production of InP single crystals by the VCZ technique has been successfully achieved. Because InP single crystal with low dislocation density has not been supplied by the other methods, despite the needs for low dislocation density InP substrates. Recently, vertical crucible techniques, such as VB and VGF, have been developed (Young *et al.* 1998, Kawase *et al.* 2002). Competition between these techniques has begun.

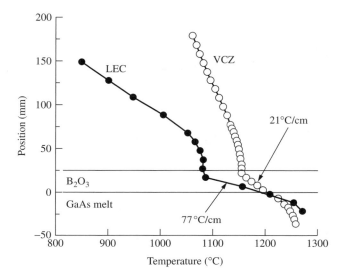

Figure 16.4 Temperature profile in a VCZ puller measured at the typical growth condition of GaAs in comparison with LEC.

16.3.4   VB AND VGF TECHNIQUES

Gault *et al.* (1986) showed that VGF was a very promising technique to grow III-V compound semiconductor single crystals with very few dislocations. Detailed research by Hoshikawa *et al.* (1989) on the LE-VB technique of GaAs has made vertical crucible methods one of the most popular growth techniques for GaAs.

Figure 16.5 is an example of the growth system of a VB technique. Presynthesized GaAs polycrystals and a seed crystal are placed in a fused-silica or PBN crucible. In mass production, a PBN crucible is generally used, because the fused-silica crucible has insufficient strength at the growth temperature. Boric oxide is essential when using PBN crucibles, because polycrystals are nucleated when the crystal contacts the crucible (Hoshikawa *et al.* 1989). A thin boric oxide layer on the inner surface of the crucible suppresses the direct contact between crystal and crucible. Preoxidation is suggested to form a uniform boric oxide layer on the crucible surface. After GaAs polycrystals are melted in the crucible, a single crystal is grown as the temperature of the melt is decreased (Bourret and Merk 1991).

In the VB technique, the crucible or the heater is moved vertically under a fixed temperature profile. In the VGF technique, the temperature profile is continuously changed toward the lower temperature with a fixed relative position between the crucible and the heater. In analogy to HB and LEC growth there are two methods to suppress dissociation of the GaAs melt. In the first, the crucible

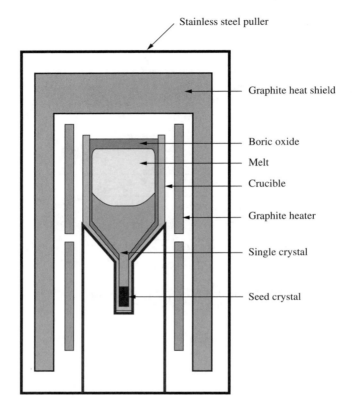

Figure 16.5 Schematic drawing of the VB technique.

with raw material is sealed in a fused-silica ampoule to keep the equilibrium arsenic pressure, and in the second method the GaAs melt is encapsulated by boric oxide and pressurized by argon or nitrogen in a water-cooled stainless steel puller (LE-VGF, LE-VB). Large-diameter GaAs with low dislocation density and low residual strain can be grown by VB and VGF techniques, such as 3-inch diameter Si-doped n-type and 4- to 6-inch diameter semi-insulating substrates.

16.4 APPLICATIONS AND REQUIREMENTS FOR GaAs SINGLE CRYSTALS

Applications and requirements for GaAs and InP substrates are shown in Table 16.3. As mentioned above, there are two categories of applications, namely for photonic and for electronic devices. As for photonic devices, GaAs substrates are used from red to infrared light emitting diodes and laser diodes. N-type substrates with carrier concentration from 10^{17} to 10^{19} cm^{-3} are typically used.

Table 16.3 Applications and requirements for GaAs and InP substrates (O – mass production, Δ – R&D)

	Device	Application	GaAS	InP	Requirement
Photonic	Light emitting diode	Lamp, personal computer	O	–	Low price
	Laser diode	Optical communication CD player, DVD player	O	O	Low dislocation density
	Photodetector	Optical communication	O	O	Low dislocation density
	Solar cell	Artificial satellite	O	–	Low price
Electronic	Transistor (MES-FET, HEMT, HBT)	Mobile communication, Optical communication	O	O	Electrical properties (Resistivity, Mobility), uniformity
	IC	Computer	O	Δ	Large diameter, Low price

Since dislocations in the substrate degrade the lifetime of photonic devices, especially of laser diodes, substrates with dislocation density less than $500\,\mathrm{cm}^{-2}$ are preferred.

Recently, requirements for GaAs electronic devices have been increasing as key devices for portable telephones, because of the low power consumption in the high-frequency region. In this field, there is severe competition in the cost performance with Si devices. Development of a large-diameter wafer process is the key to winning the competition.

16.5 GROWTH OF LARGE SINGLE CRYSTALS

Competition for low-cost mass production of semiconductor devices has required large-diameter substrates, which are difficult to produce. Though successful growth of 4-inch single crystals by HB and HGF methods has been reported, mass production of those crystals has not been achieved due to the cost-reduction problems. The cross section of the crystal is half-circular or rectangular, which causes large losses by crystal machining. Additionally, it is very difficult to grow large-diameter crystals more than 4-inch diameter, because of asymmetric thermal environment in the growth atmosphere. Therefore, large-diameter crystals of 4- to 6-inch have been manufactured by LEC, VCZ, VB and VGF techniques.

Since the temperature difference between the center and the periphery of the crystal increases with the diameter, precise control of temperature profile becomes

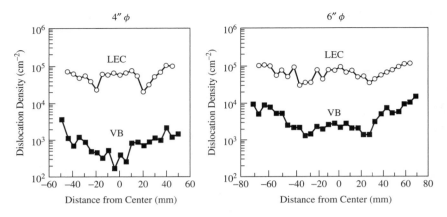

Figure 16.6 Radial profiles of dislocation density on 4-inch and 6-inch semi-insulating GaAs substrates grown by VB and by LEC.

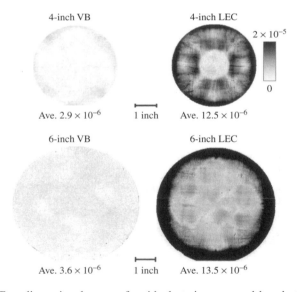

Figure 16.7 Two-dimensional maps of residual strain measured by photo-elastic effect measurement on 4-inch and 6-inch semi-insulating GaAs substrates grown by the VB technique and by LEC.

more important to grow larger-diameter crystals. Wafers of LEC-grown large diameter crystals have a significant risk of breakage due to their increased residual strain. In contrast, VB- and VGF-grown wafers have no problem of breakage because of the smaller residual strain. However, when the crystals with large diameter and great length are grown, polycrystallization is sometimes caused by direct contact between crystal and crucible. This is one of the serious problems in VB and VGF techniques.

Figures 16.6 and 16.7 show radial profiles of dislocation density and two-dimensional maps of residual strain measured by the photoelastic effect measurement in 4-inch and 6-inch semi-insulating GaAs substrates grown by the VB technique in comparison with LEC. Dislocation densities of VB substrates are one order of magnitude lower, and residual strain is less than half that of LEC substrates. It is proven that VB substrates with low residual strain have a lower probability of slip-line generation during the epitaxial growth process (Kawase *et al.* 1992, 1999a). Large crystals of more than 4-inch diameter were exclusively produced by the LEC technique, until mass production by VB and VGF techniques was achieved. The shares of VB and VGF substrates, having low dislocation density and low residual strain, is increasing.

16.6 GROWTH OF LOW-DISLOCATION-DENSITY GaAs CRYSTAL

Dislocations are generated by plastic deformation when the thermal stress in the crystal exceeds the critical resolved shear stress, that is, $\tau_T > \tau_{CRSS}$ where τ_T is thermal stress caused in a crystal and τ_{CRSS} is the critical resolved shear stress of GaAs. There are two approaches to reduce dislocation density.

One is by increasing τ_{CRSS} by doping with impurities. It is well known that silicon, zinc and indium are impurities that increase τ_{CRSS} effectively. 3-inch GaAs crystals with dislocation density less than $10\,cm^{-2}$, nominally dislocation-free crystals, were grown by Si doping (Hagi *et al.* 1996). Growth of undoped or semi-insulating crystals with such low dislocation density has not been achieved, due to the smaller τ_{CRSS} than Si-doped crystals. The lowest dislocation density of undoped or semi-insulating crystals grown by VB or VGF, which has been reported up to now, is about $1000\,cm^{-2}$ (Kawase *et al.* 1996).

Another method to reduce dislocation density is decreasing τ_T. τ_{CRSS} depends on the temperature, and as the lowest value at the melting point. Therefore, it is important to make τ_T near the solid/liquid interface smaller. τ_T depends on the temperature difference in the radial direction. In the case of the LEC technique, the largest τ_T is found mainly near the solid/liquid interface and near the surface of the B_2O_3 encapsulant. The temperature gradient in the B_2O_3 layer is much larger than in the atmosphere above the B_2O_3 layer, because of the small thermal conductivity and small thermal convection in the B_2O_3. Consequently, the temperature gradient changes steeply along the axial direction at the boundaries between the atmosphere and the B_2O_3, where a large temperature difference occurs in the radial direction resulting in large τ_T. The radial temperature gradient can be reduced in the VCZ technique, due to the small axial temperature gradient in the B_2O_3 layer. Consequently, τ_T near the boundaries between the atmosphere and the B_2O_3 is much smaller than that of the LEC technique.

It is believed that dislocation density has a strong correlation with excess thermal stress τ_{EX}. τ_{EX} is defined as the sum of the positive differences between the

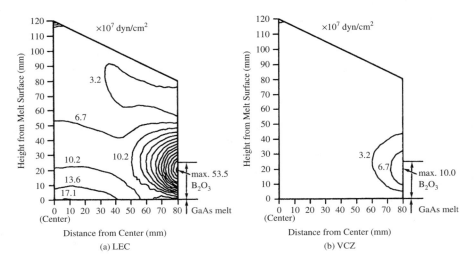

Figure 16.8 Calculated results of τ_{EX} in 6-inch undoped GaAs during crystal growth by VCZ and LEC techniques.

absolute value of each of the resolved shear stress components to twelve equivalent $\langle 110 \rangle$ slip-directions on $\{111\}$ slip-planes, and the critical resolved shear stress τ_{CRSS}. τ_{EX} caused in the crystal can be estimated by computer simulation. Figure 16.8 shows the calculated results of τ_{EX} in 6-inch undoped GaAs caused during the crystal growth by VCZ and LEC techniques. τ_{EX} caused near the surface of the B_2O_3 layer in the VCZ crystal is much smaller than that of the LEC crystal, which is believed to be why dislocation densities of VCZ crystals are much lower than those of the LEC crystals.

In the case of VB and VGF techniques, the thermal stress caused near the solid/liquid interface is dominant, because the B_2O_3 layer has little effect on the solid/liquid interface. Large τ_T is caused by the large temperature difference in the radial direction, when the solid/liquid interface becomes excessively convex or concave. Making the solid/liquid interface flatter is most important to decrease τ_T near the solid/liquid interface. Figure 16.9 shows the calculated results of τ_{EX} in Si-doped GaAs crystals grown by the VB technique with flatter convex (a) and with concave (b) interfaces. τ_{EX} in the crystal with a flatter interface (a) is much smaller than that in the crystal with concave interface (b). Figure 16.10 shows the experimental results of dislocation density. Dislocation density of a crystal with a flatter interface is much lower than that of a crystal with a concave interface. From τ_{EX} estimated by computer simulation and the experimental results of dislocation density, it is certain that making the solid/liquid interface flatter is most important to reduce dislocation density in the VB technique.

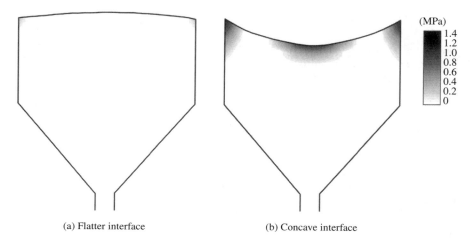

(a) Flatter interface   (b) Concave interface

Figure 16.9   Shows the calculated results of τ_{EX} in Si-doped GaAs crystals grown by the VB technique with (a) flat convex and (b) concave interface.

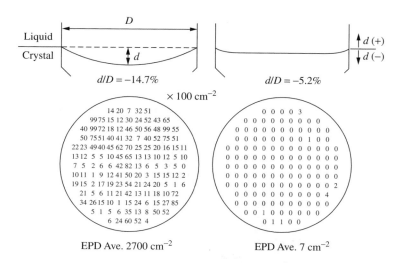

EPD Ave. 2700 cm^{-2} EPD Ave. 7 cm^{-2}

Figure 16.10   Experimental results of dislocation density. The left figure shows a crystal with concave interface, the right figure a crystal with flatter interface.

16.7   CONTROL OF QUALITY AND YIELD OF GaAs CRYSTALS

Suppression of twinning and lineage is the key for the mass production of GaAs crystals, because these defects decrease the yield of the crystals.

16.7.1   TWINNING

Twinning is probably caused in the cone part between the seed and the cylindrical body. Once twinning is initiated, the crystal grows in an undesired direction. For example, in the case of $\langle 100 \rangle$ growth of GaAs, the growth direction changes from $\langle 100 \rangle$ to $\langle 122 \rangle$. Therefore twinning is one of the most unfavorable defects.

The mechanism of twinning has not been clarified yet, however, it has been suggested that twinning was related to growth stability of the $\{111\}$ facet. It is expected that growth of the $\{111\}$ facet is enhanced under a low temperature gradient and a high concentration doping. Figure 16.11 shows the striation profiles of a $\{111\}$ facet around a twin of a S-doped InP. Since twinning depends on the stacking-fault energy, it is a serious problem in InP with low stacking-fault energy.

It is empirically known that twinning possibly occurs due to the following causes.

(a)   Large temperature fluctuation in the melt;
(b)   Contact of scum on the melt surface to the solid/liquid interface (in LEC and VCZ);
(c)   Wetting of the melt with the boat or crucible (in HB, HGF, VB and VGF).

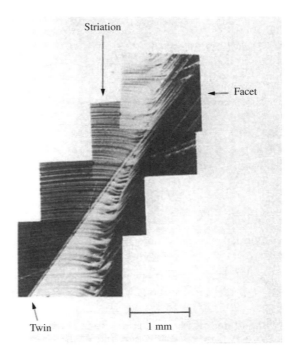

Figure 16.11   Striation profiles of $\{111\}$ facet around twin of a S-doped InP crystal.

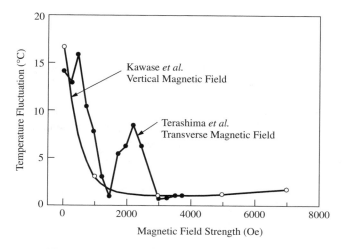

Figure 16.12 The relation between the amplitudes of temperature fluctuations in GaAs melt and the strength of magnetic field.

It is suggested to apply magnetic fields to suppress temperature fluctuations in the melt. Figure 16.12 shows the relation between the amplitude of temperature fluctuations in GaAs melt and the strength of magnetic field (Terashima *et al.* 1986, Kawase *et al.* 1986). The amplitude is reduced to 10 % by applying a magnetic field of 0.1 T. A magnetic field is applied in mass production of InP (Hosokawa *et al.* 1998).

Twinning sometimes occurs when scum on the melt surface contacts the solid/liquid interface in pulling techniques, such as LEC and VCZ. It is expected that the occurrence of scum is related to the crucible material, water content in B_2O_3, and melt composition.

Wetting of the melt to the boat or the crucible sometimes causes twinning in HB, HGF, VB, and VGF techniques. Wetting is one of the most important problems to be overcome. Sandblasting the inner surface of the fused-silica boat or crucible is effective to suppress wetting. This technique has led to the successful mass production by HB and HGF techniques. A PBN crucible is commonly used in VB and VGF techniques. Though GaAs melt originally wets the crucible, direct contact of the melt to the crucible can be protected by insertion of a B_2O_3 layer between them. However, twinning and polycrystallization will occur, if enough thickness of the B_2O_3 layer can not be obtained. Preparing a B_2O_3 layer of sufficient thickness and good uniformity is very important to realize good productivity (reduced twinning and polycrystallization) in mass production. A uniform B_2O_3 layer (Bourret *et al.* 1991) can be formed on the inner surface of the crucible by oxidizing PBN itself according to

$$4BN + 3O_2 \longrightarrow 2B_2O_3 + 2N_2$$

16.7.2   LINEAGE

Formation of lineage and the resulting polycrystallization decrease the yield
of undoped or semi-insulating GaAs crystals. It is suggested that lineage is
formed when dislocations propagate and interact perpendicular to the concave
solid/liquid interface. It is believed that polycrystallization is caused by growth
of grains with different crystallographic orientations from the original crystal,
when the lineage reaches the solid/liquid interface. It is reported that in LEC
growth when the center of the circle contacting the solid/liquid interface at the
peripheral region comes inside the surface of the crystal, formation of lineage
and the resulting polycrystallization are caused (Shibata *et al.* 1993). Therefore,
the shape of the solid/liquid interface must be convex to achieve a high yield of
single crystal.

The relation between latent heat Q_L, heat flux out of the crystal Q_{OUT}, heat
flux into the crystal from the melt Q_{IN}, and heat flux into the crystal in the radial
direction Q_R is shown in Figure 16.13. The shape of the solid/liquid interface
depends on the balance of heat fluxes as shown in Equation (16.3).

$$Q_R = Q_{OUT} - (Q_L + Q_{IN}) \tag{16.3}$$

The shape of the solid/liquid interface becomes convex when $Q_R > 0$, concave when $Q_R < 0$. Therefore, Q_L, Q_{IN}, Q_{OUT} must be controlled to satisfy

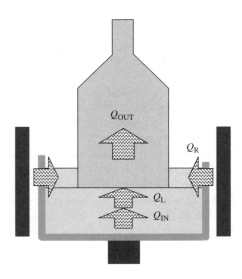

Figure 16.13   Relation among latent heat Q_L, heat flux out from the crystal Q_{OUT}, heat
flux into the crystal from the melt Q_{IN}, and heat flux into the crystal along the radial
direction Q_R.

the following relation in order to realize a convex interface.

$$Q_R = Q_{OUT} - (Q_L + Q_{IN}) > 0 \qquad (16.4)$$

Decreasing Q_L, which is realized by decreasing growth speed, is undoubtedly advantageous to achieve a convex interface, however, low growth speed decreases the productivity of crystals. Therefore, it is favorable to control Q_{IN} and Q_{OUT} in order to a achieve a convex interface.

16.8 CONTROL OF THE ELECTRONIC QUALITY OF GaAs

16.8.1 ABSOLUTE VALUE OF RESISTIVITY

The resistivity of semi-insulating GaAs substrates is one of the most important properties, especially for electronic devices fabricated by the ion-implantation techniques, because the performance of the devices strongly depends on the electronic properties of the substrate. The semi-insulating property of GaAs is realized when the deep donor EL2 compensates the concentration difference between shallow acceptors and shallow donors. This relation is expressed by the following relation, where N_{DD}, N_{SA}, N_{SD} are concentrations of deep donor, shallow acceptor, and shallow donor, respectively,

$$N_{DD} > N_{SA} - N_{SD} \qquad (16.5)$$

Possible shallow donor impurities are Si from the fused-silica crucible and S from graphite components in the growth chamber. Concentrations of these impurities can be minimized by using a PBN crucible instead of a fused-silica crucible, and by using highly purified graphite components. The resistivity of semi-insulating GaAs is substantially changed by the concentrations of carbon doped as shallow acceptor impurity, and EL2. EL2 is an intrinsic defect, which is generated during the heat treatment of 800–950 °C. GaAs crystals are subjected to a complicated thermal history during the growth. Consequently, the concentration of EL2 in as-grown crystals becomes extremely inhomogeneous along the growth direction. Therefore, the concentration of EL2 has to be adjusted by post-growth annealing at 800–950 °C.

Carbon is unintentionally doped into GaAs melt through CO gas generated from graphite components in the growth chamber. The carbon concentration in the crystal can be adjusted by the water content in the B_2O_3, however, decrease of carbon concentration from seed to tail can not be avoided due to the large effective segregation coefficient of carbon ($k_{eff} > 1$). The carbon concentration correlates with the CO partial pressure in the LEC puller, therefore it can be controlled by adjusting the CO partial pressure (Doering et al. 1990). Uniform carbon concentration from seed to tail has been realized in mass production using this technique.

16.8.2   UNIFORMITY OF MICROSCOPIC RESISTIVITY

Uniformity of microscopic resistivity on GaAs substrates is required to decrease
the variation of threshold voltage of FET (field effect transistor) across the wafer.
Figure 16.14 shows a photograph of etch pits revealed by KOH etching and a
two-dimensional map of resistivity in a microscopic region on a VB substrate.
The distribution of resistivity corresponds to the cellular structure of etch pits,
therefore to an inhomogeneous distribution of EL2 around dislocations.

EL2 is generated at 800 to 950 °C and reaches the equilibrium concentration
after heat treatment for several hours. It is suggested that EL2 is related to an
arsenic atom substituted on a site of gallium. The concentration of EL2 in an
as-grown crystal is usually lower than the equilibrium concentration, because the
duration at 800–950 °C during growth is not sufficient. Usually, since the generation speed of EL2 is higher around the network of dislocations than elsewhere,
the concentration of EL2 around the network is higher than elsewhere in the
crystal. The resulting inhomogeneous distribution of microscopic resistivity is
reduced by annealing the as-grown crystals at 800–950 °C to improve the homogeneity of EL2 by generating sufficient EL2 inside the network of dislocations.
By this post-growth annealing, substrates with good uniformity are manufactured
in mass production.

Improved homogeneity of microresistivity can be achieved by dissolving the
precipitates of arsenic. Arsenic precipitates nucleate selectively on dislocations
and grow larger by absorbing arsenic atoms due to the strain field around dislocations at temperatures below 1100 °C during crystal growth. As a result, the
concentration of excess arsenic inside the network of dislocations decreases so
that the EL2 concentration inside the network can not reach the equilibrium
concentration even after annealing at 800–950 °C. Consequently, the inhomogeneous distribution of resistivity remains even after post-growth annealing. In an
improved process, the arsenic precipitates around the dislocations are melted by
annealing over 1100 °C. After that, the crystal is cooled down to 500 °C and then
annealed at 850 °C for several hours.

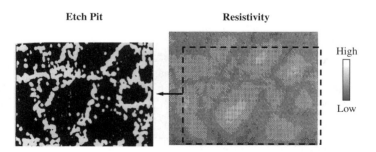

Figure 16.14   A photograph of etch pits revealed by KOH etching and a two-dimensional
map of resistivity in a microscopic region of a VB substrate.

16.9 TREND OF GROWTH METHODS FOR GaAs

A comparison of diameter and dislocation density of Si-doped n-type GaAs substrates and semi-insulating GaAs substrates grown by the major growth techniques is shown in Figure 16.15. In the case of photonic devices, a low dislocation density is required, but large diameters of substrates are not so critical, because the tip size of the device is very small. Therefore, it is expected that HB and HGF substrates will be used for small-diameter 2-inch substrates. However, VB and VGF substrates are exclusively used for large-diameter 3-inch substrates. The ratio of VB and VGF substrates is expected to increase with increase of the requirement for large-diameter substrates.

In the case of semi-insulating substrates for electronic devices, VB and VGF techniques have made remarkable progress recently. The requirement for VB and VGF substrates, by suppressing slip-line generation, is increasing, as wafer size is increasing from 4- to 6-inch diameter. LEC substrates still cover about half of the wafer market, whereas the share of VB and VGF substrates is increasing steeply. It is certain that VB and VGF techniques will become the major growth techniques of semi-insulating GaAs single crystals.

16.10 InP

InP substrates are mainly used for photonic devices, such as laser diodes and photodetectors, in optical communication. Since dislocations in the substrate degrade the lifetime of laser diodes, S-doped substrates with very low dislocation density are preferred. Sn-doped and Fe-doped substrates with higher dislocation density are usually used for photodetectors and for electronic devices.

Since InP melt has a high equilibrium vapor pressure of about 2.8 MPa, a water-cooled stainless steel puller is essential for polycrystal synthesis and

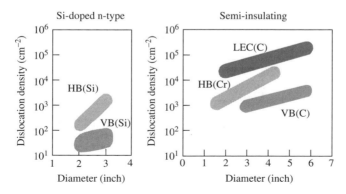

Figure 16.15 Comparison of diameters and dislocation densities of Si-doped n-type GaAs substrates and semi-insulating GaAs substrates for the major growth techniques.

single-crystal growth. Direct synthesis, which is the preferred synthesis method of GaAs, is not possible. Figure 16.16 shows one of the typical synthesis techniques of InP. A fused-silica boat with solid indium and solid phosphorus is sealed in a fused-silica ampoule. The ampoule is placed in a stainless steel high-pressure chamber, and InP is synthesized by heating the ampoule.

Another synthesis technique of InP is shown in Figure 16.17. InP is synthesized in a stainless-steel high-pressure chamber, which is pressurized above the evaporation pressure of InP melt. Phosphorus vapor is injected into the In melt encapsulated with boric oxide, by which phosphorus and indium are reacted.

Single crystals are grown by LEC, VCZ, VB and VGF techniques. A comparison of dislocation densities of 4-inch Fe-doped semi-insulating InP crystals produced by the major growth techniques is shown in Table 16.4. The ratio of semi-insulating InP substrates with low dislocation density manufactured by VCZ, VB and VGF techniques is increasing.

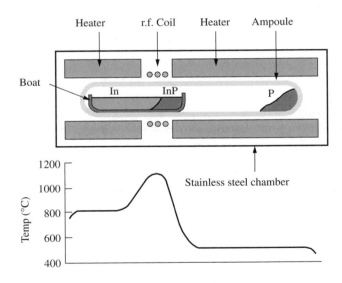

Figure 16.16 A typical synthesis technique of InP with temperature program: Boat method.

Table 16.4 Comparison of dislocation densities of 4-inch Fe-doped semi-insulating InP crystals produced by the major growth techniques

Growth technique	LEC	VCZ	VB/VGF
Dislocation density	50 000~ 100 000	15 000~ 30 000	2 000~ 20 000

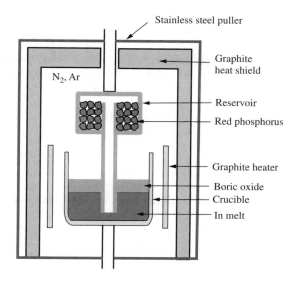

Figure 16.17   A synthesis technique of InP: Injection method.

16.11   SUMMARY

Recently, the qualitative and quantitative requirements for the III-V semiconductors GaAs and InP have abruptly increased for applications as photonic devices for optical communication and electronic appliances, and as electronic devices for high-frequency mobile communication. However, the III-V compounds have inherent disadvantages, such as higher production cost and lower resistance against breakage than Si. Innovation of crystal-growth technology, such as VCZ, VB and VGF, have made it possible to manufacture 6-inch large-diameter GaAs single crystals with low dislocation density and low residual strain. Production costs have been reduced by about 50 % in recent years. However, progress of performance and cost reduction of Si devices is also very fast. Competition between GaAs and Si devices in cost and performance is severe. Continuous effort on cost reduction and improvement of performance is essential to increase the share of GaAs technology.

REFERENCES

Akai, S., Fujita, K., Nishida, Y., Kito, N., Sato, Y., Yoshitake, Y. and Sekinobu, M., (1983) Proceedings of Symposium on Opto-electronics Epitaxy and Device Related Process, San Francisco, p. 41.
Bourret, E. D. and Merk, E. C., (1991) *J. Cryst. Growth* **110** 395.

Doering, P. J., Freidenreich, B., Tobin, R. J., Pearah, P. J., Tower, J. P. and Ware, R. M., (1990) Proceedings of the International Conference on Semi-insulating III-V Materials, Toronto, p. 173.

Gault, W. A., Monberg, E. M. and Clemans, J. E., (1986) *J. Cryst. Growth* **74** 491.

Hagi, Y., Kawarabayashi, S., Inoue, T., Nakai, R., Kohno, J., Kawase, T. and Tatsumi, M., (1996) Proceedings of Conference on Semiconducting and Insulating Materials, Toulouse, p. 279.

Hagi, Y., Nakai, R., Ohtani, S., Kawarabayashi, S., Nanbu, K., Sawada, S., Morimoto, T., Yao, H., Ueda, T., Sakurada, T. and Nakajima, S., (2001) Digest of Papers, International Conference on Compound Semiconductor Manufacturing Technology, Las Vegas, p. 64.

Hashio, K., Sawada, S., Tatsumi, M., Fujita, K. and Akai, S., (1997) *J. Cryst. Growth* **137** 33.

Hoshikawa, K., Nakanishi, H., Kohda, H. and Sasaura, M., (1989) *J. Cryst. Growth* **94** 643.

Hosokawa, Y., Yabuhara, Y., Nakai, R. and Fujita, K., (1998) Proceedings of 13th International Conference on Indium Phosphide and Related Materials, p. 34.

Kawase, T., Kawasaki, A. and Tada, K., (1986) Proceedings of the International Conference on GaAs and Related Compound, Las Vegas, p. 27.

Kawase, T., Araki, T., Miura, Y., Iwasaki, T., Yamabayashi, N., Tatsumi, M., Murai, S., Tada, K. and Akai, S., (1990) *Sumitomo Electric Tech. Rev.* **29** 228.

Kawase, T., Wakamiya, T., Fujiwara, S., Hashio, K., Kimura, K., Tatsumi, M., Shirakawa, T., Tada, K. and Yamada, M., (1992a) Proceedings of the International Conference on Semi-insulating III-V Materials, Ixtapa, p. 85.

Kawase, T., Fujiwara, S., Hashio, K., Kimura, K., Tatsumi, M., Shirakawa, T. and Tada, K., (1992b) Proceedings of the International Conference on GaAs and Related Compound, Karuizawa, p. 13.

Kawase, T., Hagi, Y., Tatsumi, M., Fujita, K. and Nakai, R., (1996) Proceedings of Conference on Semiconducting and Insulating Materials, Toulouse, p. 275.

Kawase, T., Yoshida, H., Sakurada, T., Hagi, Y., Kaminaka, K., Miyajima, H., Kawarabayashi, S., Toyoda, N., Kiyama, M., Sawada, S. and Nakai, R., (1999a) Proceedings of Conference on Gallium-Arsenide Manufacturing Technology, Vancouver, p. 125.

Kawase, T., Tatsumi, M. and Fujita, K., (1999b) *J. Korean Assoc. Cryst. Growth* **9** 535.

Kawase, T., Hosaka, N., Hashio, K., Matsushima, M., Sakurada, T. and Nakai, R., (2002) Technical Digest of GaAs IC Symposium, Monterey, p. 143.

Nakai, R., Hagi, Y., Kawarabayashi, S., Miyajima, H., Toyoda, N., Kiyama, M., Sawada, S., Kuwata, S. and Nakajima, S., (1998) Technical Digest of GaAs IC Symposium, Atlanta, p. 243.

Shibata, M., Suzuki, T., Kuma, S. and Inada, T., (1993) *J. Cryst. Growth* **128** 439.

Suzuki, T., Akai, S., Koe, K., Nishida, Y., Fujita, K. and Kito, N., and Sumitomo, (1978) *Electric Tech. Rev.* **18** 105.

Tada, K., Tatsumi, M., Nakagawa, K., and Kawase, T., (1987) Proceedings of the International Conference on GaAs and Related Compound, Heraklion, p. 439.

Tatsumi, M., Kawase, T., Araki, T., Yamabayashi, N., Iwasaki, T., Miura, Y., Murai, S., Tada, K. and Akai, S., (1989) Proceedings of the International Conference on InP and Related Materials, Oklahoma, p. 18.

Terashima, K., Washizuka, S., Nishio, J., Shimada, H., Yasuami, S. and Watanabe, M., (1986) Proceedings of the International Conference on Semi-insulating III-V Materials, Hakone, p. 59.

Young, M., Liu, X., Zhang, D., Zhu, M. and Hu, X. Y., (1998) Proceedings of 13th International Conference on Indium Phosphide and Related Materials, p. 30.

17 CdTe and CdZnTe Growth

R. TRIBOULET

Laboratoire de Physique des Solides de Bellevue CNRS F-92195 MEUDON CEDEX, FRANCE

17.1 INTRODUCTION

For the past forty years and since the encyclopedic and pioneering work of de Nobel, CdTe has attracted great interest and found several industrial applications. It has been the main topic of several international conferences, among which were the Strasbourg conferences of 1972, 1977, 1992 and 1995, and its story has been reviewed in the excellent monograph of Zanio (1978). CdTe shows unique properties making it important and very suitable for several applications. Its band gap of 1.5 eV, just in the middle of the solar spectrum, makes it an ideal material for photovoltaic conversion, and several companies now commercialize CdTe-based solar cells. Its high average atomic number of 50, wide band gap, and good transport properties are very convenient for nuclear detection. Its high electro-optic coefficient combined with its low absorption coefficient allows high-performance electro-optic modulators and photorefractive devices. It can show n- or p-type conductivity, making diode technology and field effect transistors possible, and a semi-insulating state as well. CdTe-based semimagnetics like CdMnTe have exciting properties that so far have not been exploited.

CdTe is also present at the industrial level through its ternary alloys CdHgTe, a major material for infrared detection, and with CdZnTe substrates for the epitaxial deposition of CdHgTe layers. Furthermore, CdTe-based nuclear detectors have great potential, mainly for medical purposes.

The considerable efforts for more than forty years in crystal growth of CdTe will be reviewed in close relationship with its intrinsic properties, mainly those that are related to the characteristics of the Cd–Te chemical bond.

17.2 PHASE EQUILIBRIA IN THE Cd–Te SYSTEM

Crystal growth of CdTe requires precise knowledge of the existence regions of the solid, liquid and gas phases with respect to temperature, pressure and composition. For this purpose, the temperature versus composition, $T-x$, component

Crystal Growth Technology, Edited by H. J. Scheel and T. Fukuda
© 2003 John Wiley & Sons, Ltd.   ISBN: 0-471-49059-8

pressure or total pressure versus temperature, $p-T$ or $P-T$, and finally $p-T-x$ or $P-T-x$ diagrams have been experimentally determined and theoretically modeled. Deviations from stoichiometry of the crystals strongly affect their semiconducting properties, thus requiring the knowledge of the extent of the homogeneity region and of the position of the stoichiometric line inside the three-phase boundary.

The first liquidus data were obtained either from visual (subjective) observation of the onset of initial freezing (Kobayashi 1911, de Nobel 1959), or by thermal analysis (Kulwicki 1963, Lorenz 1962a) and differential thermal analysis (Steininger et al. 1970). Such measurements as a function of component partial pressure established the first $p-T$ diagrams (de Nobel 1959, Lorenz 1962a), while Brebrick (1971) measured the optical density of the vapor to determine the component partial pressures.

$p-T$ diagrams were first calculated using a regular-associated-solution model, which postulates the existence of stable CdTe complexes in the liquid phase and shows an infinite interchange energy at equiatomic composition (Jordan and Zupp 1969). The $T-x$ diagram was calculated from an improved approach in which liquid-mixing functions are expressed in an association model by using the Redlich–Kister polynomia form of species interactions (Marbeuf et al. 1985).

An extent of the homogeneity region of CdTe of about 10^{17} cm^{-3} on both Cd and Te sides was found from electrical measurements interpreted by simple point-defect models, with the solidus of retrograde shape (de Nobel 1959, Whelan and Shaw 1968, Matveev et al. 1969, Zanio 1970, Smith 1970, Wienecke et al. 1993). Overestimated values of stoichiometry deviations have been reported by several authors. The deviation from stoichiometry was suggested to be as large as 2×10^{19} cm^{-3} on the Cd side and 5.7×10^{19} cm^{-3} on the Te side from density measurements (Kiseleva et al. 1971). From calculations based on measurements of the Te vapor pressure above CdTe crystals with various limiting deviations from stoichiometry, the existence of a homogeneous region amounting to ~1 % at 800 °C was stated (Medvedev et al. 1972). More than 10^{20} cm^{-3} on both sides were estimated from a special mass spectrometry method and optical vapor density measurements (Ivanov 1996).

From new calorimetric investigations, the enthalpy of formation of CdTe $\Delta_f H^0$ and the standard entropy S_{298}^0 at 298.15 K were found to be $-100\,708$ J/mol and 93.3 J/mol K, respectively (Yang Jianrong et al. 1995). The Cd–Te phase diagram was recently revisited from total-vapor-pressure scanning experiments (Greenberg 1996). The determination of the composition X_S, X_V and X_L of the solid, vapor and liquid phases in equilibrium allowed solidus, vaporus and liquidus curves of the Cd–Te system to be drawn. Very precise $P-T$ (Figure 17.1) and $T-X$ (Figure 17.2) projections of the Cd–Te diagram were constructed, as well as several isobaric sections of the $P-T-X$ phase diagram allowing the crystallization of nonstoichiometric CdTe under various $P-T$ conditions to be followed from different matrices (liquid, vapor or both) or annealing of the prepared material.

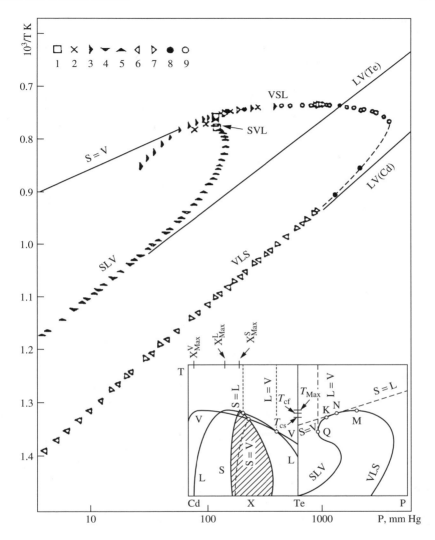

Figure 17.1   *P–T* Projection of the Cd–Te diagram. The inset shows *P–T* and *T–x* projections of the CdTe melting region. (Reprinted from J. H. Greenberg, *J. Cryst. Growth* **161** (1996) 1–11, copyright (1996) with permission from Elsevier Science.)

Several significant conclusions can be drawn from the study of Greenberg: CdTe shows congruent fusion (cf) with a maximum melting point of 1092 °C; at T_{cf}, CdTe is Te-saturated; at T_{cs} (cs: congruent sublimation), crystalline CdTe is also Te-saturated; $T_{cs} < T_{max}$ by 41 K; stoichiometric CdTe is in equilibrium with a Te-rich melt and a virtually pure Cd vapor; there is no constant congruent sublimation composition; the congruent sublimation line lies on the Te-rich side,

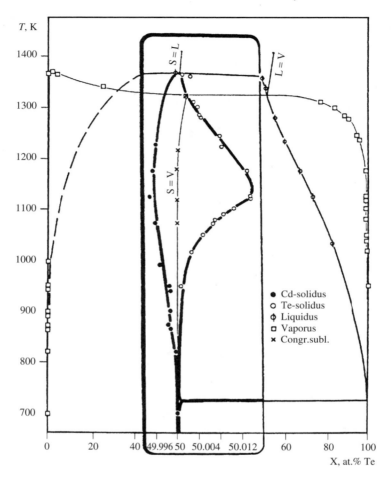

Figure 17.2 $T-x$ Projection of the Cd–Te diagram. Near-solidus region is given on an enlarged scale. (Reprinted from J. H. Greenberg, *J. Cryst. Growth* **161** (1996) 1–11, copyright (1996) with permission from Elsevier Science.)

rather far from stoichiometry; the maximum Te nonstoichiometry is of about 4×10^{18} cm^{-3}, the maximum Cd nonstoichiometry of about 10^{18} cm^{-3}. Most of the results of Greenberg were confirmed by Fang and Brebrick (1996) from recent optical density measurements. They nevertheless found the solidus curve $X_{Te} - 1/2$ extending to 10^{-4} to compare with 1.5×10^{-4} in the case of Greenberg.

Recent results throw a new light on the stoichiometric-line determination: a singly ionizable Te antisite model gives a good fit of the high-temperature electrical measurements (Brebrick and Fang 1996).

Greenberg showed how to apply his $P-T-X$ phase-equilibrium data to prepare crystals with controlled composition, either stoichiometric or with a certain deviation from stoichiometry, according to different technologies, vapor phase growth, vertical, horizontal and high-pressure Bridgman (Greenberg 1999). In vapor growth, the growth rate is linked to the composition of the vapor, and the maximum rate corresponds to the stoichiometric composition. CdTe physical vapor deposition (PVD) is recommended at temperatures well below the max-imum congruent sublimation temperature $T_{cs} = 1324$ K: the composition of the crystals becomes less temperature dependent below 1173 K, while at higher tem-peratures small deviations in the vapor composition or small pressure variations will favor the vapor–liquid–solid mechanism with detrimental consequences. For the conventional CdTe PVD temperature range (1123–1223 K) the composition of the crystals changes from 50.00083 at. % Te to 50.0015 at. % Te. In the vertical Bridgman method, the growth occurs at the near-maximum melting tem-perature from the liquid+solid state of the system. The crystals grown from the vapor–liquid–solid (VLS) equilibrium are Te-rich and, moreover, at the congru-ent melting point. To reduce the Te excess, the pressure should be increased above a pressure of 1.36 atm (up to about 2 atm according to Rudolph *et al.* 1994) either using a Cd reservoir with independent temperature control, in order to enrich the liquid in Cd, or under an inert gas overpressure of over 100 atm (Raiskin and Butler 1988). In the horizontal Bridgman method, the growth occurs from the VLS state, allowing a flexible control of the composition. The three-phase VLS equilibrium being univariant, the crystallization temperature as well as the com-position of the crystals become fixed when a vapor pressure is chosen. The vapor is essentially pure cadmium. CdTe crystals can be grown from Cd-rich melts at vapor pressures higher than the one corresponding to T_{MAX}, $P(T_{MAX})$, or from Te-rich melts for pressures lower than $P(T_{MAX})$. The results of Greenberg can also be used to adjust post-growth annealing conditions.

17.3   CRYSTAL GROWTH VERSUS Cd–Te CHEMICAL BOND CHARACTERISTICS

The fairly high ionic character of the II-VIs chemical bond affects not only most of the physical properties of these semiconductors but also their crystal growth. The higher the ionicity, the lower is the thermal conductivity. The low CdTe thermal conductivity of 55 mW cm^{-1}K^{-1} makes it difficult to control the solid/liquid interface in both melt and vapor growth. Pretransition phenomena on both sides of the melting point, in both liquid and solid states, have been found to occur in CdTe as the result of the ionic character of the Cd–Te chemical bond.

This bond causes highly associated melts close to the melting point, and the highly organized particles influence the nucleation process and growth kinetics. The associated-solution model for liquid CdTe expresses the strong interaction between Cd and Te atoms in the liquid state (Jordan 1970). The melting pro-cess of CdTe was assumed to be a solid semiconductor–liquid semiconductor transition from experimental investigations of the melt parameters (Glasov *et al.*

1969) and from the positive slope of the conductivity versus temperature in the region of the solid phase (Rud' and Sanin 1971). Above the melting point, the liquid-phase conductivity was found again to increase exponentially with temperature from eddy-current measurements (Wadley and Choi 1997). Using the eddy-current technique also to measure the variation of the electrical conductivity versus temperature, transitions occurring between 1025 and 1050 °C were suggested to be associated with allotropic phase transitions (Rosen *et al.* 1995). The mole fraction of associated species in the melt just above the CdTe melting point, or association coefficient, was estimated to be 0.908 from calculations (Marbeuf *et al.* 1985), 0.95 (Yu and Brebrick 1993) and 0.92 from new calorimetric data (Yang Jianrong *et al.* 1995). The enthalpy of formation of the melts, determined from calorimetric measurements, shows very strong departures from ideality (Amzil *et al.* 1997). The authors explain this result by the existence of short-range ordering corresponding to CdTe associates. From neutron-diffraction experiments, it was suggested that the structure of liquid CdTe could be described by a four-coordinated random network, very different from the structure of group IV elements or III-V compounds (Gaspard *et al.* 1987). From DTA measurements, endothermic effects above the melting point were observed (Figure 17.3) (Shcherbak 1998, 1999). A high degree of structural ordering was concluded by the author. Because of the presence in the melt of such highly organized particles, no supercooling is observed at small superheating, while a large supercooling occurs at superheating values higher than 9–10 K (Figure 17.4) (Rudolph and Mühlberg 1993, Shcherbak *et al.* 1996, Shcherbak

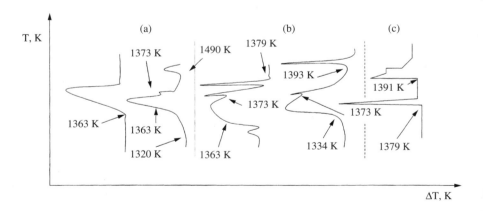

Figure 17.3 Typical differential thermograms of the pure and doped CdTe melting process: (a) rapid heating to T_m ($\tau = 10 - 15\,\text{min}$) of the sample equilibrated at room temperature; (b) thermocycling in the 950–1150 °C temperature range; (c) slow heating rate ($V_h < 1\,\text{K/min}$) of a CdTe sample previously excited by thermocycling. (Reprinted from L. Shcherbak, *J. Cryst. Growth* **197** (1999) 397–405, copyright (1999) with permission from Elsevier Science.)

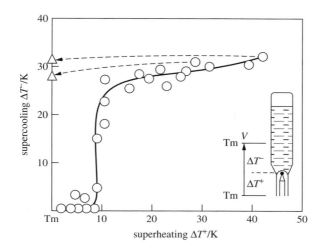

Figure 17.4 Influence of the degree of superheating ($\Delta T^+ = T_m + T$) on the degree of supercooling ($\Delta T^- = T_m - T$) for stoichiometric CdTe measured with a thermocouple at the tip of vertical Bridgman ampoules moved with a constant velocity $v = 1 \text{ mm Min}^{-1}$: Δ, supercooling of former highly superheated melts subsequently cooled and kept for 5 h at the melting point T_m. (Reprinted from P. Rudolph and M. Muhlberg, *Mater. Sci. Eng.* **B16** (1993) 8–16, copyright (1993) with permission from Elsevier Science.)

1999). There is thus some duality of supercooling–superheating. At small superheating, parasitic nucleation occurs from the highly organized particles present in the melt: if initial single-crystal growth is often observed (Figure 17.5(a)), grain boundaries and twins are originated during further growth. Higher superheating results in a polycrystalline first-to-freeze region followed by a single-crystal part (Figure 17.5(b)). But such high superheating impedes seeding: large superheating values are necessary for the destruction of associated melt complexes, but as a result seed melting becomes unavoidable.

In the solid state, the higher the ionicity, the higher is the tendency to the hexagonal structure and thus to twinning. The Cd–Te chemical bond shows a fairly high ionicity of 0.55. The tendency to twinning and to hexagonal structure appears usually in such ionocovalent compounds when ionicity exceeds 0.5. The high CdTe twinning tendency was explained according to the classical inverse relationship between stacking-fault energy and ionicity (Vere *et al.* 1983). The existence of a hexagonal structure in CdTe crystals at high temperature was reported by several authors (Appell 1954, Höschl and Konak 1963, Kendall 1964). A phase transition in the 893–920 °C temperature range was pointed out (Leybov *et al.* 1984). This tendency to twinning and to hexagonal structure in CdTe appears also through the behavior of CdMnTe alloys in which a transition in the solid state was found between a phase of hexagonal structure at high temperature and a cubic one at lower temperature (Triboulet *et al.* 1990a

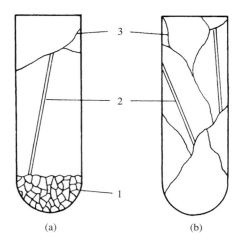

(a)   (b)

Figure 17.5 Typical growth structures of (a) a high ($\Delta T^+ = 27$ K) and (b) a low ($\Delta T^+ = 3$ K) superheated unseeded CdTe crystal: 1, polycrystalline tip region; 2, twin lamella; 3, large grain boundary. (Reprinted from P. Rudolph and M. Muhlberg, *Mater. Sci. Eng.* **B16** (1993) 8–16, copyright (1993) with permission from Elsevier Science.)

and Triboulet 1991). The transition temperature decreases with increasing Mn content in agreement with increasing ionicity with the incorporation of Mn in CdTe (Perkowitz *et al.* 1988). The transition line converges at the CdTe melting point, suggesting that the CdTe structure could very easily oscillate between hexagonal and cubic phases. Solid-phase transformations in the 900–1092 °C temperature range were reported from the electromotive-force method and from In diffusion and annealing experiments (Ivanov 1996).

Because of a stronger interaction between unlike particles arising from the considerable ionic contribution to the bond energy, the CdTe liquidus, like the other II-VI liquidus curves, shows a hyperbolic shape near the congruent melting point.

It has been shown that the higher the ionicity of the bond, the smaller is the formation energy of vacancies, which are the majority defects in II-VIs (Verié 1998). This explains the large II-VI nonstoichiometry, which is an additional complicating factor for crystal growth. The higher the ionicity, the smaller is the energy of creation of dislocations and of stacking faults. Thus the II-VI crystal lattice is very sensitive to any strain and easily becomes defective.

All these factors render the melt growth of large high-quality crystals difficult. As examples, the wide homogeneity range and the retrograde solidus necessitate a careful control of the stoichiometry to adjust the electronic properties of the crystals and to avoid the presence of precipitates. The highly associated melts lead to some duality between superheating and supercooling, making seeding extremely difficult. The low thermal conductivity makes it difficult to control the shape of the growth interface and to apply Czochralski pulling.

In order to overcome these difficulties, different mechanisms have been proposed: growth in a reduced temperature range to avoid the phase transitions in the solid state; growth under high-superheating conditions; growth under forced convection regimes such as ACRT, micro- or macro-gravity conditions; growth under vibrational stirring with the vibrations aimed at breaking the CdTe associates; growth under electric or magnetic field, etc.

17.4 CRYSTAL GROWTH

Because of these difficulties, just about all the techniques of growth of semiconductor materials have been applied to CdTe and CdZnTe, and even now, 'novel' methods are still proposed for CdTe growth!

The growth techniques of CdTe can be divided into growth from stoichiometric and off-stoichiometric melts, and into growth from the vapor phase either by sublimation or by chemical vapor transport. The melting point of $1092\,°C$ is lower than the softening point of silica, so that melt growth from silica tubes has become popular.

An important step of melt growth is the synthesis of the CdZnTe compound from the elements, which is achieved either in a separate ampoule (ex situ) or in the growth ampoule (in situ), and either by melting the elements together or by vapor transport of Cd into molten Te containing the Zn (Glass *et al.* 1998).

For growth from stoichiometric melts, vertical Bridgman in evacuated silica ampoules has been widely used. Without control of the Cd partial pressure, the growth from stoichiometric charges generally leads to Te-rich melts because of the loss of Cd by incongruent evaporation into the free ampoule volume over the charge (Route *et al.* 1984, Oda *et al.* 1985, Kennedy *et al.* 1988, Mühlberg *et al.* 1990, Bruder *et al.* 1990, Rudolph and Mühlberg 1993, Casagrande *et al.* 1993, 1994). Attempts to reduce the Cd loss with boric oxide as liquid encapsulation were reported (Carlsson and Ahlquist 1972). Excess Cd can be added to the charge to compensate for the Cd loss (Khattak and Schmid 1989, Brandt *et al.* 1990, Rudolph and Mühlberg 1993). But a better control of the stoichiometry can be achieved under a defined Cd partial pressure, using a Cd extra source, by the so-called 'modified' Bridgman or gradient-freeze techniques (Triboulet and Rodot 1968, Kyle 1971, Triboulet and Marfaing 1973, Bell and Sen 1985, Sen *et al.* 1988, Song *et al.* 1988, Mochizuki *et al.* 1988, Becker *et al.* 1989, Azoulay *et al.* 1992, Rudolph *et al.* 1994, Price *et al.* 1998, Szeles and Driver 1998, Glass *et al.* 1998, Höschl *et al.* 1998).

To minimize thermal stress, very small axial temperature gradients are now used, generally lower than 10 K/cm and even down to 1 K/cm. Such small gradients led to the vertical gradient freeze technique (VGF) which does not require any mechanical movement of the charge. Using an axial gradient of 1 K/cm with control of the Cd vapor pressure, $\langle 111\rangle$-oriented 100 mm diameter CZT single crystals were grown (Asahi *et al.* 1996). According to a very similar arrangement called the 'self-seeding technique', in which an axial temperature gradient of about 3 °C/cm opposite in direction to that typical of the classical vertical Bridgman technique allows the growth to be initiated from the free melt surface, 90-mm diameter ⟨111⟩-oriented CdTe single crystals were grown (Ivanov 1998).

From the early work of Raiskin and Butler (1988), the vertical Bridgman technique under inert gas pressure, also called high-pressure Bridgman (HPB) has become very popular for the growth of semi-insulating crystals, mainly CdZnTe, suitable for nuclear detectors (Doty *et al.* 1992, Butler *et al.* 1993, Kolesnikov *et al.* 1997, Fougères *et al.* 1998, Szeles and Driver 1998).

The influence of various parameters on the basic Bridgman and gradient freeze processes has been studied. The role of the residual atmosphere and of the gas pressure in the ampoules has been stressed. A residual hydrogen pressure in the growth tube affects the compensation state of the crystals (Brunet *et al.* 1993), and, in accordance, the resistivity of CdTe crystals was found to increase as the concentration of oxygen incorporated from the atmosphere into CdTe decreased (Yokota *et al.* 1989). The classical Bridgman process was extended by applying the accelerated-crucible-rotation technique (ACRT) of Scheel (Capper *et al.* 1993). By ACRT the size of the largest grains increased by a factor of 2 to 5, the density of small Te precipitates decreased, large precipitates were absent, and the X-ray linewidths decreased. A coupled vibrational stirring method was applied to the vertical Bridgman growth of CdTe (Lu *et al.* 1990). The application of low amplitude (0–100 Hz) mechanical vibrations to the ampoules during growth was found to move fluids rapidly without turbulence and at velocities significantly faster than ACRT. The results were an improved chemical homogeneity but a degradation of the microcrystalline quality. An enlarged grain size was obtained with the heat-exchanger method (HEM) for growth of 3-inch diameter CdTe ingots by controlled directional solidification and solid/liquid interface shape (Khattak and Schmid 1989). Asymmetrical Bridgman, initially proposed for growing compositionally uniform HgCdTe single crystals (Bagai and Borle 1989) was found also to be very effective for the growth of large CdTe single crystals (Bagai *et al.* 1994, Aoudia *et al.* 1995). In this method, the growth ampoule is held asymmetrically in the cylindrical furnace, allowing the solid/liquid growth interface to be flattered.

Horizontal Bridgman (HB), historically one of the first methods used for CdTe crystal growth, is generally achieved under Cd overpressure to adjust the stoichiometry through the solid–liquid–vapor equilibrium (Medvedev *et al.* 1968, Matveev *et al.* 1969, Lay *et al.* 1988, Cheuvart *et al.* 1990, Brunet *et al.* 1993). Nevertheless, as in the VB case, the control of the melt composition through the Cd partial pressure control is sometimes neglected, and the composition of the crystals depends on the initial composition of the charge and on the free volume over the charge (Khan *et al.* 1986, Liao *et al.* 1991). Different variants exist depending on how the temperature gradient is applied to the charge: relative movement of furnace and charge (Liao *et al.* 1991, 1993, 1994), electronic displacement of the temperature gradient (Cheuvart *et al.* 1990, Brunet *et al.* 1993) or the 'gradient freeze technique'.

CdTe zone refining has been achieved either by horizontal zone refining or by a vertical sealed-ingot zone-refining technique.

Using silica crucibles the purification and growth of CdTe by horizontal zone refining under controlled cadmium vapor pressure was reported (de Nobel 1959). After forty passes, crystals with a residual impurity content of about 1 ppma were obtained from an initial charge containing about 0.01 % impurities by weight. A maximum room-temperature electron mobility of about 850 cm^2/ V s was reached. By using graphite boats instead of quartz boats and 15 passes, room-temperature electron mobilities of 990 cm^2/ V s were measured (Matveev et al. 1969). The use of vitreous carbon as the most suitable crucible material for horizontal CdTe zone refining was suggested (Prokof'ev 1970).

The transport of impurities through the vapor and back diffusion through the hot solid, which affect horizontal zone melting, are eliminated by the vertical sealed-ingot zone refining method (Heumann 1962) which allows, furthermore, a narrow molten zone to be maintained, without having to heat the solid, and a sharp temperature gradient between melt and solid to be realized (Lorenz and Halsted 1963, Cornet et al. 1970, Woodbury and Lewandowski 1971, Triboulet and Marfaing 1973). The effectiveness of zone-refining purification has been demonstrated from chemical analysis, photoluminescence and electrical measurements. Electron mobilities exceeding 1.5×10^5 cm^2/ V s at low temperature and room-temperature mobilities as high as 1100 cm^2/ V s were measured (Triboulet and Marfaing 1973). It was suggested that the abnormally high mobility values measured at low temperature could be explained by the presence of microinhomogeneities and by donor–acceptor pairing (Caillot 1977). Such zone-refined crystals were shown, furthermore, to be highly compensated and in a metastable state as a result of the large temperature gradients in the zone melting process. Small crystals with low etch pit density ($\sim 10^3$ cm^{-2}) were obtained by floating-zone melting in space, the evaporation of Cd being controlled by adding excess Cd to the charge combined with heating the ampoule walls (850–950 °C) (Chang et al. 1993).

Attempts to pull CdTe under encapsulation (LEC) were not successful (Meiling and Leombruno 1968, Vandekerkof 1971, Mullin and Straughan 1977, Hobgood et al. 1987). Beside severe heat transfer problems associated with the low CdTe thermal conductivity, difficulties arise from the insufficient wetting of B$_2$O$_3$ to the molten CdTe leading to a continuous Cd loss, and from some solubility of CdTe in the encapsulant. Large ingots of 50 mm diameter were nevertheless grown by LEC, but exhibited a high degree of polycrystallinity (Hobgood et al. 1987).

In order to reduce the contamination resulting from high-temperature growth, CdTe solution growth has been used to lower the growth temperature. The growth from nonstoichiometric melts using a kind of vertical gradient freeze configuration, with a reticular diaphragm in the liquid to reduce convective currents and thus to stabilize growth conditions, yielded small CdTe crystals both from Cd-rich and Te-rich solutions at 800–850 °C and 750–780 °C, respectively (Medvedev et al. 1972). Large CdTe crystals (45 mm diameter) of high purity were grown by vertical Bridgman by normal freezing of a slightly Te-rich solution with the growth interface kept at a fixed position by suitably lowering the furnace temperature versus crystallized length (Schaub *et al.* 1977).

CdTe grown by the traveling-heater method (THM) is used for fabricating γ-detectors, an application that has become of great interest during the last decade. Two kinds of source material have been reported for the THM growth of CdTe, presynthesized ingots and a mixture of the constituting elements according to the 'cold THM' principle. Feed material with the required density close to 100 % is generally prepared by casting CdTe in the form of an ingot by vertical Bridgman. This can be achieved either from a stoichiometric Cd+Te mixture (Bell *et al.* 1970, Tranchart and Bach 1976, Schoenholz *et al.* 1985) or from a Te-rich solution, as in the ZnTe case (Triboulet and Didier 1975, Ohmori *et al.* 1993). A purification of the presynthesized CdTe ingot is sometimes performed by vertical zone refining before THM growth (Triboulet *et al.* 1974, Taguchi *et al.* 1978, Mochizuki and Matsumoto 1986). According to the cold-THM principle, a composite source material is constituted of a Cd rod of appropriate diameter surrounded by Te pieces and powder (Figure 17.6) (Triboulet *et al.* 1990b, El Mokri *et al.* 1994). The movement of the solvent zone through this composite charge, at a temperature of about 750 °C, induces the fractional and progressive synthesis of the compound in Te solution, its growth and purification. The same CTHM process has been applied to other tellurides like HgTe, ZnTe and PbTe. In order to increase the THM solution-zone refining purification effect, a multipass THM process was developed (Triboulet and Marfaing 1981). In this technique, the starting Te solvent consists of a Te column a certain fraction of which is taken up at each pass. This arrangement enables several passes to be carried out in the same ampoule without any handling of the material between each pass. Very pure crystals were obtained by this MTHM technique as proven by electrical, chemical and optical measurements.

Imposing a forced convection regime in the THM molten zone improved the radial and axial distribution of Te inclusions and $\mu\tau$ products when using a magnetic rotating field (Salk *et al.* 1994), and larger crystals and faster growth rates could be achieved with the accelerated-crucible-rotation technique (El Mokri *et al.* 1994).

25 cm^3 CdTe boules containing 20 cm^3 single crystals were obtained using a convection-assisted solution-growth system in a vertical-gradient-freeze arrangement (Zanio 1974). A sharp temperature gradient, inducing solution circulation and restricting Cd depletion at the interface, was used in combination with a cold finger at the bottom of the ampoule to initiate nucleation. Growth proceeds with a convex solid/liquid interface due to the radial temperature gradient imposed by the cold finger.

In order to grow n-type crystals and to avoid the large deviation from stoichiometry of the Te-rich side, Cd was used as the solvent for the CdTe THM growth at 1000 °C with a growth rate of 1 mm/day (Triboulet *et al.* 1985), while

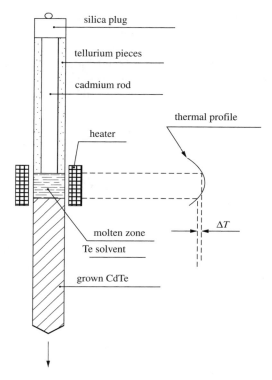

Figure 17.6 Principle of cold-traveling-heater method (CTHM). (Reprinted from R. Triboulet *et al.*, *J. Cryst. Growth* **101** (1990) 216–220, copyright (1990) with permission from Elsevier Science.)

the possibility of using such heterosolvents as In (Wald 1976) or $CdCl_2$ (Tai and Hori 1976) was also demonstrated.

The growth from Cd-rich solutions was achieved by solvent evaporation (SE) either by heating the solution from 970 to 1025 °C at constant Cd pressure (Lorenz 1962b) or by reducing the Cd pressure isothermally at 1040 °C (Lunn and Bettridge 1977). In this last vertical SE technique, temperature instabilities were found to be due to Cd droplets falling back from the upper Cd reservoir in the solution. A modified SE technique was proposed in which the positions of the CdTe/Cd solution and Cd reservoir were reversed (Mullin *et al.* 1982, Vere *et al.* 1985).

Small platelets and rods were obtained by flux growth at 900 °C under the driving force of a temperature gradient using either Bi or Sn as the solvents (Rubenstein 1966, 1968).

While millimeter-size CdTe crystals were grown by the hydrothermal technique at about 350 °C in alkaline solutions (Kolb *et al.* 1968), the growth of crystals from gels was unsuccessful because of the Te ion instability (Blank and Brenner 1971).

An exotic way to produce CdTe powder for the fabrication of solar-cells by screen-printing was developed in Matsushita Electronics (Ikegami). By stirring metallurgical Cd and Te powders mixed with distilled water in an agate ball mill at room temperature, CdTe powder was obtained after several hours.

Vapor phase growth of CdTe has become popular due to the difficulties in melt growth. Using a vertical configuration CdTe crystals were grown by sublimation or physical vapor transport (PVT) in closed tubes without seed either from a charge of composition adjusted by weighing before loading it in the growth tube (Yellin et al. 1982, Yellin and Szapiro 1984) or under controlled Cd or Te partial pressure to adjust a composition of the charge leading to optimum growth rate (Durose et al. 1985).

In classical PVT experiments, CdTe boules of limited thickness are obtained, because the initial thermal conditions are lost after the growth of a 2-to-3 cm thick crystal due to the low CdTe thermal conductivity. In order to overcome this problem of limited thickness, a new technique called sublimation THM (STHM), in which the molten solvent zone is replaced by an empty space, was proposed (Triboulet 1977). By the movement of the charge relative to the heater, a temperature difference appears in STHM between the sublimation interface and growth interface, and the empty space is made to migrate through the solid source material, allowing continuous growth over long distances. According to this STHM principle, CdTe crystals were grown without control of the pressure regime in the ampoule and without a seed (Triboulet 1977) or with a seed using a monoellipsoid mirror furnace (Laasch et al. 1994). According to a modified STHM technique, a capillary affixed to the ampoule top and staying at room temperature allows the total pressure in the ampoule to be controlled, in order to reach the conditions of minimum total pressure corresponding to congruent sublimation (Figure 17.7) (Triboulet and Marfaing 1981).

PVT growth was also carried out in a vertical configuration in semiclosed systems, with a heat sink at the bottom of the ampoules constructed according to a scheme previously employed (Markov and Davydov 1971), either with a seed and the charge at the top of the ampoule, with either Ar or H_2 or NH_4I as residual atmosphere (Golacki et al. 1982, Zhao et al. 1983) or by in situ nucleation with the charge at the bottom of the ampoule under vacuum without a heat sink (Grasza et al. 1992). In such a semiclosed system, the ampoule was left under Ar until the furnace reached the operating temperature and then evacuated to $10^{-3}-10^{-5}$ atm to initiate the growth on (111)A seeds (Kuwamoto 1984). This Markov–Davydov technique was recently reactivated (Laasch et al. 1997) and used to obtain chlorine-doped semi-insulating crystals (Kunz et al. 1998). The authors showed the influence of the heat-sink temperature on the characteristics of the crystals. In such PVT growth, the source temperature ranges generally between 700 and 1040 °C, with a 5–10 °C difference between source and seed temperatures. CdTe crystals were also grown according to a novel 'multitube' technique, very close to the Markov–Davydov method, at a growth temperature of 700 °C and a growth rate of ~5 mm/day (Aitken et al. 1999).

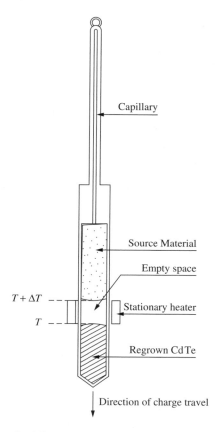

$T + \Delta T$
T

Figure 17.7 Principle of sublimation-traveling-heater method (STHM) with capillary to control the total pressure. (Reprinted from R. Triboulet and Y. Marfaing, *J. Cryst. Growth* **51** (1981) 89–96, copyright (1981) with permission from Elsevier Science.)

PVT was employed in a horizontal configuration either without a seed under controlled partial pressure of Cd or Te (Höschl and Konak 1963, 1965, Akutagawa and Zanio 1971, Igaki and Mochizuki 1974, Igaki *et al.* 1976, Mochizuki 1981, 1985), or without control of component partial pressure without seed (Wiedemeier and Bai 1990, Mycielski *et al.* 1999), or with seed selection in a capillary by visual inspection (Buck and Nitsche 1980, Geibel *et al.* 1988). In an 'improved' PVT, a three-step closing process of the ampoule was proposed to avoid any oxidation of the source and any condensation of volatile impurities evaporated from the silica tube while sealing it (Yang *et al.* 1997).

The influence of the source material composition on the growth rate was studied by measurements of the mass flux (Wiedemeier and Bai 1990). The composition of the charge was adjusted by heat treatment in order to remain close to the congruent sublimation condition and determined either by EDXRF and reflectivity in the region of free excitons (Mycielski *et al.* 1999) or using an optical absorption technique (Su *et al.* 1998).

Using seeded physical vapor transport, also called sublimation and physical vapor transport (SPVT), $Cd_{1-x}Zn_xTe$ crystals with $x = 0.2$ and 40 g in weight were grown from a pretreated source material with excess Cd (or Cd + Zn) in the ampoule (when desired) ($T_{source} = 1000\,°C$, $\Delta T = 20-40\,°C$, growth rate $= 1$ to 4 mm/day) (Palosz *et al.* 1997) and CdTe crystals of 45–50 mm diameter 250–300 g in weight under Ar pressure (0.2 atm) in a semiclosed system allowing the conditions of congruent sublimation ($T_{source} = 880\,°C$, $\Delta T \sim 40\,°C$, growth rate$\sim$0.8 mm/day) to be reached (Boone *et al.* 1994).

Single crystals shaped with (110) and (111) planes of approximately 10 mm size were obtained by an evaporation-condensation method on the surface of a polycrystalline source material, at 820–900 °C (Szczerbakow and Golacki 1993). In an investigation of the vapor phase chemical transport of CdTe using I_2 as chemical agent, it was shown that no transport is possible in closed tubes in the hot-cold direction, but only vapor-phase transport controlled by the source sublimation (Paorici and Pelosi 1977). Thin platelets were observed growing on the source under small temperature gradients. This growth was explained in terms of a reverse (cold–hot) iodine transport associated with a reduced sublimation tendency of the facetted crystallized material with respect to the unfacetted powdered charge.

CdTe single crystals were also grown in horizontal open systems either from Cd and Te_2 vapors using H_2 as a carrier gas (Lynch 1962) or from a polycrystalline charge using N_2 or Ar as a carrier gas (Corsini-Mena *et al.* 1971).

17.5 BRIDGMAN GROWTH MODELING AND INTERFACE-SHAPE DETERMINATION

The difficulties met in the CdTe melt growth have prompted the idea that analytical modeling could be gainfully employed to the optimization and design of crystal-growth processes. Several papers deal with numerical simulation of crystal growth in both vertical and horizontal configurations.

The first analytical calculation of heat transfer in a vertical Bridgman–Stockbarger configuration was reported by Chang and Wilcox (1974). Identical heat conductivities in both melt and solid were assumed, and the effect of ends was neglected. Given these approximations, the isotherm shape was shown to depend on the temperature difference between the hot and cold plateau of the furnace, the container geometry, and the degree of heat transfer taking into account the value of the Biot number Bi. The larger the Bi, the smaller is the isotherm curvature, the narrower is the transition region from the hot to the cold part of the crystal, and the steeper is the temperature gradient at the growth axis. This model was later extended (Fu and Wilcox 1980) taking into account different heat-exchange coefficients for hot and cold zones and inserting a layer of insulation between heater and cooler. This adiabatic zone is shown not only to dramatically decrease the sensitivity of the interface shape to perturbations in the system, but also to decrease the radial temperature gradient inside the ampoule. Later, a booster heater adjacent to this adiabatic zone was included (Naumann 1982).

Electrical analogues were used to model the thermal behavior of a Bridgman crystal growing system (Ickert and Rudolph 1978, Jones *et al.* 1982). This last study demonstrates that during the critical early stages of growth, the isotherm shapes in the crystal and in the melt are influenced by the end losses from the ampoule. Changing the shape of the ampoule base or changing the conductivity of the ampoule body should allow the isotherm shape to be adjusted. The actual growth-rate/extraction-rate ratio is shown to depend on the furnace characteristics. To obtain a convex isotherm, a shallow temperature gradient should be used and end losses maximized (Jones *et al.* 1984).

In an analytical treatment of both axial temperature profile and radial temperature variations using a one-dimensional model, the effects of charge diameter, charge motion, thickness and thermal conductivity of a confining crucible, thermal conductivity change at the crystal/melt interface, noninfinite charge length, length of an insulating zone between hot and cold regions, thermal coupling between charge and furnace, and generation of latent heat were considered (Jasinski *et al.* 1983, 1984). A concave growth interface shape could be the result of the presence of a crucible for semiconductors. Use of silica, with its relatively low thermal conductivity, is seen to have only a moderate effect when compared to boron nitride.

Demonstrating the power of global calculation of heat transfer applied to the numerical simulation of crystal growth in a vertical Bridgman furnace, natural convection was shown to have little impact upon the temperature field, and the impact of material properties upon the global power input was clearly emphasized (Crochet *et al.* 1989). The same group used the steady-state finite-element method (FEM) to calculate the transient thermal flows and the liquid/solid interface in horizontal Bridgman growth (Crochet *et al.* 1987, Dupont *et al.* 1987, Wouters *et al.* 1987). Furthermore, they showed that three-dimensional effects dominate the flow and that the release of latent heat of fusion has a major influence upon the shape of the interface.

With the aim of getting a better understanding of fluid flow in both vertical and horizontal Bridgman, the interest of lowest-possible thermal gradients was shown (Favier 1990). Favier pointed out that the horizontal Bridgman configuration is the least stable from the fluid point of view, due to Marangoni convection caused by the free surfaces. Thus, the vertical Bridgman configuration appears more favorable from the fluid-flow point of view, and Favier also stressed the importance of furnace and crucible design.

Besides these general studies, numerous papers have been specifically dedicated to the Bridgman growth modeling of CdTe, which differs from many materials by its low thermal conductivity and its small thermal-conductivity difference between melt and solid. These factors render the control of temperature distribution and interface shape more difficult.

The FEM model was employed to calculate the temperature and stress fields in vertical Bridgman growth of CdZnTe (Parfeniuk *et al.* 1992). From their simulation, the authors recommended the following operating conditions: low cold-zone furnace temperature (1060–1065 °C), low hot-zone furnace temperature (1110 °C), and a moderate axial gradient ($\sim$15 K cm^{-1}).

The effect of the temperature distribution in the growth ampoule on the crystallization process was studied using a finite-element approach (Sterr *et al.* 1993). The experiments carried out with a predicted convex interface shape resulted in the growth of large crystals.

From finite-element modeling also for the CZT vertical Bridgman growth, large radial gradients were shown to dominate the temperature field in the solid, while convection flattens the radial temperature distribution in the melt (Kuppurao *et al.* 1995a). Concave interface shapes are predicted to arise from the thermal conductivity mismatch between solid and liquid. The shape of the solid/liquid interface is shown to be very sensitive to the growth rate due to the significant release of latent heat. The same FEM approach was used to account for zinc segregation (Kuppurao *et al.* 1995b). The system is shown to be far from the diffusion-controlled limit and thus far from reaching a steady state. Lowering the growth rate in the system is predicted to increase axial segregation, while decreasing radial segregation. From this FEM approach, the same authors (Kuppurao *et al.* 1996) proposed an interrupted growth strategy in an unseeded vertical Bridgman system employing a series of relatively fast growth periods followed by pauses. Providing a means to mix the Cd rejected at the growing interface into the bulk, this strategy allows the use of faster growth rates for grain selection while minimizing the risk of constitutional supercooling. Using the same FEM for heat transfer, melt convection and interface position, Kuppurao and Derby (1997) proposed an ampoule design with a shallow cone sitting upon a composite support made of a highly conducting core and a less conducting outer sheath to promote axial heat transfer while inhibiting radial thermal flow. The result should be an interface shape convex toward the melt.

The importance of the ampoule-wall characteristics was stressed from a numerical simulation using a commercial computational code (Martinez-Thomas *et al.* 1999). The three heat-transfer modes, conduction, convection and radiation, and a rather complex geometry of the entire system consisting of the enclosure wall, ampoule and different graphite covers were taken into account. It was found that the graphite cover decreases the axial temperature gradient in the melt, this effect being intensified by the conical tip. Another effect of the heat-conducting cover is to increase the growth rate at the onset of the growth process in the tip in such a way that, depending on the graphite thickness, this velocity can be several times larger than the ampoule translation rate; the growth rate then reaches the stationary state after a time similar to that needed in the uncovered ampoule case. Similarly, at the onset of growth the presence of graphite increases the concavity of the interface at the tip of the ampoule, this effect being reduced as the growth proceeds in the cylindrical zone (Figure 17.8).

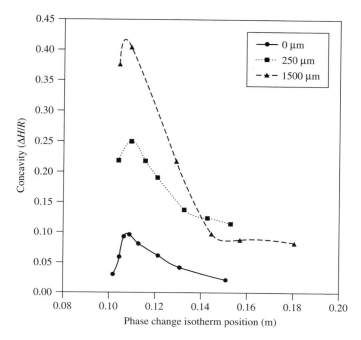

Figure 17.8 Concavity of the solid/liquid interface as a function of the axial 1365-K isotherm position. (Reprinted from C. Martinez-Thomas *et al.*, *J. Cryst. Growth* **197** (1999) 435–442, copyright (1999) with permission from Elsevier Science.)

The importance of the tip geometry, conical or flat, and of three graphite cover thicknesses on the walls of the ampoule were stressed again recently by the same group (Martinez-Thomas and Munoz). Among several very significant conclusions, it was shown that a graphite cover makes the growth rate greater than the ampoule translation rate, increases the concavity of the liquid/solid interface (mainly in the tip zone), reduces the axial thermal gradient in the melt near the interface, increases the radial thermal gradient in the tip, and produces a multicellular convective flow in the melt. Besides a flat tip, a conical tip was shown to increase again the growth rate and the interface concavity produced by the graphite cover, mainly in the early growth stages, to retard the arrival of the axial temperature gradient in the melt to an asymptotic value as the growth proceeds, to reinforce the radial thermal gradient when the interface is in the tip of uncovered ampoules, and to weaken this gradient in graphitized ampoules.

The ampoule base shape has been also taken into account in vertical Bridgman crystal-growth systems through the heat transfer between ampoule and furnace. It was found that both the flat base and the semispherical base system produced slightly concave solid/melt interface, which is undesirable (Ouyang and Shyy 1997).

In a recent CZT vertical Bridgman growth simulation, a configuration in which the heat exchange of the ampoule with the environment occurred only by radiation was studied (Lakeenkov *et al.* 1999). At the onset of the growth, heat removal through the ampoule bottom dominates, providing a convex shape of the crystallization isotherm. Portions concave towards the crystal may appear on convex isotherms as a result of the large bend height of convex crystallization isotherms.

Although the horizontal Bridgman growth modeling has attracted less attention, a recent study (Edwards and Derby 1997) deals with the understanding of CdTe and CdZnTe growth according to the horizontal Bridgman shelf growth process originally developed by Texas Instruments in the framework of the DARPA program entitled Infrared Materials Processing (IRMP) (Liao 1993, 1994). A calculation using FEM clearly shows that shelf growth naturally arises from simple heat-transfer effects in the studied low-gradient system (Figure 17.9). It is demonstrated that the shelf shape can be dramatically altered by process modifications like faster rates due to the increased rate of latent heat release, or the form and magnitude of the horizontal thermal gradient. A FEM analysis of thermal stresses presents the benefits of the shelf-growth configuration. The growing crystal is protected from adverse crucible sticking, and the low thermal gradients used to enhance shelf growth act to minimize thermal stresses.

Besides these theoretical simulations, different strategies have been used to visualize the convection in the Bridgman technique and the liquid/solid growth interface.

The influence of operating parameters on thermal convection in the Bridgman–Stockbarger technique was studied with a transparent furnace and melt (Neugebauer and Wilcox 1988). Under stabilized thermal conditions, convection was found to be nearly absent in the melt near the interface, but it always exists in the upper portion of the melt because of the temperature roll-off at the top of the heater. Under vertically destabilized thermal conditions caused by a short booster heater between the main heater and cooler, there is a significant convection throughout the melt.

Numerous studies deal, furthermore, with the interface shape observation and calculation in the CdTe vertical Bridgman growth.

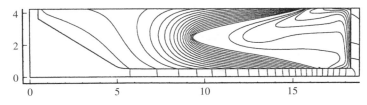

Figure 17.9   Representative basic case showing temperature isotherms corresponding to a pull rate of 1 mm/h and the gradient zone positioned between $x = 16$ cm and $x = 18$ cm. (Reprinted from K. Edwards and J.J. Derby, *J. Cryst. Growth* **179** (1997) 120–132, copyright (1997) with permission from Elsevier Science.)

[111]In was used as a radiotracing dopant for melt/solid interface-shape demarcation (Route et al. 1984). The interface shape was found to be near-optimal and varied from marginally concave to slightly convex, and attempts to modify them were found to be successful but produced only slight changes. Surprisingly, the grain structure was not significantly improved. Actual crystal growth rates were found to be as much as 35 % lower than the imposed mechanical translation rate, suggesting complex boundary conditions.

While decantation was sometimes used to show the interface shape (Shetty et al. 1996), a quenching procedure was proposed to mark the interface shape (Pfeiffer and Mühlberg 1992). Although the quenching conditions were not capable of freezing-in the interface shape because of the low heat conductivity of CdTe, a small excess of Te was used to produce a constitutional-supercooled zone in the melt marking the interface shape. The experimentally observed interface shapes were compared with those obtained from thermal modeling. The interface curvatures observed were shown to depend strongly on the axial position. Noticeable influences of the end effects were observed leading to the proposal of longer ampoules in order to reduce end effects.

In some studies, the equicomposition contour of Zn in doped crystals was used to reveal the solid/liquid interface shape (Tanaka et al. 1987, Azoulay et al. 1992). But assumptions that equated the isoconcentration contours of Zn with the shape of the solid/liquid interface were shown later to be in great error due to the large degree of radial segregation across the interface (Kuppurao et al. 1995).

Beside these indirect and sometimes questionable techniques used to reveal the interface shapes, an exploration of eddy-current sensing of solid/liquid interfaces in real time during crystal growth has been proposed (Rosen et al. 1995, Shetty et al. 1996, Dharmasena and Wadley 1997). Eddy-current diagnostics is a nondestructive, noncontact and remote electromagnetic technique sensitive to changes in electric conductivity. It is motivated by the difference between the electrical conductivities of most liquid and solid semiconductors. Even in the difficult case of CdTe, the electrical conductivity of which is low compared to GaAs, which is an ideal material for such a technique, it was shown that the interface location could be determined to ±1 mm and its curvature estimated with sufficient precision to be of use to characterize the vertical Bridgman growth process (Wadley and Dharmasena 1997). Concave interface shapes in qualitative agreement with the shape of decanted ingots were found (Shetty et al. 1996).

17.6 CdZnTe PROPERTIES

17.6.1 PROPERTIES AT MACROSCOPIC AND MICROSCOPIC SCALE

Cadmium telluride is now replaced by cadmium zinc telluride (CZT) for most applications, among which are epitaxial substrates and nuclear detectors. In CdZnTe substrates a perfect lattice match to any CdHgTe composition can be achieved by adjusting the Cd/Zn ratio. Furthermore, Zn in CdTe increases the band gap, and thus resistivity, and the energy of defect formation, facilitating the fabrication of nuclear detectors with improved performance. The role of Zn in CdTe can be considered at both macroscopic and microscopic scales.

The Zn–Te bond length and ionicity are smaller and the Zn–Te binding energy is higher than in Cd–Te, which strengthens the CdTe lattice by the incorporation of Zn as shown by the increased shear modulus, which is a key signature of material stability (Sher *et al.* 1985). This leads to the concept of increased stability of Zn-containing alloys. Thus, the addition of Zn in CdTe reduces the density of both dislocations and subgrains (James and Stoller 1984, Bell and Sen 1985, Tranchart *et al.* 1985, Yoshikawa *et al.* 1987). The solid-solution hardening in CZT was modeled and experimentally verified by plastic deformation and microhardness experiments (Guergouri *et al.* 1988, Imhoff *et al.* 1991).

At the microscopic scale, the local bond-length mismatch in the lattice leads both to structural distortions from ferroelectric origin (Marbeuf *et al.* 1990) and to a miscibility gap in the solid below 428 °C, as a result of the very strong repulsive mixing enthalpy in the solid. This last effect was theoretically predicted and experimentally verified in both thin films and bulk crystals (Figure 17.10) (Marbeuf *et al.* 1992, Ruault *et al.* 1994). Thus, slowly cooled CdZnTe crystals show phase separation and an extreme mechanical fragility. Phase separation can be avoided by sufficiently fast post-growth cooling (Ruault *et al.* 1994).

17.6.2   SEGREGATION

In addition to the CdTe growth problems discussed in Section 17.3, the segregation of Zn during the CdZnTe growth is another issue. The Zn segregation coefficient in CdTe was estimated to be in the 1.16–1.35 range (Steininger *et al.* 1970, Tanaka *et al.* 1989, Toney *et al.* 1996) and gives rise to inhomogeneous distribution along both axial and radial directions in the boules. Various approaches to avoid Zn segregation, which strongly affects the crystal uniformity required for any application (substrates or nuclear detectors) and thus the production yield, are described in the following.

With the segregation coefficients of Zn and Se in CdTe being respectively larger than 1 (1.35) and lower than one (0.9, Strauss and Steininger 1970), it has been shown possible to grow Zn and Se codoped crystals by the vertical-gradient-freeze method with a very uniform lattice constant and high structural perfection (Tanaka *et al.* 1989).

Another approach used the thermodynamic proportionality of the Zn solute partial pressure with its concentration in the Cd+Zn solution. Under such a Cd/Zn partial pressure control, the axial Zn concentration of VGF-grown $Cd_{0.96}Zn_{0.04}Te$ ingots was found to be uniform within ±3 % from X-ray diffraction and electron microprobe analysis (Azoulay *et al.* 1992).

Ingots showing a good Zn axial homogeneity were grown by a modified vertical Bridgman technique using a (Cd, Zn) alloy source in communication with the melt, whose temperature was gradually changed from 800 to 840 °C during growth (Lee

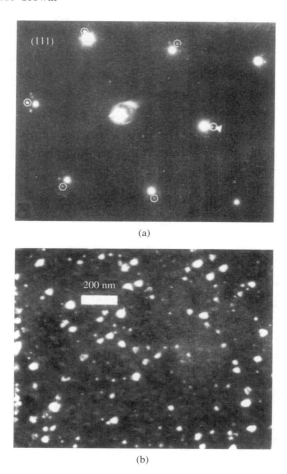

(a)

(b)

Figure 17.10   (a) Diffraction pattern of an as-grown $Cd_{0.96}Zn_{0.04}Te$ crystal showing the (001)ZnTe//(111) matrix. Spots from ZnTe are circles. Extra spots are due to double diffraction. (b) Dark-field image of the superstructure spot indicated in (a), showing coherent ZnTe precipitates. (Reprinted from M.-O. Ruault *et al.*, *J. Cryst. Growth* **143** (1994) 40–45, copyright (1994) with permission from Elsevier Science.)

et al. 1995). In a similar approach in vertical Bridgman growth of $Cd_{0.96}Zn_{0.04}Te$ a replenishing melt was supplied from a second crucible immersed in the melt from which the crystal grew (Tao and Kou 1997). The replenish crucible had a long small-diameter passageway between the two melts to suppress diffusion. The melts were encapsulated with B_2O_3 under pressurized Ar to prevent evaporation.

2-inch diameter $Cd_{0.96}Zn_{0.04}Te$ and $Cd_{0.8}Zn_{0.2}Te$ ingots of excellent radial and axial uniformity (Figure 17.11) were obtained by cold THM, which benefits from the additional purification effect of THM solution zone refining (Triboulet 1994).

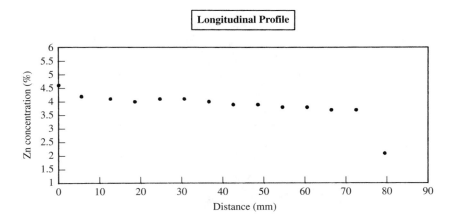

Figure 17.11 Longitudinal composition profile of a $Cd_{0.96}Zn_{0.04}Te$ ingot grown by cold THM. (Reprinted from R. Triboulet, *Prog. Cryst. Growth Charac.* **28** (1994) 85–144, copyright (1994) with permission from Elsevier Science.)

17.6.3 INDUSTRIAL GROWTH

Only a few of the growth techniques described in Section 17.4 are actually used for the industrial production of CdTe and CdZnTe crystals for substrates and nuclear detectors.

The traveling heater method using Cl-doped Te as a solvent (Eurorad, Acrotec), the high-pressure vertical-Bridgman (eV, Digirad), and the modified horizontal-Bridgman (Imarad) techniques are used for the production of CdTe and CdZnTe nuclear detectors. For substrate fabrication, horizontal Bridgman according to the 'shelf-growth process' initially developed by Texas Instruments (TI) (II-VI Inc., JME, TI) and vertical Bridgman either classical (AEG, BAE Systems, Hughes Res. Lab., JME, Sofradir) or gradient freeze (Japan Energy) are essentially employed. CdTe substrates are also produced by SPVT (Eagle Picher).

17.7 PROPERTIES AND DEFECTS OF THE CRYSTALS

Some properties of CdTe and CdZnTe crystals are displayed in Table 17.1.

Several kinds of extended defects are frequently observed in CdTe, precipitates and inclusions, dislocations and subgrain boundaries, twins and stacking faults.

Inclusions and precipitates, generated during the growth from nonstoichiometric melts, differ essentially by their mode of generation. Many papers on inclusion analysis in CdZnTe are discussed in a review on II-VI melt growth (Rudolph 1998). Inclusions are formed by the capture of melt droplets at the diffusion-boundary layer enriched by the excess component, Cd, Zn or more frequently Te.

Table 17.1

Material	Technique	Section	Resistivity (Ω cm)	R (mm/h)	G (K/cm)	EPD (cm^{-2})	FWHM (arcsec)	$\mu\tau_e$ (cm^2/V)	$\mu\tau_h$ (cm^2/V)	References
CdTe	PVT	Ø36 mm	$\sim 10^7$	$\sim 4 \times 10^{-5}$	–	–	–	10^{-4}	10^{-5}	Akutagawa and Zanio 1971
CdTe	PVT	Ø20 mm	10^9	–	–	10^3–10^4	–	–	–	Buck and Nitsche 1980
CdTe	PVT	Ø23 mm	$\geq 10^8$	$\sim 4 \times 10^{-5}$	–	2×10^3–10^4	–	–	–	Yellin et al. 1982
CdTe	HB	Ø50 mm (sr)	1.0×10^7	–	–	10^4–10^5	9.5–15	–	–	Khan et al. 1986
CdTe	VGF	Ø30–60 mm	–	–	1.5–25	$\sim 10^4$	9.1	–	–	Tanaka et al. 1987
CdTe	PVT	Ø22 mm	10^2	–	30	10^5	–	–	–	Geibel et al. 1988
CdTe	HGF	Ø4 cm (sr)	–	2–3	–	2–8×10^4	10–37	–	–	Lay et al. 1988
CdZnTe	VMB	Ø2 inch	–	1	<12	$\sim 10^4$	8–14	–	–	Sen et al. 1988
CdZnTe	VB	Ø47 mm	10^2	1.5–2	15	3–8×10^4	11.4	–	–	Bruder et al. 1990
CdZnTe	VBACRT	Ø50 mm	–	1	–	3.3–5.8×10^4	16–26	–	–	Capper et al. 1993
CdZnTe	HPVB	Ø100 mm	$> 10^{11}$	–	–	$\leq 10^4$	10–15	6×10^{-4}	3×10^{-5}	Butler et al. 1993
CdZnTe	Mod. HB	40×40 mm^2	–	3–6	~ 2	–	20–30	–	–	Brunet et al. 1993

(continued overleaf)

Table 17.1 (*continued*)

Material	Technique	Section	Resistivity (Ω cm)	R (mm/h)	G (K/cm)	EPD (cm^{-2})	FWHM (arcsec)	$\mu\tau_e$ (cm^2/V)	$\mu\tau_h$ (cm^2/V)	References
CdTe	VB	Ø64 mm	8.3×10^3	1	3–8	2.5×10^4	9.4	–	–	Casagrande et al. 1993
CdTe:Cl	THM	Ø32 mm	$>10^8$	$\sim 8 \times 10^{-2}$	25	–	–	–	–	Ohmori et al. 1993
Cd$_{Zn}$Te	HB	Ø4 inch (sr)	–	~ 4–8	–	$\sim 10^4$	–	–	–	Liao
CdTe$_{Se}$	VB	Ø37 mm	–	1	8	8×10^3	5.6	–	–	Hähnert et al. 1994
Cd$_{Zn}$Te	VB	Ø64 mm	–	–	3–5	$\leq 5 \times 10^4$	9.2	–	–	Casagrande et al. 1994
CdTe	SPVT	Ø45–50 mm	5–10	4×10^{-5}	4.5	3–7×10^4	10–25	–	–	Boone et al. 1994
CdTe:Cl	STHM	Ø15 mm	8×10^8	8×10^{-2}	115	–	–	–	–	Laasch et al. 1994
Cd$_{Zn}$Te	VGF	Ø100 mm	30–37	1.8	1	4–6×10^4	8–13	–	–	Asahi et al. 1996
Cd$_{Zn}$Te	HPVB	Ø140 mm	1.7–4×10^{10}	–	–	–	–	0.5–5×10^{-3}	0.2–5×10^{-5}	Szeles et al. 1998
Cd$_{Zn}$Te	mod. HB	45 × 45 mm^2	$\sim 5 \times 10^9$	–	–	10^3–10^4	–	1–3×10^{-3}	–	El Hanany 2000
CdTe:Cl	THM	Ø15–25 mm	3×10^9	–	–	–	–	10^{-3}	10^{-4} – 10^{-5}	Hage-Ali and Siffert 1995

The density of Te inclusions was correlated to the growth rate (Dinger and Fowler 1977). The fundamental problem of inclusions as a form of growth instability and the relation to growth parameters was discussed earlier (Scheel and Elwell 1972) and also in the book of Elwell and Scheel (1975). The size of inclusions ranges usually from 1 μm to several tens of micrometers. They can show, depending on the growth conditions, a polyhedral, hexagonal, triangular, square or star shape, and can be surrounded by screw dislocations disturbing the crystal quality. The inclusion density can reach values as high as $10^7 \, cm^{-3}$ at the ingot ends.

Precipitates are formed at lower temperature in the solid due to the retrograde shape of the solidus on both the Cd and Te side. Due to the frequently Te-rich growth conditions, Te precipitates are more generally observed, but Cd precipitates can be found for growth under Cd-rich conditions. Several studies deal with the crystallographic shapes of Te precipitates. They were shown from X-ray diffraction to present the same structural phase as observed in elemental Te under high pressure (Shin et al. 1983). Te precipitates having well-defined crystallographic shapes with facets parallel to the {111} planes were observed by transmission infrared microscopy (Wada and Suzuki 1988). The same density found for precipitates and etch pits suggests that there exists a strong correlation between the presence of precipitates and dislocations. Compared to the microstructure of CdTe and CdZnTe, Te precipitates with a density and size lowest and largest were observed in CdSeTe by TEM (Rai et al. 1991). The origin of the often-reported large stress fields in and around Te precipitates and associated punching of dislocation loops in star-like patterns was explained from a thermodynamic calculation (Yadava et al. 1992). The occurrence of varying precipitate morphologies is also explained, and the possibility of nucleating the high-pressure monoclinic Te phase is considered.

Cd vacancies, V_{Cd}, are considered as a deep double-acceptor level in the band gap, at $E_{VB} + 0.47 \, eV$ (Meyer and Stadler 1996). V_{Cd} cannot be easily detected in CdTe because the dominant residual impurities, like Ag and Cu, are present at a rather high level, close to $10^{17} \, cm^{-3}$, and are incorporated in Cd substitutional sites. V_{Cd} participates nevertheless in a very significant defect, the metal vacancy–donor pair defect, also called the A-center. The chlorine A-center in CdTe, with a binding energy of 125 meV, participates with other intrinsic defects in the compensation behavior and explains the high resistivity of chlorine-doped crystals. From a statistical-thermodynamic derivation of the equilibrium equations for two-defect models (Brebrick and Fang 1996), the presence of singly ionizable Te antisite donors is assumed in Cd-rich CdTe. The singly ionizable antisite model gives the best overall fit of the electrical high-temperature measurements reported so far.

17.8   PURITY, CONTAMINATION, DOPING

From chemical analysis measurements by glow-discharge mass spectrometry and Zeeman-corrected graphite furnace atomic absorption, the main elements detected in CdZnTe substrates (grown in three companies, TI, II-VI and JME) by their standard horizontal and vertical Bridgman processes) were found to be Li, Na, Mg, Al, Si, Cl, Se and Cu. These impurities were attributed to the crystal-growth process environment (crucible, silica, furnace, residual atmosphere) more than to the raw materials, except for Se originating from Te (Tower *et al.* 1995). Cd was suspected to be the main source of Cu contamination for residual levels in the 1–40 ppba range (Ard 1995). In another study, such acceptors as Li, Na, K, As, Cu, Ag, Sb, Bi, the donors Al, Cl, Ga, In, Tl, and the neutral impurities Si, Ca, Cr, Co, Sn and Pb were identified as classical residual impurities in as-grown p-type vertical Bridgman CdTe crystals (Rudolph *et al.* 1992). Contamination by residual acceptors like Cu, Li, P and Na from silica ampoules during CdTe Bridgman growth were found by electrical and optical measurements (Triboulet *et al.* 1995). Surface oxides on the high-purity elements, residual gases, and the hydrocarbon vapors from the evacuation system were identified as contamination sources by infrared-absorption measurements (Bagai *et al.* 1994). Furthermore, it was found in the same study that both Zn and Se substitution enhances, and hydrogen purging reduces, the uptake of these contaminants.

Although the electrical activity of impurities can be hampered by self-compensation from native defects or other residual impurities and by gettering of impurities into Te precipitates, p-type and n-type doping of CdTe is easy to achieve. The elements of the first and fifth columns of the periodic chart act as acceptors, and those of the third and seventh columns act as donors. Mainly Al, Ga, In, Cl and I have been used as donors, and Li, Cu, Ag, N, P and As as acceptors. Some of these elements show special behavior in CdTe, depending on the site they occupy in the crystalline lattice, like the amphoteric behavior of Li or Ag (Lyubomirsky *et al.* 1996). A complete review of the impurity doping of CdTe was given elsewhere (Astles 1994). The maximum doping levels achievable in bulk CdTe are of some 10^{17} cm^{-3} for holes (As, P, Li), and some 10^{18} cm^{-3} for electrons (Al, I, In), thus very close to the deviations from stoichiometry on both Cd and Te sides.

17.9 CONCLUSIONS AND PERSPECTIVES

Over forty years numerous efforts have been dedicated to the growth and materials science of CdTe and its solid solutions, which have become key issues for many industrial and scientific applications. The achieved crystal perfection demonstrates that CdTe has now come to maturity. All these studies now open the way to CdTe-based gamma-ray cameras for medical purposes, which is likely to become a major application and thus boost again the interest for this attractive class of material.

REFERENCES

Aitken, N. M., Potter, M. D G., Buckley, D. J., Mullins, J. T., Carles, J., Halliday, D. P., Durose, K., Tanner, B. K., Brinkman A. W. (1999) *J. Cryst. Growth* **198/199**, 984.

Akutagawa, W., Zanio, K. (1971) *J. Cryst. Growth* **11**, 191.

Amzil, A., Mathieu, J.-C., Castanet, R. (1997) *J. Alloys Compd.* **256**, 192.

Aoudia, A., Rezpka, E., Lusson, A., Schneider, D., Marfaing, Y., Triboulet, R. (1995) *Opt. Mater.* **4**, 241.

Appell, J. (1954) *Naturfors.* **9**, 265.

Ard, C. K. oral communication at the(1995) US Workshop on the Physics and Chemistry. of HgCdTe and other IR Materials.

Asahi, T., Oda, O., Taniguchi, Y., Koyama, A. (1996) *J. Cryst. Growth* **161**, 20.

Astles, M. G. (1994) *Properties of Narrow Gap Cd-based Compounds*, ed. Capper P, INSPEC, London, **10**, 494.

Azoulay, M., Rotter, S., Gafni, G., Tenne, R., Roth, M. (1992) *J. Cryst. Growth* **117**, 276.

Bagai, R. K., Borle, W. N. (1989) *J. Cryst. Growth* **94**, 561.

Bagai, R. K., Yadava, R. S D., Sundersheshu, B. S., Seth, G. L., Anandan, M., Borle, W. N. (1994) *J. Cryst. Growth* **139**, 259.

Becker, U., Zimmermann, H., Rudolph, P., Boyn, R. (1989) *Phys. Stat. Sol. (a)* **112**, 569.

Bell, R. O., Hemmat, N., Wald, F. (1970) *Phys. Stat. Sol. (a)* **1**, 375.

Bell, S. L., Sen, S. (1985) *J. Vac. Sci. Technol.* **A3**, 112.

Blank, Z., Brenner, W. (1971) *J. Cryst. Growth* **11**, 255.

Boone, J. L., Cantwell, G., Harsch, W. C., Thomas, J. E., Foreman, B. A. (1994) *J. Cryst. Growth* **139**, 27.

Brandt, G., Emen, H., Moritz, R., Rothemund, W. (1990) *J. Cryst. Growth* **101**, 296.

Brebrick, R. F. (1971) *J. Electrochem. Soc.* **118**, 779.

Brebrick, R. F., Rei. Fang (1996) *J. Phys. Chem. Solids* **57**, 451.

Bruder, M., Schwartz, H.-J., Schmitt, R., Maier, H., Möller, M.-O. (1990) *J. Cryst. Growth* **101**, 266.

Brunet, P., Katty, A., Schneider, D., Tromson-Carli, A., Triboulet, R. (1993) *Mater. Sci. Eng.* **B16**, 44.

Buck, P., Nitsche, R. (1980) *J. Cryst. Growth* **48**, 29.

Butler, J. F., Doty, F. P., Apotovsky, B., Lajzerowicz, J., Verger, L. (1993) *Mater. Sci. Eng.* **B16**, 291.

Caillot, M. (1977) *Rev. Phys. Appl.* **12**, 241.

Capper, P., Harris, J. E., O'Keefe, E., Jones, C. L., Ard, C. K., Mackett, P., Dutton, D. (1993) *Mater. Sci. Eng.* **B16**, 29.

Carlsson, L., Ahlquist, C. N. (1972) *J. Appl. Phys.* **43**, 2529.

Casagrande, L. G., Larson, D. J., Di Marzio, D., Wu, J., Dudley, M., Black, D. R., Burdette, H. (1994) *Proc. SPIE* **2228**, 21.

Casagrande, L. G., Marzio, D. D., Lee, M. B., Larson, D. J., Dudley, M., Fanning, T. (1993) *J. Cryst. Growth* **128**, 576.

Chang, C. E., Wilcox, W. R. (1974) *J. Cryst. Growth* **23**, 135.

Chang, W.-M., Wilcox, W. R., Regel, L. (1993) *Mater. Sci. Eng.* **B 16**, 23.

Cheuvart, P., El Hanani, U., Schneider, D., Triboulet, R. (1990) *J. Cryst. Growth* **101**, 270.

Cornet, A., Siffert, P., Coche, A. (1970) *J. Cryst. Growth* **7**, 329.

Corsini-Mena, A., Elli, M., Paorici, C., Pelosini, L. (1971) *J. Cryst. Growth* **8**, 297.

Crochet, M. J., Dupret, F., Ryckmans, Y., Geyling, F. T., Monberg, E. M. (1989) *J. Cryst. Growth* **97**, 173.

Crochet, M. J., Geyling, F. T., Van Schaftingen, J. J. (1987) *Int. J. Numer. Methods Fluids* **7**, 29.

De Nobel, D. (1959) *Philips Res. Rep.* **14**, 361.

Dharmasena, K. P., Wadley, H. N G. (1997) *J. Cryst. Growth* **172**, 303.

Dinger, R. J., Fowler, I. L. (1977) *Rev. Phys. Appl.* **12**, 135.

Doty, F. P., Butler, J. F., Schetzina, J. F., Bowers, K. A. (1992) *J. Vac. Sci. Technol.* **B10**, 1418.

Dupont, S., Marchal, J. M., Crochet, M. J., Geyling, F. T. (1987) *Int. J. Numer. Methods Fluids* **7**, 49.

Durose, K., Russell, G. J., Woods, J. (1985) *J. Cryst. Growth* **72**, 85.

Edwards, K., Derby, J. J. (1997) *J. Cryst. Growth* **179**, 120, 133.

El Hanany, U. (2000) private communication.

El Mokri, A., Triboulet, R., Lusson, A., Tromson-Carli, A., Didier, G. (1994) *J. Cryst. Growth* **138**, 168.

Elwell, D., Scheel, H.-J. (1975) *Crystal Growth from High-Temperature Solutions*, Academic Press, London New York.

Favier, J. J. (1990) *J. Cryst. Growth* **99**, 18.

Fang, Rei., Brebrick R. F. (1996) *J. Phys. Chem. Solids* **57**, 443.

Fougeres, P., Hage-Ali, M., Koebel, J.-M., Siffert, P., Hassan, S., Lusson, A., Triboulet, R., Marrakchi, G., Zerrai, A., Cherkaoui, K., Adhiri, R., Bremond, G., Kaitasov, O., Ruault, M.-O., Crestou, J. (1998) *J. Cryst. Growth* **184/185**, 1313.

Fu, T. W., Wilcox, W. R. (1980) *J. Cryst. Growth* **48**, 416.

Gaspard, J. P., Bergman, C., Bichara, C., Bellisent, R., Chieux, P., Goffart, J. (1987) *J. Non-Cryst. Solids* **97–98**, **1283**.

Geibel, C., Maier, H., Schmitt, R. (1988) *J. Cryst. Growth* **86**, 386.

Glasov, V. M., Chizhevskaya, S. N., Glagoleva, N. N. (1969) *Liquid Semiconductors*, Plenum Press, New York.

Glass, H. L., Socha, A. J., Parfeniuk, C. L., Bakken, D. W. (1998) *J. Cryst. Growth* **184/185**, 1035.

Golacki, Z., Gorska, M., Makowski, J., Szczerbakow, A. (1982) *J. Cryst. Growth* **56**, 213.

Grasza, K., Zuzga-Grasza, U., Jedrzejczak, A., Galazka, R. R., Majewski, J., Szadkowski, A., Giodzicka, E. (1992) *J. Cryst. Growth* **123**, 519.

Greenberg, J. H. (1996) *J. Cryst. Growth* **161**, 1.

Greenberg, J. H. (1999) *J. Cryst. Growth* **197**, 406.

Guergouri, K., Triboulet, R., Tromson-Carli, A., Marfaing, Y. (1988) *J. Cryst. Growth* **86**, 61.

Hage-Ali, M., Siffert, P. (1995) *Semicond. Semimet.* **43**, 292.

Hähnert, I., Mühlberg, M., Berger, H., Genzel, C. (1994) *J. Cryst. Growth* **142**, 310.

Heumann, F. K. (1962) *J. Electrochem. Soc.* **109**, 345.

Hobgood, H. M., Swanson, B. W., Thomas, R. N. (1987) *J. Cryst. Growth* **85**, 510.

Höschl, P., Konak, C. (1965) *Phys.Stat. Sol.* **9**, 107.

Höschl, P., Ivanov Yu, M., Belas, E., Franc, J., Grill, R., Hlidek, P., Moravec, P., Zvara, M., Sitter, H., Toth, A. L. (1998) *J. Cryst. Growth* **184/185**, 1039.

Höschl, P., Konak, C. (1963) *Czech. J. Phys.* **13**, 850.

Ickert, L., Rudolph, P. (1978) *Krist. Tech.* **13**, 107.

Igaki, K., Mochizuki, K. (1974) *J. Cryst. Growth* **24/25**, 162.

Igaki, K., Ohashi, N., Mochizuki, K. (1976) *Jpn. J.Appl. Phys.* **15**, 1429.

Ikegami, S., Matsushita Electric Industrial Co., private communication.

Imhoff, D., Zozime, A., Triboulet, R. (1991) *J. Phys. France* **1**, 1841.

Ivanov Yu, M. (1996) *J. Cryst. Growth* **161**, 12.

Ivanov Yu, M. (1998) *J. Cryst. Growth* **194**, 309.

James, R. W., Stoller, R. E. (1984) *Appl. Phys. Lett.* **44**, 56.

Jasinski, T., Rohsenow, W. M., Witt, A. F. (1983) *J. Cryst. Growth* **61**, 339.

Jasinski, T., Witt, A. F., Rohsenow, W. M. (1984) *J. Cryst. Growth* **67**, 173.

Jones, C. L., Capper, P., Gosney, J. J. (1982) *J. Cryst. Growth* **56**, 581.

Jones, C. L., Capper, P., Gosney, J. J., Kenworthy, I. (1984) *J. Cryst. Growth* **69**, 281.

Jordan, A. S. (1970) *Metall. Trans.* **1**, 239.

Jordan, A. S., Zupp, R. R. (1969) *J. Electrochem. Soc.* **116**, 1285.

Kendall, E. J M. (1964) *Phys. Lett.* **8**, 237.

Kennedy, J. J., Amirtharaj, P. M., Boyd, P. R., Qadri, S. B., Dobbyn, R. C., Long, G. G. (1988) *J. Cryst. Growth* **86**, 93.

Khan, A. A., Allred, W. P., Dean, B., Hooper, S., Hawkey, J. E., Johnson, C. J. (1986) *J. Electron. Mater.* **15**, 181.

Khattak, C. P., Schmid, F. (1989) *Proc. SPIE* **1106**, 47.

Kiseleva, K. V., Klevkov, U. V., Maximovsky, S. N., Medvedev, S. A., Senturina, N. N. (1971) *Proc. Intern. Symp. on CdTe*, Strasbourg, eds Siffert P, Cornet A, XII–1.

Kobayashi, M. Z. (1911) *Anorg. Chem.* **69**, 1.

Kolb, E. D., Caporasso, A. J., Laudise, R. A. (1968) *J. Cryst. Growth* **3**, **4**, 422.

Kolesnikov, N. N., Kolchin, A. A., Alov, D. L., Ivanov Yu, N., Chernov, A. A., Schieber, M., Hermon, H., James, R. B., Goorsky, M. S., Yoon, H., Toney, J., Brunett, B., Schlesinger, T. E. (1997) *J. Cryst. Growth* **174**, 256.

Kulwicki, B. (1963) Ph.D Dissertation, Univ. of Michigan, Ann Arbor, Michigan.

Kunz, T., Laasch, M., Meinhardt, J., Benz, K. W. (1998) *J. Cryst. Growth* **184/185**, 1005.

Kuppurao, S., Brandon, S., Derby, J. J. (1995a) *J. Cryst. Growth* **155**, 93.

Kuppurao S, Brandon S, Derby J. J. (1995b) *J. Cryst. Growth* **155**, 103.

Kuppurao, S., Brandon, S., Derby, J. J. (1996) *J. Cryst. Growth* **158**, 459.

Kuppurao, S., Derby, J. J. (1997) *J. Cryst. Growth* **172**, 350.

Kuwamoto, H. (1984) *J. Cryst. Growth* **69**, 204.

Kyle, N. R. (1971) *J. Electrochem. Soc.* **118**, 3304.

Laasch, M., Kunz, T., Eiche, C., Fiederle, M., Joerger, W., Kloess, G., Benz, K. W. (1997) *J. Cryst. Growth* **174**, 696.

Laasch, M., Schwarz, R., Rudolph, P., Benz, K. W. (1994) *J. Cryst. Growth* **141**, 81.

Lakeenkov, V. M., Ufimtsev, V. B., Shmatov, N. I., Schelkin Yu, F. (1999) *J. Cryst. Growth* **197**, 443.

Lay, K. Y., Nichols, D., McDevitt, S., Dean, B. E., Johnson, C. J. (1988) *J. Cryst. Growth* **86**, 118.

Lee, T. S., Lee, S. B., Kim, J. M., Kim, J. S., Suh, S. H., Song, J. H., Park, I. H., Kim, S. U., Park, M. J. (1995) *J. Electron. Mater.* **24**, 1057.

Leybov, V. A., Inanov Yu, M., Vanyukov, A. V. (1984) *Electron. Technik. Mater.* **10**, 28.

Liao, P. K., oral communications in the 1993 and 1994 US Workshop on the Physics and Chemistry. of HgCdTe and other IR Materials.

Liao, P. K., Colombo, L., Castro, C. A., Dean, B. E., Johnson, C. J., Boyd, P. R., Mozer, W., Comtois, R. (1991) in:, *Proc. IRIS Mater.* **245**.

Lorenz, M. R. (1962a) *J. Phys. Chem. Solids* **23**, 939.

Lorenz, M. R. (1962b) *J. Appl. Phys.* **33**, 3304.

Lorenz, M. R., Halsted, R. E. (1963) *J. Electrochem. Soc.* **110**, 343.

Lu, Y. C., Shiau, J.-J., Feigelson, R. S., Route, R. K. (1990) *J. Cryst. Growth* **102**, 807.

Lunn, B., Bettridge, V. (1977) *Rev. Phys. Appl.* **12**, 151.

Lynch, R. T. (1962) *J. Appl. Phys.* **33**, 1009.

Lyubomirsky, I., Lyakhovitskaya, V., Triboulet, R., Cahen, D. (1996) *J. Cryst. Growth* **159**, 1148.

Marbeuf, A., Ferah, M., Janik, E., Heurtel, A. (1985) *J. Cryst. Growth* **72**, 126.

Marbeuf, A., Druilhe, R., Triboulet, R., Patriarche, G. (1992) *J. Cryst. Growth* **117**, 10.

Marbeuf, A., Mondoloni, C., Triboulet, R., Rioux, J. (1990) *Solid State Commun.* **75**, 275.

Markov, E. V., Davydov, A. A. (1971) *Neorg. Mater.* **4**, 575.

Martinez-Thomas, C., Munoz, V., private communication, to be published.

Martinez-Thomas, C., Munoz, V., Triboulet, R. (1999) *J. Cryst. Growth* **197**, 435.

Matveev, O. A., Prokof'ev, S. V., Rud' Yu, V. (1969) *Izv. Akad. Nauk, Neorg. Mater.* **5**, 1175

Medvedev, S. A., Klevkov Yu, V., Kiseleva, K. V., Sentyurina, N. N. (1972) *Izv. Akad. Nauk, Neorg. Mater.* **8**, 1210.

Medvedev, S. A., Maksimovskii, S. N., Klevkov Yu, V., Shapkin, P. V. (1968) *Izv. Akad. Nauk, Neorg. Mater.* **4**, 2025.

Meiling, G. S., Leombruno, R. (1968) *J. Cryst. Growth* **3**, **4**, 300.

Meyer, B. K., Stadler, W. (1996) *J. Cryst. Growth* **161**, 119.

Mochizuki, K. (1981) *J. Cryst. Growth.* **51**, 453.

Mochizuki, K. (1985) *J. Cryst. Growth.* **73**, 510.

Mochizuki, K., Matsumoto, K. (1986) *Mater. Lett.* **4**, 298.

Mochizuki, K., Masumoto, K., Miyazaki, K. (1988) *Mater. Lett.* **6**, 119.

Mühlberg, M., Rudolph, P., Genzel, C., Wermke, B., Becker, U. (1990) *J. Cryst. Growth* **101**, 275.

Mullin, J. B., Jones, C. A., Straughan, B. W., Royle, A. (1982) *J. Cryst. Growth* **59**, 135.

Mullin, J. B., Straughan, B. W. (1977) *Rev. Phys. Appl.* **12**, 105.

Mycielski, A., Szadkowski, A., Lusakowska, E., Kowalczyk, L., Domagala, J., Bak-Misiuk, J., Wilamowski, Z. (1999) *J. Cryst. Growth* **197**, 423.

Naumann, R. J. (1982) *J. Cryst. Growth* **58**, 554 and 569.

Neugebauer, G. T., Wilex, W. R. (1988) *J. Cryst. Growth* **89**, 143.

Oda, O., Hirata, K., Matsumoto, K., Tsuboya, I. (1985) *J. Cryst. Growth* **71**, 273.

Ohmori, M., Iwase, Y., Ohno, R. (1993) *Mater. Sci. Eng.* **B16**, 283.

Ouyang, H., Shyy, W. (1997) *J. Cryst. Growth* **173**, 352.

Palosz, W., George, M. A., Collins, E. E., Chen, K.-T., Zhang, Y., Hu, Z., Burger, A. (1997) *J. Cryst. Growth* **174**, 733.

Paorici, C., Pelosi, C. (1977) *Rev. Phys. Appl.* **12**, 155.

Parfeniuk, C., Weinberg, F., Samaraserka, I. V., Schevzov, C., Li, L. (1992) *J. Cryst. Growth* **119**, 261.

Perkowitz, S., Sudharsanan, R., Wrobel, J. M., Clayman, B. P., Becla, P. (1988) *Phys. Rev. B* **38**, 5565.

Pfeiffer, M., Mühlberg, M. (1992) *J. Cryst. Growth* **118**, 269.

Price, S. L., Hettich, H. L., Sen, S., Currie, M. C., Rhiger, D. R., Mc Lean, E. O. (1998) *J. Electron. Mater.* **27**, 564.

Prokof'ev, S. V. (1970) *Izv. Akad. Nauk, Neorg. Mater.* **6**, 1077.

Rai, R. S., Mahajan, S., McDevitt, S., Johnson, C. J. (1991) *J. Vac. Sci. Technol. B* **9**, 1892.

Raiskin, E., Butler, J. E. (1988) *IEEE Trans. Nucl. Sci.* **35**, 81.

Rosen, G. J., Carlson, F. M., Thompson, J. E., Wilcox, W. R. (1995) *J. Electron. Mater.* **24**, 491.

Route, R. K., Wolf, M., Feigelson, R. S. (1984) *J. Cryst. Growth* **70**, 379.

Rubenstein, M. (1966) *J. Electrochem. Soc.* **113**, 623.

Rubenstein, M. (1968) *J. Cryst. Growth* **3**, **4**, 309.

Ruault, M.-O., Kaitasov, O., Triboulet, R., Crestou, J., Gasgnier, M. (1994) *J. Cryst. Growth* **143**, 40.

Rudolph, P. (1998) *Recent Development of Bulk Crystal Growth*, ed. and Isshiki,, Research Signpost, Trivandrum, 127.

Rudolph, P., Mühlberg, M. (1993) *Mater. Sci. Eng.* **B16**, 8.

Rudolph, P., Mühlberg, M., Neubert, M., Boeck, T., Möck, P., Parthier, L., Jacobs, K., Kropp, E. (1992) *J. Cryst. Growth* **118**, 204.

Rudolph, P., Rinas, U., Jacobs, K. (1994) *J. Cryst. Growth* **138**, 249.

Rud' Yu, V., Sanin, K. V. (1971) *Sov. Phys. Semicond.* **5**, 1385.

Salk, M., Fiederle, M., Benz, K. W., Senchenkov, A. S., Egorov, A. V., Matioukin, D. G. (1994) *J. Cryst. Growth* **138**, 161.

Schaub, B., Gallet, J., Brunet-Jailly, A., Pelliciari, B. (1977) *Rev. Phys. Appl.* **12**, 147.

Scheel, H. J., Elwell, D. (1972) *J. Cryst. Growth* **12**, 153.

Schoenholz, R., Dian, R., Nitsche, R. (1985) *J. Cryst. Growth* **72**, 72.

Sen, S., Konkel, W. H., Tighe, S. J., Bland, L. G., Sharma, S. R., Taylor, R. E. (1988) *J. Cryst. Growth* **86**, 111.

Shcherbak, L. (1998) *J. Cryst. Growth* **184/185**, 1057.

Shcherbak, L. (1999) *J. Cryst. Growth* **197**, 397.

Shcherbak, L., Feichouk, P., Panchouk, O. (1996) *J. Cryst. Growth* **161**, 16.

Sher, A., Chen, A.-B., Spicer, W. E., Shih, C. K. (1985) *J. Vac. Sci. Technol.* **A3**, 105.

Shetty, R., Ard, C. K., Wallace, J. P. (1996) *J. Electron. Mater.* **25**, 1134.

Shin, S. H., Bajaj, J., Moudy, L. A., Cheung, D. T. (1983) *Appl. Phys. Lett.* **43**, 68.

Smith, F. T J. (1970) *Metall. Trans.* **1**, 617.

Song, W.-B., Yu, M.-Y., Wu, W.-H. (1988) *J. Cryst. Growth* **86**, 127.

Steer Ch,, Hage-Ali, M., Koebel, J.-M., Siffert, P. (1993) *Mater. Sci. Eng.* **B16**, 48.

Steininger, J., Strauss, A. J., Brebrick, R. F. (1970) *J. Electrochem. Soc.* **117**, 1305.

Strauss, A. J., Steininger, J. (1970) *J. Electrochem. Soc.* **117**, 1420.

Su, C.-H., Sha, Y.-G., Lehoczky, S. L., Liu, H.-C., Fang, R., Brebrick, R. F. (1998) *J. Cryst. Growth* **183**, 519.

Szczerbakow, A., Golacki, Z. (1993) *Mater. Sci. Eng.* **B16**, 68.

Szeles, C., Driver, M. C. (1998) *Proc. SPIE* **3446**, 2.

Taguchi, T., Shirafuji, J., Inuischi, Y. (1978) *Jpn. J. Appl. Phys.* **17**, 8.

Tai, H., Hori, S. (1976) *J. Jpn. Inst. Met.* **40**, 722.

Tanaka, A., Masa, Y., Seto, S., Kawasaki, T. (1987) *Mater. Res. Soc. Symp. Proc.* **90**, 111.

Tanaka, A., Masa, Y., Seto, S., Kawasaki, T. (1989) *J. Cryst. Growth* **94**, 166.

Tao, Y., Kou, S. (1997) *J. Cryst. Growth* **181**, 301.

Toney, J. E., Brunett, B. A., Schlesinger, T. E., Van Scyoc, J. M., James, R. B., Schieber, M., Goorsky, M., Yoon, H., Eissler, E., Johnson, C. (1996) *Nucl. Instrum. Methods. Phys. Res.* **A 380**, 132.

Tower, J. P., Tobin, S. P., Kestigian, M., Norton, P. W., Bollong, A. B., Schaake, H. F., Ard, C. K. (1995) *J. Electron. Mater.* **24**, 497.

Tranchart, J.-C., Bach, P. (1976) *J. Cryst. Growth* **32**, 8.

Tranchart, J.-C., Latorre, B., Foucher, C., Le Gouge, Y. (1985) *J. Cryst. Growth* **72**, 468.

Triboulet, R. (1977) *Rev. Phys. Appl.* **12**, 123.

Triboulet, R. (1991) *Mater. Formn.* **15**, 30.

Triboulet, R. (1994) *Prog. Cryst. Growth Charact.* **28**, 85.

Triboulet, R., Aoudia, A., Lusson, A. (1995) *J. Electron. Mater.* **24**, 1061.

Triboulet, R., Didier, G. (1975) *J. Cryst. Growth* **28**, 29.

Triboulet, R., Marfaing, Y. (1973) *J. Electrochem. Soc.* **120**, 1260.

Triboulet, R., Marfaing, Y. (1981) *J. Cryst. Growth* **51**, 89.

Triboulet, R., Marfaing, Y., Cornet, A., Siffert, P. (1974) *Nature Phys. Sci.* **245**, 140.

Triboulet, R., Legros, R., Heurtel, A., Sider, B., Didier, G., Imhoff, D. (1985) *J. Cryst. Growth* **72**, 90.

Triboulet, R., Heurtel, A., Rioux, J. (1990a) *J. Cryst. Growth* **101**, 131.

Triboulet, R., Pham Van, K., Didier, G. (1990b) *J. Cryst. Growth* **101**, 216.

Triboulet, R., Rodot, H. (1968) *C.R. Acad. Sci. B* **266**, 498.

Vandekerkof, J. (1971) *Proc. Int. Symp. CdTe Mater. Gamma Ray Detect.*, Strasbourg, France, eds Siffert, P., Cornet, A., VIII.

Vere, A. W., Cole, S., Williams, D. J. (1983) *J. Electron. Mater.* **12**, 551.

Vere, A. W., Steward, V., Jones, C. A., Williams, D. J., Shaw, N. (1985) *J. Cryst. Growth* **72**, 97.

Verie, C. (1998) *J. Cryst. Growth* **184/185**, 1061.

Wada, M., Suzuki, J. (1988) *Jpn. J. Appl. Phys.* **27**, L972.

Wadley, H. N G., Choi, B. W. (1997) *J. Cryst. Growth* **172**, 323.

Wadley, H. N G., Dharmasena, K. P. (1997) *J. Cryst. Growth* **172**, 313.

Wald, F. V. (1976) *Phys. Stat. Solidi (a)* **38**, 253.

Whelan, R. C., Shaw, D. (1968) *Phys. Stat. Solidi* **29**, 145.

Wiedemeier, H., Bai, Y.-C. (1990) *J. Electron. Mater.* **19**, 1373.

Wienecke, M., Berger, H., Schenk, M. (1993) *Mater. Sci. Eng.* **B13**, 219.

Woodbury, H. H., Lewandowski, R. S. (1971) *J. Cryst. Growth* **10**, 6.

Wouters, P., Van Schaftingen, J. J., Crochet, M. J., Geyling, F. T. (1987) *Int. J. Numer. Methods Fluids* **7**, 131.

Yadava, R. D S., Bagai, R. K., Borle, W. N. (1992) *J. Electron. Mater.* **21**, 1001.

Yang Jianrong,, Silk, N. J., Watson, A., Bryant, A. W., Argent, B. B. (1995) *Calphad* **19**, 399.

Yang, B., Ishikawa, Y., Doumae, Y., Miki, T., Ohyama, T., Isshiki, M. (1997) *J. Cryst. Growth* **172**, 370.

Yellin, N., Eger, D., Shachna, A. (1982) *J. Cryst. Growth* **60**, 343.

Yellin, N., Szapiro, S. (1984) *J. Cryst. Growth* **69**, 555.

Yokota, K., Yoshikawa, T., Inano, S., Katayama, S. (1989) *Jpn. J. Appl. Phys.* **28**, 1556.

Yoshikawa, M., Maruyama, K., Saito, T., Maekawa, T., Takigawa, H. (1987) *J. Vac. Sci. Technol.* **A5**, 3052.

Yu, T.-C., Brebrick, R. F. (1993) *J. Phase Equilibria* **14**, 271.

Zanio, K. (1970) *J. Appl. Phys.* **41**, 1935.

Zanio, K. (1974) *J. Electron. Mater.* **3**, 327.

Zanio, K. (1978) *Semicond. Semimet.* **13**.

Zhao, S. N., Yang, C. Y., Huang, C., Yue, A. S. (1983) *J. Cryst. Growth* **65**, 370.

Part 4

Oxides and Halides

18 Phase-Diagram Study for Growing Electro-Optic Single Crystals

SHINTARO MIYAZAWA

SHINKOSHA Co. Ltd., 3-4-1 Kosugaya, Sakae-ku, Yokohama, Kanagawa 247-0007, Japan

ABSTRACT

A phase diagram of compounds is the 'compass' in growing single crystals with high quality, especially when crystal properties depend on crystal stoichiometry. A retrospective of how the congruent composition of $LiTaO_3$ was determined is presented in detail as a representative of practical electro-optic crystals. Re-examination of the phase diagram of quadratic electro-optic $Bi_{12}TiO_{20}$ is described as a representative of incongruently melting compounds for reproducible top-seeded solution growth (TSSG). From these two experimental approaches to the phase-relation studies, hints for establishing a detailed phase diagram of a crystal to be grown will be given.

18.1 INTRODUCTION

A large number of electro-optic oxides have been synthesized in the past two decades and most of their single crystals are now grown by the Czochralski (Cz) pulling technology. In Cz-growth, single crystals are usually grown from a melt by adding calcined ceramics with a weight equal to a previously grown crystal. The crystal quality required for the types of devices and its operational conditions and density have to be achieved by the proper control of the process parameters. In micro- and opto-electronics, the bulk crystal defines the device yield and performance.

Frequently a poor reproducibility of crystal quality and crystal composition is observed. Here it is quite important to know the causal relationship between a crystal/melt composition and material properties, i.e., the stoichiometry of a crystal to be grown, in order to grow single crystals reproducibly for practical applications. Although most stoichiometric compounds on phase diagrams are shown in general as a line compound without any solid-solution ranges, the most noticeable aspect in the stoichiometry is that the crystal composition does

Crystal Growth Technology, Edited by H. J. Scheel and T. Fukuda
© 2003 John Wiley & Sons, Ltd. ISBN: 0-471-49059-8

not coincide necessarily with its starting melt composition. For example, the stoichiometry of $LiNbO_3$ was well established (Prokhorov and Kuz'minov 1990) and that of GaAs was treated in relation to its semi-insulating property (Terashima *et al.* 1986, Miyazawa *et al.* 1988). In both cases, the phase-diagram study was a 'key' issue for reproducible growth of single crystals in industrial crystal-growth technology.

In this chapter, studies on phase relations of electro-optic $LiTaO_3$ and $Bi_{12}TiO_{20}$ are chosen as representatives of congruently and incongruently melting compounds, respectively. First, the determination of the congruently melting composition of $LiTaO_3$ will be described in detail. With the increasing importance of $LiTaO_3$ single crystals in device applications, there is specific interest in $LiTaO_3$ with sto-ichiometric composition, of which the optical/dielectric properties are different from crystals grown at congruent composition (Furukawa *et al.* 1999, Kitamura *et al.* 1999). Here, a retrospective of how the congruent point of $LiTaO_3$ was established (Miyazawa and Iwasaki 1971) can give hints for the growth of electro-optic single crystals of high optical quality. Secondly, we will focus on the quadratic electro-optic crystal $Bi_{12}TiO_{20}$, which melts incongruently, and the new phase relation for reproducible growth of single crystals (Miyazawa and Tabata 1998). $Bi_{12}TiO_{20}$ is still useful for electro-optic sensing devices (Shinagawa *et al.* 1998) and for photorefractive hologram-memory applications.

The following methods have been applied to establish phase diagrams for crystal growth:

   (i)   Differential thermal analysis (DTA);
  (ii)   Lattice constant measurement; X-ray diffractometry;
 (iii)   Chemical analysis of compositions and precision analysis of stoichiometry;
 (iv)   Physical property measurements;
  (v)   Optical property measurements;
 (vi)   Measurement of characteristic properties closely correlated with composition;
(vii)   Crystal-growth experiments.

In the case of $LiTaO_3$, the ferroelectric–paraelectric phase transition temperature, the Curie temperature (T_c), was quite useful, and therefore methods (vi) and (vii) were combined and method (v) was applied in order to verify the optical quality of $LiTaO_3$ grown at the congruent composition. In the case of $Bi_{12}TiO_{20}$, meth-ods (i) and (ii) were applied with topic (vii) because distinct properties related to composition are lacking.

18.2   PHASE-RELATION STUDY OF $LiTaO_3$

$LiTaO_3$ has long been considered one of the most promising materials for electro-optic device applications. More recently, in the new optical-fiber communication system called WDM (wavelength division multiplex) electro-optic $LiNbO_3$ crystals are used as external high bit-rate optical modulators in the form of optical waveguides. The isostructural $LiTaO_3$ could be an alternative and is widely used for SAW (surface-acoustic wave) devices. SAW applications require reproducible growth of large single crystals with the same acousto-optic properties and the congruent composition opened up its practical uses (Fukuda *et al.* 1979). However, optical-grade $LiTaO_3$ single crystals are not available commercially, which is an obstacle for its wider use.

In the case of $LiTaO_3$, the commercial DTA instruments cannot be used because of the high melting point of $LiTaO_3$ (about $1650\,°C$). However, the ferroelectric Curie temperature T_c sensitively depends on the composition as reported for the cases of $LiNbO_3$ (Bridenbaugh *et al.* 1970), $LiTaO_3$ (Ballman *et al.* 1967) and $Sr_xBa_{1-x}Nb_2O_6$ (Ballman and Brown 1967), and thus can be used to determine the composition of single crystals.

Earlier reported T_c values of $LiTaO_3$ are $665\,°C$ (Shapiro *et al.* 1964), $660 \pm 10\,°C$ (Levinstein *et al.* 1966), $630\,°C$ (Shapiro *et al.* 1969), $618\,°C$ (Glass 1968) and so on. In our experiments, the T_c of a single crystal grown from a stoichiometric melt $(Ta_2O_5/(Li_2O + Ta_2O_5) = 50.0$ in mol %) was usually measured as about $620\,°C$, while that of the calcined ceramics with the stoichiometric composition was about $660\,°C$. A practical phase relation of $LiTaO_3$ to explain this discrepancy has not been established so far, except for $LiNbO_3$.

18.2.1 PRELIMINARY STUDIES BY X-RAY DIFFRACTOMETRY

The congruently melting composition in the binary Li_2O-Ta_2O_5 system was determined by

(i) Comparison of the X-ray powder diffraction pattern of stoichiometric ceramics with that of a bulk polycrystal prepared by melting the stoichiometric ceramics at $\sim1650\,°C$ (calcined at $1450\,°C$ for 4 h). Results are shown in Figure 18.1, where (a) is the diffraction pattern of the calcined sample and (b) that of the bulk sample. The diffraction peaks at $2\Theta = 15.5°$, $33.5°$, $36.7°$ of sample B (indicated by arrows) are not found in (a). If the stoichiometric composition coincided with the congruently melting one, both diffraction patterns should be identical.

(ii) Comparison of the X-ray powder diffraction patterns of ceramics with different ratios of $Li_2O/Ta_2O_5 = 1/3$, $1/1$, $2/1$ and $3/1$ (calcined at $1450\,°C$ for 4 h) showed increasing intensities of diffraction peaks at $2\Theta = 15.5°$, $35.5°$ and $36.7°$ with increasing Li_2O/Ta_2O_5 ratio. On the other hand, these characteristic diffraction peaks were not observed for the sample with $Li_2O/Ta_2O_5 = 1/3$. Thus, the congruently melting composition does not

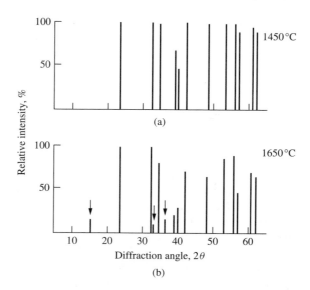

Figure 18.1 X-ray diffraction patterns of stoichiometric LiTaO$_3$; ceramics calcined at 1450 °C for 4 h (a), and bulk crystals prepared by melting the ceramics at ~1650 °C (b).

coincide with the stoichiometric one, and will be located on the Ta$_2$O$_5$-rich side of the phase diagram.

18.2.2 DETERMINATION OF THE CONGRUENTLY MELTING COMPOSITION

The phase relations in the Li$_2$O-Ta$_2$O$_5$ system in the vicinity of LiTaO$_3$ were determined by measuring the Curie temperatures of single crystals (sample A) pulled from melts with various Ta$_2$O$_5$/Li$_2$O ratios and those of the crystals picked up randomly from the solidified melts that remained in the same crucibles (sample B). Samples A were prepared from the first-grown part of the single-crystal boules, which corresponds to the primary crystallization. The composition of melts was varied from Ta$_2$O$_5$/(Li$_2$O + Ta$_2$O$_5$) = 49.0 to 54.25 mol %. The starting materials were prepared by mixing Li$_2$CO$_3$ and Ta$_2$O$_5$ (purities ≥ 99.99 %) in the required ratios, and the crystals were grown using a conventional r.f.-heating Cz-pulling apparatus with an Ir crucible (50 mm diameter, 50 mm depth and 1 mm wall thickness). Crystal pulling was performed along the [001] axis. The best pulling conditions were 5 mm/h of pulling rate, 40–60 rpm of seed rotation rate, temperature stability within 0.5 °C, ~80 °C/cm temperature gradient above the melt, and a N$_2$ atmosphere. The size of the grown crystals was 10–15 mm in diameter and 25–35 mm in length. The Curie temperatures were determined by measuring the temperature dependence of the dielectric constants at 10 kHz using an admittance bridge.

The measured Curie temperatures of samples A and B are plotted in Figure 18.2 as a function of the nominal starting melt composition. The Curie temperatures of samples A change almost linearly with the nominal melt composition, and the results agree well with Ballman *et al.* (1967). The filled circles represent the Curie temperatures of samples B and are dispersed over a wide temperature range. It is noted, however, that the Curie temperatures of samples B at $Ta_2O_5/Li_2O + Ta_2O_5 = 51.25$ mol % are dispersed over a relatively narrow temperature range. For a congruently melting composition, the Curie temperature of single crystals grown from this melt should coincide with that of crystals picked up from the residual melt solidified in a crucible, and thus Curie temperatures of samples B should not be dispersed. Therefore, it can be concluded that the congruently melting composition lies close to $Ta_2O_5/(Li_2O + Ta_2O_5) = 51.25$ mol %, which agrees well with the preliminary studies and T_c of 608 °C.

When melts of composition $Ta_2O_5/(Li_2O + Ta_2O_5) = 49.0$ mol % were cooled slowly, white precipitates, which were identified by X-ray diffraction as a mixture of $Ta_2O_5 \cdot 3Li_2O$ and $Ta_2O_5 \cdot Li_2O$, were observed in the crucible. Melts of composition $Ta_2O_5/(Li_2O + Ta_2O_5) = 54.0$ mol % precipitated a mixture of $Ta_2O_5 \cdot$

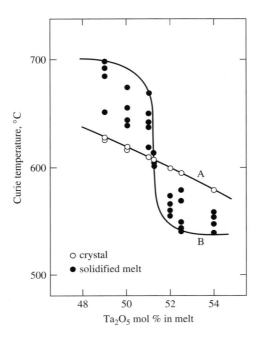

Figure 18.2 Curie-temperature variations as a function of the nominal starting melt composition; single crystal grown from melt (A), and bulk samples solidified in a crucible (B).

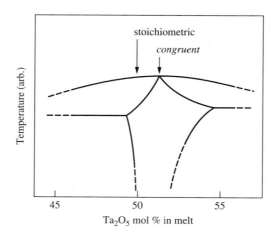

Figure 18.3   Detailed phase relation of Li_2O-Ta_2O_5 in the vicinity of $LiTaO_3$.

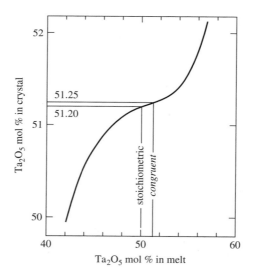

Figure 18.4   Crystal composition as a function of the nominal melt composition.

Li_2O and $3Ta_2O_5 \cdot Li_2O$ on slow cooling. From these experimental results, a tentative phase relation in the vicinity of $LiTaO_3$ is drawn in Figure 18.3, where the temperature scale is arbitrary. This was the first report on the stoichiometry of $LiTaO_3$ (Miyazawa and Iwasaki 1971). The compositional relation between the starting melt and the grown crystal is shown in Figure 18.4. The distribution coefficient k_o of $LiTaO_3$ at the stoichiometric melt is calculated to be about 1.02.

Lerner *et al.* (1968) investigated the binary system Li_2O-Nb_2O_5 around $LiNbO_3$ in detail, and found that the stoichiometric composition of $LiNbO_3$, $Li/Nb = 1.0$

in molar ratio, did not coincide with the congruently melting composition. Byer and Young (1970) grew LiNbO$_3$ single crystals of high optical homogeneity from the melt of the congruently melting composition, which was defined as Li/Nb = 0.941 in molar ratio. Recently, Chen *et al.* (1999) showed that the melt density of the congruent composition does not vary much with time, while that of the stoichiometric one increases by about 0.3 % from 3.588 g/cm^3 to a stable value of 3.600 g/cm^3 after 3 h of aging. This aging effect was explained by the existence of two phases, LiNbO$_3$ and Li$_3$NbO$_4$, before complete melting. Referring to this experimental result, the congruent melt of LiTaO$_3$ is also suitable for a practical growth technology.

18.2.3   OPTICAL QUALITY OF THE CONGRUENT LiTaO$_3$

A qualification of the optical quality may assist to verify the phase/stoichiometry of the crystal. Usually, the conoscopic observation of a thin plate normal to the optic axis of a uniaxial crystal in a polarizing microscope is convenient to characterize the optical homogeneity (Shubnikov 1960). Inhomogeneous birefringence will disturb the conoscopic figure. Conoscopic interference patterns of single-crystal boules were observed with a simple optical setup shown in Figure 18.5, where the laser beam is slightly divergent so that a conoscopic interference is observed. Specimens were cut from crystals grown from melts with compositions of Ta$_2$O$_5$/(Li$_2$O + Ta$_2$O$_5$) = 50.0, 51.25 and 52.5 mol %, and their z-faces were polished within λ/10 of flatness and 30 min of parallelism. The specimens were 15–17 mm in diameter and about 20 mm in length along the z-axis, the growth direction.

Three conoscopic figures are shown in Figure 18.6. The conoscopic figure of specimen (a), grown from the stoichiometric melt, seems to be slightly distorted, having hanks of veins, which are origins of inhomogeneous birefringence. The conoscopic figure (c) of a specimen grown from the melt of composition Ta$_2$O$_5$/(Li$_2$O + Ta$_2$O$_5$) = 52.5 mol % looks optically biaxial. It is assumed that the specimen (c) is more distorted than the specimen (a). These observations

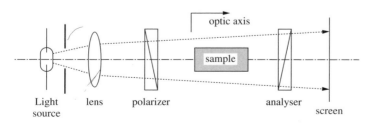

Figure 18.5   Schematic drawing of the apparatus used for observing the conoscopic interference pattern of a crystal sample of ~20 mm thickness.

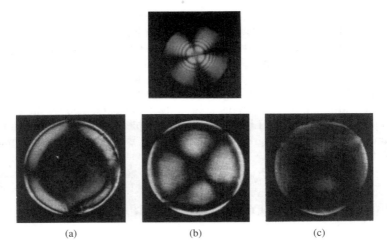

Figure 18.6 Conoscopic interference patterns of c-axis single crystals grown from melt with compositions of (a) 50 mol % Ta_2O_5; (b) 51.25 mol % Ta_2O_5; (c) 52.5 mol % Ta_2O_5. The inserted pattern is a typical microscopic conoscope pattern of a thin specimen without any inhomogeneities.

confirm that crystals grown from noncongruent melts have compositional variations and related birefringence variations along the pulling direction, the z-axis. On the other hand, the conoscopic figure (b) of a specimen grown from the congruent melt determined as $Ta_2O_5/(Li_2O + Ta_2O_5) = 51.25$ mol % shows an optically uniaxial pattern, quite similar to the inserted microscopic conoscopic pattern, which proves much lower optical inhomogeneity.

As single crystals grown from the congruent melt have no variation of composition through the crystal boule, as evidenced by identical Curie temperatures of the top and bottom portions (within an experimental error of ± 1 °C), they are expected to have high optical homogeneity. This was confirmed by measuring the dependence of the light extinction ratio (ER) on the light-beam diameter. Based on the theoretical approach (Sugibuchi *et al.* 1968, Miyazawa 1995a), ER can be expressed as

$$ER = r^2(\pi/\lambda)^2 \Delta_r n^2 l^2 \tag{18.1}$$

where r is a diameter of the beam, λ is a wavelength (μm), $\Delta_r n$ is a radial variation of refractive index in a beam, and l is a sample thickness. According to Equation (18.1), ER is proportional to r^2, when the refractive-index variation $\Delta_r n$ is constant.

The apparatus using crossed polarizers is shown in Figure 18.7. A specimen without antireflection coating was set so that its crystallographic z-axis makes an angle 45° to the polarization direction of an incident laser beam. Half-wave ($\lambda/2$) and quarter-wave ($\lambda/4$) plates were adjusted first to minimize the transmitted

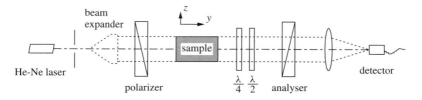

Figure 18.7 Schematic drawing of an apparatus used for measuring the light-extinction ratio.

light intensity and then to make it maximum, and the extinction ratio was defined as the ratio of the minimum to the maximum intensities. The specimens were annealed beforehand at 1100 °C for 4 h and poled by the field-cooling method, because as-grown crystals have ferroelectric multidomain structures that cause optical scattering (Miyazawa and Iwasaki 1971). Their dimensions were 9 mm wide along the y-axis, 15 mm long along the z-axis, and 7 mm thick for the x-axis, along which the light was transmitted.

The result is demonstrated in Figure 18.8, in which values of two different portions of the crystal were plotted. On increasing the laser beam diameter from 1 mm up to 6 mm, the light-extinction ratios decrease almost linearly. This linear tendency clearly obeys Equation (18.1). On the other hand, the ER of a crystal grown from the stoichiometric melt does not obey a linear relation of Equation (18.1), which indicates a relatively large birefringence variation in the boule as shown in Figure 18.6(a). A rapid decrease of the extinction ratios (in dB) with light-beam diameters larger than 6 mm is probably due to strain near the crystal surface. From the obtained extinction ratios, a natural birefringence variation in a crystal was estimated to be less than 8.9×10^{-6}/cm from the simple relation deduced from Equation (18.1)

$$\Delta_r n = (\lambda l / \pi)(r \sqrt{\text{ER}})^{-1} \tag{18.2}$$

Devices such as light modulators require birefringence variations of less than 5×10^{-5}/cm. Therefore, the congruent LiTaO$_3$ of this study may find practical applications.

18.2.4   CONCLUSION

The phase diagram in the vicinity of LiTaO$_3$ within the system Li$_2$O-Ta$_2$O$_5$ was investigated by measuring the ferroelectric Curie temperatures of the grown crystals and their melts with various compositions of Ta$_2$O$_5$/Li$_2$O. It was found that the congruently melting composition is located at about 51.25 mol % Ta$_2$O$_5$, and the composition of single crystals grown from the stoichiometric melt contained less than 51.25 mol % Ta$_2$O$_5$, thus the distribution coefficient of Ta$_2$O$_5$ was determined as 1.02.

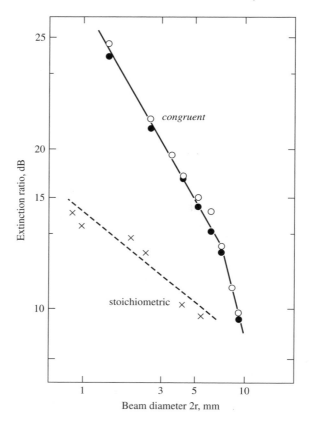

Figure 18.8 Measured light-extinction ratio as a function of beam diameter for crystals grown from congruent and from stoichiometric melts.

The optical homogeneity of the congruent crystals was proved to be much higher than those grown from stoichiometric melts, thus making them useful for practical devices such as SAW and optical waveguide devices.

18.3 PHASE-RELATION STUDY OF $Bi_{12}TiO_{20}$

Single-crystal $Bi_{12}TiO_{20}$ (BTO) has the largest electro-optic constant r_{41} among sillenites, making it useful for EOS (electro-optic sampling) probe heads (Shinagawa *et al.*, 1998) and also for photorefractive applications.

This section concerns firstly the phase relations in a Bi_2O_3-rich region near the stoichiometric $Bi_{12}TiO_{20}$ composition (14.28 mol % TiO_2) in the Bi_2O_3-TiO_2 binary, and secondly BTO crystal growth using the TSSG method. According to the Bi_2O_3-TiO_2 binary system reported by Bruton (1974), as shown on the left in Figure 18.9, BTO is a line compound melting incongruently at $873 \sim 875\,^{\circ}C$, thereby decomposing into liquid and $Bi_4Ti_3O_{12}$. Therefore, BTO single crystals are usually grown using TSSG from Bi_2O_3-rich solution of less than about 12 mol % TiO_2 (Bruton *et al.* 1974, Rytz *et al.* 1990, Okano *et al.* 1991, Miyazawa 1995b). However, a standard method for growing reproducibly BTO single crystals has not been established. Bruton's phase diagram was revisited experimentally, and crystal growth was carried out by a new technology, which suppresses a few problems in BTO crystal growth. Ultrahigh-purity raw materials of Bi_2O_3 (99.9995 %) and TiO_2 (99.995 %) were used throughout this work.

18.3.1 RE-EXAMINATION OF PHASE DIAGRAM

Differential thermal analysis (DTA) of binary compounds with different TiO_2 concentrations close to the stoichiometric composition was carried out combined with TSSG crystal growth in order to re-examine the peritectic reaction and the eutectic temperatures.

Figure 18.9 shows the essential part of the phase diagram determined by the DTA experiments. On the DTA curves of ceramics with different mol % TiO_2 and using heating and cooling rates of $1\,^{\circ}C/min$, strong exothermic and endothermic peaks were obtained, while peaks corresponding to the liquidus line were weak. The eutectic and peritectic temperatures were determined to be $823 \pm 3\,^{\circ}C$ and $855 \pm 2\,^{\circ}C$, and thus were higher and lower, respectively, than those reported by Bruton (1974) shown in Figure 18.9.

In order to verify the peritectic composition, TSSG-pulling was carried out from solutions containing 10.25 to 11.25 mol % TiO_2. Polycrystalline boules consisting of Bi_2O_3, BTO and $Bi_4Ti_3O_{12}$ were obtained from 11.00 mol % TiO_2 solutions, while single-phase BTO crystals were obtained from 10.50 mol % TiO_2 solutions. A boule grown from a 10.75 mol % TiO_2 solution consisted of a first-grown $Bi_4Ti_3O_{12}$ polycrystalline phase and a monocrystalline BTO phase with a lattice constant of 10.1743_9, which resulted from a gradual change in solution composition passing the peritectic composition along the liquidus curve. Therefore, the peritectic composition must lie close to the 10.75 mol % TiO_2 concentration.

18.3.2 LATTICE-CONSTANT VARIATIONS OF THE $Bi_{12}TiO_{20}$ PHASE

To examine the existence width of the BTO phase, mixed binary compounds with 13.30–14.70 mol % TiO_2 were calcined at $830 \pm 5\,^{\circ}C$ for 20 h in air, very close to the determined eutectic temperature described above, resulting in sufficient precipitation of the BTO phase. The lattice constants of the BTO phase in each ceramic sample were evaluated by X-ray powder diffractometry at room temperature with Si powder as reference by means of the Nelson–Riley approximation method. The accuracy was estimated to be better than $1 \times 10^{-4}\,\text{Å}$.

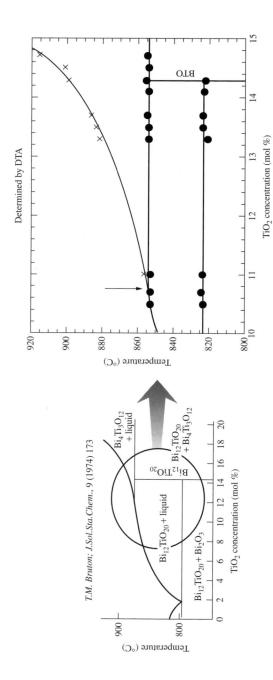

Figure 18.9 Section of revised phase diagram based on DTA and crystal-growth experiments. A crystal boule grown from the solution (10.75 mol %) contained $Bi_4Ti_3O_{12}$ phase and $Bi_{12}TiO_{20}$ single crystal (see text). The left diagram was reported by Bruton (1974).

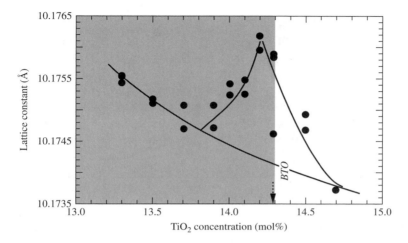

Figure 18.10 Lattice-constant variation of the BTO phase in calcined ceramics as a function of nominal TiO$_2$ concentration.

Figure 18.10 shows these lattice constants as a function of the nominal TiO$_2$ concentration (Miyazawa and Tabata 1998). X-ray diffraction patterns of samples with a TiO$_2$ of less than the stoichiometric 14.286 mol % TiO$_2$ (hypoperitectic region) showed the γ-Bi$_2$O$_3$ phase, while those of the samples exceeding 14.286 mol % TiO$_2$ (hyper-peritectic region) showed relatively small signals of Bi$_4$Ti$_3$O$_{12}$. The lattice constant of BTO decreased monotonously with TiO$_2$ concentration and seemed to reach a minimum at around 13.80 $\sim$ 13.85 mol % TiO$_2$. From this minimum, the lattice constant increased rapidly to a maximum value and again decreased through the stoichiometric compound (14.28$_6$ mol % TiO$_2$). This lattice-constant variation suggests the existence of a nonstoichiometric solid-solution range. Also, we noticed that the lattice constant of the stoichiometric compound was smaller than that of the BTO phase in the 14.20 mol % TiO$_2$ compound. Such a nonlinear relationship between the solution composition and the lattice constant of crystals was also discussed in Bi$_{12}$SiO$_{20}$ (BSO) (Brice *et al.* 1977), where the composition of a crystal grown from $x \sim 11.6$ melt composition in the formula of Bi$_x$SiO$_{1.5x+2}$ showed the minimum (Hill and Brice 1974).

According to the phase diagram shown in Figure 18.9, BTO crystallizes as the primary phase from a Bi$_2$O$_3$-rich solution in the hypoperitectic region and the crystal composition is independent of the solution compositions. However, a rapid increase of lattice constants in the hypoperitectic region shown in Figure 18.10 suggests that the stoichiometry of the primarily crystallized BTO must be influenced by the composition of the starting solution.

Then, single crystals were grown along the [100] using TSSG from solutions containing from 4.0 up to 10.75 mol % TiO$_2$. The growth apparatus and the essential growth conditions will be described later. Three to five single crystals

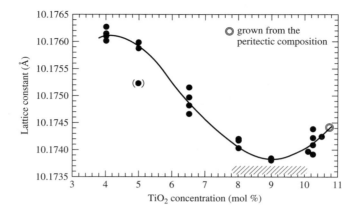

Figure 18.11 Lattice constant of BTO single crystals versus TiO_2 concentration in solution. Each crystal was grown from a solution by adding a stoichiometric BTO ceramics equal in weight to that of the previously grown crystal. The shaded bar indicates usable starting solutions for obtaining homogeneous crystals (see text).

weighing about 40 g, typically 15–20 mm in diameter and about 25 mm in length, were grown in succession from one starting solution. For successive growth from the initial solution, calcined stoichiometric BTO ceramics, equal in weight to that of the previously grown crystal, were added to the residual solution. The first crystallized fractions of the grown crystals were characterized by their lattice constants as described before.

Figure 18.11 shows that the lattice constants of the successively grown BTO single crystals decreased monotonically with the TiO_2 concentration to the minimum of crystals from a 9.0 mol % TiO_2 solution. Noticeably, the lattice constants of crystals grown from 8.0 and 10.25 mol % TiO_2 solutions are almost the same, and larger than those grown from a 9.0 mol % TiO_2 solution. This means that the crystal compositions grown from 8 and 10.25 mol % TiO_2 solutions must be very similar, suggesting that the solidus curve must be retrograde, regardless of point defects such as vacancies, oxygen deficiency and/or Bi_{Ti}^{3+}-antisites (Oberschmid 1985).

18.3.3   NEW PHASE DIAGRAM

From the above experimental results and considerations, the relevant part of the phase diagram in the hypoperitectic region close to the stoichiometric composition of BTO (14.28_6 mol % TiO_2) is drawn and shown in Figure 18.12 with a plausible retrograde solid solution close to stoichiometric BTO. We assume the turning point of the retrograde solidus curve to be around $13.80 \sim 13.85$ mol % TiO_2, which corresponds to 9 mol % TiO_2 of the liquidus line, because the minimum value in Figure 18.10 lies at around $13.80 \sim 13.85$ mol % TiO_2. This

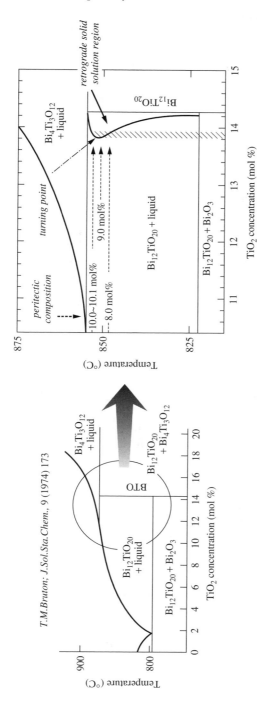

Figure 18.12 Section of revised Bi_2O_3-TiO_2 binary phase relation in the hypoperitectic region of BTO. From the solutions of 10.10–8.0 mol % TiO_2, the crystal composition lies in the shaded area, 13.85 ± 0.05 mol % TiO_2.

retrograde characteristic is verified when relatively long single crystals are grown as described later.

From Figure 18.11, we can expect high reproducibility of the crystal composition grown from the 9 mol % TiO_2 solution. Such BTO single crystals were applied for practical EOS devices (Shinagawa *et al.* 1998).

18.3.4 GROWTH OF LONG SINGLE CRYSTALS

In the case of TSSG-pulling, the starting solution composition changes gradually along the liquidus line as the crystal grows. The corresponding variation of the lattice constant was investigated by growing long single crystals.

The problem in TSSG-pulling of BTO is the evaporation of Bi (Okano *et al.* 1991). Also, the BTO crystals sometimes contain strong striations caused by temperature fluctuations in the solution. To eliminate both the evaporation of the Bi constituent and temperature fluctuations, a double-crucible configuration shown in Figure 18.13 was developed (Miyazawa 1995b). The outer Pt crucible was heated by r.f. induction. The inner Pt crucible was placed onto alumina powder in the outer heating crucible. This configuration avoided direct heating of the solution in the inner crucible, thus allowing suppression of Bi evaporation. In fact, evaporation of the Bi constituent was negligibly small even after 7 days of the growth run. Another expected advantage of this double-crucible configuration is the suppression of striations (Miyazawa 1995b). With this arrangement, a pulling rate of 0.25 mm/h and a crystal rotation rate of $10 \sim 35$ rpm, relatively long crystals were successively grown.

First, a 55 mm long, $\sim$15-mm diameter single crystal of 148-g weight was grown from a fresh 9.0 mol % TiO_2 solution of 373.72 g weight. Lattice constants at the top and bottom were 10.1738_7 and 10.1747_0 Å, respectively. The lattice constant of the primarily crystallized top corresponds well to that of crystals grown from the 9.0 mol % TiO_2 solution, and that of the bottom is close to that of a crystal grown from a 6.7 mol % TiO_2 solution, as is deduced from Figure 18.11. This tendency implies that the TiO_2 concentration in the solution gradually decreases along the liquidus line as the crystal grows, and the corresponding TiO_2 concentration in the crystal increases along the solidus line, resulting in an increase in the lattice constant, as shown in Figure 18.10.

Second, a 53-mm long crystal of 91 g weight was grown from a 10.10 mol % TiO_2 solution. Table 18.1 summarizes lattice constant and composition variations (Bi/Ti atomic ratio analyzed by ICP) along the crystal length at intervals of $5-7$ mm from the shoulder part. It is noted that the lattice constant varies according to the retrograde solidus line through the turning point, i.e. the crystal composition varies with a change in the solidus curve.

In referring to Figure 18.11, crystals grown from solutions with TiO_2 concentrations of between $\sim$7.5 and $\sim$10.0 mol % have lattice constant deviations of less than $\sim 1.0 \times 10^{-4}$ Å. Consequently, a starting solution with $10.0-10.10$ mol %

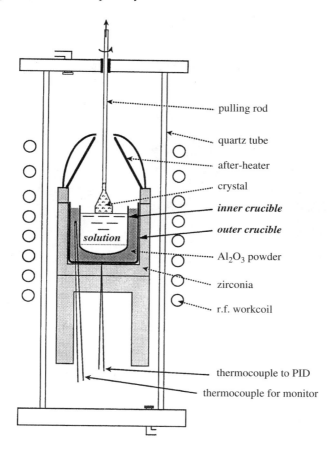

Figure 18.13 Schematic drawing of a double-crucible furnace to reduce Bi_2O_3 evaporation (Miyazawa 1995.).

Table 18.1 Lattice-constant variation in a large $Bi_{12}TiO_{20}$ crystal

Portion	Lattice constant (Å)	Bi/Ti atomic ratio (ICP)
(1)	10.1739_7	–
(2)	10.1736_6	12.13_3
(3)	10.1729_8	12.25_6
(4)	10.1736_3	12.29_7
(5)	10.1739_7	12.25_6
(6)	10.1744_3	12.21_5

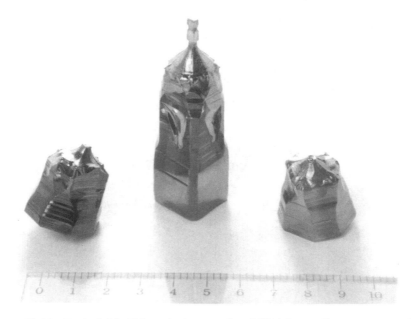

Figure 18.14 Typical $Bi_{12}TiO_{20}$ single crystals of 75 ± 2 mm diameter grown from $9.0-10.10$ mol % TiO_2 solutions.

TiO_2 is very practical for growing relatively homogeneous single crystals whose lattice constants are within $\pm 1 \times 10^{-4}$ Å over the whole crystal boule, when the solidified fraction is less than 45 %. Figure 18.14 shows crystals grown by these principles, which were used in practical EOS devices (Shinagawa *et al.* 1998).

18.3.5 CONCLUSION

For establishing a standard growth method of sillenite $Bi_{12}TiO_{20}$ (BTO) crystals, we revised the phase diagram of the hypo-eutectic Bi_2O_3-TiO_2 region and characterized the TSSG-grown BTO crystals by their lattice constants. This study proved the existence of a solid-solution region with a retrograde solidus curve. A starting solution with a $10.0-10.1$ mol % TiO_2 concentration and a solidified fraction of the crystal of less than about 45 % will result in a relatively homo-geneous single crystal whose lattice constant deviation is less than 1×10^{-4} Å. In addition, a new furnace configuration was designed that could suppress both evaporation of Bi_2O_3 and striations.

18.4 SUMMARY

The phase diagram is a 'compass' for growing single crystals with high crystalline quality. An important aspect in the phase relation is that the crystal composition does not coincide necessarily with its starting stoichiometric melt composition, that is, a 'congruently melting' composition does not coincide with a 'stoichiometric' composition. Therefore, a detailed phase diagram should be established in cases where a compound has an existence range showing solid solubility with one or both constituents.

This chapter reviewed how phase relations have been clarified of congruently melting $LiTaO_3$ and incongruent $Bi_{12}TiO_{20}$ as representatives of electro-optic single crystals. The phase relations of $LiTaO_3$ were first reported in 1971 (Miyazawa and Iwasaki). Since then the determined congruent composition is used widely in industrial production. The first experimental results showing a retrograde solid solution of incongruently melting $Bi_{12}TiO_{20}$ were presented.

These examples have shown that a detailed investigation of phase relations, complemented by systematic crystal-growth experiments, may be lengthy, but must be encouraged for the following reasons: First, crystals with great potential for applications may be grown reproducibly to best performance and sometimes may be developed only from sufficient knowledge of the phase relations. Secondly, defect control and important properties like electrical conductivity, optical absorption, and ferroelectric transition temperature depend sensitively on stoichiometry and thus on phase-relation-based growth processes. Finally, the production of high-quality crystals at highest yield is only possible when the growth conditions are scientifically derived from the phase relations.

ACKNOWLEDGMENT

The author would like to express his thanks to Dr H. J. Scheel for a critical reading this text.

REFERENCES

Ballman, A. A., Levinstein, H. J., Capio, C. D. and Brown, H., (1967) *J. Am. Ceram. Soc.* **50**, 657.
Ballman, A. A. and Brown, H., (1967) *J. Crystal Growth* **1**, 311.
Byer, R. L. and Young, J. F., (1970) *J. Appl. Phys.* **41**, 2320.
Brice, J. C., Hight, M. J., Hill, O. F. and Whiffin, P. A. C., (1977) *Philips Tech. Rev.* **37**, 250.
Bridenbaugh, P. M., Carruthers, J. R., Dziedric, J. M. and Nash, F. R., (1970) *Appl. Phys. Lett.* **17**, 104.
Bruton, T. M., (1974) *J. Solid State Chem.* **9**, 173.
Bruton, T. M., Brice, J. C., Hill, O. F. and Whiffin, P. A. C., (1974) *J. Cryst. Growth* **23**, 21.
Chen, X. M., Wang, Q., Wu, X. and Lu, K. Q., (1999) *J. Cryst. Growth* **204**, 163.
Fukuda, T., Matsumura, S., Hirano, H. and Ito, T., (1979) *J. Cryst. Growth* **46**, 179.
Furukawa, Y., Kitamura, K., Suzuki, E. and Niwa, K., (1999) *J. Cryst. Growth* **197**, 889.
Glass, A. M., (1968) *Phys. Rev.* **172**, 172.

Hill, O. F. and Brice, J. C., (1974) *J. Mater. Sci.* **9**, 1252.

Kitamura, K., Furukawa, Y., Niwa, K., Goplan, V. and Mitchell, T. E., (1999) *Appl. Phys. Lett.* **73**, 3073.

Lerner, P., Legras, C. and Dumas, J. P., (1968) *J. Cryst. Growth* **3/4**, 231.

Levinstein, H. J., Ballman, A. A. and Capio, C. D., (1966) *J. Appl. Phys.* **7**, 4585.

Miyazawa, S. and Iwasaki, H., (1971) *J. Cryst. Growth* **10**, 276.

Miyazawa, S., Watanabe, K., Osaka, J. and Ikuta, K., (1988) *Rev. Phys. Appl.* **23**, 727.

Miyazawa, S., in *KOUGAKU KESSHOU* (Optical Crystals, Japanese), BAIFUKAN, Tokyo, 1995a, 271.

Miyazawa, S., (1995b) *Opt. Mater.* **4**, 192.

Miyazawa, S. and Tabata, T., (1998) *J. Cryst. Growth* **191**, 512.

Oberschmid, R., (1985) *Phys. Stat. Sol. (a)* **89**, 263.

Okano, Y., Wada, H. and Miyazawa, S., (1991) *Jpn. J. Appl. Phys.* **30**, L1307.

Prokhorov, A. M. and Kuz'minov, Yu. S., (1990) in *Physics and Chemistry of Crystalline Lithium Niobate* (transl. from Russian by T. M. Pyankova and O. A. Zilbert), Adam Hilger, English edition.

Rytz, D., Wechsler, B. A., Nelson, C. C. and Kirby, K. W., (1990) *J. Cryst. Growth* **99**, 864.

Shapiro, Z. I., Fedulov, S. A. and Venevtsev, Yu. N., (1964) *Fiz. Tverd. Tela.* **6**, 316.

Shapiro, Z. I., Fedulov, S. A., Venevtsev, Yu. N. and Rigerman, I. G., (1969) *Sov. Phys.-Cryst.* **10**, 725.

Shinagawa, M., Nagatsuma, T. and Miyazawa, S., (1998) *IEEE Trans. Instrum. & Meas.* **47**, 235.

Shubnikov, A. V., (1960) in *Principles of Optical Crystallography* (transl. from Russian), Consultants Bureau, New York.

Sugibuchi, K., Tsuya, H. and Fujino, Y., (1968) *Appl. Phys. Lett.* **13**, 107.

Terashima, K., Nishio, J., Okada, A., Washizuka, S. and Watanabe, M., (1986) *J. Cryst. Growth* **79**, 46.

19 Melt Growth of Oxide Crystals for SAW, Piezoelectric, and Nonlinear-Optical Applications

KIYOSHI SHIMAMURA[a], TSUGUO FUKUDA[b], VALERY I. CHANI[c]

[a] *Kagami Memorial Laboratory for Materials Science and Technology, Waseda University 2-8-26 Nishiwaseda, Shinjuku 169-0051, Japan*
[b] *Institute of Multidisciplinary Research for Advanced Materials, Tohoku University 2-1-1 Katahira, Aoba-ku, Sendai 980–8577, Japan*
[c] *Materials Science and Engineering, McMaster University 1280 Main Street West, Hamilton, Ontario, Canada*

19.1 INTRODUCTION

Bulk oxide crystals are important for acoustic, piezoelectric and optical device applications. Growth methods used for commercial production of these materials are based on liquid–solid phase transitions and can be classified into melt- and flux-growth techniques depending on congruent or incongruent melting behavior. In the first case, melt and crystal compositions are almost equal. Therefore, in general, the segregation phenomenon is weak. In such a case, crystal-growth techniques based on simple solidification of isocompositional liquids are useful. Thus, selection of a suitable growth technique depends mainly on the phase diagram of the corresponding mixture (congruency, volatility, etc.) (Fukuda *et al.* 1998). The Czochralski (CZ) method (Figure 19.1) is the basic technique for mass production of large and perfect crystals which melt congruently.

The crystal formation of incongruently melting compounds is possible only from melts with composition differing from that of the crystal. In this case, the segregation phenomenon has to be studied precisely, because the main problems of such processes are determined by the incorporation rates of all cations into the crystal. Although the top-seeded solution growth (TSSG) technique allows the growth of such crystals, the reproducibility of TSSG growth and the properties of the crystals grown are much lower than those of the CZ method.

Single-crystal fibers have become the subject of intense study in recent years because of their remarkable characteristics (Feigelson 1986). We have developed a superior growth technique, the micro-pulling-down (μ-PD) method. Figure 19.2

Crystal Growth Technology, Edited by H. J. Scheel and T. Fukuda
© 2003 John Wiley & Sons, Ltd. ISBN: 0-471-49059-8

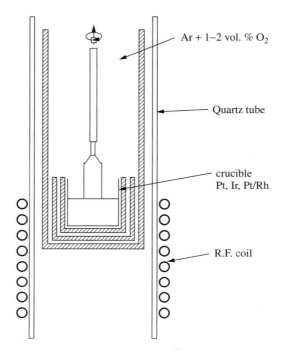

Figure 19.1 Schematic of the CZ technique.

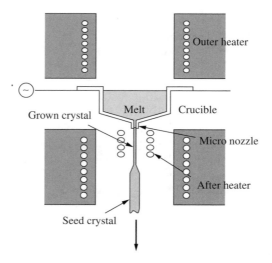

Figure 19.2 Schematic of the μ-PD technique.

shows a schematic of the μ-PD method. This method enables us to fabricate new excellent fibers of oxides, semiconductors and ceramic-matrix composites with high melting temperatures and/or incongruent-melting composition. Fiber crystals with a small diameter in the μm to mm range grown by the μ-PD method exhibit characteristic features for a wide field of advanced applications (Chani *et al.* 1998(a); Chani *et al.* 1998(b); Chani *et al.* 1999; Imai *et al.* 1997).

The development of new nonlinear-optical crystals also has become of considerable interest because of the progress of optoelectronic technologies. Fiber-form single-crystalline materials are of particular interest because of their unique characteristics, such as compact size, higher doping concentration, and, in some cases, novel properties due to their hybrid structure. The μ-PD method has an important role in this field and gives us new excellent optical fiber crystals.

Although there are alternative techniques to grow fiber single crystals, the μ-PD method has several advantages including high growth stability and growth of small-sized uniform crystals free from degradation. Moreover, this technique allows fabrication of single crystals of incongruently melting materials from melts that contain small (5–10 mol %) amounts of flux.

19.2   LiTaO₃ FOR SAW DEVICES

$LiNbO_3$ (LN) and $LiTaO_3$ (LT) single crystals are preferred for current electric and optical devices (Ballman 1965; Bordui *et al.* 1991, 1995; Byer *et al.* 1970; Carruthers *et al.* 1971; Fukuda *et al.* 1981; Matthias and Remeika 1949; Miyazawa and Iwasaki 1971). Recently LN and LT achieved a large share of the market, next to quartz single crystals, as materials for surface-acoustic-wave (SAW) devices. Research to industrialize LN and LT started in 1960. In 1965 these crystals with 'cm' sizes were grown at about the same time in the USA and Russia with the CZ technique. After this success, investigations to grow large-size and high-quality crystals for industrial applications have started. From the early 1970s, these crystals have come into prominence as a material for IF filters for color TV. At that time, intensive research on these materials for applications in SAW devices was started. Later, a lot of effort was put into growing large-size, high-quality, and low-cost LN and LT. In the 1990s, LT became the most important candidate for the new digital communication technology including digital TV such as CATV, CS and BS, handheld phones, etc. SAW devices made with LT are promising for next-generation devices operating in the high-frequency range around 2 GHz. Table 19.1 shows the main areas of applications for SAW devices made of different materials. It can be seen that LT is more advantageous for advanced technologies than LN.

Figure 19.3 shows the current trend of SAW crystals. The ideal material should have both a small frequency-temperature coefficient and a large electromechanical coupling factor. This means that crystals close to the upper-right corner would be ideal and thus explains the rapidly increasing market for LT.

Table 19.1 Material selection for SAW devices

Material	Main areas of application
LiTaO$_3$	TV-IF filter
(X-112°Y)	UHF filter for BS receiver
	VHF oscillator for RF modulator
	Vestigial sideband filter
	Medium bandwidth filter
LiTaO$_3$	SAW filter for cordless telephone and handheld phone
(36°Y-X)	Wide bandwidth filter
LiNbO$_3$	TV-IF filter
(128°Y-X)	UHF filter for BS receiver
	Vestigial sideband filter
	Wide bandwidth filter
LiNbO$_3$	SAW filter for handheld phone
(64°Y-X)	Wide bandwidth filter
Li$_2$B$_4$O$_7$	SAW resonator
(45°X-Z)	SAW filter for pagers and cordless telephone
	Narrow bandwidth filter
Quartz	Oscillator, stable device
(ST-X)	SAW filter for pager and handheld phone
	Narrow bandwidth filter

Although LN growth has been very well investigated for a long time, this is not the case with LT: the knowledge about the precise congruent-melting composition and the solid-solution range is lacking. However, in view of recent requirements for LT, a large effort has started to be paid in the growth of LT single crystals.

Figure 19.4 shows the diameter trend of SAW wafers made of LN and LT. The increase of diameters of LT from 2 inches to 4 inches was rather rapid, in almost 10 years. However, after 1985, the size has remained at 4 inches. The realization of 5-inch LT is now under investigation.

For the growth of LT single crystals, Ir or Pt/Rh crucibles are normally used. Since the melting temperature of LT is reported to be 1562 °C, Pt crucibles cannot be used. When LT single crystals are grown using a Pt/Rh crucible in an air atmosphere, the crystals are colored brown because of a very small amount of Rh from the crucible material. However, this coloration and the existence of Rh in LT does not affect the SAW properties. The general growth conditions for production are: pulling rate 5–7 mm/h, crystal rotation 8–10 rpm, the ratio of crystal and crucible diameter 0.6–0.7, and crystal length 80–120 mm. Although these conditions for LT are almost the same as for LN, the average temperature gradient for LT between the melt surface and 5 mm above is different from that used for LN growth. Figure 19.5 shows as-grown LN single crystals.

LT has a rather wide solid-solution range, and the stoichiometric composition and the congruent-melt composition are different. There have been several

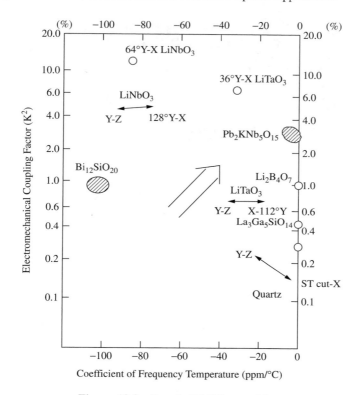

Figure 19.3   Trend of SAW materials.

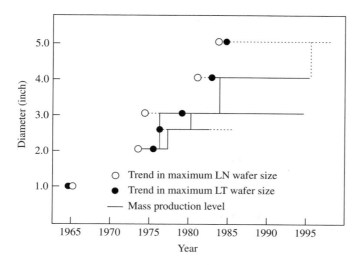

Figure 19.4   Diameter trend of SAW wafers.

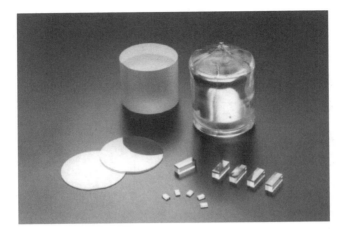

Figure 19.5   View of 4-inch LN single crystals (NGK Insulators, Ltd.)

reports that the congruent-melt composition of LT is around 48.7 Li_2O mol %, and recent investigations estimate it to be 48.3–48.4 Li_2O mol %. Knowing the exact congruent composition is very important for reproducible growth of LT single crystals with high uniformity. One way to find out is to measure the Curie temperature (T_c) which has an almost linear dependence with the composition.

For mass production, a slightly Li_2O-enriched composition is used in order to compensate for the evaporation loss during the growth process. Recent commercially available LT single crystals are guaranteed to have compositional variation among crystal boules within ±0.12 Li_2O mol %. This corresponds to a T_C variation within ±2.5 °C, which satisfies the SAW velocity variation within ±1 m/s.

Other problems observed during the growth of LT are unstable growth such as spiral phenomena and cracking. At temperature gradients smaller than 30 °C/cm, spirals are sometimes observed. However, if the temperature gradient is steep, cracking occurs. Therefore, the heating zone should be well designed, which in most cases is based on experience. Further investigation in this field is expected.

The required LT wafer specifications for the current SAW devices are: diameter 3–4 inches, mirror surface should be flat within 5 μm, and the roughness of the back surface should be less than 0.2–2.2 μm. The flatness of the back surface is also important because it causes diffuse reflection and thus reduces efficiency.

19.3 LANGASITE-FAMILY CRYSTALS FOR PIEZOELECTRIC APPLICATIONS

Progress in electronic technology requires new piezoelectric crystals with high thermal frequency stability and large electromechanical coupling factors. In the design of devices such as filters with a wide bandwidth, high stability, and small insertion loss, crystals with properties intermediate between those of quartz and LT are required.

Recently, langasite ($La_3Ga_5SiO_{14}$; LGS), and its aliovalent analogues, $La_3Nb_{0.5}Ga_{5.5}O_{14}$ (LNG) and $La_3Ta_{0.5}Ga_{5.5}O_{14}$ (LTG) single crystals, have received attention as new candidates for piezoelectric applications (Detaint et al. 1994; Kawanaka et al. 1998; Shimamura et al. 1996; Silvestrova et al. 1993; Takeda et al. 1996).

The LGS crystal has a $Ca_3Ga_2Ge_4O_{14}$-type structure with the space group P321. There are four kinds of cation sites in this structure. Therefore, this structure can be described by the chemical formula $A_3BC_3D_2O_{14}$. In this formula, A and B represent the decahedral (twisted Thomson cube) site coordinated by 8 oxygen, and an octahedral site coordinated by 6 oxygen, respectively. C and D represent tetrahedral sites coordinated by 4 oxygens; the size of the C site is slightly larger than that of D site. Different isovalent and aliovalent substitutions in a given structure are quite interesting in themselves, and could, perhaps, also result in useful physical properties (optical, piezoelectric, etc).

Single crystals were grown with a conventional RF-heating CZ technique with a Pt crucible (50 mm in diameter and height) or an Ir crucible (100 mm in diameter and height). The growth atmosphere was either air or a mixture of Ar plus 1 or 2 vol % O_2, in order to avoid the evaporation of gallium suboxide from the melt during growth. The chemical composition of the grown crystals was measured by quantitative X-ray fluorescence analysis.

For the characterization of piezoelectric properties, we actually made resonators and filters. Wafer-like samples ($9 \times 9 \times 1$ mm) were cut from grown crystals perpendicular to the X-axis, which corresponds to the [100] crystallographic direction. Using filters and resonators, electromechanical coupling factors were measured by the resonance method.

High-quality LGS, LNG and LTG single crystals were grown by the CZ technique. Figure 19.6 shows the as-grown LTG single crystals of approximately 2 and 3 inch diameter. These crystals were obtained at pulling and rotation rates of 1 mm/h and 10 rpm, respectively. Grown crystals were free of cracks and inclusions.

Lattice parameter and concentration of each oxide, i.e. La_2O_3, Ga_2O_3, SiO_2, Nb_2O_5 and Ta_2O_5, were almost constant within the estimated errors throughout the crystal boule. The uniformity of lattice parameter and chemical composition suggests that the stoichiometric compositions of the three compounds are close to the congruently melting compositions. This agrees with the fact that many cracks and inclusions were found in crystals grown from starting melts of which the compositions varied from the stoichiometric one.

Figure 19.7 shows the equivalent series resistance versus vibration modes of resonators of LGS compared with that of quartz. The equivalent series resistance of the LGS resonator was small even in the higher-order overtone mode, compared with that of quartz. The quartz resonator cannot oscillate in higher

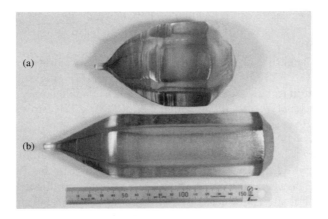

Figure 19.6 $La_3Ta_{0.5}Ga_{5.5}O_{14}$ single crystals of (a) 3 and (b) 2 inch diameter.

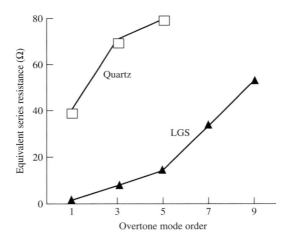

Figure 19.7 Equivalent series resistance versus vibration mode of resonators using LGS and quartz single crystals.

overtone mode because of the large equivalent series resistance, whereas the LGS resonator, even with a rough surface, is suitable for higher-order overtone use. The frequency deviation in the temperature range from −20 to 70 °C is about 100–150 ppm. The equivalent series resistance is almost constant in this temperature range.

Using LGS single crystals, we made monolithic-type filters (10.4 and 21.4 MHz) as seen in Figure 19.8. The electrical properties of these filters include low input and output impedance, small size and low attenuation, compared with filters made of quartz (see Table 19.2).

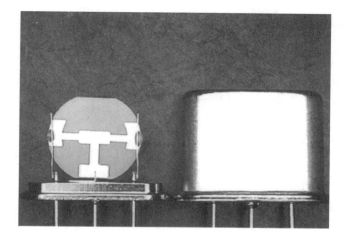

Figure 19.8   21.4 MHz filter made of LGS (size $6.9 \times 5.8 \times 2.2\,\mu m^3$).

Table 19.2   Electrical properties of LGS monolithic-type filters compared with quartz

Frequency (MHz)	Material	Passband width (3 dB) (kHz)	Attenuation (dB)	Input–output impedance (kΩ//pF)
10.7	Quartz	3 ± 15	2.5	5.5 k//-1.0
	LGS	3 ± 15	2.0	2.5 k//1.7
	Quartz	3 ± 15	2.0	2.0 k//0.5
21.4	LGS	3 ± 15	2.0	800//4.0
	LGS	3 ± 30	2.0	1.3 k//1.7

A 71-MHz-wide bandwidth SMD filter for a GSM base station was also made, which exhibited superior properties. The main electrical characteristics are:

(1)   Low input and output impedance ($940\,\Omega$//0.5 pF, versus several kΩ for quartz filters);
(2)   Interstage requires only capacitors (for quartz filters, a transfer is required);
(3)   Electrode gap may be wide (approximately $100\,\mu m$, versus several μm for quartz filters).

Because of these advantages, smaller and lighter filters, requiring less adjustment, can be achieved with LGS than with conventional discrete quartz filters. Actually, the volume and weight of a 71-MHz LGS filter are 1/5 and 1/3 of a quartz filter respectively.

In addition, more productive filters requiring relatively lower accuracy of electrode dimensions can be achieved. A 4-pole filter made of 2-pole filters in cascade could obtain characteristics such as 71 MHz center frequency, 1.6 dB insertion

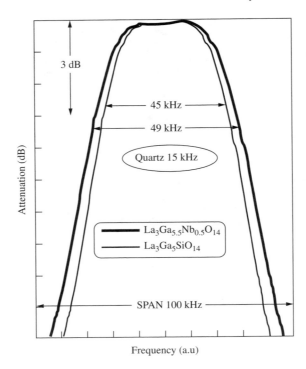

Figure 19.9 Bandwidth characteristics of filters made of LNG single crystals, compared with those of quartz and LGS.

loss, ± 89 kHz bandwidth at 3 dB, and 1.1 ms group delay distortion. These characteristics are almost the same as those of current discrete-type filters.

Figure 19.9 shows the bandwidth characteristic of LNG filter those of, compared to quartz and LGS ones. The bandwidth of the LNG filter was not only three times wider than that of quartz one, but also was a little wider than LGS one. This shows that LNG is superior to quartz and LGS for filter-device applications.

Since LNG, LTG and LGS have no phase transitions from room temperature to the melting point, low melting points, and higher hardness than LT and quartz, the growth of high-quality crystals with easy processing can be expected. Furthermore, LNG, LTG and LGS are not ferroelectric and thus do not require poling, and they are not easily soluble in any acid or base, these being further advantages. Their electromechanical coupling factors k and thermal frequency stability are between those of LT and quartz. The above characteristics indicate that LGS-family crystals are more promising for piezoelectric devices than lithium tetraborate ($Li_2B_4O_7$), which has similar piezoelectric properties (Detaint *et al.* 1994). $Li_2B_4O_7$ is easily soluble in all acids and bases, and is not stable in air.

Figure 19.10 shows the piezoelectric modulus $|d_{11}|$ increasing with the lattice parameter a of langasite-type crystals. This can be explained as follows:

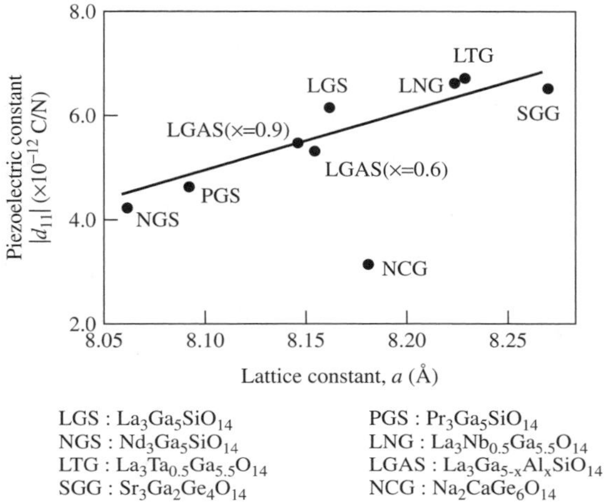
8.0
6.0
4.0
2.0
8.05
8.10
8.15
8.20
8.25
Piezoelectric constant
$|d_{11}|$ ($\times10^{-12}$ C/N)
Lattice constant, a (Å)
LTG
LGS
LNG
SGG
LGAS(x=0.9)
LGAS(x=0.6)
PGS
NGS
NCG
LGS : $La_3Ga_5SiO_{14}$
NGS : $Nd_3Ga_5SiO_{14}$
LTG : $La_3Ta_{0.5}Ga_{5.5}O_{14}$
SGG : $Sr_3Ga_2Ge_4O_{14}$
PGS : $Pr_3Ga_5SiO_{14}$
LNG : $La_3Nb_{0.5}Ga_{5.5}O_{14}$
LGAS : $La_3Ga_{5-x}Al_xSiO_{14}$
NCG : $Na_2CaGe_6O_{14}$

Figure 19.10 Dependence of piezoelectric constant on the lattice parameter.

increasing the lattice parameter means enlarging the size of the polyhedron composed of a cation and the nearest oxygens. By such an enlargement, the cations can move more easily within the crystal. Thus, even if the applied stress is relatively small the polarization could increase easily.

The piezoelectric constants of LNG and LTG are the largest of the langasite-type crystals studied. Therefore, we conclude that substitution of the B site is most effective for improvement of the piezoelectric properties. This tendency serves as a guide for the development of improved compounds with langasite-type structure.

19.4 NONLINEAR-OPTICAL CRYSTALS FOR BLUE SHG

Various nonlinear-optical niobate crystals are excellent for second-harmonic generation (SHG) of Nd:YAG and near-infrared GaAlAs lasers (Reid 1993).

$KNbO_3$ (KN) with a perovskite structure has large optical nonlinearity. $K_3Li_2Nb_5O_{15}$ (KLN) crystals with a tungsten-bronze structure have higher mechanical and chemical stability. However, since they melt incongruently, sufficiently large crystals could not be grown with the conventional crystal-growth techniques such as TSSG. The μ-PD technique has been applied to grow those fiber single crystals (Chani *et al.* 1999; Imai *et al.* 1997).

For the growth of oxide fibers with a comparatively low melting temperature (<1550 °C), the μ-PD method with Pt crucibles and resistive heating were used. Raw materials were melted in a crucible with a micronozzle at the bottom. Subsequently, the molten material was passed through the micronozzle to form a rod-shaped fiber single crystal.

Figure 19.11   View of as-grown KLN fiber crystal (scale in mm).

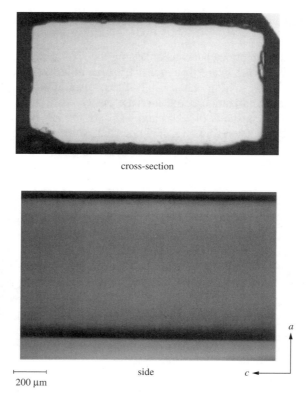

cross-section

side

a

c

200 µm

Figure 19.12   Microscopic image of KLN fiber crystal.

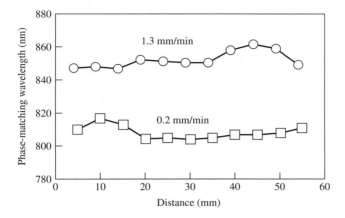

Figure 19.13   Variation of phase-matching wavelength for the KLN fiber grown along the ⟨100⟩ direction with pulling down rates of 1.3 mm/min and 0.2 mm/min.

Inclusion- and crack-free KNbO$_3$ fiber crystals can be grown with the μ-PD technique from melts enriched with K$_2$O.

The KLN fiber crystals were grown on KLN ⟨100⟩ in air atmosphere with the pulling rate ranging from 0.1 to 1.0 mm/min. A view of the as-grown KLN crystal is given in Figure 19.11, and its microscopic image is given in Figure 19.12. The {100} facets are larger than those of {001}.

Figure 19.13 shows the phase-matching wavelength along the growth axis. The deviation of the phase-matching wavelength for both crystals is almost the same, about 10 nm, which indicates that the grown crystals have high compositional uniformity. This is probably because natural convection was restricted in the micronozzle and the effective distribution coefficient of each constituent of the KLN melt approaches unity due to high growth rates.

Figure 19.14 demonstrates that the phase-matching wavelength of KLN can be controlled by the pull rate in the μ-PD process. With increased pulling-down rate the Li/Nb ratio decreases, and the phase-matching wavelength becomes longer. Figure 19.15 shows the dependence of the NCPM (noncritical phase-matching) wavelength on the Nb$_2$O$_5$ molar fraction. When the concentration of Nb$_2$O$_5$ increases, the NCPM wavelength becomes longer. These results show that we can control the optical property of KLN easily and still maintain the quality of the crystals. This is very important for actual device applications. The fiber growth process is economic and suitable to develop new materials and measure their properties.

19.5   SUMMARY

The crystals described have a significant potential in future for acoustic, piezo-electric and optical applications. Although significant progress has been made

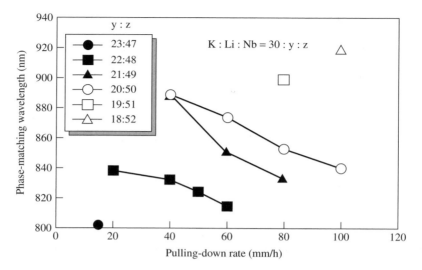

Figure 19.14 Relation between the phase-matching wavelength, pulling-down rate and Li:Nb ratio.

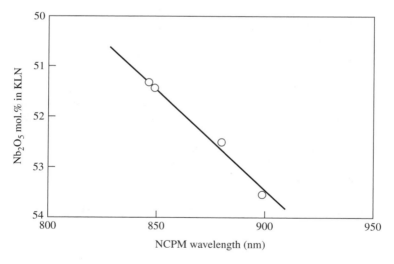

Figure 19.15 Relation between the NCPM wavelength and the Nb_2O_5 molar fraction.

in recent years in the bulk, thin-film and fiber growth techniques to prepare the single crystals described in this report, there is considerable scope for the development of new methods and new materials for next-generation devices (Fukuda *et al.* 1998).

REFERENCES

Ballman, A. A., (1965) *J. Am. Ceram. Soc.* **48**, 112.

Bordui, P. F., Norwood, R. G., Bird, C. D., Calvert, G. D., (1991) *J. Cryst. Growth* **113**, 61.

Bordui, P. F., Norwood, R. G., Bird, C. D., Carella, J. T., (1995) *J. Appl. Phys.* **678**, 4647.

Byer, R. L., Young, J. F., Feigelson, R. S., (1970) *J. Appl. Phys.* **41**, 2320.

Carruthers, J. R., Pererson, G. E., Grasso, M., (1971) *J. Appl. Phys.* **42**, 1846.

Chani, V. I., Nagata, K., Imaeda, M., Fukuda, T., (1998a) *Ferroelectrics* **218**, 9.

Chani, V. I., Nagata, K., Kawaguchi, T., Imaeda, M., Fukuda, T., (1998b) *J. Cryst. Growth* **194**, 374.

Chani, V. I., Shimamura, K., Fukuda, T., (1999) *Cryst. Res. Technol.* **34**, 519.

Detaint, J., Schwartzel, J., Zarka, A., Capelle, B., Philippot, E., *Proc. 1994 IEEE Intern. Frequency Control Symp.* p. 58.

Feigelson, R. S., (1986) *J. Cryst. Growth* **79**, 669.

Fukuda, T., Matsumura, S., Yasuami, S., Hirano, H., Fukuda, T., (1981) *Jpn. J. Appl. Phys.* **20**, 159.

Fukuda, T., Chani, V. I., Shimamura, K., in *Recent Development in Bulk Crystal Growth* ed. M. Isshiki, Research Signpost, India, 1998, pp. 191–229.

Imai, K., Imaeda, M., Uda, S., Taniuchi, T., Fukuda, T., (1997) *J. Cryst. Growth* **177**, 79.

Kawanaka, H., Takeda, H., Shimamura, K., Fukuda, T., (1998) *J. Cryst. Growth* **183**, 274.

Matthias, B. T., Remeika, J. P., (1949) *Phys. Rev.* **76**, 1886.

Miyazawa, S., Iwasaki, H., (1971) *J. Cryst. Growth* **10**, 276.

Reid, J. J. E., (1993) *Appl. Phys. Lett.* **62**, 19.

Shimamura, K., Takeda, H., Kohno, T., Fukuda, T., (1996) *J. Cryst. Growth* **163**, 388.

Silvestrova, I. M., Pisarevsky, Yu. V., Bezdelkin, V. V., Senyushenkov, P. A., 1993 *IEEE Intern. Frequency Control Symp.* p. 351.

Takeda, H., Shimamura, K., Kohno, T., Fukuda, T., (1996) *J. Cryst. Growth* **169**, 503.

20 Growth of Nonlinear-optical Crystals for Laser-frequency Conversion

TAKATOMO SASAKI, YUSUKE MORI,
and MASASHI YOSHIMURA
Department of Electrical Engineering, Graduate School of Engineering, Osaka
University 2-1 Yamada-oka, Suita, 565–0871, Osaka, Japan

20.1 INTRODUCTION

Nonlinear-optical (NLO) crystals are important for laser-frequency conversion. We will introduce the growth of KDP and of organic NLO crystals LAP and DAST from low-temperature solutions, and KTP and CLBO grown from high-temperature solutions, all to large size of several centimeters.

20.2 CRYSTALS GROWN FROM LOW-TEMPERATURE SOLUTIONS

20.2.1 GROWTH OF LARGE KDP (POTASSIUM DIHYDROGEN PHOSPHATE) CRYSTALS OF IMPROVED LASER-DAMAGE THRESHOLD

KDP crystal is suitable for second- and third-harmonic generation of large glass-laser systems for fusion experiments because of easy crystal growth and its relatively high laser-damage threshold.

Twelve years ago we grew crystals with $40 \times 40 \times 70 \, \text{cm}^3$ size for laser-fusion experiments [1]. An example is shown in Figure 20.1. The maximum growth rate in the c-direction was $1–2 \, \text{mm/day}$, and it took about one year to grow a crystal of that size from a $1.5 \, \text{m}^3$ solution within a growth vessel shown in Figure 20.2. Supersaturation was achieved by slow cooling (approximately from $60 \, ^\circ\text{C}$ to room temperature). It was important to prevent spontaneous nucleation during the long growth term: spurious crystals on the wall and at the bottom of the vessel stopped the growth of the large crystals. By cleaning the solution we could reduce spurious crystals. First, almost all crystals cracked after three or four months from the start of the growth when the pyramidal sectors of the crystal were set upward as shown in Figure 20.3(a). Then we found that after three to four months small spurious crystals settled on the pyramidal sectors and grew independently from the original main crystal, thereby causing the fatal

Crystal Growth Technology, Edited by H. J. Scheel and T. Fukuda
© 2003 John Wiley & Sons, Ltd.   ISBN: 0-471-49059-8

Figure 20.1   KDP crystals (size $45 \times 45 \times 70 \, \text{cm}^3$, growth period one year).

Figure 20.2   Growth vessel for large KDP crystal (slow-cooling method).

crack. By setting KDP downward as shown in Figure 20.3(b) we obtained the large crack-free crystals shown in Figure 20.1.

Recently, the growth of large crystal (size $45 \times 45 \times 45 \, \text{cm}^3$) from a seed of $1 \, \text{cm}^3$ within one month was reported [2, 3], this corresponding to a fast growth rate of 15 mm/day along the c-axis. In this case the solution temperature was held

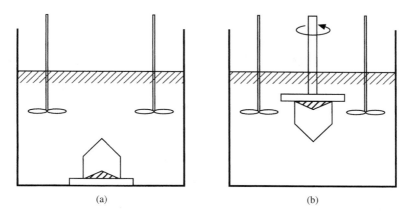

(a)   (b)

Figure 20.3 The schematics of the growth vessel. (a) pyramidal sectors set upward and (b) pyramidal sectors set downward.

for a long time significantly above the saturation temperature, which drastically reduced the rate of cluster generation and spontaneous nucleation. This facilitates the growth of large crack-free crystals.

In an attempt to increase the threshold for laser damage we found the harmful effect of organic impurities and of microbe fractions [4, 5]. When organic traces were reduced by photochemical reactions with the combination of UV irradiation and oxidants (H_2O_2), in addition to conventional ultrafiltration UF, the three times higher laser-damage threshold of 20–22 J/cm was achieved. This is shown in Figure 20.4 with the samples A, B, C grown in conventional UF-cleaned solutions, whereas samples D and E were grown from solutions with additional UV/oxidant-cleaning. Figure 20.5 shows the correlation of total organic carbon TOC with the threshold of laser damage.

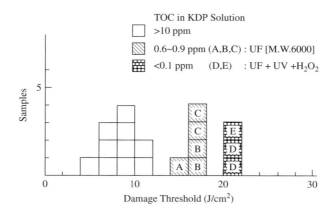

Figure 20.4 Histogram of bulk laser-damage threshold measurement of KDP crystals.

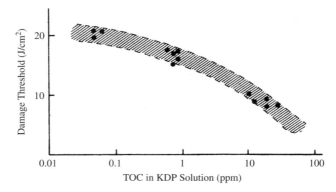

Figure 20.5 Relation between damage threshold of KDP crystals and TOC in the growth solution.

20.2.2   GROWTH AND CHARACTERIZATION OF ORGANIC NLO CRYSTALS

In the past twenty years there has been increasing interest in organic nonlinear optical (NLO) materials because of their large second- or third-order hyperpolarizabilities compared to inorganic NLO materials. It was difficult to realize devices because most organic crystals are too soft to cut and polish. We have found that organic crystals with ionic bonding and organo-metallic crystals are preferable for applications because of their higher bond strengths. Here the growth and characterization of two typical ionic crystals, LAP (L-arginine phosphate monohydrate) [6] and DAST (4-dimethylamino-N-methyl-4-stilbazolium tosylate) [7] are reported.

20.2.2.1   Growth and Stimulated Brillouin Scattering (SBS) properties of LAP crystal

LAP crystals were grown from a slowly cooled aqueous solution, which was prepared by mixing orthophosphoric acid and L-arginine in water. The solutions were filtered through a 0.2 μm membrane before starting crystal growth. Figure 20.6 shows the LAP crystal (size $7 \times 8 \times 18\,\text{cm}^3$, $a \times b \times c$) which had an extremely high bulk laser-damage threshold [8] and was thought to be good for frequency conversion of high-power lasers. Recently it was assumed that the high damage threshold is due to the stimulated-Brillouin-scattering (SBS) effect [9]. The evaluated organic nonlinear crystal DLAP (deuterated LAP) is a new solid-state phase conjugator (PC), which showed a low SBS threshold of 1.6 mJ at 1064 nm, high reflectivity of about 80 %, and stable performance without damage. DLAP crystals, also grown from a solution, were synthesized by a similar reaction as LAP in D_2O. In contrast to toxic liquids or high-pressure gases commonly used

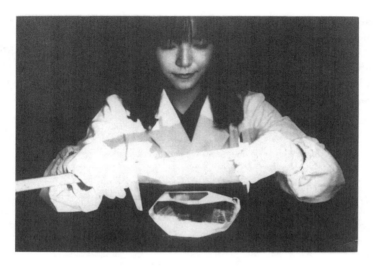

Figure 20.6 Photograph of LAP crystal with dimensions of $7 \times 8 \times 18\,\text{cm}^3$.

for SBS-PC, DLAP is harmless for environment and easy to handle. Crystals in large sizes ($10\,\text{cm}^3$) are grown and fabricated into SBS-PC devices easily. Hence, it should be possible to construct a compact all-solid-state laser system with high beam quality.

20.2.2.2 Growth and characterization of DAST crystal

DAST possesses an extremely large electro-optic coefficient of $r_{111} = 77\,\text{pm/V}$ [10]. However, the difficulties in growing reasonably large and high-quality DAST crystals prevents its real applications such as optical sampling and electro-optical (EO) sampling devices through frequency-conversion and EO effects.

DAST was synthesized by the condensation of 4-methyl-N-methyl pyridinium tosylate (which was prepared from 4-picoline and methyl toluenesulphonate) and 4-dimethylamino-benzaldehyde in the presence of piperidine, this was done in a dry nitrogen atmosphere in order to avoid orange hydrate-crystals DAST.H_2O. DAST crystals were grown from methanol solution by a slow-cooling method. The seed crystals of $1-2$ mm size were prepared by spontaneous nucleation. The (001) surface of DAST seed crystal was attached to a Teflon rod by using a silicone bond and then put into the solution. DAST crystals were grown with various growth rates ranging from 0.14 mm/day to 1.2 mm/day, controlled by temperature lowering rates between $0.04-1.2\,°\text{C/day}$. The seed crystal was rotated at a rate of 20 rpm while reversing the rotation direction every 3 min. Figure 20.7 shows a typical DAST crystal of dimensions $5.8 \times 3.4 \times 1.0\,\text{mm}^3$ grown at a rate of 0.14 mm/day for 11 days. The as-grown (001) surface of this sample is flat as observed from the largest plane in Figure 20.7. The cutoff wavelength of 600 nm results in a dark red color in the crystalline state.

Figure 20.8 shows the increasing Vickers hardness of the (001) DAST surface with decreasing FWHM of the X-ray diffraction (XRD) rocking curve. It is remarkable that the Vickers hardness increases with crystalline perfection. We think that the molecular arrangement of the crystal is closer packed for

Figure 20.7   Photograph of DAST crystal with dimensions of $5.8 \times 3.4 \times 1.0\,mm^3$ grown at a rate of 0.14 mm/day.

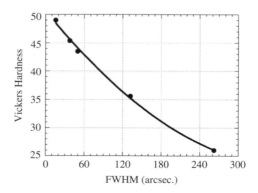

Figure 20.8   Vickers hardness of (001) DAST as a function of FWHM of XRD rocking curve.

narrower XRD rocking curve, resulting in the lager Vickers hardness. The maximum Vickers hardness of 49 of the DAST crystal is similar to that of organic salt crystals, such as LAP, tris-hydroxymethyl-amino-methane phosphate (THAMP) and tris-hydroxymethyl-amino-methane sulphate (THAMS) with values of 56, 58 and 54, respectively [11]. This means that DAST has moderate hardness for mechanical processing.

20.3 CRYSTALS GROWN FROM HIGH-TEMPERATURE SOLUTIONS

Crystals growing immersed in the solution have the best structural perfection due to low temperature gradients during growth. Top-seeded solution growth TSSG is applied because this allows monitoring of the growth process.

20.3.1 GROWTH AND OPTICAL CHARACTERIZATION OF KTP (POTASSIUM TITANYL PHOSPHATE) CRYSTAL [12–14]

KTP is a useful crystal to get efficient green light by frequency doubling of Nd:YAG laser. It has high optical nonlinearity and large temperature and angular tolerance. It is not water-soluble and is mechanically hard. It can be grown by the TSSG method at about $950\,^{\circ}C$ in the $P_2O_5 - K_2O - TiO_2$ ternary system. This crystal is not a line compound, it has an existence width. Therefore, the solid-solution composition and thus the refractive-index change with growth temperature and solution composition during growth.

KTP crystals were grown from about 21 solution in $K_6P_4O_{13}$ (K6) flux in a platinum crucible of 150 mm diameter and 150 mm height placed in a vertical cylindrical furnace with five zone heaters. The temperature difference of the solution in the crucible was less than $1\,^{\circ}C$. The solution was saturated at approximately $950\,^{\circ}C$ and slowly cooled at a constant rate of $30\,^{\circ}C/day$, whereas the seed rotation of 60 rpm was reciprocated every 30 s. After a growth period of 30 days, a crystal with dimensions of $31 \times 42 \times 84\,mm^3$ was obtained. This is shown in Figure 20.9(a). It had liquid inclusions parallel to the crystal faces (mainly (011) faces). In order to obtain a clear crystal without inclusions, real-time measurement of the weight of the crystal during the growth was carried out with an electronic load cell to estimate the bulk growth rate. It was found that the growth rate of the crystal became too fast and liquid inclusions were formed during a certain growth period by excessively high supersaturation at constant cooling rate. By changing the cooling rate during growth a KTP crystal $32 \times 42 \times 87\,mm^3$ ($a \times b \times c$) in size without liquid inclusion was successfully obtained and is shown in Figure 20.9(b).

The optical homogeneity in the grown crystal was measured by interferograms with He-Ne lasers. A sample of 7 mm thickness was cut from the grown crystal

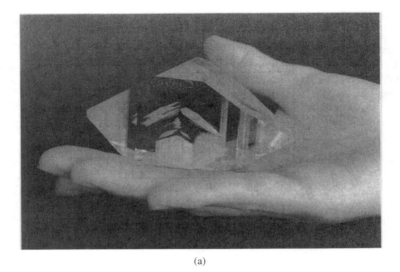

(a)

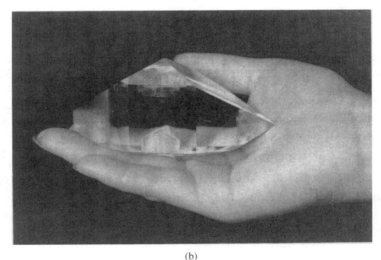

(b)

Figure 20.9 Photographs of KTP crystals: (a) obtained at constant cooling rate of 3 °C/day (size 31 × 42 × 84 mm³, a × b × c)); (b) obtained at various rates controlled between 0 °C/day and 3 °C/day (size 32 × 42 × 87 mm³).

(Figure 20.9(b)) in a direction of phase matching for second-harmonic generation of Nd:YAG laser ($\theta = 90°$, $\phi = 24°$). Figure 20.10(a) shows a transmission interferogram of wave-front distortion taken with circularly polarized light, and Figure 20.10(b) shows a schematic drawing of the growth sector structure of the sample. The central portion of the plate (denoted (110) in Figure 20.10(b)) shows

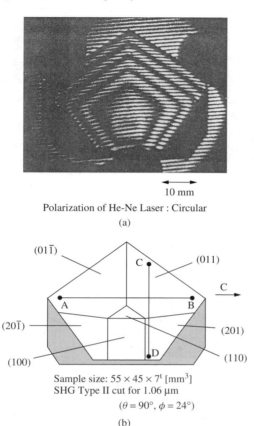

10 mm

Polarization of He-Ne Laser : Circular

(a)

Sample size: $55 \times 45 \times 7^t$ [mm^3]
SHG Type II cut for 1.06 μm
$(\theta = 90°, \phi = 24°)$

(b)

Figure 20.10 (a) He-Ne laser interferogram of wave-front distortion in transmission of a sample cut for second-harmonic generation (thickness 7 mm, circular-polarized wave); (b) schematic drawing of growth-sector structure of the sample.

a part near the seed that was formed at the initial growth stage. The interference fringes are considered as isophase lines of the light wave passing through the sample. This means that the refractive index of each growth sector changed gradually along the growth direction. Figure 20.11 shows the wave-front distortion in transmission of the KTP plate taken with linear-polarized waves. Figure 20.11(a) is refractive-index change n_z for z-polarized wave and (b) is n_{xy} for xy-polarized wave. From a quantitative analysis of the interference fringes, it was found that n_z decreased by 1.6×10^{-4} and n_{xy} increased by 7.1×10^{-6} from the seed to the edge of the grown crystal, corresponding to a change in phase-matching angle of 0.38.

The growth atmosphere affects the optical transmission of KTP crystals [15]. A colorless KTP crystal was obtained in an oxygen-deficient ambient (in this

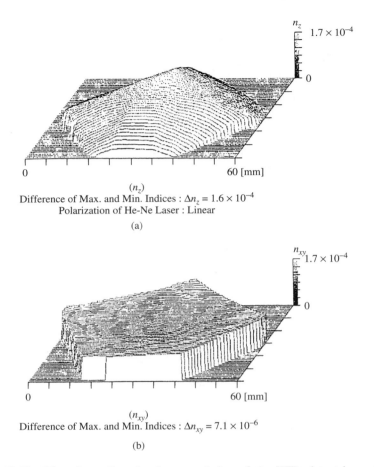

(n_z)
Difference of Max. and Min. Indices : $\Delta n_z = 1.6 \times 10^{-4}$
Polarization of He-Ne Laser : Linear

(a)

(n_{xy})
Difference of Max. and Min. Indices : $\Delta n_{xy} = 7.1 \times 10^{-6}$

(b)

Figure 20.11 Wave-front distortion in transmission of the KTP plate taken with linear-polarized waves. (a) z-polarized wave; (b) xy-polarized wave.

case nitrogen ambient). Figure 20.12 shows the improvement of absorption coefficient in the region from 400 to 550 nm with decreasing oxygen concentration. There seems to be a correlation between the absorption coefficient and Pt impurity concentration suggesting that more Pt ions (Pt^{4+}) from the Pt crucible are incorporated at higher oxygen concentration.

20.3.2   GROWTH AND NLO PROPERTIES OF CESIUM LITHIUM BORATE CLBO

Many efforts were devoted to develop ultraviolet (UV) lasers for industrial and medical applications. Frequency conversion of solid-state laser radiation

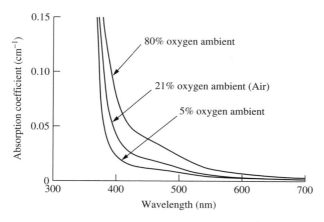

Figure 20.12   Optical absorption spectra of KTP crystals grown in various atmospheres.

in nonlinear-optical (NLO) crystals is an effective method for obtaining UV radiation with high beam quality and stability. Borate crystals are promising for UV generation because of their wide band gap and adequate optical nonlinearity. Barium borate (BBO), lithium borate (LBO), potassium beryllium boro-fluoride (KBBF), and strontium beryllium borate (SBBO) have been developed by Chen's group in China. Recently, the new borate crystal $CsLiB_6O_{10}$ (CLBO) has been discovered and developed by the present authors [16, 17]. CLBO crystals can generate the fourth and fifth harmonics of Nd:YAG laser radiation due to their relatively large birefringence. Here, we review the growth and the NLO- and frequency-conversion properties of CLBO crystals and relate the latter to other NLO crystals such as BBO and LBO.

CLBO has a tetragonal structure with space group $\bar{I}42d$ and cell dimensions $a = 10.494(1)$ Å and $c = 8.939(2)$ Å for $Z = 4$. The structure comprises isolated Cs cations and a network of chains formed from B_3O_7 groups and Li cations. It consists of a six-membered ring, where two of the boron atoms are three-fold coordinated and the third boron atom in the ring is four-fold coordinated by oxygen atoms. The borate network surrounds eight-coordinated Cs ions and four-coordinated Li ions.

The pseudobinary $(Cs_2O + 3B_2O_3) - (Li_2O + 3B_2O_3)$ and $(Cs_2O + Li_2O) - (2B_2O_3)$ phase diagrams were investigated in the ternary $Cs_2O - Li_2O - 3B_2O_3$ system using differential thermal analysis and X-ray diffractometer as shown in Figures 20.13(a) and (b) [18]. These figures show a stoichiometric congruent melting point at 848 °C, meaning that it is easier to grow CLBO crystal than BBO and LBO crystals. Phase diagrams show that CLBO crystals can be grown from fluxes that are either poor or rich in B_2O_3. This is an advantage for flux growth of CLBO compared with that of LBO, which grows only from a flux rich in B_2O_3, which has a high viscosity. Therefore, a low-B_2O_3 flux was used to grow large crystals of CLBO. The starting charges was Cs_2CO_3, Li_2CO_3 and

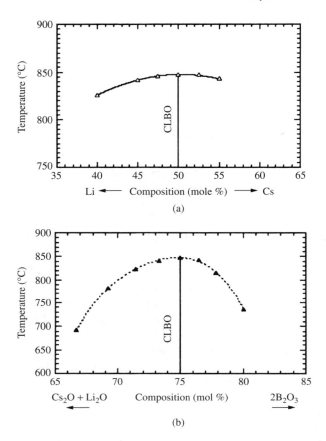

Figure 20.13 The pseudobinary $(Cs_2O + 3B_2O_3)$-$(Li_2O + 3B_2O_3)$ and $(Cs_2O + Li_2O)$-$(2B_2O_3)$ phase diagrams investigated in the ternary Cs_2O-Li_2O-$3B_2O_3$ system.

B_2O_3 with a ratio of 1:1:5.5 (73.3 % B_2O_3). From the saturation temperature 845°C the solution was cooled at a rate of ~0.1°C/day to 843.5 °C. The growing crystal was rotated at a rate of 15 rpm, reversing the rotation direction every 3 min. Figure 20.14 shows a well-facetted CLBO crystal with dimensions of $14 \times 11 \times 11 \, \text{cm}^3$ grown under this condition for three weeks [19].

The transmission range of the CLBO crystal was from 180 nm to 2750 nm. The absorption edge of CLBO at 180 nm is shorter than the reported value of BBO (189 nm) but longer than that reported for LBO (160 nm). The refractive indices were determined by the method of minimum deviation. The Sellmeier equations predict the limits of type-I and type-II PM wavelengths to be 472 and 640 nm, respectively. The Sellmeier equations also predict the fifth-harmonic generation (5HG) of Nd:YAG laser radiation in CLBO crystal by mixing the first- and fourth-harmonics radiations. Table 20.1 shows values of angular, spectral and

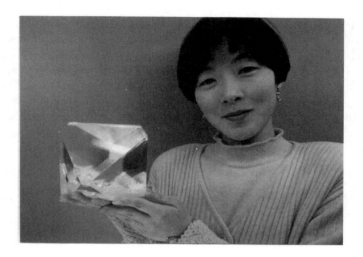

Figure 20.14   Photograph of large CLBO crystal with dimensions $14 \times 11 \times 11$ cm^3.

Table 20.1   NLO properties of CLBO and BBO for 266-nm and 213-nm generation

Wave-length (nm)	Crystal	PM angle (deg.)	d_{eff} (pm/V)	Angular band-width (mrad cm)	Spectral band-width (nm cm)	Temperature band-width (°C cm)	Walk-off angle (deg)
532 + 532	CLBO	62	0.85	0.49	0.13	8.3	1.83
= 266	BBO	48	1.32	0.17	0.07	4.5	4.80
1064 + 266	CLBO	67	0.88	0.42	0.16	4.6	1.69
= 213	BBO	51	1.26	0.11	0.08	3.1	5.34

temperature bandwidths, and walk-off angles for various frequency-conversion processes in CLBO and BBO for 266-nm and 213-nm generation. Despite its smaller nonlinear coefficient, CLBO possesses a smaller walk-off angle and larger angular and spectral bandwidths than BBO. In addition, the temperature acceptance of CLBO exceeds that of BBO. As shown in Table 20.1, CLBO exhibits better NLO properties than BBO, and this is supported by the following experimental results.

A high average power solid-state UV laser with pulse rates of 100 Hz and 1000 Hz was realized by CLBO. Powers of 9.7 W at 266 nm and 4.0 W at 213 nm were detected for a pulse rate of 100 Hz. The conversion efficiency from 1064 nm to 213 nm was 11.8 % [20]. At 1000 Hz, we obtained power of 2.5 W at 266 nm and 1.0 W at 213 nm from a fundamental of 8.8 W, corresponding to an efficiency of 11.4 % from 1064 nm to 213 nm [21]. Moderate birefringence gives CLBO a smaller walk-off angle for achieving better spatial overlapping of the mixing beams compared to BBO. Thus, higher conversion efficiency and better beam pattern of harmonics generation can be achieved. Its relatively large angular, spectral and temperature acceptance bandwidths at 532 nm favor CLBO for stable 266-nm generation of high-power Nd:YAG lasers. All these properties and a shorter absorption edge (down to 180 nm) eventually lead to a high conversion efficiency and stable 213-nm generation. Recently, we have obtained 20 W of 266 nm in collaboration with Mitsubishi Electric Co. Ltd [22].

20.4 CONCLUSIONS

The growth of various NLO crystals from low- and high-temperature solutions has been discussed with particular emphasis on the current research in the authors' laboratory. We have succeeded in growing large KDP crystals up to $40 \times 40 \times 70 \, cm^3$ by preventing spurious crystals from attaching to the growth surface during the long crystal-growth period. We can grow organic NLO crystals LAP and DAST with high crystallinity by optimizing the growth conditions. It was found that LAP exhibited the stimulated Brillouin scattering and seems to be good as a solid-state phase conjugator. A clear KTP crystal of size $32 \times 42 \times 84 \, mm^3$ ($a \times b \times c$) was grown using K6 flux. The refractive indices vary along the growth direction, which resulted from the compositional deviation of the solution. The effect of growth ambient on optical properties of KTP has been investigated. A colorless KTP crystal could be obtained by growing the crystal in nitrogen atmosphere. We have discovered and developed CLBO, which possesses advantages in growth and UV generation. A CLBO crystal with dimensions of $14 \times 11 \times 11 \, cm^3$ could be grown in three weeks. A maximum output of 20 W at 266 nm was obtained by using CLBO.

REFERENCES

[1]   T. Sasaki and A. Yokotani: *J. Cryst. Growth* **99** (1990) 820.
[2]   N. P. Zaitseva, J. J. De Yoreo, M. R. Dehaven: *J. Cryst. Growth* **180** (1997) 225.
[3]   N. P. Zaitseva, L. N. Rashkovich, S. V. Bogatyreva: *J. Cryst. Growth* **148** (1997) 276.
[4]   A. Yokotani, T. Sasaki, K. Yoshida, T. Yamanaka and C. Yamanaka: *Appl. Phys. Lett.* **48** (1986) 1030.
[5]   Y. Nishida, A. Yokotani, T. Sasaki, K. Yoshida, T. Yamanaka and C. Yamanaka: *Appl. Phys. Lett.* **52** (1998) 420.
[6]   A. Yokotani, T. Sasaki, K. Fujioka, S. Nakai and C. Yamanaka: *J. Cryst. Growth* **99** (1990) 815.

[7] H. Adachi, Y. Takahashi, Y. Mori and T. Sasaki: *J. Cryst. Growth* **198/199** (1999) 568.

[8] A. Yokotani, T. Sasaki, K. Yoshida and S. Nakai: *Appl. Phys. Lett.* **55** (1989) 2692.

[9] H. Yoshida, H. Fujita, M. Nakatsuka, T. Sasaki and K. Yoshida: *Appl. Opt.* **36** (1997) 7783.

[10] F. Pan, G. Knopfle, Ch. Bosshard, S. Follonier, R. Spreiter, M. S. Wong and P. Gunter, *Appl. Phys. Lett.* **69** (1996) 13.

[11] T. Sasaki, A. Yokotani, K. Fujioka, Y. Kitaoka and S. Nakai, *Int. Workshop on CGOM* **V-02** (1989) p. 246.

[12] T. Sasaki, A. Miyamoto, A. Yokotani and S. Nakai: *J. Cryst. Growth* **128** (1993) 950.

[13] A. Miyamoto, T. Sasaki, A. Yokotani, S. Nakai, Y. Okada and N. Ohnishi: Technol. Rep. Osaka University **43** (1993) 61.

[14] A. Miyamoto, Y. Mori, Y. Okada, T. Sasaki and S. Nakai: *J. Cryst. Growth* **156** (1995) 303.

[15] A. Miyamoto, Y. Mori and T. Sasaki: *Appl. Phys. Lett.* **69** (1996) 1032.

[16] Y. Mori, I. Kuroda, S. Nakajima, T. Sasaki and S. Nakai, *Jpn. J. Appl. Phys.* **34** (1995) L296.

[17] T. Sasaki, Y. Mori, I. Kuroda, S. Nakajima, K. Yamaguchi and S. Nakai, *Acta Crystallogr. C* **51** (1995) 2222.

[18] Y. Mori, I. Kuroda, S. Nakajima, T. Sasaki and S. Nakai, *J. Cryst. Growth* **156** (1995) 307.

[19] Y. Mori, I. Kuroda, S. Nakajima, T. Sasaki and S. Nakai, *Appl. Phys. Lett.* **67** (1995) 1818.

[20] Y. K. Yap, M. Inagaki, S. Nakajima, Y. Mori and T. Sasaki, *Opt. Lett.* **21** (1996) 1348.

[21] U. Stamm, W. Zschocke, T. Schroder, N. Deutsch, D. Basting, *Technical Digest of CLEO'97* **11** (1997) CFE7.

[22] T. Kojima, S. Konno, S. Fujukawa, K. Yasui, K. Yoshizawa, Y. Mori, T. Sasaki, M. Tanaka, Y. Okada: *OSA TOPS Adv. Solid State Lasers* **Vol. 26** (1999) 93.

21 Growth of Zirconia Crystals by Skull-Melting Technique

E. E. LOMONOVA, V. V. OSIKO

Laser Materials and Technology Research Center General Physics Institute of the Russian Academy of Sciences Moscow, Russia

21.1 INTRODUCTION

Pure zirconium dioxide (ZrO_2) exists in three polymorphic modifications. The ZrO_2 melt crystallizes at 2680 °C in cubic modification (Fm3 m, fluorite) which at 2370 °C transforms into the tetragonal ($P4_2$/nmc) and at 1160 °C into the monoclinic ($P2_1C$) modification (baddeleyite). Thus it is impossible to grow single crystals of pure zirconia directly from the melt. Fortunately, the cubic modification can be stabilized by incorporation of special dopants such as MgO, CaO, Y_2O_3. The addition of these stabilizers in concentrations of 10 or higher (up to 40) mole % forms numerous vacancies in the oxygen sublattice of ZrO_2 and thus prevents the transformation of the cubic phase on cooling: cubic zirconia (CZ) then remains stable down to room temperature.

The high melting point and extreme chemical reactivity of the melt do not allow melting and crystallization of CZ in conventional metallic or graphite crucibles. In the late 1960s the new technique of melting and solidification of refractory dielectrics called direct RF melting in a cold container or 'skull melting' was proposed. The physical reason for this technique is the specific temperature dependence of electroconductivity of dielectrics: the electroconductivity rises with temperature up to the melting point, and then sharply increases in the melt. Typically the jump of electroconductivity reaches two orders of magnitude. According to Wenckus of Zirmat Corporation (USA) [1] the skull-melting process may be thought of as a very-high-temperature analog of microwave cooking where the microwave energy heats the contents in the dish, but not the dish itself. Skull-melting utilizes a water-cooled, copper crucible-like structure to surround the RF-heated molten zirconia that is contained by a thin shell or 'skull' of its own composition. Thus the contained melt is not in contact with, or contaminated by, the copper cold crucible. The melt is then slowly cooled by lowering within the RF coil until complete solidification occurs. In this way the molten charge is converted to a mass of elongated CZ crystals. The crystallized ingot consists of columnar-shaped single crystals, the weight of the separate single crystals reaches

Crystal Growth Technology, Edited by H. J. Scheel and T. Fukuda
© 2003 John Wiley & Sons, Ltd.   ISBN: 0-471-49059-8

10–15 kg [1]. The yttrium-stabilized (YCZ) single crystals are characterized by a unique set of physical and chemical properties. They are water-clear and transparent in the range of 0.26–7.5 μm, with the refractive index $n_0 = 2.17$–2.21, and dispersion of the refractive index 0.06. It is relatively hard (Mohs hardness 8.5) and chemically extremely inert. Furthermore, YCZ possesses significant electroconductivity of the ion type at elevated temperatures. Due to such a combination of properties the YCZ crystals are very attractive in jewelry as imitation diamonds, substrates, selective electrodes and for various technical components utilizing their mechanical and optical properties. The additional advantage of YCZ is that many dopants can be incorporated in these crystals, thereby varying their spectroscopic, electrical, and mechanical properties.

Another new material based on zirconia solid solutions is so-called partially stabilized zirconia (PSZ). It forms when the concentration of stabilizing oxide in zirconia is less than 6 %. PSZ prepared in the solidification process by the skull technique represents a new kind of composite material. An important feature is that the initial material is formed as cubic single crystals, which, in the course of cooling, partially transform into the tetragonal phase. The final product has all the features of a composite nanostructured material, is composed of nanodomains, specially oriented in the bulk volume, and preserves the shape of the initially grown cubic crystal. Thus the composite material is formed by phase transformation of initial single crystals instead of by the conventional process of sintering powders.

PSZ differs very strongly in its mechanical properties from cubic zirconia single crystals. It displays the unique combination of high toughness and shock resistance with a low friction coefficient and thermal stability. The specific combination of mechanical properties meets many requirements of practical applications as scalpels, perforators, and drills in medical instrumentation, cutting and abrasive instruments for materials processing, draw plates, rollers, mills, thread-leading tools and bearings.

21.2 PHYSICAL AND TECHNICAL ASPECTS OF THE DIRECT RADIO-FREQUENCY MELTING IN A COLD CONTAINER (SKULL MELTING)

The theoretical and experimental investigation of the direct radio-frequency melting in a cold container with respect to dielectric materials with the electric resistivity at room temperature 10^4–10^{14} ohm cm have been accomplished by French, American, and Soviet authors in the late 1960s and early 1970s [2–5].

The laws of decrease of the amplitudes of the magnetic and electrical fields in the solid body are:

$$H_{\text{ampl}} = H_0 \exp(-Z/\Delta) \quad \text{and} \quad E_{\text{ampl}} = H_0 \sqrt{2} \exp(-Z/\Delta)/\gamma \Delta$$

In this equation, H_0 is the amplitude of the magnetic field on the surface of the body, Z is the distance from the surface (depth), γ is the specific electrical conductivity of the body, and Δ characterizes the reduction of E and H amplitudes; it is equal to the depth at which H_{ampl} diminishes e-times, and is called the penetration depth. It is shown that

$$\Delta = 5.03 \times 10^3 (\rho/\mu f)^{1/2} \text{ (cm)}$$

where ρ is the specific electrical resistivity of the material, μ is its magnetic permeability, and f is the frequency of the electromagnetic field. The value of the active energy flow in the body at a depth Z is also dependent on Δ:

$$S_a = H_0^2 \exp(-2Z/\Delta)/2\gamma\Delta$$

Substituting Z for $Z = 0$ and $Z = \Delta$ in this formula we have

$$S_{0a} - S_{\Delta a} = 0.864 H_0^2/2\gamma\Delta = 0.864 S_{0a}$$

for the energy absorption in a layer of the material with a thickness equal to Δ. Due to this it is assumed in technical calculations that the RF field energy is practically completely absorbed in a surface layer with a thickness Δ.

To estimate the possibilities of RF melting for zirconia it is necessary that the electrical properties of this material both in the solid phase and in the melt are known. As the great majority of dielectric materials are nonmagnetic so that $\mu = 1$. Unfortunately, the available published information is rather insufficient. The data on the electric resistivity of some solid and molten oxide materials show that they possess electric resistivity within $0.01 - 100$ ohm cm at temperatures close to their melting points. Also it is known that melting increases the conductivity of these materials [6, 7]. The increase is in some cases very high. At room temperature the electric resistivity of these compounds is $10^4 - 10^{14}$ ohm cm, i.e. they are typical dielectrics. Figure 21.1 a shows the temperature dependence of the resistivity of aluminum oxide, a typical dielectric, in solid (corundum) and molten states directly measured in air [6]. It is seen that as a result of melting, the resistivity decreases more by than two orders of magnitude and is 0.1 ohm cm for the melt. This means that for the aluminum oxide melt a frequency from 500 kHz to 10 MHz should be used. In this case the value of penetration depth is quite reasonable and is equal to 2.2–0.5 cm.

Let us now discuss some distinguishing features of the RF melting process of oxide materials in a cold container, which result from the electric properties of this type of material. Figure 21.1(b) shows the absorption of the energy of the RF field by a two-layered cylinder; the external shell of the cylinder consists of the solid phase (alumina) and the internal volume of a melt of the same composition as the shell. This diagram illustrates that in the case of the material

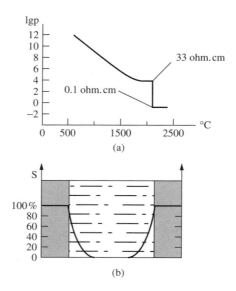

Figure 21.1 (a) Specific electrical resistivity versus temperature and (b) radial distribution of RF field energy, absorbed in two layered cylinder, consisting of solid (outer layer) and molten (internal volume) phases of aluminum oxide at the frequency of electromagnetic field of 5 MHz.

with decreasing dependence of electric resistance on temperature, the solid phase is in fact 'transparent', as regards the electromagnetic field it does not prevent field penetration into the melt, and is essentially not heated (Figure 21.1(b)). Thus the direct induction heating of the melts in a cold container is well suited for the materials with conductivity increasing with temperature, and at the same time it is less suited for metals.

The energy efficiency of the melting process, i.e. the conditions of optimum absorption of RF energy by the melt, has been studied [8]. It was shown that the maximum power released in the melt is mainly determined by the inductor capacity factor $\cos \varphi$. The dependence of $\cos \varphi$ on the parameter $m = \sqrt{2} r / \Delta$ (where r is the melt radius and Δ is the penetration depth) has a marked maximum, which corresponds to the optimum heating conditions (Figure 21.2). This makes it feasible to determine the optimal frequency of RF heating, inductor-to-melt height ratio, and inductor position with respect to the melt surface. The procedure of skull melting and crystallization of refractory oxides consists of four steps:

1. Local heating of the material from room temperature up to that of melting, which can not be achieved directly by RF-heating, (so-called melt seeding); after local melting the seeding melt drop becomes able to absorb RF energy independently and to grow in volume.

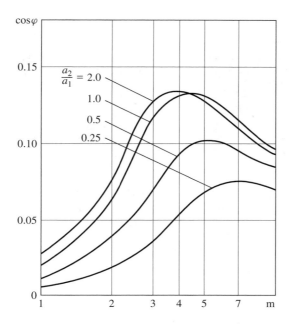

Figure 21.2 The dependence of $\cos\varphi$ on parameter $m = \sqrt{2}r/\Delta$ and on a_2/a_1 ratio: a_1 – RF coil height; a_2 – melt height.

2. Melting the charge with subsequent pouring the initial powder material into the cold crucible until the desired volume of the melt is reached.
3. Establishment of optimal electrical parameters of RF-generator and heat equilibrium conditions in the melt and solid shell (skull) system.
4. Crystallization of the melt.

Usually a small portion of the metal, which forms the oxide, is incorporated in the charge in order to initiate the seed melting process. In the case of YCZ crystal growth it can be zirconium or yttrium (or the metal of any other stabilizing component). The pieces of seed metal absorb RF energy and are rapidly heated. Additional heat is provided due to exothermic oxidation of the metal. The overheated metal melts the surrounding charge, which begins to absorb RF energy independently. The rest of the metal oxidizes in the air atmosphere and admixes itself to the charge without any contamination of the melt Figure 21.3(a).

When the volume of the melt reaches the desired level it is necessary to achieve thermal and spatial equilibrium between melt and solid polycrystalline shell or skull (Figure 21.3(b)). Phase equilibrium means that the temperature of any point on the interface is equal to the melting temperature, and does not change in the course of time. The quantity of heat, given off by the melt to the surface of the solid phase (assuming that complete stirring of the melt takes

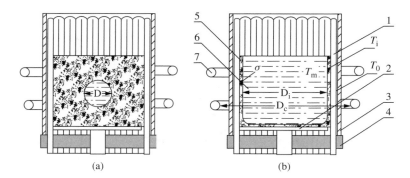

(a)   (b)

Figure 21.3 Illustration of (a) the initial melting and (b) melt–solid equilibrium in a cold container: (1) water-cooled tubing, (2) water-cooled bottom, (3) insulating screen, (4) insulating: bottom, (5) skull, (6) melt, (7) RF-coil. The meaning of the letters is given in the text.

place), can be expressed by the equation

$$Q_1 = \alpha(T_m - T_i)F \cdot \tau$$

where α is the heat-emission coefficient, T_i is the temperature of the solid/liquid interface, F is the surface emitting heat, τ is the time. The quantity of heat transported through a solid layer with thickness σ and thermal conductivity λ, to the heat carrier, can be expressed by

$$Q_2 = (\lambda/\sigma)(T_i - T_0)F \cdot \tau$$

For the stationary regime the temperature of the interface T_i is equal to the temperature of melting T, and $Q_1 = Q_2$, so that $\lambda/\sigma\alpha = (T_m - T)/(T - T_0)$.

$T_m - T = \Delta T$ is the value characterizing the overheating of the melt. If $\Delta T \to 0$, then $\sigma \to \infty$. This means that for phase equilibrium the melt should be overheated. If we assume λ, α and T_0 independent of σ, then

$$\sigma \cdot \Delta T = \text{const.}$$

It can be seen from the equations that an increase or reduction of T_m results in a reduction or increase of the thickness of the solid phase layer. Furthermore, any processes in the solid shell resulting in an alteration of its thermal conductivity (for example, sintering) would disturb the phase equilibrium and therefore change the temperature of the melt or the thickness of the solid layer. Convection in the melt determines the value of the heat-emission coefficient and affects the phase equilibrium as well. The condition of overheating, necessary for the existence of the melt itself in the scheme described, distinguishes the cold container from conventionally heated crucibles.

After the thermal equilibrium and melt homogenization are achieved, the system is ready for crystallization. The crystallization process usually consists of slowly lowering the container with the melt towards the RF coil.

The obvious advantages of the skull melting and crystallization are:

1. The temperature of the process is not limited and may reach 3000 °C or higher.
2. There is no contact of the melt with the crucible material; this provides the high 'purity' of the process.
3. Freedom of choosing any gas atmosphere.

However, skull melting also has some shortcomings, which are not as evident as the advantages:

1. It is impossible to control the crystallization process by direct changing the temperature of the melt.
2. The melt in a cold container is in contact with polycrystalline skull of the same chemical composition. Each grain of the skull is a potential seed for the crystallizing melt. Thus it is impossible to grow one single crystal in the melt volume.

21.3   RF-FURNACES FOR ZIRCONIA MELTING AND CRYSTALLIZATION

Several types of furnaces for skull melting and crystallization of zirconia have been developed since the mid-1960s. Table 21.1 contains some data on the 'Crystal' series of furnaces that are commercially available in Russia. Russia is the only country at present that produces specialized equipment for skull melting of zirconia and other oxides.

Table 21.1 Commercially available installation for CZ and PSZ skull melting and crystallization

Commercially available installation	'Crystal 401'	'Crystal 403'	'Crystal 405'
Total electrical power, kW	100	250	250
Power of oscillations, kW	60	160	160
Frequency of oscillations, MHz	5.28	1.76	1.76
Crucible diameter, m	0.2	0.4	0.5
Crucible height, m	0.3	0.7	1.0
Productivity, kg h^{-1}	0.3–0.5	2–3	4–5

The most important part of the skull furnace is the cold container. The construction of the cold container must meet a number of special requirements [2, 4, 5]. First it should be 'transparent' to RF energy. This means that the container should be sectioned to avoid circular currents. Also it should be made of a material with high thermal conductivity and be intensively cooled by water flow. Figure 21.4 shows a photograph of a typical cylindrical container. It is seen that the container consists of electrically isolated loops fabricated of heavy-walled copper tubes. The copper water-cooled bottom is also sectioned. The loop elements and the bottom are mounted in an electrically insulating micalex holder. The cylindrical container is surrounded by the RF coil of the generator. In some furnaces the diameter of the container reaches 1 m and the height reaches 2 m [1].

The tube RF generators have frequencies from 400 kHz to 5.28 MHz and power up to several hundreds of kW, which depends on the diameter of the container and the mass of the charge. In Figure 21.5 a photograph of a 'Crystal-403' furnace is shown, which is the most common apparatus for skull melting of zirconia in Russia (see also Table 21.1).

As was mentioned above, it is impossible to control the melting and crystallization processes in a cold container directly by measuring the temperature of the melt or observing the solid/liquid interface. Usually, the electrical parameters of the RF generators are used to control the process. Figure 21.6 shows the sequence of electric parameters of a 'Crystal-401' furnace during different stages

Figure 21.4 Photograph of the cold container of 'Crystal-405' skull-melting furnace.

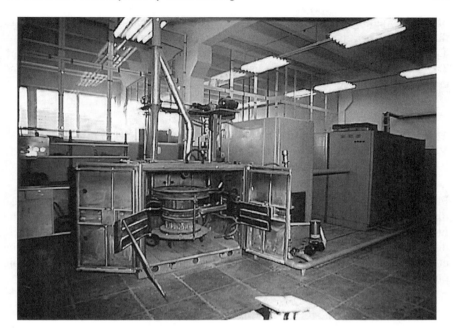

Figure 21.5 Photograph of 'Crystal-403' skull-melting furnace.

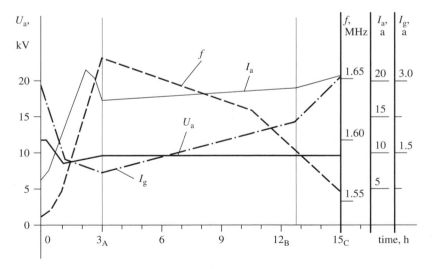

Figure 21.6 The sequence of electric parameters of 'Crystal-401' skull-melting furnace in real process of YCZ melting and crystallization: I_a – anodic current; f – frequency; I_c – grid current; U_a – anodic voltage; O-A – homogenization process; A-B – growth process; B-C – annealing.

of a real melting-crystallization process for YCZ. It can be seen that most of the electrical parameters are very sensitive to the changing of generator loading. The most suitable parameters for feedback are frequency and anodic current. These are the parameters that are usually used for the control of the melting and crystallization processes.

21.4 PHASE RELATIONS IN ZIRCONIA SOLID SOLUTIONS. Y-STABILIZED (YCZ) AND PARTIALLY STABILIZED (PSZ) ZIRCONIA

Figure 21.7 presents the zirconia-rich section of the ZrO_2-Y_2O_3 phase diagram [9]. It shows the composition fields where the cubic, tetragonal, and monoclinic phases of ZrO_2-Y_2O_3 are the only stable phases, and the fields where different phases coexist. It can be seen that at any concentration of yttria the cubic phase crystallizes first. Further phase transformations of the grown cubic phase under cooling conditions depend primarily on the yttria concentration.

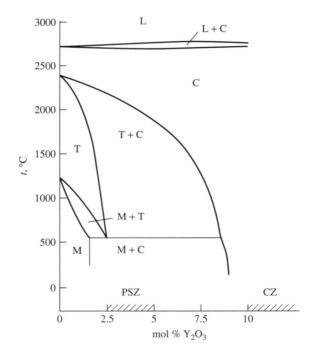

Figure 21.7 Sketch of the ZrO_2-Y_2O_3 phase diagram. C, T, and M- are cubic, tetragonal, and monoclinic phases, respectively, L is liquid phase (the melt). The regions of the compositions corresponding to YCZ and PSZ crystals are marked on the abscissa axis.

Then it should be mentioned that, according to the phase diagram, the crystals crystallized from melt are not stable, in a thermodynamic sense, at room temperature whatever the chemical composition is.

Let us consider now the real phase-transformation processes and phase composition in the course of YCZ and PSZ crystal growth.

Generally, the crystals with 1–8 mole % of Y_2O_3 consist of several phases. At 8–40 mole % of Y_2O_3 crystals can be monophase and can consist of pure fluorite phase. But there follow some very important details. When the crystals with 8 mole % of Y_2O_3 are grown in a small container (with the capacity of about 1 l) at relatively high growth rates of 15–20 mm h^{-1}, then we get perfect optically clear crystals with fluorite structure. Even after 45 days of annealing at 1200 °C in air such crystals do not decompose and remain water-clear. On the other hand, crystals of the same composition grown at low growth rates (1–2 mm h^{-1}) or in large multilitre containers become opaque indicating the partial phase separation of the solid solution. This feature can be observed when the concentration of Y_2O_3 remains less than 14 mole %. At 14 mole % of Y_2O_3 crystals remain clear at any reasonable growth and annealing conditions. Thus it should be assumed that the partial decay of ZrO_2-Y_2O_3 solid solutions at Y_2O_3 concentrations from 8 to 14 mole % strictly depends on how long the crystals remain in the high-temperature region, where diffusion processes can take place.

At concentrations of Y_2O_3 from 8 to 2 (2.3) mole % the crystals are obviously biphased and consist of a cubic matrix with inclusions of tetragonal phase particles dispersed in the volume. The lower the Y_2O_3 concentration, the higher is the concentration of tetragonal inclusions. As was shown by Raman-scattering investigations [10], there are two varieties of tetragonal phases. T tetragonal phase forms in the matrix of a cubic solution as a result of diffusional decay (chemical segregation) in the course of cooling of the grown crystal. Normally the Y_2O_3 concentration in T phase is lower than its initial concentration in the protocubic phase. As the cooling process proceeds, segregation is reduced. Further transition excludes diffusion processes, and leads to the formation of the T' tetragonal phase, which differs from the T phase by a much lower tetragonal distortion of the cubic structure. Obviously the chemical composition of T' phase is exactly the same as the composition of surrounding cubic matrix. The last region of compositions, which is not yet discussed, is the 1 to 2.3 mole % of Y_2O_3. Different methods of phase analysis confirm the presence of the monoclinic phase together with the tetragonal one, at these compositions.

Two composition regions are of special interest: ZrO_2–10 to 40 mole % of Y_2O_3, which corresponds to YCZ crystals, and ZrO_2–2.5 to 5.0 mole % of Y_2O_3 – the composition region of PSZ crystals.

As was mentioned in Section 21.1 many other oxides besides yttria can be used as stabilizing components in CZ and PSZ. Corresponding phase diagrams can be found in [11, 12]. There are some differences between yttria and other stabilizers, but the main effects on the phase transformations are identical.

21.5   GROWTH PROCESSES OF YCZ AND PSZ CRYSTALS

The starting components ZrO_2 and stabilizing oxide are usually of 'high-purity' grade. Magnesium, calcium, strontium, barium, scandium, yttrium, lanthanum, and rare earth oxides in concentrations up to about 40 % can serve as stabilizing compounds. Yttrium oxide is used most frequently. Table 21.2 shows typical concentrations of contaminating impurities in ZrO_2 and Y_2O_3 powders, which are used to grow YCZ and PSZ crystals. Silicon and aluminum are the most undesirable impurities preventing the growth of perfect crystals. The other impurities are less critical, though most of them change the optical, electrical, and mechanical properties of the crystals and influence to some extent the melting and crystallization processes.

The oxide powders are thoroughly mixed and put into a cold container. Metallic yttrium or zirconium in the form of chips is placed as a compact pile into the volume of the oxide mixture. The chemical purity of the metal should be, as a rule, as high as that of the basic charge materials, i.e. not less than 99.99 %.

Then the RF generator is switched on, and the metallic 'seed' begins to heat itself and the surrounding charge. After the charge adjacent to the metal is melted and its volume has exceeded the critical volume the melting process begins to progress gradually, the molten zone increases in volume, and simultaneously intensive oxidation of the metal takes place. When the melt reaches the top surface of the charge, additional portions of the powder or previously melted pieces are gradually poured into it. In this way the desired volume of the melt can be reached gradually, and the melt−solid shell system equilibrium can be established. This stage of the process is reflected in Figure 21.8(a). The solid shell or skull is formed of sintered material. The top layer of sintered material, which forms a vault over the melt, plays a very important role in preventing heat losses.

Table 21.2   Chemical purity of the initial materials

Admixtures, wt. %	ZrO_2	Y_2O_3
Silicon	$<5 \times 10^{-3}$	5×10^{-4}
Aluminum	$<1 \times 10^{-3}$	−
Iron	$<5 \times 10^{-4}$	1×10^{-3}
Chromium	$<1 \times 10^{-4}$	5×10^{-4}
Titanium	5×10^{-3}	−
Manganese	1×10^{-4}	1×10^{-3}
Vanadium	1×10^{-4}	−
Cobalt	1×10^{-4}	5×10^{-4}
Praseodymium + Neodymium	−	4×10^{-3}
Terbium + Holmium	−	2×10^{-3}
Dysprosium	−	2×10^{-3}
Sulfates	1×10^{-4}	−
Chlorides	2×10^{-4}	−

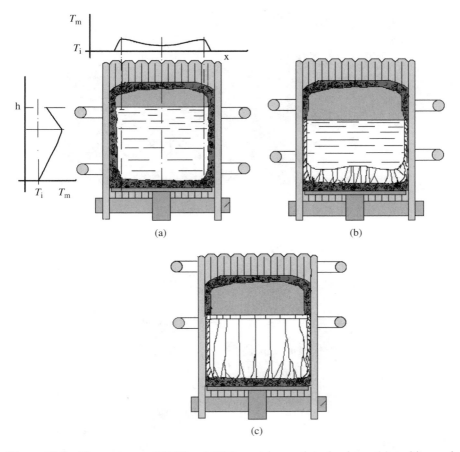

Figure 21.8 Three stages of YCZ and PSZ crystal-growth technology: (a) melting and homogenization process; (b) seeding and the selection of growth rates; (c) complete solidification of the melt.

Figure 21.8(a) simultaneously shows the schematic diagram of radial and vertical temperature distribution in the melt. After the thermal equilibrium in the system is reached, the crystallization of the melt by slow lowering of the container with the melt towards the RF coil can be started. In large furnaces, where the weight of the charge reaches several tons, the RF coil is moved towards the container [13]. The practical rates of crystallization depend on the composition of the melt and are in the range of 3 to 15 mm h^{-1}. As was mentioned above, crystal growth in the skull process differs in an essential way from that of the conventional Bridgman 'hot' crucible technique. The melt is in constant contact with the polycrystalline shell, where each grain is a potential seed crystal. Nevertheless, as crystallization proceeds, the number of seeded crystals is gradually reduced

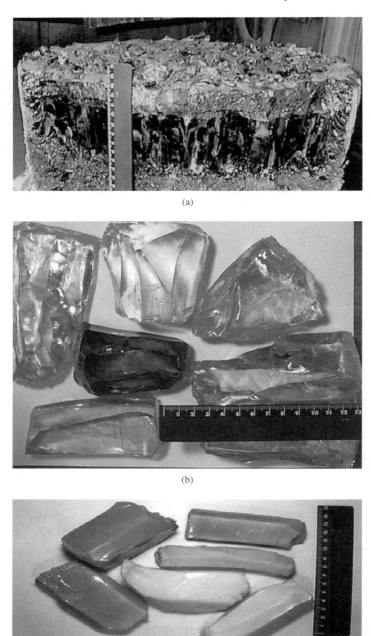

(a)

(b)

(c)

Figure 21.9   YCZ and PSZ crystals: (a) cross section of the ingot; (b) single-crystalline blocks of YCZ; (c) crystalline blocks of PSZ.

due to selection of growth rates so that a restricted number of crystals remains and grows. This stage of the process is shown in Figure 21.8(b). As a result of crystallization and of competition of the growing crystals the ingot becomes composed of elongated single-crystalline blocks with their long axes parallel to each other.(Figure 21.8(c)). In the last stage of the process the RF power is gradually decreased during several hours (depending on the mass of the charge). When the RF generator is switched off, and the grown crystals are cooled enough, the ingot is extracted from the container and is separated into single-crystalline blocks (Figure 21.9).

21.6 STRUCTURE, DEFECTS, AND PROPERTIES OF YCZ AND PSZ CRYSTALS

In this section we will briefly describe the structure, typical intrinsic and growth defects, and the most important properties of YCZ and PSZ crystals.

The structure of YCZ crystals. YCZ crystals have cubic structure of the fluorite type (space group O_h^5-Fm3m). The lattice parameter varies from 5.14 to 5.22 Å, depending on the Y_2O_3 concentration (5.14 for 10 % and 5.22 for 40 % of Y_2O_3 in YCZ crystal) [7] (Figure 21.10(a)).

Stabilizing components incorporated into the ZrO_2 crystal lattice produce oxygen vacancies due to valence difference. Each CaO molecule (or other stabilizer of MO formula) gives one oxygen vacancy. In the case of Y_2O_3 (or any other of the formula M_2O_3) each Y_2O_3 molecule (or each two atoms of Y) gives again one vacancy. If we consider the crystalline lattice of YCZ in terms of structure elements after Kröger [14], and mark the ions and vacancies occupying lattice sites as Ca^{2+} in the cation site, Ca_{Zr}; Y^{3+} in the cation site, Y_{Zr}; oxygen vacancy, V_0, then $nV_0 = nCa$ and $nV_0 = 0.5n$ Y are the expressions

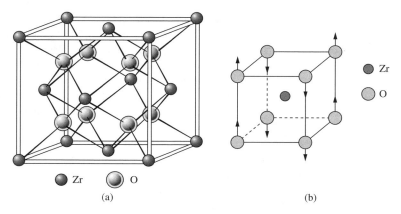

Figure 21.10 Crystal structure of zirconia solid solutions: (a) cubic (fluorite) phase; (b) the distortion of cubic lattice leading to the formation of tetragonal structure.

that characterize the main correlation between structure elements. In the case of ZrO_2-Y_2O_3 solid solutions the concentration of oxygen vacancies in the lattice reaches 7.8×10^{21} cm^{-3} at the concentration of 40 mole % Y_2O_3. It is necessary to emphasize that oxygen vacancies play an important role in the stabilization of cubic structure of YCZ and affects most physical properties of these crystals.

The structure of PSZ. According to the data presented in Section 21.4 the crystals with 2.5–5 mole % of Y_2O_3 consist of two phases: cubic and tetragonal. Initially grown crystals of cubic fluorite structure with O^5_h space group begin to disintegrate with the formation of tetragonal phase with D^{15}_{4h} space group at cooling. The phase transformation is accompanied by chemical segregation of solid-solution components. The tetragonal structure of ZrO_2-Y_2O_3 can be represented as slightly distorted fluorite structure. The character of the distortion is indicated in Figure 21.10(b). The residual cubic matrix is enriched with additional Y_2O_3 and in the course of subsequent cooling can transform into the T′ phase. The concentration of tetragonal (T and T′) phases in the matrix and the contribution of diffusion-dominated and diffusion-less transformation processes depend on the Y_2O_3 concentration and on the growth conditions.

The unique feature of the PSZ formation is that phase transformations do not destroy the grown crystal mechanically. The tetragonal phases appear in the form of tiny domains with dimensions of 8–50 nm [15] dispersed in the crystal volume. The dimensions and orientation of the domains depend on the composition and thermal prehistory of the crystal. Figure 21.11(a) gives an example of a tetragonal domain structure in ZrO_2-Gd_2O_3 8 mole % grown in a 1-l container with the growth rate of 15 mm h^{-1} and subsequent rapid cooling of the grown crystals, revealed by means of polarized Raman spectroscopy. The Raman spectra for crystals with different Gd_2O_3 concentrations in Figure 21.11(a) show how the cubic structure gradually transforms into the tetragonal one with decreasing Gd_2O_3 concentration [10]. The interpretation of polarized Raman spectroscopy data leads to the conclusion that the 4-fold axes of tetragonal domains are oriented along 4-fold axes of the initial cubic lattice (Figure 21.11(b)). It is interesting that these crystals were water-clear and looked like real single crystals, though their Raman spectra were characteristic of the tetragonal structure. These features certify that in this case we have a pure C (cubic) $\rightarrow$ T′ transformation. Figure 21.12 presents the structure of a ZrO_2-Y_2O_3 3 mole % PSZ crystal. It can be concluded that its domain structure is more complicated than in the previous case. The structure is formed with close-packed elongated and regularly oriented tetragonal domains. This picture is characteristic of the compositions where chemical segregation processes take place and the T-phase forms. Visually the crystals of such compositions are opaque (milky-white), have a shiny surface and are completely free of cracks and pores. The special nanosized domain structure of PSZ provides a set of outstanding mechanical properties that will be discussed below.

Growth defects in YCZ crystals. [16] The majority of YCZ single crystals are water-clear with high optical perfection and with gradients of the refractive index

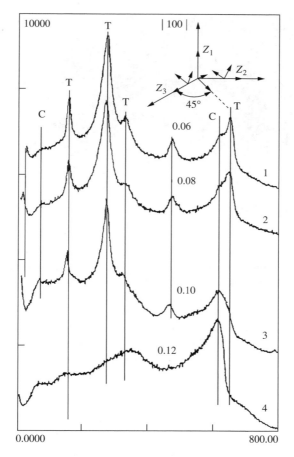

Figure 21.11 Raman spectra of $(1 - x)$ ZrO$_2$-xGd$_2$O$_3$ crystals. (1) $x = 0.12$, (2) $x = 0.10$, (3) $x = 0.08$; (4) $x = 0.06$. C and T mark peaks corresponding to cubic and tetragonal phases. Orientation of 4-fold axes of tetragonal domains relative to 4-fold axes of initial cubic lattice is indicated on the upper part of the figure.

not higher than 5×10^{-3}. Nevertheless some of the samples contain defects, the most characteristic of which are discussed below.

Growth striations located perpendicular to the growth direction are typical of most of the solid solutions and are caused mainly by two factors: the noncongruent nature of liquid–solid transition, and growth-rate fluctuations. Although the Y$_2$O$_3$ distribution coefficient is 1.09 [7], i.e. very close to unity, it still provides the possibility of nonuniform Y$_2$O$_3$ distribution. A typical value of the refractive-index gradient in the striations for YCZ is 10^{-2}.

Light scattering microscopic and submicroscopic solid phase inclusions with dimensions from 0.03 to 10 μm can be randomly dispersed in the volume of the

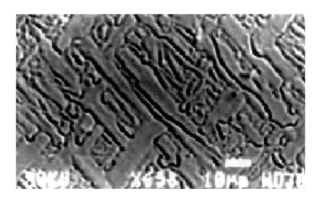

Figure 21.12   Electron-microscopic images of ZrO_2-3 mole % Y_2O_3 crystals.

crystal. The most common cause for light-scattering inclusions is the precipitation of silicates (or aluminates) of yttrium (or any other stabilizing component). Such inclusions can often be observed when the silicon or aluminum contamination in the initial charge is higher than 5×10^{-3} mole %.

Cellular structure appears sometimes in the central and top parts of the ingot, and generally is connected with insufficient purity of the initial chemicals. Silicon and aluminum cause the cellular structure due to their high chemical affinity to the stabilizing agents. Thus, in the melt, silicates and aluminates are formed that are nonisomorphic with zirconia and therefore cause constitutional supercooling and cellular growth.

Mechanical stresses cause refractive-index gradients of values up to 8×10^{-4}. The mechanical stresses originating from the high temperature gradients during the growth and cooling process of YCZ crystals can be largely reduced by high-temperature annealing in vacuum at $2100\,^{\circ}C$ with subsequent slow cooling.

Dislocations are revealed by chemical etching. The typical densities of etch pits in the [111] plane of YCZ crystals are as high as $10^5\,cm^{-2}$. By annealing, the dislocation density could be decreased by one order of magnitude.

Properties of YCZ And PSZ crystals. Table 21.3 contains some data about general physical and thermodynamic properties of YCZ [7]. It should be noted that YCZ crystals have extremely high melting point, high chemical stability, high refractive index and its dispersion, wide optical transparency window, and high hardness. This set of properties ensures some interesting applications of YCZ described below. Many elements of the periodic chart can be incorporated in ZrO_2 in the form of oxides as stabilizing and/or doping agents with the corresponding effects on the spectroscopic, electric, acoustic, and mechanic properties.

Table 21.4 presents the most important properties of PSZ crystals with emphasis on mechanical properties that significantly differ from those of YCZ. PSZ displays a unique combination of high fracture toughness and bending strength with low friction coefficient and high thermal stability. On the other hand, PSZ

Table 21.3   Some properties of CZ crystals

Melting point ($ZrO_2:Y_2O_3$ 10 mol %), °C	2780
Density ($ZrO_2:Y_2O_3$ 10 mol %), kg m^{-3}	5910
Refractive index ($ZrO_2:Y_2O_3$ 10 mol %), D_{Na}	2.173
Dispersion of refractive index	0.06
Optical transparency window, mcm	0.26–7.5
Thermal expansion coefficient,	$11-13 \times 10^{-6}$
°C^{-1}(20–1000 °C)	
Heat capacity, 20 °C, kcal mol^{-1} grad^{-1}	0.015
Heat conductivity, 20 °C, kcal m^{-1} h^{-1}grad^{-1}	2.2
Hardness,	
Mohs	8.5
Vickers, GPa	15.29
Fracture toughness, ($ZrO_2:Y_2O_3$ 10 mol %),	2–3
K_{1C}, MPa m$^{1/2}$	
Electroconductivity, ohm^{-1} m^{-1}	
20 °C	10^{13}
1000 °C	10^4

Table 21.4   Some properties of PSZ crystals

Melting point ($ZrO_2:Y_2O_3$ 3 mol %), °C	2680
Density ($ZrO_2:Y_2O_3$ 3 mol %), kg m^{-3}	6080
Thermal expansion coefficient, °C^{-1} (20–1200 °C)	$9.3-11.4 \times 10^{-6}$
Heat conductivity, 20 °C, kcal m^{-1} h^{-1} grad^{-1}	7.48
Hardness,	–
Mohs	8.5
Vickers, GPa ($P = 200$ g)	12–15
Young's modulus, GPa	190–360
Flexural strength, MPa	–
20 °C	600–1400
1400 °C	700
Fracture toughness, ($ZrO_2:Y_2O_3$ 3 mol %), K_{1C},	6–14
MPa m$^{1/2}$	
Temperature resistance, ΔT, °C	400
Gliding friction coefficient on steel	0.04

loses the attractive optical properties of YCZ. The main reason for the unique properties of PSZ is the specific nanodomain structure of this composite material.

21.7   APPLICATIONS OF YCZ AND PSZ CRYSTALS

The advantages of YCZ and PSZ crystals in practical applications directly follow from their properties. The following applications are more or less developed at present.

Jewelry, the main application for pure and doped YCZ crystals [13, 17, 18], exploits the attractive optical and mechanical properties of YCZ (see Table 21.3): high optical transparency, high refractive index and dispersion, combined with sufficiently high hardness, and stability towards different chemical agents and temperature fluctuations. The most impressive is the large-scale application of colorless YCZ crystals as a diamond imitation [17, 18]. The optical properties of YCZ and diamond are so close to each other that it is not easy to distinguish YSZ from diamond when it is mounted in jewelry. Special physical methods of identification have been developed and described for this purpose [19]. The most successful is the method developed by Wenckus that is based on the significant difference in thermal conductivity between YCZ and diamond [20].

An additional and important advantage of CZ as a gemstone is its isomorphic capacity. Table 21.5 gives some data on the typical dopants in jewelry CZ with the corresponding colors. Figure 21.13 presents cut stones of pure and doped YCZ crystals for jewelry.

Optical components: windows, lenses, prisms, filters and laser elements [7] find applications due to the combination of optical characteristics, mechanical toughness, and chemical stability. In particular, YCZ windows are successfully used in the chemical industry for online monitoring of corrosive liquids [13].

Substrates for semiconductor and superconductor films. Some of the crystallographic planes of YCZ are compatible with silicon, some $A^{III}B^{V}$ semiconductors,

Table 21.5 Colors produced by specific dopants, added to CZ

Dopant	Color
Ce	Orange
Pr	yellow, chrisolite
Nd	Lilac
Sm	Yellow
Eu	Rosy
Tb	Yellow
Ho	Champagne
Er	Pink
Ti	Yellow
V	Olive
Cr	yellow-green
Mn	pink-brown
Fe	Yellow
Co	amethyst, violet, blue, sky-blue
Ni	Peachy
Cu	Aqua
Mo	yellow-green
Pr + Co	emerald-green
Ce + Nd	wine-red

Figure 21.13   Jewelry cut stones of YCZ.

and some of the high-T_c superconductors. It was shown in [21] that YCZ substrates practically do not contaminate the epitaxial layers with oxygen, zirconium, or yttrium even at elevated temperatures [22]. Oriented polished YCZ substrates up to 100 mm diameter are commercially available [13].

The oxygen-ion electric conductivity of YCZ crystals (see Table 21.3) is utilized in ion-selective electrodes based on YCZ [13].

The mechanical properties listed in Table 21.4, especially the combination of high hardness and shock resistance, high wear and tear resistance, low friction coefficient with high chemical and thermal resistance, characterize PSZ crystals as a unique construction material. At elevated temperatures, up to 1000 °C in air atmosphere, PSZ preserves most of its mechanical properties. Additionally, PSZ is compatible with bio tissues.

Figure 21.14(a) demonstrates some articles and components fabricated from PSZ crystals. First we consider supersharp scalpels. The special composite structure allows the working edge of the scalpel to be cut to a width of 1000 Å. Figure 21.14(b) compares scanning electron micrographs of two working edges: one fabricated from special steel and the other from PSZ. It can be seen, that PSZ edge is much smoother, compared with the steel edge, and has no defects. Figure 21.14(a) contains a photograph of different scalpels, which are certified in Russia for cardiosurgery, neurosurgery, vascular surgery, embryonal surgery, facial surgery and ophthalmology applications. Several years experience of PSZ scalpels application confirms their outstanding cutting ability and reliability. The other applications of PSZ are not as developed as the application in scalpels, but still are impressive.

These are: cutting and abrasive instruments for processing of some metallic and nonmetallic materials; draw plates, rollers, mills, and mortars; thread-leading tools; bearings; supporting prisms and bio-inert medical implants (Figure 21.14).

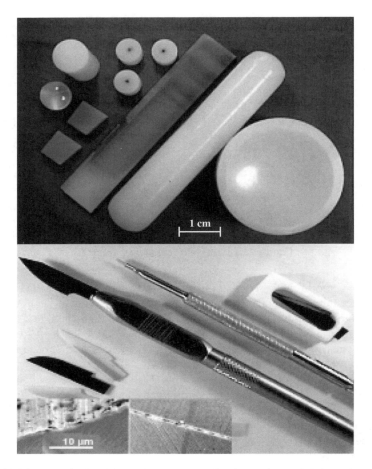

Figure 21.14 Technical articles and components fabricated from PSZ. Scanning electron micrographs of the working edges of two scalpels fabricated from special steel (left) and from PSZ (right).

21.8 CONCLUSION

According to American estimates the current world market for YCZ crystals is close to 600 tons per year [13]. This means that the production scale of this material is one of the largest among man-made crystals. J. F. Wenckus – the leading American expert in YCZ production and skull-melting process – notes [13] that well over 98 % of the CZ crystals produced in 1998 were fabricated into gemstones, mostly as diamond imitations. The yield of finished gems is approximately 15 % of the rough YCZ. As an example the Ceres Corporation of the US, one of the leading CZ producers, has a current YCZ output of approximately 20 tons per month using 110-cm diameter skull furnaces holding charges of 2.5

to 3.5 tons [1]. About 50 % of the charge is transformed into high-quality crystals. Some of the grown crystalline blocks reach 130 to 150 mm in cross section weighing 10 to 15 kg. However, most gem cutters reject YCZ crystals larger than 25 mm in cross section, as their cutting equipment does not accommodate larger crystals. Therefore this crystal-growth process has been specially suited to produce crystals of that size. As to the large crystals, they are well suited for cutting optical windows and substrates [13]. K. Nassau – analyzing the sources of new synthetics – mentioned that 'the most recent diamond imitations – yttrium-aluminum garnet, gadolinium-gallium garnet, and cubic zirconia–all originated as spin-offs from technological research'. And he posed a question: 'Is it then likely that this source will produce another material, that might provide a diamond imitation that is superior to cubic zirconia? The answer to this question is yes, but a qualified yes.' YCZ crystals are very close to diamond in the optical properties, in hardness, and wearability, so that improvements can not be expected. 'All in all – Nassau concludes – it would seem that cubic zirconia is likely to be with us for quite some time' [18]. About two decades have passed since this prediction, and though many new artificial crystals appeared during that time, there is not one that would replace YCZ crystals in the market of diamond imitations.

The production of PSZ is much smaller than that of YCZ. Research and development activity in the field of PSZ and its applications is widespread due to the expected good future for this material.

After over three decades of development of the skull-melting process itself and its most significant products – YCZ and PSZ crystals – one may speculate about the following prospects for the future:

1. The improvement of the energy efficiency of the skull-melting process, which presently does not exceed 50 % of the energy input, with the rest of the energy dissipating in the RF generator. For this reason the energy consumption per unit of grown crystals reaches $150 \, kW \, h \, kg^{-1}$. As was shown in [23] the partial elimination of this shortcoming can be achieved by increasing the capacity of the container. The specific energy consumption decreases approximately linearly with the increase of the container diameter.

2. The impossibility of growing single crystals of a size comparable with the volume of the initial melt. The reasons for this shortcoming were discussed above. Some improvements in crystal dimensions have been achieved by increasing the container capacity. But still the crystal-to-melt volume relation remains very low. Another approach to increase the crystal dimensions uses seeds placed on the bottom of the container [16]. In this case the initial stages of seeding and growth-rate competition are prevented.

3. Difficulties in temperature and liquid/solid interface position control during the process. Electrical parameters such as frequency and anodic current of the RF generator provide some possibilities to control the process, but it is still insufficient in order to have comprehensive information about the crystallization process.

4. The large scale of YCZ production and relatively low (15%) yield of the final product (cut stones) was an ecological threat as much waste is generated in YCZ gem production. Fortunately, in recent years this problem was solved almost completely by recycling defective crystals and scraps in subsequent skull processes. Even the fine powder of YCZ remaining after diamond cutting and polishing of crystals, can be filtered, pressed and sintered into very dense and thermally stable ceramic components.

We hope, in conclusion, that all of the enumerated problems will be successively solved by the next generation of crystal-growth technologists.

ACKNOWLEDGMENTS

We wish to thank the following for help in the preparation of this manuscript and for helpful discussions: Dr N. R. Studenikina (LMTRC, Moscow, Russia), Dr S. V. Lavrishchev (LMTRC, Moscow, Russia), Dr J. F. Wenckus (Zirmat Corp, Waltham/Mass., USA), Dr V. V. Kochurikhin (LMTRC, Moscow, Russia), Dr. A. Gaister (LMTRC, Moscow, Russia).

REFERENCES

[1] V. V. Osiko, J. F. Wenckus, Growth of Stabilized Zirconia Crystals by Skull Melting, Book of Lecture Notes, First Int. School on Crystal Growth Technology, Beatenberg, Switzerland, Sept. 5–16, 1998, editor H. J. Scheel, p. 580–590.

[2] V. I. Aleksandrov, V. V. Osiko, V. M. Tatarintsev, Synthesis of Laser Materials from the Melt by Direct Radio Frequency Heating in a Cold Container, Otchet, FIAN, Moscow, 1968 (in Russian).

[3] Y. Roulin, G. Vitter, C. C. Deportes, New Device for Melting without a Crucible, Fusion of High Melting Oxides in a Multitubular Furnace, Revue Internationale des Hautes Temperatures et des Refractaires 6, 1969, 153–157 (in French).

[4] V. I. Aleksandrov, V. V. Osiko, V. M. Tatarintsev, The Melting of Refractory Dielectric Materials by RF-Heating, Pribory i tekhnika experimenta 5, 1970, 222–225 (in Russian).

[5] J. F. Wenckus, Study, Design, and Fabrication of a Cold Crucible System, Scientific Report, Ceres Corp, No 1 (15 May 1974).

[6] V. I. Aleksandrov, V. V. Osiko, V. M. Tatarintsev, Electroconductivity of Al_2O_3 in Solid and Molten States, Izvestiya Akademii Nauk SSSR, seriya Neorganicheskiye Materialy 8 (1972) 956 (in Russian).

[7] V. I. Aleksandrov, V. V. Osiko, A. M. Prokhorov, V. M. Tatarintsev, Synthesis and Crystal Growth of Refractory Materials by RF-Melting in a Cold Container, in Current Topics in Materials Science 1, editor E. Kaldis, North Holland Publishing Company (1978), 421–480.

[8] Yu. B. Petrov, Induktionnaya Plavka Okislov, Energoatomizdat, Leningrad 1983, 103 pages (Russian).

[9]   H. G. Scott, Phase Relationships in the Zirconia – Yttria System, *J. Mater. Sci.* **10** (1975) 1527–1535.

[10] Yu. K. Voronko, M. A. Zufarov, B. V. Ignatyev, E. E. Lomonova, V. V. Osiko, A. A. Sobol, Raman Scattering in ZrO_2-Gd_2O_3 and ZrO_2-Eu_2O_3 Single Crystals with Tetragonal Structure, *Optika i Spektroskopiya* **51** (1981) 569 (in Russian).

[11] V. S. Stubican, J. R. Hellman, Phase Equilibria in Some Zirconia Systems, in Advances in Ceramics 3, editors A. H. Heuer and L. W. Hobbs, The American Ceramic Society, Columbus/OH 1981, 25–37.

[12] A. Rouanet, Diagrams de Solidification et Diagrams de Phases de Haute Tempera-ture. Des Systemes Formes par la Zircone avec les Sesquioxydes des Lanthanides, in Les Elements des Terres Rares, tome 1, Colloques Internationaux du Centre National de la Recherche Scientifique 1, **No. 180**, Paris-Grenoble 5–10 Mai 1969, 221–235 (in French).

[13] J. F. Wenckus, private communication 1998.

[14] F. Kröger, Khimiya nesovershennykh krystallov, (Mir, Moscow 1969) (in Russian).

[15] A. F. Shchurov, I. F. Perevozchikov, Mechanical Properties of Stabilized Zirconia Crystals, *Neorganicheskiye Materialy* **33, No. 9** (1997) 1087–1092.

[16] V. I. Aleksandrov, M. A. Vishnjakova, V. F. Kalabukhova, E. E. Lomonova, V. A. Panov, Growth of Zirconia Single Crystals by Direct Crystallization in a Cold Container, *Proc. Indian Nation. Sci. Acad* **57.A, No 2** (1991) 133–144.

[17] K. Nassau, Cubic Zirconia, the Latest Diamond Imitation and Skull Melting, *Lapidary J.* **31** (1977) 900–904, 922–926.

[18] K. Nassau, Cubic Zirconia: an Update, *Lapidary J.* **35, No. 6** (1981) 1194–1200, 1210–1214.

[19] K. Nassau, Distinguishing Diamond from Cubic Zirconia: Old and New Tests for the Identification of Diamond, *Gems and Gemology* **16**, (1978–79b), 111–117.

[20] K. Nassau, A Test of the Ceres Diamond Probe, *Gems and Gemmology* **16**, (1978–79a), 98–103.

[21] V. G. Shengurov, V. N. Shabanov, A. N. Buzynin, V. V. Osiko, E. E. Lomonova, Heteroepitaxial Silicon Structures on Phyanite Substrates, *Microelectronika* **6** (1996) 204–209 (in Russian).

[22] A. N. Buzynin, V. V. Osiko, E. E. Lomonova, Yu. N. Buzynin, A. S. Usikov, Epi-taxial Films of GaAs and GaN on Phyanite Substrate, Widebandgap Semiconductors for High Power, High Frequency, and High Temperatures, *Proc. Mater. Res. Soc. Symp.* **512** (1998), 205–210.

[23] V. I. Aleksandrov, N. A. Iofis, V. V. Osiko, A. M. Prokhorov, V. M. Tatarintsev, Phyanites and the Prospects of their Practical Application, *Vestnik Akademii Nauk SSSR* **6** (1980) 65–72 (in Russian).

22 Shaped Sapphire Production

LEONID A. LYTVYNOV

Institute for Single Crystals, Kharkov, Ukraine

22.1 INTRODUCTION

In 1938 Stepanov suggested for crystal growth to impose a preset shape on the melt column providing the necessary feeding flux (Stepanov, 1959).

The theory of crystal-shape stability in meniscus-controlled growth was described by Surek and co-workers (Surek *et al.* 1977). Theoretical aspects of capillary phenomenon in shaped sapphire growth were studied by Tatartchenko and co-workers (Tatartchenko and Brener, 1976).

Due to the increasing demand for shaped high-temperature oxides (especially sapphire) new, more efficient modifications of methods growth were designed (LaBelle, 1969, 1980. Antonov *et al.* 1980. Dobrovinskaia *et al.* 1994).

Since the profiles are grown at high speed, their structure is more defective than crystals, grown under milder conditions. This is why a number of methods for improvement of the structural perfection were developed.

22.2 CRYSTAL-GROWTH INSTALLATION

Specialized machines were created for growing the various products: model 'Specter' for sapphire tape, 'Zarya' for sapphire pipes with minimal tolerance, and multipurpose machines.

Sapphire is grown at a speed of 30–100 mm/h. So that the growth time is comparable with the time of the machine preparation, heating, and cooling.

Advanced installation allows reduction in the loss time, and growth of crystals in a sequence without opening the chamber. As an example, a scheme of such a setup, model 'Krystall-610' (Byndin *et al.* 1990) is shown in Figure 22.1.

The chamber has a seed-inserting device, and a pulling shaft with an eccentric disc. The seeds are held in cone-traps in this disc. The growth process is controlled by a balance and TV system. The length of the pulling shaft allows shaped crystals to be pulled up to 1.5 m. After growth the crystal is removed by sliding through a slot of a turning disc into the collecting pocket.

Replenishment of melt is required for higher productivity.

Crystal Growth Technology, Edited by H. J. Scheel and T. Fukuda
© 2003 John Wiley & Sons, Ltd.   ISBN: 0-471-49059-8

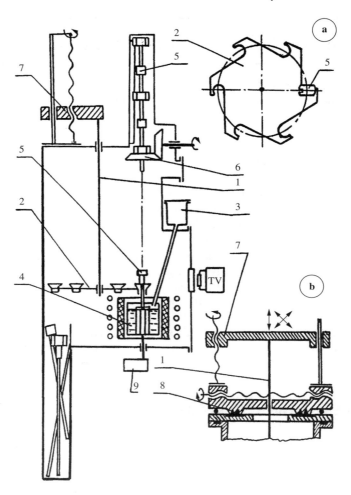

Figure 22.1 (a) Installation for group crystal growth; (b) Lift system with coordinate mechanism: 1. pull rod; 2. disc; 3. feeder; 4. crucible; 5. seed; 6. seed inserting device; 7. screw pair; 8. magnetic lock; 9. balance.

For the growth of complicated profiles a special additional device is used (Figure 22.1(b)). The shaft, while preserving a possibility of vertical displacement, gets additional degrees of freedom in the horizontal plane. The growing crystal can be moved from one die to another shaping element.

22.3 GROWING OF CRUCIBLES

Small crucibles ($D = 2$–5 mm) are obtained (Figure 22.2) by replacing the growing crystal from one die to another. Then the 'bamboo-like' long crystal is sliced

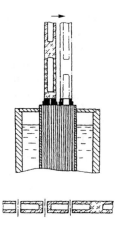

Figure 22.2   Growth of 'bamboo-like' crystals for cutting of small crucibles.

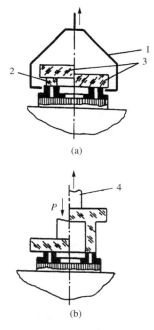

Figure 22.3   Device for seeding by crystal disc: 1. support; 2. distance crystal ring; 3. seed disc; 4. crystal rod.

into multiple crucibles, as is shown in the lower part of the drawing. This method is highly productive.

Larger crucibles are grown from one die (Figure 22.3). In this case the seeding takes place by a thin disc, which becomes the bottom of the future crucible.

The seeding disc is suspended on the metal support hooks (Figure 22.3(a)). The distance ring improves the contact of the disc with the melt. After this ring has been melted the disc goes down onto the die.

This approach can be used not only for crucible growth. Usually the seeding of tubes with $D > 40$ mm from the seed rod takes several hours. The seeding with a thin disc, which overlaps at once the cross section of die, allows this time to be reduced to several minutes.

Another method uses the seeding disc and crystal-supporting rod with the same crystallographic orientation (Lytvynov *et al.* 1989). In Figure 22.3(b) the two-stage seeding is shown: first a sapphire rod intergrows with the disc under pressure P and then, using this rod, a sapphire crucible is pulled up.

The durability of the joint disc-rod depends on the temperature and temperature gradient in the connection zone, value of pressure and time of endurance test under loading (Table 22.1).

Table 22.1 shows that if temperature and loading pressure are sufficient (2270 K and 2 kg/mm^2) we obtain a rather durable joint, which reaches 0.6–0.7 of the tensile strength of sapphire and can be used for pulling the crystal. Temperature plays the main role in the achievement of high enough ratio values of tensile strength for compound (σ_{comp}) to $\sigma_{Al_2O_3}$. At a premelting temperature 2270 K the connected components are plastic and a required ratio $\sigma_{comp}/\sigma_{Al_2O_3}$ is reached.

During a crucible growth, a rarefaction ($P_1 > P_2$, Figure 22.4) takes place inside a closed cavity space. The difference in pressure inside and outside of the close cavity is the reason for wall distortion. This effect is especially noticeable when growing thin-wall items and those of large diameter.

Arches made in the dice connect the space inside and outside the profile and $P_1 = P_2$. These arches do not impede the melt lifting towards the crystallization front.

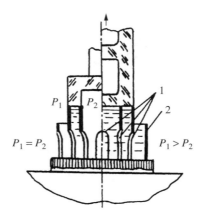

Figure 22.4 Device for changing pressure in the growing closed vessel: 1. arch; 2. ring.

Table 22.1 Parameters of the process of two-stage seeding and seed compound strength with crystal holder

Item	Crystal holder type	Seed disc Thickness, mm	Diameter, mm	T, K	dT/dr, K/cm	P, kg/mm^2	Time of endurance test under loading, min	$\sigma_{comp.}/\sigma_{Al_2O_3}$
Sapphire crucible $D = 5$ mm, wall thickness – 1 mm	Sapphire rod $D = 1.5$ mm	1	8	1770	50	5.0	5	0.05
				1820	50	5.0	5	0.16
				2070	150	3.5	3	0.30
				2220	300	2.0	1	0.40
Sapphire crucible $D = 28$ mm, wall thickness – 2 mm	Sapphire pipe $D = 6$ mm, wall thickness – 2 mm	2.5	30	2270	300	2.0	1	0.70
				1770	50	5.0	10	0.01
				1820	50	5.0	5	0.02
				2070	150	3.0	3	0.30
				2220	300	2.0	2	0.45
				2270	300	2.0	2	0.60

But the difference in pressure can be neglected and used for the growth of 'bamboo-like' crystals. For this purpose ring 2 (Figure 22.4) is periodically turned. Arches can be closed or opened. When arches are closed the melt due to rarefaction moves inside and the bottom is formed.

22.4   GROWTH OF COMPLICATED SHAPES

For some requirements it is necessary to have the sapphire profiles with variable shape of the inner channel (cone channel for variation of the metallic wares, for the decrease of turbulence of the hydraulic flow at the exit from the nozzle, etc.).

Profiles with variable shape of the inner capillary channel can be grown at changing growth rate by changing the voltage supplied to the heater (Zhilin *et al.* 1998). A synchronous variation of these parameters in a certain sequence, i.e. pulling of the crystals with the increasing (decreasing) or constant rate allows a variable cross section to be obtained inner diameter channel.

It was experimentally found that if the pulling rate (ΔV) is increased by the value $K \Delta V$ one can obtain the uniformly variable inner cross section of the article, e.g. a cone-like one. With this, the larger the initial inner diameter of the article the higher the variation of pulling rate that is necessary. This dependence is defined by the coefficient K, which is a function of the difference between inner diameters at the beginning (d_b) and at the end (d_e) of pulling.

The coefficient K was found experimentally since the accuracy of the thermophysical calculations of the systems operating at $2100\,^{\circ}$C is insufficient and it is impossible to find the coefficient with a prespecified accuracy. The variation of the rate ΔV was also selected experimentally for the whole range of the grown capillary tubes, in particular, $d_b = 0.7$–1.2 mm and $d_e = 0.9$–1.4 mm, respectively. At a length of 100 mm the variation of the rate is about 5 mm/h. On increasing the pulling rate and, consequently, increasing the inner diameter, the outer diameter of the article becomes less since the temperature inside the capillary is always higher than outside due to a re-emission of the light flow in the capillary. The heat is scattered to the environment mainly at the side surface, which leads to the decrease of the outer diameter of the article. To avoid the decrease of the outer diameter simultaneously with the increase of the pulling rate one should decrease the value of power supplied to the crucible. The size of the outer diameter depends on the variation of power supplied during formation of the crystal. Depending on application, the outer diameter of capillary tube is usually in the range of 2–10 mm. Shown in Figure 22.5 are the examples of the articles that can be grown by the described method.

To obtain more complicated shapes variable shaping was suggested (Borodin *et al.* 1998). The variation ability is provided by a set of capillary systems of different lengths (Figure 22.6). Each of the systems serves for the growth of one element of the profile. All the capillary systems are fixed in one position. By

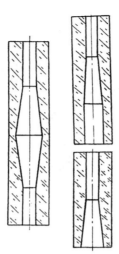

Figure 22.5   Profiles with variable shape of inner cross section.

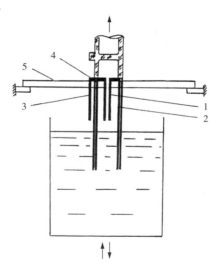

Figure 22.6   Variable shaping: 1. capillary for growth of bottom; 2. capillary for growth of tube; 3. capillary for growth of ribs; 4. dice; 5. support table.

lifting or lowering the crucible one or another system is switched. The complicated shapes were obtained by this approach (Figure 22.7).

The rotation and alternation of rotation direction gives additional opportunities for forming of fantastic shapes, for example, rods with twisted ribs or holes twisted in different directions in the articles (Figure 22.8).

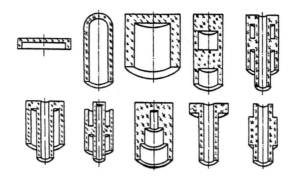

Figure 22.7   Shaped articles grown by variable shaping.

Figure 22.8   Article grown by alteration of rotation.

Large sapphire products can be made in the same way as their metal analogs by growing individual profiles followed by their joining (intergrowth) under high temperature and pressure. It is the analog of the welding process in conformity with crystals.

Provided the intergrown block-free planes coincide along crystallographic axes, a vacuum-tight connection is achieved (Dobrovinskaia *et al.* 1994).

22.5   DICE

The principal scheme of any modification of the process includes the die (sharper) which can be wetted or not wetted with the melt. The majority of high-temperature oxide melts wet the die.

In the classical variant the capillary system was terminated with a die. But for high-temperature conditions this scheme is not convenient: at 2000 °C metals are plastic and the edges of the dice are easily bent at seeding. In this case it is necessary to change the whole capillary system.

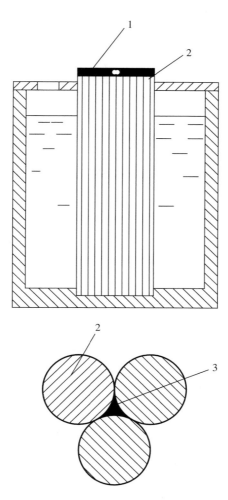

Figure 22.9 Crucible with divided capillary and die: 1. die; 2. wires; 3. capillary channel.

It is more convenient to divide capillaries and the die (Figure 22.9). The capillary system is made in the form of a block, which consists of the compressed thin rods. The spaces between the rods are distinctive capillary channels. Such capillaries are more efficient than round ones. This block is universal, long-lived and virtually eternal: it is replaced from crucible to crucible. The die is mounted on the top of the block and can be easily interchangeable.

It is possible to prevent the thin-wall dice or their edges from being damaged at seeding by automatic seeding. Automatic seeding (or, more exactly, natural seeding without any action) occurs on bringing the seed to the die for a distance a little shorter than the thermal elongation of all heated elements (Figure 22.10).

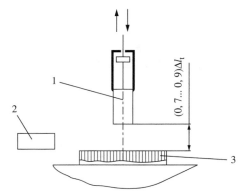

Figure 22.10 Seeding by thermal elongation of heated elements: 1. seed; 2. TV control; 3. capillary.

After heating, natural seeding takes place under thermal elongation. This is the most preferred seeding.

Under growth of hexagonal profiles die parts (or elements) should be matched with the crystallographic symmetry of a seed. So, if the seed's planes $(11\bar{2}0)$ are not parallel to the die faces it will lead to decreasing of the outer form of crystal symmetry. As a result, the configuration of such crystals has only those elements of their own (crystallographic) symmetry, which coincide with the symmetry elements of the die.

It is rather difficult to tread a screw on the inner surface of sapphire articles even by a diamond. Using the die as an element of screw and rotation it is possible to grow a profile with outer and inner screws.

In a combined version of the Stepanov method and the method of 'cold' crucible, a massive die is put into the original raw material and fixed out of a crucible. The die, heated by the HF-field, melts the raw material.

22.6   GROUP GROWTH

Group growth is used for mass production of profiles. For a simultaneous growth of several profiles on a capillary system top a set of dice is placed at a minimal distance from each other.

For instance, on the top $D = 30\,\text{mm}$ one can place in four lines, 12 dice for growing rods $d = 4.3\,\text{mm}$, so that all rods could be observed by a TV control system.

The seed holder copies the location of the dice. The seeds 'swim' in their nests (Figure 22.11), compensating a tolerance of dice and their arrangement. The seeds are sharpened for better of the dice sockets.

In any type of group growth, the crystals are in different temperature conditions. For the decrease of temperature gradient buffer profiles are grown between

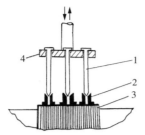

Figure 22.11   Group growth of rods: 1. seed; 2. die; 3. capillary; 4. seed holder.

the crystals (Figure 22.12). Their task is to redistribute temperature and to heat the neighboring crystals.

Related to the group method is also a coaxial one. Coaxial growth allows simultaneous growth of different sizes of hollow-type profiles, which are formed one in another (Figure 22.13). The seed has a fork-form and dice are made

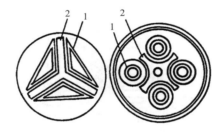

Figure 22.12   Dice (1) with buffer profiles (2).

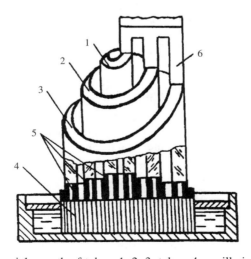

Figure 22.13   Coaxial growth of tubes: 1, 2, 3. tubes; 4. capillaries; 5. dice; 6. seed.

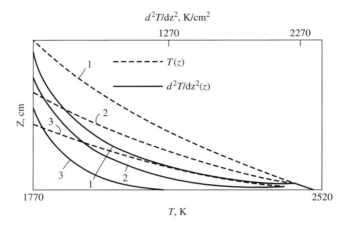

Figure 22.14 Heat distribution in 3-coaxial tubes system: 1. inner tube; 2. middle tube; 3. outer tube.

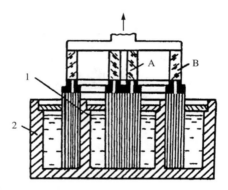

Figure 22.15 Coaxial growth of different kinds of crystals: 1. inner crucible; 2. outer crucible.

with different heights. The inner profile, screened by the external ones, grow in more uniform temperature conditions, under a lower temperature gradient (Figure 22.14). This is why it has better structural and optical perfection.

Coaxial growth allows two kinds of crystals to grow simultaneously, including crystals of decomposing components, which can be grown in a closed space from the inner crucible (Figure 22.15). The tube A is screened and protected by the inert sapphire profile B.

22.7 LOCAL FORMING

This variant of the Stepanov method (Borodin *et al.* 1982) allows sapphire tube to grow with a diameter that is bigger than the crucible's (Figure 22.16).

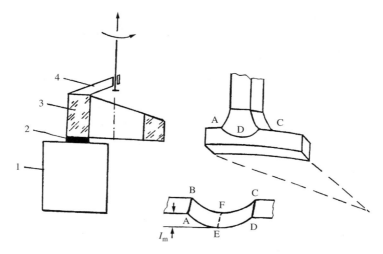

Figure 22.16 Local shaping: 1. crucible; 2. die; 3. growing crystal; 4. seed.

The die is only part of the form; it controls only the width of melt film and the radius of the seed's rotation determinates the final shape. Every dot of the seed moves along a spiral line with an inclination, which is a result of the velocity of pulling V, value of radius r and frequency of rotation n. At the beginning the meniscus is being repeatedly torn up and then after closing of the profile it becomes constant.

The plane ABCD is the outline of moistening. During the growth each part is being sequentially brought into the melt and then is being taken out. Moving through the melt the surface of the former crystallized layer remelts to the depth l_m and then the layer $l_c = l_m + V/n$ is being built up.

The crystallization front is a surface ABEF. At any point on it $V = 2\pi r n$.

Under local forming crystal structure is improved. The layers that are being formed during one rotation are being added to conventional growth layers. A relief of the outer surface has a circulating character, which depends on V/n. The distances between accumulation of pores concur with the relief period. Remelted layers can cause opening of the pores of the previous layer that leads to pore agglomeration.

This variant of the method is used for growth of constructional crystals.

22.8   SAPPHIRE PRODUCTS FOR MEDICINE

A new unexpected field of sapphire application is medicine.

Inorganic crystals as implants have a number of advantages compared to their competitors, especially metals (Lytvynov, 1988). In numerous investigations it is shown that sapphire is one of the most inert and biocompatible materials. Unlike metal implants, sapphire, which possesses electrical neutrality, is not transferred by electrochemical reactions to other parts of the body and does not give rise to disturbances in the immune system.

The principal difference of crystal implants is their ability, being crystallographically suitable, to intergrow with the crystalline fiber component of bone tissue, forming a strong unit.

It is noticed that in spite of evident inertness the sapphire implants show osteogenic activity. Study of this event has led us to approach the problem of forming the implant boarder–bone tissue, taking into consideration crystallographic correspondence (Δ) between the crystal lattice of implants with that of crystal fibers (mineral part of the bone tissue) which are inside microfibrils.

$$\Delta = 2(a_{A_1} - a_{B_1})(a_{A_1} + a_{B_1}) + 2(a_{A_2} - a_{B_2}) + 2(\Phi_A - \Phi_B)/(\Phi_A + \Phi_B)$$

where a_{A_1} and a_{A_2}, a_{B_1} and a_{B_2} are periodicity of the location of metal atoms along the first and second chains in the substances A and B. Φ_A, Φ_B are the angles between these chains.

This correspondence played a particular role in the condition of implant adaptation in the body (Lytvynov, 1994). Estimation of the misfit degree for the crystal lattice parameters of hydroxilapatite of bone and the most neutral crystal of different syngonia show (Table 22.2) that sapphire has the minimal misfit $\Delta = 0.2$. Such misfit satisfies the condition of heteroepitaxy.

But implant biocompatibility does not guarantee their functional adaptation in a body. For osteogenes stimulation besides tight insertion in a bone socket it is necessary to activate the surface and create retention points. The surface activation is achieved by making a subsurface layer with a high density of dislocations, point defects, pores, microcracks, which decrease the energy of crystal-nucleus formation and accelerate jointing. Besides constructional hollows on the surface of implants it is desirable to have macropores $100-300\,\mu$m, in which bone tissue can grow through.

Several groups of sapphire implants were elaborated, tested and worked out in practice (Lytvynov, 1998a). The main groups are dental implants, maxillofacial

Table 22.2 Correspondence Δ between closely packed chains of Ca and Al atoms in (A) $Ca_5(PO_4)_3OH$, and (B) Al_2O_3

Flat index		Chain direction index		Periodicity of chains, Å		Angles between chains, deg		Δ
A	B	A	B	A	B	A	B	
(1100)	(1210)	[0001][11$\bar{2}$3]	[0001][20$\bar{2}$1]	6.88 3.89	6.50 3.50	53.9	51.8	0.202
(2111)	(10$\bar{1}$2)	[1123][2201]	[2201][02$\bar{2}$1]	3.89 3.89	3.50 3.50	88.8	85.7	0.246

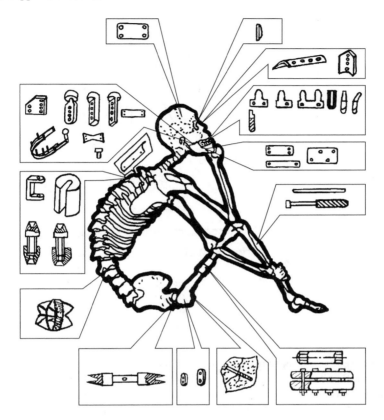

Figure 22.17   Sapphire medical implants.

implants, vertebrals, lenses for eyes, implants for rinoseptoplastic, orthophedy, traumatology (Figure 22.17).

The next medical application is surgery. Success in microsurgery is to a great extent defined by the sharpness of the cutting instrument which in its turn depends on the hardness of the material. Diamond is very expensive, but the next best is sapphire. The probability of formation of keloid and hypertrophic postoperational scars and remission depends on the sharpness of the blade (Lytvynov,1998b).

The blades are made with different profiles and angles of edge (Figure 22.18). The hardness, corrosion resistance, chemical inertness in the medium of human body allow use of sapphire microscalpels in all areas of microsurgery, and big abdominal scalpels in common surgery. Sapphire scalpels ensure less traumatic incision, lower pressure is required for pricking and cutting through tissue, and losses of endothelial cells are lower (Lytvynov, 1999).

For a number of operations performed under a microscope the most efficient are scalpels of blue sapphire. They are more contrasting on the background of tissue.

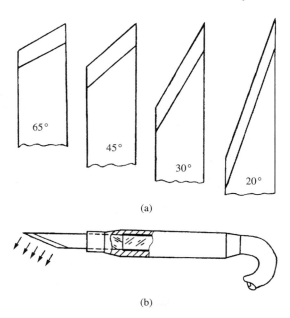

(a)

(b)

Figure 22.18 (a) Sapphire scalpels; (b) Sapphire scalpel with laser light.

Transparency of a sapphire blade leads to a unique possibility – to insert a laser beam via a flexible lightguide, connecting with the back end of the blade, directly into an incision (Figure 22.18). Besides the therapeutic effect and visualization of the sharp edge a possibility to illuminate and determine capillary vascular, nervous and other anatomic formations in the cutting area appears (Kolpakov *et al.* 1999).

22.9 IMPROVEMENT OF STRUCTURE QUALITY OF PROFILE SAPPHIRE

In shaped sapphire the nonuniformity of dislocation density (ρ) distribution over the cross section is more pronounced than in bulk crystals (Figure 22.19). Dislocation-free zones 200 μm in width can appear on the outer surfaces due to their exit to the surface. In the central part a decrease of ρ is observed. In a certain number of cases ρ-free zones are also formed at distances of about 50 μm from the inner surfaces (Dobrovinskaia *et al.* 1994).

The character of ρ-structure depends on the degree of overheating of the melt. In Figure 22.20 the distribution of ρ-density along the crystal length is shown in the growth process of which the overheating of the melt changes. In the region AB the ρ-density manifested itself because of a small overheating of the melt.

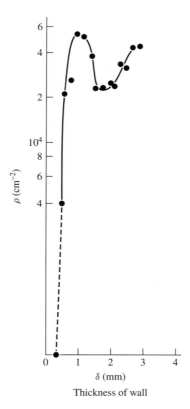

Figure 22.19 Distribution of dislocation density along the cross section of small tubes.

Table 22.3 Critical density of dislocations

Method	Critical dislocation density, cm^{-2}	
	(0001)	(11$\overline{2}$0)
Verneuil	(3–6)10^5	(5–8)10^5
Stepanov	(1–4)10^5	(3–6)10^5
Czochralski	(5–7)10^4	(6–8)10^4
Horizontal Bridgman/Bagdasarov	(2–3)10^4	(4–6)10^4
Kyropoulos	(7–9)10^3	(1–2)10^4

The dislocations, after reaching a critical density, build the grain boundaries. We checked sapphire grown by five methods and found, that a typical ρ-density critical density of dislocation ρ_{cr} might be different by two orders (Table 22.3). This difference depends on the level of residual stresses in the crystals and is determined by the growth method.

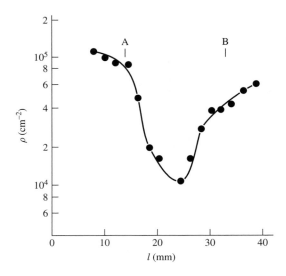

Figure 22.20 Distribution of dislocation density along the crystal length, grown with changing the overheating of the melt.

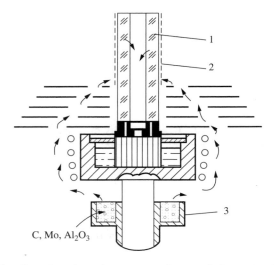

Figure 22.21 Growth and coating process: 1. crystal; 2. coating layer; 3. cap.

Applying the reflecting coating to the external surface of the crystal during its growth allows decreasing ρ-density. Since the powders C, Mo, Al_2O_3 in the cap (Figure 22.21) are evaporated, aluminum carbides and oxycarbides are formed and settle on the crystal surface. The density of coating and their distance from the die were varied by changing the design of the heating unit (Dobrovinskaia

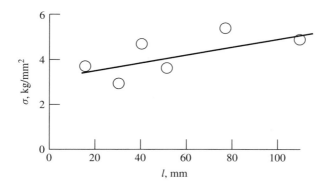

Figure 22.22   Residual stress in the sapphire plate versus the distance between the die and the absorbing coating.

et al. 1990a). With an increase of the distance between the die and the coating layer, the residual stresses become higher (Figure 22.22).

Distribution of residual stresses in the width dimension of the sapphire plate has a complex character (Figure 22.23). While in the center of the plate without an absorbing coating the value of residual stress does not exceed 4 kg/mm, it can reach 8 kg/mm in the periphery of the plate. Using a dense absorbing coating, the stress value drops several orders. The difference between the max and min of residual stress decreases by a factor of two (Dobrovinskaia *et al.* 1990b).

A specific of the Stepanov method – capillary replenishment of the melt layer, is the cause of some peculiarities of structure defects distribution. The pores appear in the places where melt streams come into collision. In multicapillary replenishment the distribution of dense pores depends on crystal length. With increasing capillary length the pressure becomes lower and square areas, where melt streams are mixed, decreased (Figure 22.24), but on further growth, damage on the die surface can occur. After that a cavitation appears and as a result, the concentration of pores increases.

Changing the capillary location and their quantity is possible to control the density and location of the pores (Figure 22.25).

An analogous dependence for residual stresses is also observed on lowering the curvature of the temperature field by introducing scattering centers (small pores) in the crystal volume (Dobrovinskaia *et al.* 1980).

Some possible patterns of the distribution of scattering centers and related distribution of subgrain boundaries are shown in Figure 22.25 for one and two capillary slots. In the places of pore aggregation a more pronounced grain-boundary structure and increased ρ-density are observed.

Control of the location of the ρ-pores gives the possibility to drive back these inclusions to the least important zones of the sapphire article or to zones, that will be removed by subsequent mechanical treatment.

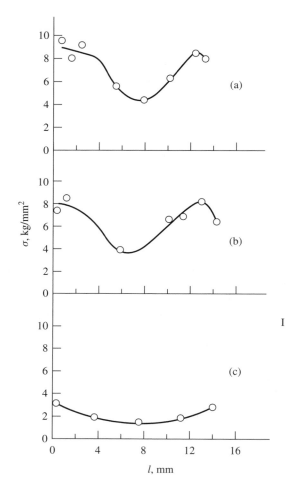

Figure 22.23 Distribution of residual stresses in the width dimension of sapphire plate crystal: (a) without absorbing coating on the crystal surface; (b) with a coating of optical density 1; (c) with a coating of optical density 2.

Figure 22.24 Pores distribution of sapphire profile at 4-capillaries replenishment: crystal length: 10, 155, 300, 450, 600 mm.

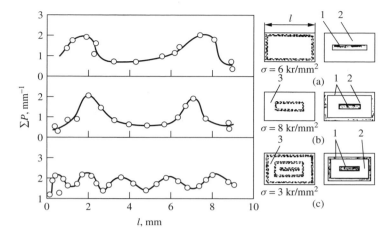

Figure 22.25   The scheme of pores distribution (a,b,c) and variation of grain-boundary length in the thickness cross section of sapphire plate crystals: 1. capillary; 2. die; 3. crystal.

Feeding of melt by one slot capillary and its movement to the remote part of the product by grooves on the die surface make it possible to avoid collision of fluxes in dangerous zones.

In the pore-free crystals, the density of dislocations and length of grain boundaries (ΣP) depends on the stability of thermal and kinetic conditions and the value of the temperature gradient in the growth zone (Figure 22.26). The instability of those conditions is caused by the variation in the crystal shape and the density of structural defects in those zones. With the decrease of the temperature gradient the correlation between instability of growth conditions and variation of the value of those defects becomes weaker. This is true for sapphire as well as for other high-temperature oxides.

The majority of crystal-growth specialists try to minimize structural defects. However, for a number of applications, defects, localized in required places, may enhance the functional property of sapphire products. Even the perennial problem in sapphire – the pores–may be useful, if it is possible to control their distribution in the item.

The Stepanov method allows localization of the high density of structural defects (pores, dislocations, vacancy, vacancies' complexes) in specified zones better than any other method. This is achieved by using a special shape of die. The die defines the location of the melt streams near the butt-end of the die or with its meniscus. Practically these collisions may be formed at any point of the profile cross section. The density of structural defects in those points increases several fold.

The sapphire products with assigned density of defects are designed for metallurgy, laser equipment, and medicine (Figure 22.27).

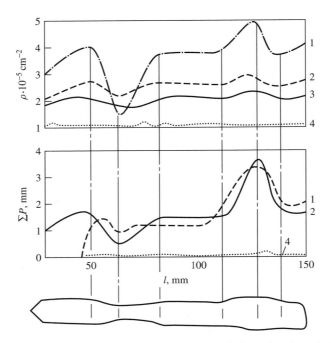

Figure 22.26 Sensitivity of dislocation density and grain-boundary length to the shape changing: 1. Al_2O_3, $dT/dz = 200\,°C/cm$; 2. Al_2O_3, $dT/dz = 150\,°C/cm$; 3. $SrTiO_3$, $dT/dz = 150\,°C/cm$; 4. Al_2O_3, $dT/dz = 50\,°C/cm$.

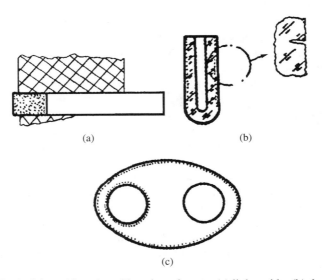

Figure 22.27 Articles with assigned location of pores: (a) light guide; (b) dental implant; (c) laser illuminator.

It is only the pores that scintillate but not those immersed in the liquid metal.

The pores, localized round the channel of the pump lamp and near side surface of the laser illuminator increase the efficiency of laser-rod pumping.

Structural defects on the surfaces of sapphire medical implants promote the growth of dense joint tissue on the implant surface. This phenomenon provides a quicker adaptation of the implant in the organism.

REFERENCES

Antonov, P. I., Nicanorov, S. P. *J. Cryst. Growth* **50/1** (1980) 3–7.

Borodin, V. A., Steriopolo, T. A., Tatartchenko, V. A. Proc. European Meeting on Crystal Growth Materials for Electronics Prague. 1982. Abstract book. p. 320.

Borodin, V. A., Sidorov, V. V., Steriopolo T. A. 12th Int. Conference on Crystal Growth. Jerusalem. 1998. Abstract book. 1998. p. 11.

Byndin, V. M., Lytvynov, L. A., Pischik, V. V. *et al.*, Patent USSR #1603847, 1990.

Dobrovinskaia, E., Lytvynov, L., Pischik V. *J. Cryst. Growth* **50/1** (1980) p. 341–344.

Dobrovinskaia, E., Lytvynov, L., Pischik, V. *J. Cryst. Growth* **104** (1990a) p. 154–156.

Dobrovinskaia, E., Lytvynov, L., Pischik, V. *J. Cryst. Growth* **104** (1990b) p. 165–168.

Dobrovinskaia, E., Lytvynov, L., Pischik, V. Naukova Dumka Kiev, 1994 p. 255. (in Russian).

Kolpakov, S. N., Lytvynov, L. A., Sofienko, G. G. XII intern. Conf. Kharkov. 1999 p. 108–109, 136–137 (in Russian).

LaBelle, H. E. US patent #3471266, 1969.

LaBelle, H. E. *J. Cryst. Growth* **50/1** (1980) p. 8.

Lytvynov L. *Izvestia Academii Nauk SSSR. Seria Fiz.* **#10 T.52** (1988) p. 1911–1913 (in Russian).

Lytvynov, L., Pischik, V. Patent USSR #1503355, 1989.

Lytvynov, L. Forum New Materials. Abstract book. Florence, 1994 p. 237.

Lytvynov, L. *Odontstomatologia de Implantes.* **6(2)** (1998a) p. 89–91 (in Spanish).

Lytvynov, L. The 5th Int. Congress of Wound Association, Tel-Aviv (1998b) p. 87.

Lytvynov, L. *Vestnic Assotsiaysii Stomatologov Ukraini.* **#1(6)** (1999) p. 9 (in Russian).

Stepanov, A. V. *J. Thechnique Fhisic* **29.3** (1959) p. 381–393. (in Russian).

Surek, T., Chalmers, B., Mlavsky A. I. *J. Cryst. Growth* **42** (1977) p. 453–465.

Tatartchenko, V. A., Brener E. A. *Izvestia AN USSR*, **seria fiz. 40** (1976) p. 1456.

Zhilin, A., Koval, Y., Lytvynov L. 12th Intern. Conf. on Crystal Growth, Jerusalem, (1998) Abstract book. p. 284.

23 Halogenide Scintillators: Crystal Growth and Performance

A. V. GEKTIN, B. G. ZASLAVSKY

Institute for Single Crystals, Lenin Av. 60, 61001 Kharkov, Ukraine

23.1 INTRODUCTION

A large number of materials have been found to scintillate, or emit light when ionizing radiation is incident upon the material. The scintillation process remains one of the most useful methods available for the detection and spectroscopy of a wide range of radiations (Birks, 1967).

The most widely used scintillation crystals are alkali halides (preferably NaI-and-CsI based crystals). These crystals combine a good detection efficiency with a high light output. General characteristics of $A^I B^{VII}$ scintillation material are presented in Table 23.1. Recently, along with alkali halide scintillators the ever-increasing application find other halogenide crystals (van Eijk, 1997; Gektin, 1996).

Each scintillator has its specific application. Alkali halide scintillators are widely used for nuclear medicine (NaI(Tl)), high-energy physics (CsI(Tl), CsI pure), geophysics (NaI(Tl), CsI(Na)), environment and security control (CsI(Tl), ^{6}LiI(Eu)) and others.

The market for scintillation detectors annually consumes tens of tons of alkali halide scintillation crystals. This being so, new application fields arise along with traditional ones. They dictate a necessity for R&D of new technologies for more perfect crystal growth. Thus, the growth of alkali halide crystals is a core matter of scientific and technological activity in production and improvement of alkali halide scintillators.

23.2 MODERN TENDENCY IN $A^I B^{VII}$ CRYSTAL GROWTH

23.2.1 R&D FOR HALOGENIDE CRYSTAL PERFECTION

Peculiarities of new applications define the tendencies in research and engineering of crystal growth and treatment. Regarding nuclear medicine application – it is

Crystal Growth Technology, Edited by H. J. Scheel and T. Fukuda
© 2003 John Wiley & Sons, Ltd.   ISBN: 0-471-49059-8

Table 23.1 Physical properties of alkali halide scintillators

	NaI(Tl)	CsI(Na)	CsI(Tl)	CsI (undoped)	CsI(CO$_3$)	LiF(W)	^{6}LiI(Eu)
Density, g/cm^3	3.67	4.51	4.51	4.51	4.51	2.64	4.08
Melting point, K	924	894	894	894	894	1133	719
Thermal expansion coefficient, K^{-1}	47.4×10^6	49×10^6	49×10^6	49×10^6	49×10^6	37×10^6	40×10^6
Cleavage plane	⟨100⟩	none	none	none	none	⟨100⟩	⟨100⟩
Hardness (Mho)	2	2	2	2	2	3	2
Hygroscopic	yes	yes	slightly	slightly	yes	no	very
Wavelength of emission maximum, nm	415	420	550	310	405	430	470
Refractive index at emission maximum	1.85	1.84	1.79	1.95	1.84	1.4	1.96
Light output % of NaI(Tl) (for gamma rays)	100	85	45	5–6	60	3.5	30–35
Primary decay time, μs	0.23	0.63	1	0.01	2	40	1.4
Afterglow (after 6 ms) %	0.3–5	0.5–5	0.1	–	0.06	–	–
Lower wavelength cutoff, nm	300	300	320	260	300	–	425

enlarging the crystal size to that of the human body and changing from flat to curve plate NaI(Tl) detectors (Bicron, 1999).

The next generation of high-energy physics experiments is based on CsI pure or CsI(Tl) calorimeters (Belle, 1995; BaBar, 1995; Ray, 1994). The most important issue for these application is related to radiation damage suppression. (Radiation damage is defined as the change in scintillation characteristics caused by prolonged exposure to intense radiation). Apart from this, deviations of uniformity along long elements of the calorimeter (300–500 mm) must not exceed 2–3 %.

For fast timing applications, undoped CsI are proposed (Kubota, 1988; Gektin, 1990). The main problem of undoped CsI preparation connects with a decrease of slow decay components, i.e. crystals have been developed that exhibit extremely low afterglow. Afterglow originates from the presence of microsecond (for HEP) or millisecond (for CT scanners) decay time components. This defines the tendency of obtaining superpure crystals.

Neutrons can be defected through their interaction with the nuclei of suitable element. In ^{6}LiI(Eu) scintillation crystal, neutrons interact with ^{6}Li nuclei to produce in α-particle and a triton which in turn produce scintillation light. In recent years new Li-based halide scintillators (for example, $LiBaF_3$) are under investigation (van Eijk, 1997).

23.2.2 TRADITIONAL CRYSTAL GROWTH METHODS

Undoubtedly, the simplest from the point of view of technical aspect growth method for alkali halide crystals is Bridgman–Stockbarger (Stockbarger, 1937) which is widely used at present. The most troubling problems of this method are connected with a direct contact of the crystal with the crucible walls. When adhesion between the crystal and crucible material is substantial, as is often the case, the difference in their thermal expansion coefficients causes additional stresses in the growing crystal. Moreover, the grown crystal cannot be extracted unless it is melted all over the surface contacting with the crucible. Sometimes (in the case of crystals with cleavage) this may result in cracking of crystals.

Another problem of this method is nonuniformity of impurity entry into the crystal lattice. It is extremely difficult to attain the uniform activator distribution through the single-crystal ingot.

One disadvantage of this technique is concerned with insufficient convectional melt mixing before the crystallization front. This leads to the appearance of inclusions and striations in the grown crystal.

Development and improvement of methods for pulling alkali halide scintillation crystals from the melt during the last 20–25 years is explained in the first place by the fact that they have a number of advantages as compared to the Bridgman–Stockbarger technique.

The most important of them are:

- absence of a direct contact of the growing crystal with the crucible walls;
- intensive compulsory mixing of the melt by the growing crystal.

Even in the simplest version (without melt replenishment and automatic control of crystallization rate) the Czochralski technique (Czochralski, 1918) allows an increase in crystal growth rate of several times owing to a higher axial and radial temperature gradients and due to an intensive mixing of the melt by the rotating crystal. Besides, this technique seems to be the most efficient when crystals are required with high structural perfection. At the same time the methods of pulling on a seed are more complicated technically; they need permanent control and correction of the main parameters, such as temperature of the melt and pulling rate. One can state with confidence that for nonautomated pulling the success of growth is defined in many aspects by the skill of the operator. When the growth process lasts for tens or hundreds of hours the need for automatic control of pulling becomes especially important.

23.2.3 AUTOMATED GROWTH PRINCIPLES AND TECHNIQUE

There are several ways of automated control of crystal pulling process. The best known of them are: a direct measuring of the growing crystal diameter using X-rays or optical sensor, the use of the melt weight gauge or weighing the growing crystal and use of a melt-level sensor. Certainly, for the automation of the process of pulling large alkali halide crystals, the duration of which is tens and hundreds hours, the melt-level sensor is the most applicable. The progress in the creation of modern highly efficient apparatus and growth techniques for alkali halide crystals of several hundreds of kilograms in mass became possible owing to the creation of a simple and reliable electric contact melt-level sensor (Belogurov, 1970). The following advantages of the electric-contact melt-level sensor should be mentioned additionally:

- high accuracy of measuring the melt level (about $200\,\mu m$) which provides a high precision of controlling the diameter of the growing crystal, this compares favorably with the use of a weight gauge;
- independence of measuring the melt level upon possible variation of the perimeter shape of the growing crystal (from circular to the faceted or ellipsoidal one).

Perhaps the only disadvantage of such sensors is the fact that it assumes a lowering of the melt level, due to evaporation, as a change of crystal growth rate. However, there are means that make this evaporation negligible, e.g. create a certain composition and pressure of gas atmosphere above the melt. Besides, at growth upwards the evaporation velocity becomes constant. This gives a systematic constant error that can be easily allowed for and it practically has no effect on the accuracy of the diameter regulation.

One of the first automated installations for growing large alkali halide crystals was created in 1981. It included the electrocontact sensor for the growth rate control. In accordance with (Goriletsky, 1981) seeded single-crystal pulling is carried out simultaneously with feeding of the raw material, which melts at the bottom of a circular gutter 1 of a platinum crucible 2 and runs down into the melt (see Figure 23.1). The crucible is suspended on a rotary support ring 3 ensuring, jointly with the rotating crystal holder, the axial symmetry of thermal fields in the melt and growing crystal. The melt-level gauge 4 with its platinum probe 5 responds to the slightest changes of melt column height in the crucible. Across a correcting circuit 6 the gauge regulates the temperature of the bottom heater 7 (at constant temperature of the lateral heater 8) so that the mass growth rate of the crystal is equal to the feeding rate $\dot{m}$. It follows from the mass balance equation that

$$d = 2(\dot{m}/\pi\rho_s V_p)^{1/2} \tag{23.1}$$

where d is the diameter of the growing crystal, V_p is the pulling rate, and ρ_s is crystal density.

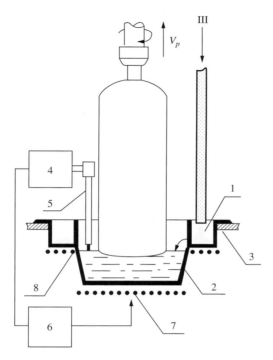

Figure 23.1 Diagram of the automated crystal pulling unit; (1) crucible circular gutter; (2) crucible; (3) supporting ring; (4) melt-level gauge; (5) probe; (6) correcting circuit; (7) bottom heater; (8) side heater. The growth unit and the source of material are enclosed in different pumped volumes that are connected with each other.

It is evident from Eq. (23.1) that d is determined by the ratio $\dot{m}/V_p$; hence, stability of the feeding rate $\dot{m}$ and pulling rate V_p automatically ensures the stability of the preset crystal diameter. Therefore, by programming the ratio $\dot{m}/V_p$ it is possible to ensure automated broadening of the crystal starting from the seed and up to the final diameter with any crystal profile. If it is not desirable to vary the axial rate of crystal growth (equal to pulling rate V_p) automated broadening can be achieved only by programming the feeding rate $\dot{m}$. A linear change of $\dot{m}$ with time is most convenient because it results in the smooth decrease of radial growth rate as crystal diameter increases; the shape of the broadened portion of the crystal should conform with a paraboloid of rotation. Suppose $\dot{m}$ changes as follows:

$$\dot{m}(t) = \dot{m}_i + \frac{\dot{m}_f - \dot{m}_i}{t_b} t \qquad (23.2)$$

where $\dot{m}_i$ and $\dot{m}_f$, are the initial and final values of the feeding rate, respectively, t_b, is the radial growth time, and t is the current time. By substituting the value of $\dot{m}$ from Eq. (23.1) into Eq. (23.2), we obtain

$$\frac{1}{4}\pi d^2 V_p \rho_s = \dot{m}_i + \frac{\dot{m}_f - \dot{m}_i}{t_b} t \tag{23.3}$$

If we designate the length of the growing crystal by l, then $t = l/V_p$, and we obtain from Eq. (23.3)

$$d = 2 \left(\frac{\dot{m}_i}{\pi V_p \rho_s} + \frac{\dot{m}_f - \dot{m}_i}{\pi V_p^2 \rho_s t_b} l \right)^{1/2} \tag{23.4}$$

showing that the dependence of d on l is parabolic.

An example of this growth method realization was described for the first time in (Goriletsky, 1981). The apparatus allowed pulling single crystalline boules with a diameter to 450 mm and to 750 mm long. As a result, KCl crystals 140 kg in weight (diameter 450 mm) and CsI(Na) 115 kg in weight (diameter 300 mm) have been grown. Now this technique allows putting of such crystals with diameter more than 500 kg weight. Figure 23.2 shows the view of this type furnaces.

Figure 23.2   'Rost'-type single-crystal growth furnaces.

23.3 MODIFIED METHOD OF AUTOMATED PULLING OF LARGE-SIZE SCINTILLATION CRYSTALS

The automatization of the pulling process intended to stabilize the mass rate of the crystal growth or its diameter is performed using a sensor of the melt level. When pulling large-diameter single crystals for the melt, the main difficulties are associated with the following circumstances:

- a sufficient radial temperature gradient on the melt surface providing stable radial growth is difficult to assure;
- insufficient process information hinders the effective use of the melt-level sensor and the automatization of radial growth;
- the high vapor pressure of volatile activator leads to its evaporation and leads to inhomogeneous distribution in upper part of the crystal.

This was the reason for the appearance of another, modified technique for the large single-crystal growth (Zaslavsky, 1999). There are three main differences of this method and apparatus from those described before.

First, it is the availability of closed feeder with melted raw material. This gives a possibility to perform additional processing of the raw material in the RAP atmosphere.

Secondly, feeding of the raw material in the crucible is made by the melt.

Thirdly, crystal pulling proceeds from a conical crucible that is not too deep.

23.3.1 PRINCIPLES OF THE METHOD

The drawbacks and difficulties mentioned above, which are inherent in pulling methods, can be eliminated provided that the free surface of the melt is as small as possible both in the initial stage (the radial growth) and in the subsequent stages, it is independent of the crystal diameter. The diameter of the initial melt surface must be comparable to that of the seed crystal and then increase along with that of the radially growing crystal. The requirement for the variable melt-surface geometry is simple to meet if on the radial growth stage the crystal is pulled from the melt under a continuous level elevation of the latter in a variable-section crucible, for example, in a conical crucible. The method concept and growth stages are presented in Figure 23.3. The initial stage of radial growth is done in the lowest part of a conical crucible where the melt surface diameter, d_1, is comparable to that of the seed (Figure 23.3(a)).

When radial growth is initiated, the crystal is pulled up at a V_p. Simultaneously, the melt level is elevated at a rate V_1 by means of feeding the raw material at a mass rate, $\dot{m}$, such that $V_p \geq V_1$. As the melt level elevates and its surface diameter increases, the single crystal radial growth is done (Figure 23.3, stages b–d); the melt temperature being adjusted so that the linear speed of the radial

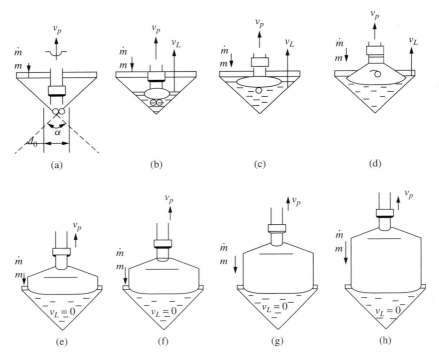

Figure 23.3 Stages of growth: (a) starting stage, (b–d) radial growth, (e–h) vertical growth.

growth is essentially equal to that of the melt surface diameter increase. Thus, the radial crystal growth from the seed diameter up to the predetermined value is performed at a minimum free surface of the melt. When the radial growth is completed, the melt level position is stabilized ($V_1 = 0$) and the crystal grows in height (Figure 23.3, stages e–h).

23.3.2   THE METHOD MODEL AND THE PROCESS CONTROL PARAMETER

To describe the growth process, let us analyze the mass balance equation, since it defines the correlation between the new control parameter and other main ones, for example, the pulling speed and the crystal diameter or cross-sectional area (Zaslavsky, 1999).

At the stage of radial growth, the feeding mass rate, $\dot{m}$, is defined as the sum of the crystal growth rate, $\dot{m}_s$ and that of the melt mass increase in the crucible $\dot{m}_1$

$$\dot{m} = \dot{m}_s + \dot{m}_1 \qquad (23.5)$$

Taking into account that the crystal pulling rate measured with respect to the melt surface is $(V_p - V_1)$, the mass balance equation can be written as

$$\pi[d_s(t)]^2(V_p - V_1)\rho_s + \pi[d_1(t)]^2 V_1\rho_1 = 4\dot{m} \qquad (23.6)$$

where $d_s(t)$ and $d_1(t)$ are the crystal diameter and the melt surface diameter, respectively, at any time t, ρ_s and ρ_1 are the crystal and melt densities, respectively.

It follows from Eq. (23.6) that the diameter of the crystal being grown in the radial direction is

$$[d_s(t)]^2 = \{4\dot{m} - \pi[d_1(t)]^2 V_1\rho_1\}/\pi(V_p - V_1)\rho_s \qquad (23.7)$$

In the case of a conical crucible with the vertex angle α

$$d_1(t) = 2t\,V_1 \tan\alpha/2 + d_0 \qquad (23.8)$$

where d_0 is the melt surface diameter at the initial moment of radial growth (Figure 23.3(a)).

At $V_1, = 0$ (radial growth is completed, Figure 23.3(e–h)) $d_s(t) = (4\dot{m}/\pi V_p\rho_s)^{1/2}$ this corresponds exactly to the case of pulling from a cylindrical crucible (see Equation (23.1)), that is, the shape of the latter is of no further significance. It is seen from Equations (23.7) and (23.8) that the crystal diameter is defined by the ratio of $\dot{m}$, V_p and V_1 on the radial growth stage and by that of $\dot{m}$ and V_p on the axial growth one.

On the other hand, axial and radial temperature gradients in the growing crystal and the melt temperature are of critical importance for the crystal diameter or the crystallization mass rate. Since the melt (or the crystal) temperature is an independent parameter defining the crystal-growth mass rate, the automated control of the crystal diameter at a constant pulling rate can be done by means of feedback, provided that a certain parameter indicative of the growth rate affects the melt temperature. Temporal variations of the crystal (Kimoto, 1973) or the melt mass (Zinnes, 1973), the melt level variations (Eidel'man, 1985) or those of the crystal diameter can be used as control parameters. It is just the melt level variation that is the preferential informative parameter when growing large alkali halide single crystals where the process duration is several hundred hours, because relatively simple and reliable electrocontact level sensors are available (Belogurov, 1970).

When realizing the method under consideration, a melt-feeding system including a feeder and an electrocontact probe is used to elevate the melt level in a conical crucible during the crystal radial growth and then throughout the course of its axial pulling. On the radial growth stage, the melt level in the crucible is elevated by lifting the probe at a preset speed. When the crystal is grown in height, the probe displacement is stopped, so the melt level remains constant.

Thus, the melt level in the crucible cannot be used as the control parameter characterizing the mass rate of the crystal growth, because it is defined only by the probe position and is independent of the crystal diameter, pulling rate or any other factor of importance.

The exact information on the mass growth rate is contained in the feeding regimen, if this regimen is controlled by the probe signal. The central point is that the limiting meniscus height, Δh (height of the melt column between the probe tip and the melt surface at the moment when it breaks), is strictly constant and depends only on the tip area and the melt properties (in particular, its surface tension). Since the feeding is done only when the tip is out of contact with the melt, i.e. when the limiting meniscus becomes broken, it is performed in a discrete regime. The melt mass, Δm_0 required to resume the probe contact with the melt surface in the crucible is defined by the Δh value and the free-melt surface area (area of ring enclosed between the crucible wall and the crystal):

$$\Delta m_0 = \pi\{[d_l(t)]^2 - [d_s(t)]^2\}\Delta h \rho_l/4 \tag{23.9}$$

Since the crystal pulling and the probe displacement are done in a continuous manner, the meniscus becomes broken in certain time intervals, when its height attains its limiting value so that the surface tension becomes unable to maintain it. When the meniscus is broken, an ordinary melt dose Δm_0 comes to the crucible, the contact is renewed to be broken again after the time period $\Delta\tau$, and so on.

The feeding mass rate can be presented as

$$\dot{m} = \Delta m_0/\Delta\tau = \pi\{[d_l(t)]^2 - [d_s(t)]^2\}\Delta h \rho_l/4\Delta\tau \tag{23.10}$$

It follows from Equation (23.10) that the constancy of $\Delta\tau$ evidences that the free melt surface $[d_l(t)]^2 - [d_s(t)]^2$ is constant or the crystal diameter $d_s(t) =$ const. (in the case when the melt level is stable, $d_l(t) =$ const.). If the melt level in a conical crucible elevates linearly in time (that is equivalent to a linear increase of $[d_l(t)]^2$ then, the constancy of $\Delta\tau$ points to a linear increase of the growing crystal cross section. Thus, $\Delta\tau$ (time intervals between the feeding operations) in an exact parameter bearing information on the growing crystal diameter convenient to govern an automated pulling system. An increase in $\Delta\tau$ evidences a diminution of the crystal diameter while, in contrast, its decrease points out that the diameter grows.

From Equations (23.7) and (23.10), a relationship between the crystal diameter and the control parameter is obtained

$$[d_s(t)]^2 = [d_l(t)]^2(\Delta h \rho_l - V_l\rho_l\Delta\tau)/[\Delta h \rho_l + (V_p - V_l)\rho_s\Delta\tau] \tag{23.11}$$

Realizing feedback between $\Delta\tau$ and the melt temperature allows to control the crystal growth mass rate or it diameter. Note that no additional sensor is necessary to measure $\Delta\tau$. The information is obtained from the electrocontact probe that governs the feeding. The time intervals are therefore measured from the cessation of feeding (restoring the probe contact with the melt) to the start of the next feeding (the contact break). Since $\Delta\tau$ is measured between each two feedings, the process information and, consequently, the control efficiency increases by several orders when a conical crucible is used. Assuming, for simplicity sake, $V_l = 0$, we obtain from Equation (23.11) that the process information is determined as

$$1/\Delta\tau = [d_s(t)]^2 V_p \rho_s / \{[d_l(t)]^2 - [d_s(t)]^2\} \Delta h \rho_l \qquad (23.12)$$

It follows from Equation (23.12) that, at a constant V_p value, the feeding frequency $1/\Delta\tau$ depends on the ratio of the crystal cross-sectional area to that of the melt free surface. So, for example, when a crystal is grown-out radially in a 500 mm diameter. cylindrical crucible on a 100 mm diameter. seed, that ratio amounts to 0.04, while it is 1.25 in the case of a conical crucible where the initial melt surface diameter $d_0 = 150$ mm. That is, on the initial radial growth stage, at an usual pulling speed, e.g., 5 mm/h, and a real limiting meniscus height amounting to 0.2 mm, the information on the current crystal diameter will come to the control system every 2400 s in the first case and every 125 s in the second one.

23.4 EXPERIMENTAL AND PRACTICAL METHOD REALIZATION

Principles of process control stated above have been realized in the industrial growth unit 'Crystal-500' (Zaslavsky, 1999). The unit block diagram is presented in Figure 23.4. The platinum feeder 1 containing the melt is placed in a sealed growth furnace 2 under the crucible 3 also made of platinum. The feeder is shaped as a ring-container of rectangular cross section, thus making it possible to arrange the bottom heater 4 in an immediate vicinity of the conical crucible vertex and to effectively control the crystallization front shape. The feeder capacity is about 270 kg of cesium iodide or about 215 kg of sodium iodide. The melt is supplied to the crucible using an inert gas (Ar or N_2) pressure. The necessary gas pressure in the feeder is provided by an electromagnetic valve 5 controlled by the melt-level control block 6. A platinum probe 7 is included into two control circuits simultaneously: the feeding control one and that intended to measure deviations of actual $\Delta\tau$ values from those set by the control system based on a personal computer 8. A signal proportional to that deviation is used in the block 9 to correct automatically the temperature and thus to eliminate the mentioned deviation. The control system so designed provides the constant crystal diameter at an accuracy of 1 % (at a diameter of 400–450 mm and pulling speeds 6.0–6.5 mm/h (Figure 23.4)). The activator distribution nonuniformity over the single-crystal volume does not exceed 10 %. Table 23.2 presents axial and radial Tl distribution in one of CsI(Tl) crystals, 420 mm in diameter and 400 mm in height. Radial distribution was measured in the layer cut from the upper area of the cylindrical crystal. Axial distribution was measured in the column cut parallel

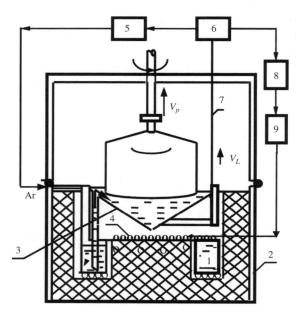

Figure 23.4 Block-scheme of 'Crystal-500' installation: (1) feeder, (2) growth chamber, (3) crucible, (4) bottom heater, (5) gas valve, (6) melt-level regulator, (7) probe, (8) PC, (9) temperature-correction block.

Table 23.2   Tl distribution in grown CsI(Tl) crystal

Radial distribution								
Distance from center (mm)	0	25	50	75	100	125	150	175	200
$C_{Tl} \times 10^3$ (mass %)	80	83	89	86	80	82	88	87	88
Axial distribution								
Height (mm)	40	80	120	160	220	260	300	340	400
$C_{Tl} \times 10^3$ (mass %)	87	86	85	89	87	85	82	90	88

to the cylinder axis. The polarographic method was used to measure the activator concentration (C_{Tl}). Figure 23.5 and 23.6 show the general view of 'Crystal-500' type growth furnace and CsI(Tl) large size single crystal.

This modified method has an extra advantages:

- the crystal is grown at a minimum and controllable free melt surface, thus providing a stable radial growth of a crystal of any prespecified diameter within limits of the crucible diameter;

Figure 23.5 View of 'Crystal-500' crystal growth furnace.

Figure 23.6 Typical CsI(Tl) crystal, grown in the 'Crystal-500' installation: seed diameter – 70 mm; crystal diameter – 450 mm.

- the whole radial growth stage, from the seed to the final prespecified value, is performed in an automated regime;
- the activator evaporation intensity is decreased substantially, for example, in growing Na(Tl) and CsI(Tl) crystals, thus providing an uniform activator distribution over the crystal volume;
- the method is versatile and makes it possible to grow single crystals of any prespecified diameter (within limits of the crucible one) due to the melt-surface diameter control, as well as to grow multicomponent crystals;
- the melt feeding makes it possible to purify it additionally in the feeder in the course of the growth process, thus eliminating both impurities insoluble in the melt and soluble ones, for example, oxygen-containing anions, as well as to use fragments, shavings, cuttings, etc., formed at crystal machining, as the raw material.

23.5 SCINTILLATOR QUALITY

The quality of scintillation single crystals is affected preferably by the host crystal purity, activator distribution, crystal transmittance and other. To assure these parameters the crystal growth technique and technology have to be variable.

23.5.1 ACTIVATED SCINTILLATORS

The largest share of halogenide scintillators is based on the activated alkali halides and the activator uniformity is the critical parameter for any application. This section is devoted to the activator distribution investigation for both described earlier automated techniques.

In terms of the impurity distribution in the growing crystal the method under consideration is similar to zone melting (Pfann, 1970), the melt contained in the crucible playing the role of the melted zone. The peculiarity lies in the fact that during crystallization the impurity transfer through the free surface of melt may occur. In this case one can use the relation valid at $V_p = $ const (Boomgaard, 1955):

$$C = C_f^* - (C_f^* - C_0) \cdot \exp\left\{-(\xi K + \eta)\frac{l}{V_p}\right\}, \qquad (23.13)$$

where $\xi = \dfrac{\dot{M_s}}{M_l}$, $C_f^* = K\dfrac{\xi C_{f-\eta}C_l}{\xi K + \eta}$, C is impurity concentration in the crystal, C_0 is initial concentration value, K is distribution coefficient, l is length of crystal, M_s is mass of melt crystal, M_l is mass of melt, C_f is impurity concentration in the make-up material, C_l is equilibrium impurity concentration in the melt (established in the melt at the absence of crystallization), η is parameter characterizing the impurity interaction with the environment. Equation (23.13) shows that the trend of impurity distribution over the length of growing crystal is governed by the sign of the expression $(C_f^* - C_0)$. When growing the undoped single crystals, if maximum purification from impurities contained in the starting material is required, the condition $(C_f^* - C_0) < 0$ should be satisfied. Therefore, in this case the relation should be held.

$$\frac{\dot{M}_s}{M_1} < \frac{\eta}{K} \frac{C_0 - kC_1}{C_f - C_0} \tag{23.14}$$

According to (Eidel'man, 1985), NaI(Tl) single crystals were grown with due account for the high rate of activator evaporation from the melt. Make-up with pure raw material was performed at the stage of radial growth. When the preset final diameter of crystal was reached, the melt was fed with charge where the activator concentration C_{f_1}, was high enough to ensure adequately rapid rise of thallium concentration in the growing crystal up to the preset optimum value $C_{opt.}$. At the moment of reaching $C_{opt.}$ activator concentration in the growing crystal the TlI concentration in make-up material was reduced to $C_{f_2} > C_{opt.}(1 + \eta/\xi K)$ and then remained unchanged. Then the thallium concentration in the melt was maintained constant in time and equal to $C_{opt.}/K$, thus ensuring uniform activator distribution in the remaining portion of the growing crystal at the preset value of $C_{opt.}$. The chemical analysis data indicated that the deviation from the average activator concentration in 600 mm long single crystals of NaI(Tl) grown in compliance with the above technique remained within $\pm 7\%$.

A modified technique is based on the feeding by the melt with dissolved activator salt. To assure uniform activator distribution, the crucible is fed with the melt containing the necessary activator concentration selected on account for its evaporation. The stability of the mass growth rate also favors uniform activator distribution. An activator nonuniformity less than 5 % by mass over the whole crystal volume is permissible to provide the light yield homogeneity and, consequently, a high-energy resolution.

As was clear from Sections 23.3 and 23.4 the modified technology of crystal growth allows suppression of the activator evaporation due to minimization of the free-melt surface at the radial stage of crystal growth and the preservation of a constant activator concentration in the feeder.

23.5.2   UNDOPED SCINTILLATORS

As follows from the Table 23.1 only pure CsI single crystals is used for the scintillation applications (Gektin, 1990; Ray, 1995; Kubota et al., 1988) due to the fast UV intrinsic luminescence. The typical scintillation decay curve is shown at Figure 23.7. Usually among fast scintillations there are some undesirable pulses (afterglow) with microsecond decay times overlapping the fast components ($t \sim 2$, 10 and 30 ns). The presence of afterglow, its intensity, spectral composition and decay time depend on crystal pulling conditions. Besides UV emission (310 nm)

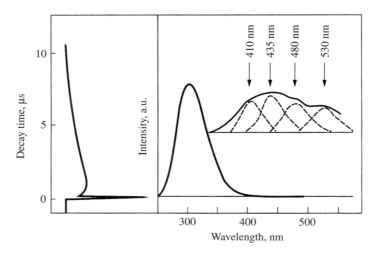

Figure 23.7 Scintillation decay curve and emission spectra (fast and slow components) of CsI single crystal, 300 K, ^{137}Cs.

luminescence spectra contain several additional bands with maxima at 410, 430, 470 and 530 nm (Figure 23.7). Their decay times are between 2 and 30 μs.

Comparing absorption, luminescence and excitation spectra with IR-spectroscopy data, the origins of harmful luminescence centers were clarified (Gektin, 1992). Scintillations with a 410-nm maximum are conditioned by the CO_3^{2-} ions traces. The luminescence at 430 nm is usually present in crystals with a high content of oxygen impurities. This emission connected with the O^{2-} or O_2^{2-}-type centers. Weak luminescence in region 460–700 nm tends to the equidistant structure that is typical for the O^{2-} centers in alkali halide crystals. The Na and Tl ions presence is absolutely prohibited, to avoid scintillation typical for the CsI(Na) and CsI(Tl). Yet one luminescence band in the blue region is observed in single crystals with an extra content of intrinsic point defects vacancies, first of all. This emission is connected with exciton transitions perturbed by structural defects: bivacancies or their aggregates.

The raw materials being used, in particular, cesium and sodium iodides, are usually rather pure and quite suitable to produce commercial CsI and NaI scintillators on which no special requirements are placed. Only a thorough high-temperature drying before melting in the feeder is required, provided that the melt hydrolysis resulting in carbonates and other oxygen-containing impurities are prevented. The process of high-temperature drying is the evacuation of the furnace volume within 14–16 h at 400 °C.

The inert gas is used in the melt-feeding system. An insufficient gas purity may cause melt contamination, therefore, it is purified additionally from oxygen and moisture using chemisorption. The total oxygen and moisture content in the inert gas does not exceed 0.001 vol %.

But the standard accuracy and raw material treatment procedure are not enough for the undoped scintillators. The crystals must meet some special requirements, e.g. enhanced radiation resistance, since the chips resulting from crystal machining are used as the raw material, an additional melt purification is performed immediately in the feeder to remove oxygen-containing impurities. To that end, a small amount of a metal-forming oxides essentially insoluble in the melt is added to the latter; such oxides settle to the feeder bottom. To prevent melt hydrolysis, a small amount of elemental iodine is introduced into the feeder. Such treatment of the melt allows growth of single crystals containing oxide anions, e.g. CO_3^{2-} in amounts lying at the sensitivity threshold of optical analysis methods (IR spectroscopy). The absorption of such crystals on the emission spectral maximum does not exceed $0.01\,\mathrm{cm}^{-1}$.

23.6 CONCLUSION

In this brief review are described both conventional and modern methods and apparatus for growing halogenide scintillation crystals. It can be seen that modifications of the methods and apparatus are focused on two main aims.

First, an increase of efficiency of the growth technique. With this, the task of growing large-size crystals is solved. Besides, the automation of the growth process allows to increase the industrial yield of crystals with high perfection of scintillation parameters.

Secondly, it is the realization of special conditions (additional purification of raw material and/or melt, control and regulation of activator distribution along the crystal growth axis, suppression of activator evaporation etc.) which ensure growth of crystals for unique applications.

The analysis of the dynamics of the requirements to the crystal quality allows a necessity of further modification of the technology for growing halogenide crystals to be forecast. Therefore, in conclusion, it should be particularly noted that despite the above-described technological developments there is an apparent perspective for perfection of techniques and apparatus for crystal growth.

REFERENCES

The BaBar Collaborations, *BaBar Technical Design Report*, SLAC-R-95-457 (1995).

The Belle Collaborations, *Technical Design Report*, KEK Report 95-1, April 95 (1995).

Belogurov Yu. P., Kostenko V. I., Nechmir V. M. and Radkevich A. V., *Prib. i Sistemy Upravlenya* **8** (1970) 48.

Bicron 1999, http://www.bicron.com.

Birks J. B., *The Theory and Practice of Scintillation Counting*, Pergamon Press, Oxford, London, Edinburgh, New York, Toronto, Sydney, Paris, Braunschweig (1967) p. 662.

Boomgaard Van Den J.: *Philips Res. Rept.*, **10** (1955) 319.

van Eijk C. W. E., New scintillators in *Proc. SCINT 97* ed. by Yin Zhiwen, CAS, Shanghai Branch Press (1997) p. 3.

Czochralski J., *Z. Physik. Chem.* **92** (1918) 219.

Eidel'man L. G., Goriletsky V. I., Nemenov V. A., Protsenko V. G., Radkevich A. V., *Crystal Res. Technol.* **20** (2) (1985) p. 167.

Gektin A. V., Shiran N. V., Charkina T. A. *et al.*, Heavy Scintillators for Scientific and Industrial Applications, Frontiers, France, (1992) p. 493.

Gektin A. V., Gorelov A. I., Rikalin V. I. *et al.*, *Nucl. Instrum. Methods* **294A, N3** (1990) p. 591.

Gektin A. V., Ivanov N. P., Nesterenko Y. A. *et al.*, *J. Cryst. Growth* **166** (1996) p. 419.

Goriletsky V. I., Nemenov V. A., Protsenko V. G. *et al.*, *J. Cryst. Growth* **52** (1981) p. 509.

Kimoto T., Wachi A., Sakurai S., Mikami M., *Trans. Soc. Instrum. Control Eng.* **9** (95) (1973) 595.

Kubota S., Sakuragi S., Nasshimoto S., Ruan J., *Nucl. Instrum. Methods* **A268, N1**, (1988) 275.

Pfann V., Zonnaya Plavka (Zone Melting) (Mir, Moscow, 1970) (Russian edn.).

Ray R. E., *The KTeV Pure CsI Calorimeter, Proc. Fifth. International Conference on Calorimetry in High Energy Physics*, World Scientific, New Jersey, Sept. 94 (1994) p. 110.

Stockbarger D. C. *Rev. Sci. Instrum.* **7** (1936) 133.

Zaslavsky B. G. *J. Cryst. Growth* **200** (1999) 476.

Zinnes A. E., Nevis B. E., Brandle C. D. *J. Cryst. Growth*, **19** (3) (1973) 187.

Part 5

Crystal Machining

24 Advanced Slicing Techniques for Single Crystals

C. HAUSER AND P. M. NASCH

HCT Shaping Systems SA, 1033 Cheseaux, Switzerland

ABSTRACT

For low-cost and high-precision slicing of hard and brittle monocrystalline materials for electronic applications, manufacturing problems are complex. Only a few methods exist, among which only the internal diameter (ID) and the multiwire slurry (MWS) cutting techniques are nowadays commonly used in production lines. Two different material-removal processes are involved. For abrasive bonded on the cutting rim of the ID blade (fixed abrasive), material is removed by a plunge grinding cutting action. In the MWS material is removed by third-party free abrasive grains transported in a liquid media (slurry). The cutting action is essentially that of a fast three-body lapping process characterized by a *rolling and indenting* cutting mechanism. Abrasive processing essentially induces mechanical and thermal effects at the locus of kerf generation. The overall depth of penetration of these damages is dependent upon the dynamics of material removal process. Subsurface depth of damage caused by free abrasive machining is reduced by as much as 50–60 % compared to that produced by fixed abrasive. The use of single crystals for optical or semiconductor applications often requires to slice the crystals with a very precise orientation relative to the crystal lattice. A model is proposed that allows simultaneously the geometrical axis of the ingot to be horizontal with the displacement of the cutting device being parallel to the targeted cutting plane of the single crystal. This property is of prime importance since it reduces cutting time and allows optimized (multi-ingot) loading of the slicing machine.

24.1 INTRODUCTION

Generally, monocrystalline materials of industrial applications such as silicon, GaAs, YAG, InP, III-V compounds, carbides, quartz, ceramics are hard (few tens of GPa) and rigid, rendering their machining and slicing particularly difficult. Crystal slicing is the first post-crystal-growth step toward producing a

Crystal Growth Technology, Edited by H. J. Scheel and T. Fukuda
© 2003 John Wiley & Sons, Ltd.   ISBN: 0-471-49059-8

polished wafer for electronic device fabrication in semiconductor and photo-voltaic industries. For low-cost production of single-crystal slices, manufacturing problems are complex because of the requirements of minimum kerf loss (i.e., high material utilization), minimum surface damage, minimum warpage and thickness variation, high production rates and yields, and machining impacts on downstream processes. The wire saw has emerged as a leading technology for ultraprecision machining of large cross-sectional area wafers in semiconductor and photovoltaic industries.

24.2   GEOMETRICAL PARAMETERS

For electronic applications, four geometrical parameters are essential: (i) the wafer's central thickness (TV); (ii) the total thickness variation (TTV); (iii) warp; and (iv) bow. Also of great importance are surface-quality parameters such as rugosity and surface damages. For photovoltaic (PV) applications, only wafer thickness and TTV are critical. The wafer's central thickness, TV, is a measure of thickness distribution within a production batch (one thickness measurement per wafer done at the center), whereas TTV is a measure of thickness distribu-tion within a single wafer (multiple thickness measurements per wafer). TTV is defined as the difference between maximum and minimum thickness. TV is generally due to tool wear and the tool-feeding mechanism. TTV is caused by tool wear and the local variation in cutting efficiency.

The warp is a 3-D topographical feature of the wafer's geometrical surface rel-ative to a mean surface. Figure 24.1 is an example of a measured warp diagram of a silicon wafer that shows, in 2-D, the topographical variations. Such data are usually acquired by passing the entire wafer surface between two capacitive

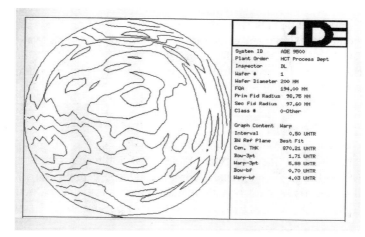

Figure 24.1   Typical warp mapping of a single-crystal silicon wafer as cut with wiresaw.

probes and measuring the potential variations associated with topographic variations. The warp is generated by instability of machine parts such as those induced by temperature variations (thermal expansion) and internal stresses and is therefore an excellent signature of cutting-process variations and crystal-manufacturing (growth) stability. Warp is an important problem to the electronic industry as it is very difficult and costly to reduce it by means of after-cut processes.

Bowage is a curvature with circular symmetry typically created by the release, during the cut, of internal stresses that were introduced in the crystal during crystallization. It is hence essentially a crystal-growth problem. Unlike TV and TTV, the warp and the bow can not be eliminated by subsequent treatment such as polishing, grinding and etching. For this reason they must be minimized at the cutting or growing stage.

Surface damage is dependent upon the material removal process (see below) and is caused by local variations of tool pressure and grit diameter. Surface roughness in the micro-range $0.5-5\,\mu m$ is dependent upon process parameters such as cutting force, vibrations, tool tensioning, cutting rate, and is related to grit size. Medium- to large-amplitude undulation such as waviness, caused by sudden variation of process control or material inhomogeneity such as hard inclusions (e.g., SiC, SiN in cast polycrystalline Si), is evidently also very detrimental to wafer quality.

24.3 SURVEY ON SLICING METHODS FOR SILICON SINGLE CRYSTAL

The small number of existing methods illustrates the difficulty for large-scale production of high-precision slicing of single crystals for electronic applications:

- Conventional internal diameter (ID)
- Outer-diameter (OD)
- Grind-slice ID (GS)
- Fixed abrasive multiblade (MBF)
- Fixed abrasive multiwire (FAST^{TM})
- Multiblade slurry (MBS)
- Multiwire slurry (MWS)

Figure 24.2 depicts the ID sawing principle. The ID sawing unit employs a stainless steel blade core with the cutting edge on the inner diameter of a centered hole in the blade. The horizontal workpiece is fed into the center hole and wafer slicing is accomplished with a vertically moving blade arrangement. In some instances, this configuration has been changed to a horizontal blade and a vertical workpiece. The blade is mounted on a ring driven by an electric motor. Very critical to the proper functioning of the saw and for reduction of wafer warpage is the uniform tensioning of the blade within the clamping head

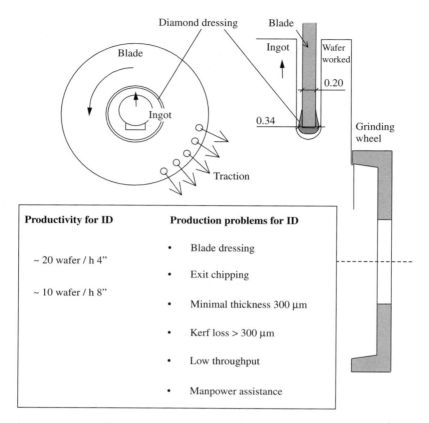

Figure 24.2   Internal-diameter (ID) sawing.

in order to assure adequate axial rigidity. Tensioning is accomplished either by means of a mechanical or hydraulic system, which stretches the blade radially, and produces the elongation required for keeping the blade in a rigid configuration. Main development efforts on conventional ID cutting are still focused into different methods for blade-deflection control such as compressed-air nozzles and rotation-speed control. The inner rim consists of a matrix of diamond particles embedded in a bond layer of nickel electroplated to the blade core. The thickness of the blade augmented by the abrasive layer determines a lower bound for the cutting width or 'kerf loss'. For a 250-μm thick steel blade, a typical kerf loss of 350-μm is obtained in ID sawing. A fluid coolant such as water and compounded coolants (lubricants and wetting agents) is directed into the cutting area to cool the blade/ingot interface and to assist in removing abraded particles (kerf). After the saw exits the ingot, the newly cut wafer is picked up by a vacuum chuck and transferred to a cassette. One of the major problems of ID slicing is the generation of wafer defects that occur where the saw exits the crystal and are therefore termed 'exit chipping' or 'saw fracture' (e.g., Dyer, 1979). Exit chipping significantly reduces the material yield.

Combination with automatic soft dressing of the blade, finer diamond grain size, and adapted cutting fluid, increases the throughput and reduces the kerf loss. Optimization of all these parameters yields a warp capability in the range 15 to 25 μm for 200 mm wafers. The 40-year old ID slicing technology is well established thanks to the extensive research devoted to it since its development during the early 1960s (e.g., Chonan *et al.*, 1993 and references therein).

Further improvement of the warp capability is possible by application of grind-slice (GS) cutting, which consists of a conventional ID saw mounted with a grinder for the ingot frontside (see Figure 24.2) (Tönshoff *et al.*, 1990; 1997). Grinding the ingot front face before slicing thus forms a reference plane that yields warp values as low as 5–12 μm. However, the GS procedure increases the cycle time of the process and consequently reduces throughput. Another big disadvantage is the higher consumption of raw material. Increased stock removal and cycle time result in a cost increase of 15 to 20 % compared to conventional ID cutting (Steffen *et al.*, 1994).

Figure 24.3 shows a schematic arrangement of a reciprocating multiblade/wire saw. A net of wires, as in the fixed-abrasive slicing technique known as FAST™ (Smith *et al.*, 1991), or a net of blades with fixed abrasive (MBF) or with slurry (MBS), is stretched between two bladeheads mounted onto a reciprocating frame. Two grooved guidance rollers are placed as close as possible to the workpiece. The ingot is placed in a feed mechanism perpendicular to the net of parallel cutting wires (or blades). Since the method has been essentially developed for photovoltaic applications, a mechanism further rocks the rectangular cross-section workpiece, changing the cutting angle to shorten the contact length by cutting the edges first. Recent advances in FAST make use of a series of

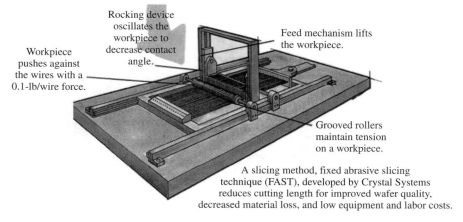

Figure 24.3 Schematic arrangement of FAST™. (Reprinted with permission from Crystal Systems, Inc., copyright (1999).)

teardrop-shaped wires with diamond abrasive particles bonded to the cutting edge (Schmid *et al.*, 1994). Kerf loss (thickness of material removal) of 230 µm is achievable with current technology and values as low as 150 µm are expected with improved profiled wires (Helmreich *et al.*, 1988). Water is used as a coolant and flushes away sawdust without hazardous waste disposal problems. FAST is not used on large-scale electronic wafer production but is a competitive technique for small-scale production.

Figures 24.4(a) and (b) describe the high-efficiency multiwire slurry saw. The wire saw consists of one 180-µm diameter steel wire moving, either unidirectionally or bidirectionally, on the surface of the workpiece (e.g., single-crystal silicon ingot) (Figure 24.4(a)). The single wire is wound on wire-guides carefully grooved with constant pitch forming a horizontal net of parallel wires or web (Figure 24.4(b)). The wire-guides are rotated by drives, causing the entire wire-web to move at a relatively high speed $(5-25\,\mathrm{m\,s^{-1}})$. A couple of high flow-rate

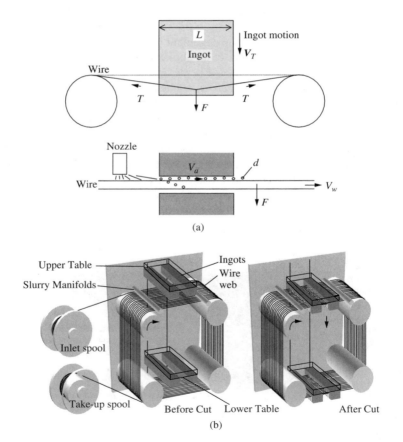

Figure 24.4 (a) principle of slurry wiresaw. (b) multiwire (web) arrangement.

nozzles are feeding the moving wires that bring, on account of surface tension, the slurry-suspended abrasion particles (e.g., SiC) into the cutting zone to produce a cut. The workpiece (or the wire-web) is moved vertically. The wire tension is maintained constant (10–30 N) during the cutting process with state-of-the-art feedback control. A wire feed reel provides the length of new wire and a wire take-up reel stores the used wire. The ingot capacity diameter of a wire saw is limited only by the shaft spacing of the wire guides of the web and the vertical travel. Latest-generation wire-saw machines can cut in a single batch up to two 200 mm diameter × 500 mm long ingots, or six 75 mm diameter × 500 mm long ingots.

The kerf loss, K, in MWS sawing is empirically obtained as

$$K = D + \tfrac{5}{2}\,\mathrm{d} \tag{24.1}$$

where D is the wire diameter and d is the free abrasive grit size. A typical kerf loss for MWS is less than 200 μm with a current wire diameter of 160 μm. The tendency in wire technology will soon yield wire diameters of 140 μm or less resulting in a concomitant reduction in kerf loss. For smaller ingots and softer materials (e.g., InP for space research and application), smaller grit size abrasive and finer wire diameter (for example 80 μm) can be used, yielding kerf losses of about 100 μm.

The primary functions of the slurry are as follows: (i) to carry the abrasive particles to the cutting zone; (ii) to flush away workpiece chips and residues (kerf); (iii) to remove heat. Most commonly used slurry are oil-based, water-soluble and water-based coolants, with generally SiC as abrasive. However, for silicon material, water-based coolant is not recommended because of a hazardous hydrogenation effect:

$$\mathrm{Si} + 2\mathrm{H_2O} \longrightarrow \mathrm{SiO_2} + 2\mathrm{H_2}^{\uparrow}$$

Since the modern multiwire saw has emerged as the technology for slicing large silicon wafers, fundamental research in modeling and control of modern MWS manufacturing process has been conducted in order to better understand the cutting mechanism and to optimize industrial processes (Kao *et al.*, 1998a; Li *et al.*, 1998; Sahoo *et al.*, 1998).

24.4 MATERIAL-REMOVAL PROCESS

The various slicing methods can be categorized according to their material-removal process. In the case of fixed-abrasive techniques (ID, OD, MBF, MWF, FAST), material is removed by diamond grains or clad that are firmly bonded on the cutting rim or wire by electroplating methods. Hence, for fixed abrasive, the abrasive impact velocity obeys

$$v_{\mathrm{abrasive}} = v_{\mathrm{tool}} \tag{24.2}$$

The cutting action is essentially that of a conventional single-edge cutting tool, which consists of a plunge-grinding process, characterized by high cutting speeds and a comparatively high total contact zone between tool and workpiece.

In the case of free-abrasive techniques (MBS, MWS), material is removed by third-party free abrasive grains transported in a liquid media (slurry). Hence, for free abrasive

$$v_{abrasive} = C v_{tool} \qquad (24.3)$$

where $C < 1$ accounts for tribologic properties (wettability, surface tension), geometry (porosity) and hydrodynamic properties (viscosity, pressure) and depends on the abrasive grit size, the slurry composition as well as the tool material and velocity. A modern high-efficiency wire saw can spool wires at a speed greater than $15\,\mathrm{ms}^{-1}$, and targeting $30\,\mathrm{ms}^{-1}$.

The cutting action in the case of free-abrasive machining (hereafter referred to as FAM) is essentially that of a fast three-body lapping process characterized by a *rolling and indenting* cutting mechanism (Figure 24.5) (Li *et al.*, 1998; Kao *et al.*, 1998b). The wire acts as a slurry carrier and applies a cyclic compressive loading. The larger abrasive grains are occasionally trapped by the asperities of the surfaces of the specimen and the tool, and are then forced to rotate by the parallel shear motion of these two bodies. During this rotation the abrasive grains transmit part of the applied compressive load from one surface to the other resulting in both surfaces being indented. The rotation of the particles enables indentation but not scratching to occur on the surfaces. The relative penetration depths into each surface will be determined by the relative hardness of the two materials. The stress pattern in the indenting zone produced by a line contact between surfaces subjected to a compressive loading and sliding friction is the superposition of the stresses due to normal and tangential forces applied by the wire. The maximum compressive stress (normal force) is found at the tip of the indenting notch where microcracking occurs. Combined with lowering of fatigue strength due to cyclic loading, these cracks propagate as the ingot is moved in a direction transverse to the wire axis. The maximum shear stress (tangential force) is attained at a small distance from the surface. Maximum subsurface shear stress, characteristic of contact mechanics, facilitates the formation of chips (kerf) on the ingot surface. The removal rate of matter (or wear rate) $\dot{Q}$ is proportional to

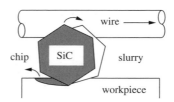

Figure 24.5 Rolling and indenting model of material removal in free-abrasive machining.

the product of the contact load pressure P times the abrasive speed $v_{abrasive}$

$$\dot{Q} = bv_{abrasive} \tag{24.4}$$

where b is a wear-like coefficient, which depends essentially upon elastic properties of involved materials (Young's modulus, fracture toughness, and hardness), tribologic parameters (contact area, film thickness), and viscosity of liquid carrier (Kao et al., 1998a; Phillips et al., 1977; Buijs and Korpel-van Houten, 1993). Further, the coefficient b depends on the concentration of abrasive grit and is proportional to the grit diameter. Note that a ratio of about 7 is generally found between the mean kerf diameter and the mean abrasive grit size.

In practice, Eq. (24.4) is expected to hold valid only below a critical pressure. Due to elastohydrodynamic interaction between the wire and the viscous slurry, a film of slurry forms in the cutting zone, whose thickness increases with increasing slurry viscosity and decreases with increasing load pressure (Kao, 1998a). Hence, at sufficiently high load pressure, one expects the slurry to be completely expelled from the interface between the tool and the workpiece, leading to a sudden loss of cutting efficiency. A result (Kao, 1998a) suggests that, within typical wiresaw operating conditions, increasing the wire speed can increase the value of that pressure threshold.

Note that wire speed also has its limitation. It can be shown indeed that the wire encounters severe vibration instabilities, but only at very high speed, typically over $300 \, m \, s^{-1}$ (Kao et al., 1998a). Clearly, this is of little concern since the wiresaw-operating velocity is currently in the range $15–25 \, m \, s^{-1}$.

On the other hand, for a given feeding rate of the raw material through the wire web, the lower the wire speed, the higher the pressure on the wire becomes. This is because the wire velocity and the feeding rate are correlated through the material removal rate. Furthermore, there is a lower limit in wire velocity below which the cutting efficiency drops, owing to insufficient hydrodynamic pressure of the slurry at the entrance slot. No slurry is then capable of reaching the ingot internal cutting zone, resulting in a loss of control of the process.

At a given permissible load pressure and wire velocity, variation in cutting efficiency is affected, in particular, by the amount of slurry reaching the cutting zone. For circular cross section ingots, the entrance angle, α, of the wire into the workpiece is less than $90°$ for the first half of the cut, whereas it is greater than $90°$ for the second half (Figure 24.6). When α is acute ($<90°$), the slurry is dragged into the cutting zone by a pinching effect. On the other hand, when α is obtuse ($>90°$), the entrance edge wipes the slurry out of the wire. By analogy, this is similar to exit chipping on ID sawing (but with a different mechanism).

One way to increase the efficiency of cut is to increase the amount of slurry reaching the cutting zone. This can be achieved, for example, by increasing the wire speed (Kao, 1998a) or by adding an ultrasonic vibration to the wire, which leads to transporting wavelets of particles into the cutting area (Figure 24.7) (Hauser, 1992). Ultrasonic assistance allows faster cutting

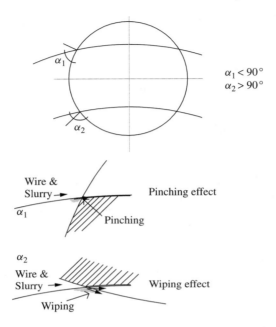

$\alpha_1 < 90°$
$\alpha_2 > 90°$

Wire &
Slurry →   Pinching effect

α_1
                              Pinching

α_2
Wire &
Slurry →   Wiping effect

Wiping

Figure 24.6   Effect of geometry on the slurry penetration into the cutting zone.

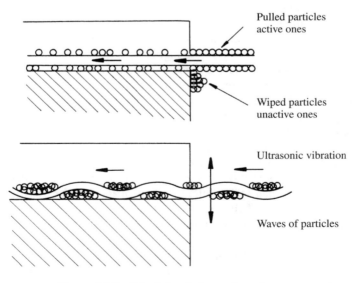

Pulled particles
active ones

Wiped particles
unactive ones

Ultrasonic vibration

Waves of particles

Figure 24.7   Principle of ultrasonic assistance.

and improved geometry, resulting in an increase in production throughput of 30–50 %, without degradation of surface roughness. Beside mechanical complications, a disadvantage of ultrasonic assistance lies in the increase in kerf loss due to vibrations, whereas wire speed below $25 \, \mathrm{m \, s^{-1}}$ has almost no influence on the vibration amplitude (Kao, 1998b; Kao et al., 1998c).

24.5 GENERAL COMPARISON OF DIFFERENT SLICING METHODS

One of the major weaknesses of fixed abrasive (diamond impregnated) sawing methods is the irreversible separation of diamond particles from the cutting edge due to high impact velocities and shear stresses inherent to the ploughing mechanism. It follows that the cutting capability of the tool decreases rapidly with time, hence increasing operating costs due to the complex blade redressing process. In the slurry-based methods, the free abrasive particles rejoin the cutting process in a continuous manner.

The MBS methods, among others, encounter dynamic instabilities due to both the single contact point between blade and workpiece and the changing wear profile during the cutting process. These hydrodynamic instabilities can lead to wafer breakage (Werner and Kenter, 1988). This problem is clearly absent in MWS where the wires are always in contact with the whole cutting area of the workpiece and continuously renewed.

The applied load pressure in MWS (and FAST) is limited by the relatively low tensile strength of the wires, severely reducing the material removal rate according to Eq. (24.4). Consequently, the permissible feed rate in wire sawing, compared to ID sawing, is drastically reduced, hampering some of the advantages of slicing hundreds of wafers at the same time. Despite this drawback, current state-of-the-art technology in wire sawing yields a throughput 5 to 30 times higher than modern ID technology.

Reduction in the production cost of silicon substrate wafers for semiconductor devices is achieved in part by increasing the wafer diameter. Silicon wafers, which started at 100 mm diameter, have now reached 200 mm. Already efforts are being put toward integrating 300 mm in production lines (Huff and Goodall, 1996) and 400 mm discussions have been started (Oishi et al., 1998). This trend poses a severe problem to ID sawing where the blade diameter must be approximately 4 times the ingot diameter. This implies that the blade thickness has to be increased to ensure proper rigidity, which in turn increases the kerf loss, not to mention additional mechanical design problems. According to Oishi et al. (1998), an ID saw for a 200-mm diameter ingot has a typical kerf loss of about $350 \, \mu\mathrm{m}$, which will rise to about $500 \, \mu\mathrm{m}$ with a 1600-mm diameter blade core essential for 400-mm ingot slicing. For comparison, a modern MWS using a $180 \, \mu\mathrm{m}$ wire diameter has a kerf loss of about $220 \, \mu\mathrm{m}$, which will remain constant for 400-mm ingots and higher. A gain greater than 50 % in material usage is then achieved for large ingot diameters with MWS compared to ID.

Table 24.1   Main advantages and disadvantages of slicing techniques

Method	Advantages	Disadvantages
Conventional ID	Mature technology	Scale up problems for wafer
–	Low cost of consumable	diameter >200 mm
–	Simple maintainability	Large kerf loss >300 μm
–	–	Low throughput
–	–	Exit chipping
–	–	Higher depth of damage
–	–	Operator-skill dependent
GS-cutting	Reduced warp	Increased cost
–	Reference plane	Low material utilization
–	Low depth of damage	Large kerf loss
–	–	Less throughput than conv. ID
FAST™	Low kerf loss	TTV
–	Low cost of consumables	Not developed for electronics
Multiblade slurry (MBS)	High cutting rate	Wafer breakage (reduced yield)
–	–	Low accuracy
Multiwire slurry (MWS)	Low kerf loss	TTV
–	Low depth of damage	High cost of consumables
–	Low warp	–
–	Decreased production cost	–
–	High throughput and yield	–

The main advantages and disadvantages of the different cutting technologies are summarized in Table 24.1. Nowadays, only the ID and MWS cutting techniques are commonly used in production lines for high-precision slicing of silicon wafers.

24.6   SURFACE DAMAGE

Because of the tight technical, physical, chemical, and geometrical requirements imposed on substrates for highly complex microelectronic components, microstructural defects can lead to altered characteristics of the substrate, in particular in the near-surface layer. Since, in general, all deviations from these requirements can be considered as defects, only the most important physical defects caused by abrasive machining will be mentioned. Abrasive processing essentially induces mechanical and thermal effects at the locus of kerf generation. Subsurface physical damage is divided into several zones (Figure 24.8).

The first zone, which is strongly damaged, is a few hundred nanometers to a few micrometers thick quasipolycrystalline layer caused by the local increase in temperature associated with elastic deformation of moving abrasive grains against

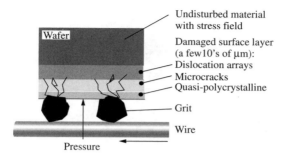

Figure 24.8 Subsurface damage structure layer.

workpiece asperities. The resulting polycrystalline structure leads to unintended changes in the electric properties of the sliced material. Also, the density and the mechanical properties are altered. Below this layer is a zone with a high density of microcracks. A few deep-reaching cracks can be found in a following dislocation-rich transition zone above a stress zone free of cracks. The high stress fields generated by abrasive machining markedly increase the dislocation concentration which, in turn, leads to mechanical and thermal near-surface residual stresses (Tönshoff *et al.*, 1994). Deeper in the material, these stresses are compensated by elastic deformation of the bulk material, i.e., strain stresses. The mismatch in thermal expansion and density between the troubled near-surface and the nondisturbed bulk substrate causes convexity or flatness errors in the wafer known as warpage and bowage.

The overall depth of penetration of these damages is within a few tens of micrometers and is dependent upon the dynamics of the material-removal process. High impact velocities of a single cutting edge on workpiece asperities, as generated by fixed abrasive, induce greater depth damages than in free-abrasive machining (FAM). Moreover, thermal effects in FAM are not as pronounced as in fixed-abrasive machining owing to a lower impact velocity of abrasive grains. A typical 200-mm ID saw has a tangential velocity at its inner rim of about $30\,\mathrm{m\,s^{-1}}$, which has to be compared to the free-abrasive velocity in any case not greater than the wire velocity currently of $15-25\,\mathrm{m\,s^{-1}}$. The depth of damage in MWS is empirically found to be about 1.5 times the abrasive grain diameter and is generally less than half the depth of damage caused by fixed abrasive machining. Reduction of the depth of damage can in principle be achieved by reducing the grit size of the abrasive, which leads to lower grain forces and to a smaller damage zone (Lundt *et al.*, 1994). However, conditionability of the grinding tool and suitability of bond type are rendered more difficult in fixed abrasive systems. Moreover, the material removal rate is proportional to the abrasive diameter in FAM (Buijs and Korpel-van Houten, 1993) through the wear coefficient, b, introduced in Eq. (24.4). Nevertheless, several published studies have shown indeed that the depth of surface damage in MWS silicon wafers is reduced by as much as $50-60\,\%$ compared to that in wafers produced by ID sawing (Kao *et al.*, 1998a;

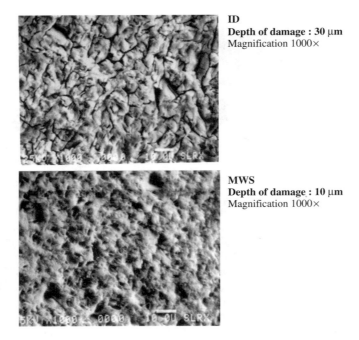

Figure 24.9 Electron micrographs of the surface of single-crystal Si wafer as obtained with ID saw (top) and MWS saw (bottom).

Werner and Kenter, 1988; Tabata *et al.*, 1984) (see Table 24.2). This is further illustrated in Figure 24.9 showing scanning electron micrographs of silicon surfaces obtained by both ID and MWS sawing. It is seen that the surface seems to be smoother in MWS, and deep cracks are observed in the ID wafer.

24.7 ECONOMICS

Minimal production costs of high-precision single-crystal substrates for microelectronic and micromechanic devices are likely to be achieved with slicing processes that provide low kerf loss, minimum surface damage, minimum warpage and thickness variation, high production rates and yields. The minimum kerf-loss requirement derives essentially from the high costs associated with the complex production technologies involved in single-crystal growth. Behind minimum surface damage as well as minimum geometrical defects requirements lies the add-on cost associated with downstream machining time and equipment necessary for surface conditioning such as lapping, etching and polishing processes.

A raw economic assessment of MWS and ID slicing techniques for monocrystalline materials is obtained by comparing basic criteria such as kerf loss (μm), cutting rate (m^2/h), material yield (m^2/kg), subsurface damage depth (μm), and

Table 24.2 Comparative survey of basic criteria of main slicing techniques for silicon

	MWS Saw		ID Saw		FAST™	
	Single crystal (8″)	Polycrystal (4″)	Single crystal (8″)	Polycrystal (4″)	Single crystal (8″)	Poly crystal (4″)
Cutting principle	FAM/fast lapping		Grinding		Grinding	
Material-removal mode	Rolling and indenting		Plunge grinding		Plunge grinding	
Subsurface depth of damage (μm)	5–15 (uniform)		15–30 (variable)		–	
Ingot diameter	200 mm and higher		up to 200 mm			up to 5″ × 5″
Maximum loadable ingots per run	2	4	1	1	Not applicable	2
Typical wafer thickness (μm)	0.9	0.2	0.9	0.35		0.35
Kerf loss (μm)	220	220	350	350		230
Wafer/hr	97	714	10	20		131
Wafer/cm	9	24	8	14		17
Cutting rate (m²/hr)	3.1	7.1	0.3	0.2		1.3
Material yield (m²/kg)	0.39	1.04	0.35	0.62		0.75
Wafer/kg	12	104	11	62		75
Loss Si (kg/kg)	0.2	0.5	0.3	0.5		0.4
Theoretical yearly throughput (300 days)	699 107	5 142 857	68 400	141 429		943 448
1 MWS saw is equivalent to (rounded up):			11	37		6
Production yield (%)	95	95	90	90		–

(FAST™ Single crystal (8″) column: NOT APPLICABLE)

The theoretical number (37) of ID saws required to match MWS yearly theoretical productivity of PV wafers is reduced to approximately 20 in practice due to actual MWS production yield, average operational and saw-loading conditions.

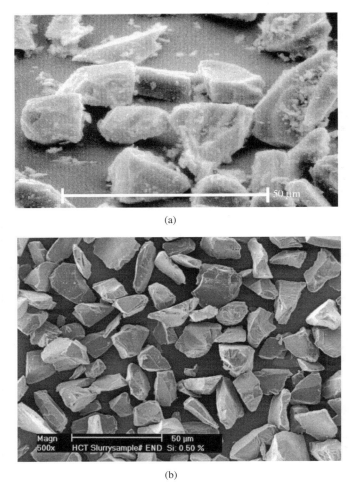

Figure 24.10 (a) Used SiC grains; note the small kerf particles attached to the grit; (b) cleaned recovered SiC grains.

yearly theoretical throughput as applied to monocrystalline silicon wafer manufacture. Such a comparative survey on operational criteria of the main silicon slicing techniques is summarized in Table 24.2, where data were compiled from internal sources and published work (Schmid *et al.*, 1994; Helmreich *et al.*, 1988; Kao *et al.*, 1998a; Werner and Kenter, 1988; Tabata *et al.*, 1984). Data for polycrystalline silicon wafers for photovoltaic (PV) applications are also shown in Table 24.2. This confirms that the multiwire saw has much higher throughput and yield with less kerf loss and surface damage.

At an equivalent production yield of 99 % and final wafer thickness of 285 μm with MWS, a saving of at least 25 % in wafering cost can be potentially achieved by reducing wire diameter and grit size from typical values of 180 μm and 15 μm down to probable limiting values of 100 μm and 5 μm, respectively (Hauser and Nasch, 1998).

The SiC abrasive alone accounts for about 35 % of the consumable costs, and the slurry (grit and coolant) for about 60 % of the total consumable costs in MWS wafer slicing. Also, the slurry loses its cutting efficiency once the sawing dust reaches 10–20 % in weight of the abrasive particles and must be changed. This means that, with a typical kerf loss of 200 μm and wafer thickness of 800 μm, the amount of waste slurry produced is equivalent to the amount of silicon sliced. In other words, approximately 10 000 tons of sludge is produced annually. The price of new slurry (SiC + coolant) being in the range of 10$/kg, one can evaluate to about 100M$ the annual cost of this consumable discarded. Fortunately, this waste is mostly avoidable. Indeed, irreversible breakdown or wear of the grit certainly occurs, but at a negligible rate. If all the grains involved in the cutting were to be used only once, this would be equivalent to coating the wire with a

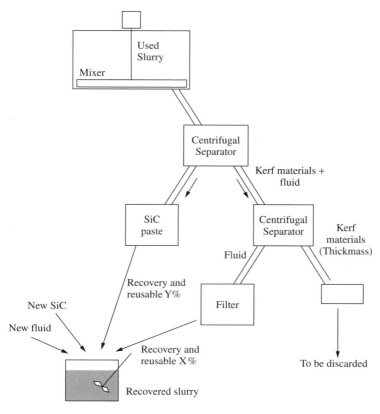

Figure 24.11 Flow chart diagram of the process for slurry recovery.

layer of abrasive grain, as in fixed-abrasive machining. For example, the volume of a single layer of $10 \mu m$ abrasive grit on a $500 km$ spool of 180-μm diameter wire would be less than 2 liters per cut. In fact, the cutting efficiency of an aging slurry is mostly hindered by the adhesion of fine sawdust particles (kerf) onto the abrasive grain surface (Figure 24.10(a)). These pollutant particles fixed onto the grain are acting as ball bearings, thus preventing the grit from cutting. Slurry recovery can be achieved by means of a special installation, whose flow sequence is shown in Figure 24.11. An example result of recovered clean grit is shown in Figure 24.10(b). It is a low-cost operation, which separates, under an appropriate centrifugal force field, the fine particles produced by the kerf from the coarser reusable abrasive grit. Such an installation has proven to give a separation efficiency of better than 85 %, with a yield of recovery of reusable grit of 85 % or higher for a cut-off size greater than $5 \mu m$. The installation also allows for recovering and clarifying the coolant. This means that the use of such a technique not only decreases significantly the cost of wafers produced by the wiresaw, but also reduces drastically the yearly amount of sludge from 10 000 t to 1500 t with an associated cost reduced to 15M$. At the same time, environmental factors are not as acute as before. Clearly, slurry management becomes one of the very important aspects of the wiresaw operation (Hauser and Nasch, 1999).

24.8   CRYSTAL ORIENTATION

The use of single crystals for optical or semiconductor applications often requires slicing the crystals with a very precise orientation relative to the crystal lattice. On the one hand, since the crystal-growth process does not allow perfect crystal orientation with respect to the geometrical axis, it is necessary to correct for growth or machining error in order for the cut to meet precise orientation requirements. On the other hand, it is sometimes required to slice the crystals with an orientation slightly modified compared to the natural crystalline orientation of the grown ingot. In both cases, one needs to precisely measure or know the actual crystal axis orientation and to position and maintain the geometrical ingot in space such that the displacement of the cutting device is parallel to the targeted cutting plane of the single crystal. There is an infinite number of such possible positions, but among these, only four positions exist that allow simultaneously the geometrical axis of the ingot to be horizontal. This property is of prime importance when using a wire saw because the ingot is then parallel to the horizontal wire-web, which reduces cutting time and allows optimized (multi-ingots) loading of the machine (Hauser, 1997).

Devices for orienting single crystals are already known and used in the semiconductor industry on internal diameter sectioners (ID saws) or on wire saws. Positioning is commonly achieved with an orientable table mounted directly on the machine. The table permits horizontal angular scan and vertical tilt (Figure 24.12). An adjustment (i.e., vertical and horizontal) of the ingot is made

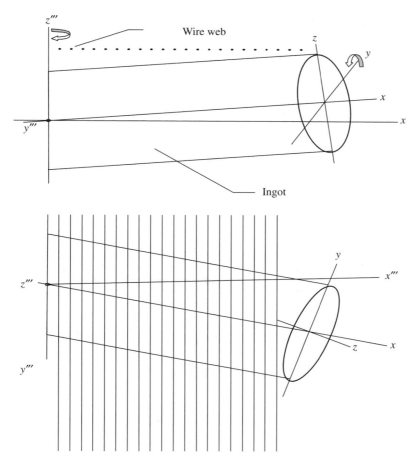

Figure 24.12 Orthogonal views of conventional positioning of single crystal.

after optical or X-ray measurement. This procedure has several disadvantages inherent to the vertically inclined position of the ingot relative to the cutting element. Among these, the decrease in productivity caused by the increase in cutting time associated with: (i) a cut that does not start simultaneously on both sides of the ingot in the case of wire saw; (ii) a length of cut that is not minimized for ID saws. Moreover, this manner of processing requires adjusting the machine table before each cut in a very precise manner. The adjustment time of the machine also contributes to reduce productivity.

To overcome these drawbacks, an orientation procedure has been devised that avoids the vertical tilt by using another combination of rotation axis (Figure 24.13). The orientation procedure positions the geometric single crystal in a plane perpendicular to the vertical cutting direction (i.e., the geometrical

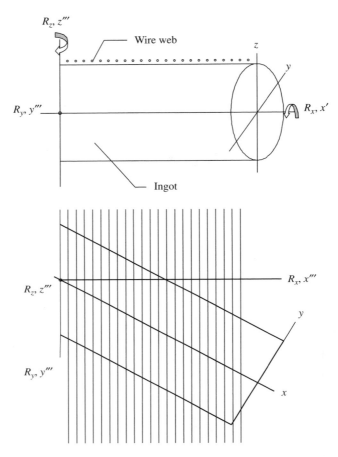

Figure 24.13 Orthogonal views of positioning of single crystal according to present method.

axis is parallel to the horizontal wire web) while bringing the targeted crystal plane of the single crystal parallel to the direction of cutting of the machine. Also, a positioning device separate from the cutting machine (Figure 24.14) was developed that allows the orientation to be done independently of the cutting machine (Hauser, 1996). The process and the device permit very precise positioning of the single-crystal to be obtained outside the machine under desirable conditions. This results in an increase in productivity due to (i) a minimized length of cut, (ii) the elimination of the machine downtime needed to adjust the cutting tool, and, (iii) the optimized loading of the saw with several ingots mounted on the same holder for a simultaneous cut.

In order for the geometrical x-axis of the ingot to be parallel to the wire web, rotation about the ingot y-axis is forbidden. Under this constraint, an algorithm

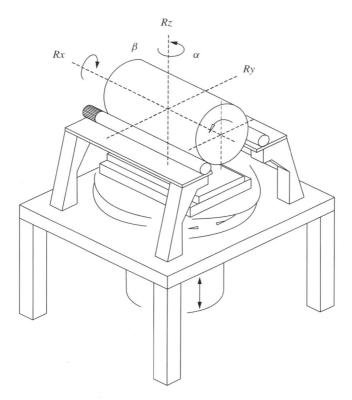

Figure 24.14 External mounting device.

has been developed that determines the rotation angles α and β about the z- and x-axis, respectively, to be imposed on the geometrical ingot in order to bring the measured or known crystal x'-axis into a predetermined position relative to the cutting plane (y''', z'''). Orienting the ingot is a two-fold procedure. The first step consists in localizing the orientation of a virtual wafer plane relatively to (x, y, z) such that the crystallographic x'-axis is seen from the wafer referential with the desired after-cut orientation (Figure 24.15-1). The second step consists in properly rotating the geometrical ingot (x, y, z) about authorized axes, x''' (Figure 24.15-2) and z''' (Figure 24.15-3), such as to bring the wafer plane into the cutting plane of the machine (y''', z''').

The wafer vector in referential i, noted $\underline{w}[i]$ ($i = 0$ for ingot referential; $i = 1$ for crystal referential; $i = 2$ for wafer referential; $i = 3$ for machine referential), is defined by a normalized vector perpendicular to the wafer plane (i.e., x'' in Figure 24.15-1) and

$$\underline{w}[2] \equiv x'' = \begin{pmatrix} 1 \\ 0 \\ 0 \end{pmatrix} \qquad (24.5)$$

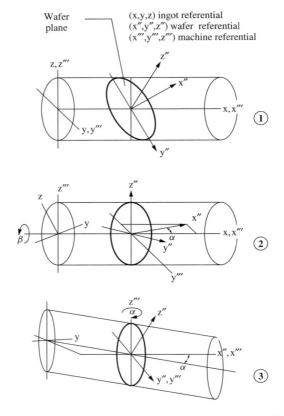

Figure 24.15 Orientation procedure (see text for details).

In the crystal referential [1], the wafer vector becomes

$$\underline{w}[1] = m[2] \cdot \underline{w}[2] \tag{24.6a}$$

In the ingot referential [0], the wafer vector becomes

$$\underline{w}[0] = m[1] \cdot \underline{w}[1] \tag{24.6b}$$

where the matrix $m[i]$ is a generalized transformation matrix of the form

$$
m[i] = \begin{pmatrix}
\dfrac{\cos \varphi_j \cos \theta_j}{N_j} & \dfrac{-\sin \varphi_j \cos^2 \theta_j}{N_j} & \sin \theta_j \\[2ex]
\dfrac{\sin \varphi_j \cos \theta_j}{N_j} & \dfrac{\cos \varphi_j}{N_j} & 0 \\[2ex]
\dfrac{-\cos \varphi_j \sin \theta_j}{N_j} & \dfrac{\sin \varphi_j \cos \theta_j \sin \theta_j}{N_j} & \cos \theta_j
\end{pmatrix} \tag{24.7}
$$

The columns of $m[i]$ give the coordinates of referential $[i]$ in referential $[j]$ when the longitudinal angle (φ_j) and azimutal angle (θ_j) are measured in referential $[j]$. Therefore $m[1]$ is characterized by the crystal growth production error angles φ_0 and θ_0, and $m[2]$ is defined by φ_1 and θ_1, which are the offset or add-off angles selected or imposed by subsequent processes or uses and are relative to the crystal lattice referential. The normalization factor is $N_j \equiv \sqrt{\cos^2 \theta_j + \cos^2 \varphi_j \sin^2 \theta_j}$. One further verifies that the transformation is orthogonal, i.e.,

$$\text{Det}(m[i]) = 1 \Longleftrightarrow m[i]^T = m[i]^{-1} \tag{24.8}$$

In the ingot referential $[0]$, the wafer vector is explicitly given by

$$\underline{w}[0] \equiv \begin{pmatrix} w_x \\ w_y \\ w_z \end{pmatrix} = m[1] \cdot m[2] \cdot \underline{w}[2]$$

$$= \begin{pmatrix} \dfrac{\cos \varphi_0 \cos \theta_0 \cos \varphi_1 \cos \theta_1}{N_o} - \dfrac{\sin \varphi_0 \cos^2 \theta_0 \sin \varphi_1 \cos \theta_1}{N_1} & \dfrac{\cos \varphi_1 \sin \theta_1}{N_1} \\ \dfrac{\sin \varphi_0 \cos \theta_0 \cos \varphi_1 \cos \theta_1}{N_o N_1} + \dfrac{\cos \varphi_0 \sin \varphi_1 \cos \theta_1}{N_o N_1} & \\ -\dfrac{\cos \varphi_0 \sin \theta_0 \cos \varphi_1 \cos \theta_1}{N_o N_1} + \dfrac{\sin \varphi_0 \cos \theta_0 \sin \theta_0 \sin \varphi_1 \cos \theta_1}{N_o N_1} & \dfrac{\cos \varphi_1 \sin \theta_1}{N_1} \end{pmatrix}$$

$$\tag{24.9}$$

The crystal vector, noted $\underline{c}[i]$, is commonly defined by a normalized vector perpendicular to the crystal plane of interest. In the crystal referential $[1]$, the direction $\langle 100 \rangle$ is obviously

$$\underline{c}[1] = \begin{pmatrix} 1 \\ 0 \\ 0 \end{pmatrix} \tag{24.10}$$

In the ingot referential $[0]$, the crystal vector becomes

$$\underline{c}[0] \equiv \begin{pmatrix} c_x \\ c_y \\ c_z \end{pmatrix} = m[1] \cdot \underline{c}[1] = \begin{pmatrix} \dfrac{\cos \varphi_0 \cos \theta_0}{N_o} \\ \dfrac{\sin \varphi_0 \cos \theta_0}{N_o} \\ \dfrac{-\cos \varphi_0 \sin \theta_0}{N_o} \end{pmatrix} \tag{24.11}$$

In the case in which it is desired to obtain wafers parallel to the $\langle 100 \rangle$ crystal direction, φ_1 and θ_1 are set to zero in Eq. (24.9), and $\underline{w}[0] = \underline{c}[0]$.

The ingot will be properly mounted on its support and ready to be sliced in a targeted crystal orientation and with its principal geometrical axis x perpendicular to the cutting plane when the wafer vector verifies

$$\Re_z(\alpha) \, o \, \Re_x(\beta) \cdot \underline{w}[0] = \begin{pmatrix} 1 \\ 0 \\ 0 \end{pmatrix} \tag{24.12}$$

where $\Re_z(\alpha) = \begin{pmatrix} \cos\alpha & -\sin\alpha & 0 \\ \sin\alpha & \cos\alpha & 0 \\ 0 & 0 & 1 \end{pmatrix}$ and $\Re_x(\beta) = \begin{pmatrix} 1 & 0 & 0 \\ 0 & \cos\beta & -\sin\beta \\ 0 & \sin\beta & \cos\beta \end{pmatrix}$

are the rotation matrices about authorized z- and x-axis, respectively.

It can be shown that the solutions of Eq. (24.12) for the mounting angles α and β are given by

$$\alpha = \text{ArcTan}\left(\frac{\sqrt{w_y^2 + w_z^2}}{w_x} \right) \tag{24.13}$$

$$\beta = \text{ArcTan}\left(\frac{w_z}{w_y} \right) \tag{24.14}$$

Note that α is equivalently obtained using the scalar product (i.e., projection of $\underline{w}[0]$ onto x''')

$$\alpha = \text{ArcCos}\left(\underline{w}[0] \cdot \begin{pmatrix} 1 \\ 0 \\ 0 \end{pmatrix} \right) = \text{ArcCos}(w_x) \tag{24.15}$$

Whereas φ_1 and θ_1 are selected offset angles, φ_0 and θ_0 need to be precisely measured. This is commonly achieved by means of X-ray diffraction methods in which the sets of lattice planes reflect X-rays of a defined wavelength according to Bragg's Law

$$n\lambda = 2d \sin\theta \tag{24.16}$$

where n is a positive integer, λ is the radiation wavelength, and d is the distance between lattice planes.

The method localizes the crystal axis through the measurements of its orthogonal projections on two planes. Referring to Figure 24.16, φ_0 (resp. θ_0) is the angle between the x-axis and the orthogonal projection of the crystal axis $\underline{c}[0]$ on the horizontal (x, y) (resp. the vertical (x, z)) plane. The (x, y) plane is the plane defined by the X-rays, and the x-axis is the ingot principal axis. Measurement of φ_0 is straightforward since it belongs to the X-ray plane, whereas measurement of θ_0 requires a 90° rotation of the ingot about its principal axis so as to bring the vertical component into the horizontal X-ray plane. The process is divided into four single steps with the notch/flat 90° apart. It is clear, however, that a

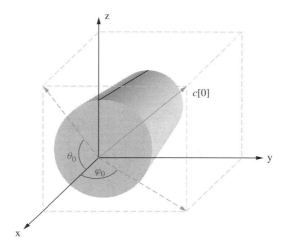

Figure 24.16 Angle definition. The dark line on the ingot represents the notch or flat reference.

3-dimensional vector is fully determined by its projections on two planes, which would imply that two of the four measurements are superfluous. In fact, two measurements $180°$ apart are done for every projection, in order to make the choice of the angle's origin irrelevant.

With known $\varphi_0, \theta_0, \varphi_1$ and θ_1 one can compute α and β using Eqs. (24.9), (24.13), and (24.14). The ingot is then rotated about its geometrical axis x or x''' of the angular value β to bring the wafer vector x'' into the horizontal machine plane (x''', y''') (Figure 24.15-2). A rotation through an angle α of the geometric single crystal about the axis z''' brings the vector x'' into a position that is colinear with the axis x''' (Figure 24.15-3). After these two rotations, the geometric single crystal x, y, z is oriented parallel to the plane (x''', y''') with an angle α relative to the normal x''' of the cutting plane corresponding to the requirements of the process ultimately used. The resulting sawing will then have the required angles φ_1 and θ_1 relative to the crystallographic axes y' and z'.

The mounting of the single crystal on the support is done with a positioning device, which permits exact measurement of the applied angles of rotation of the crystal (Figure 24.14). The single crystal is carried by two support rotating cylinders mounted on the upper portion with their principal axis oriented parallel to the x-axis. An angular measurement member, such as an encoder, permits the applied rotation β of the crystal about its principal axis x to be measured. An indexed plate is mounted about the vertical z''' axis on the lower portion of the frame. An angular-measurement system integrated in the plate permits measuring the rotation angle α applied to the single crystal (or $-\alpha$ applied to the supporting plate) about the z''' axis. Once these two rotations α and β take place, a mechanism brings together the support with the single crystal itself whilst maintaining their relative position. This can take place either by raising the indexed plate or by lowering the single crystal. Once placed into contact, the single crystal will be gripped or cemented in position to the support. The single crystal mounted on its support can then be transferred to the cutting machine. The single crystal is oriented with its geometrical axis horizontal and ready to be cut.

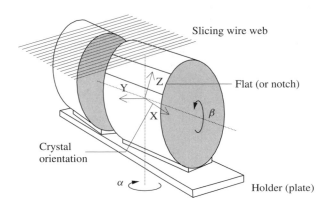

Figure 24.17   Multiple ingots mounting.

Figure 24.18   Fully automatic measuring and mounting ingot station using a robot-driven handling tool.

With these characteristics, it is possible to obtain a precise positioning and orientation of the single crystal in a suitable measuring environment, without the need to affect any adjustment of the position on the cutting machine. The downtime of the latter is thus considerably reduced so as to increase productivity. Moreover, thanks to these characteristics, productivity is further improved by allowing the mounting of multiple ingots on the same holder, each one with its own set of angles (α, β) but all being with their principal geometrical axis x horizontal (Figure 24.17) (Hauser, 1997).

Finally, external mounting of a single crystal outside the cutting machine has allowed the development of a robot-driven installation for a fully automatic orientation and mounting of multiple ingots (Figure 24.18). Such a complex application has been in a production environment for several years with a mounting accuracy within $\pm 5'$.

REFERENCES

Buijs, M., and K. Korpel-van Houten, (1993) A Model for Three-Body Abrasion of Brittle Materials, *Wear* **162–164**, 954–956.

Chonan, S., Z. W. Jiang, and Y. Yuki, (1993) Stress Analysis of a Silicon-Wafer Slicer Cutting the Crystal Ingot, *J. Mech. Design* **115**, 711–717.

Dyer, L. D., (1979) Exit Chipping in I.D. *Sawing of Silicon Crystals*, Texas Instruments, Inc.

Hauser, C., (1992) High Speed Slicing with Ultrasonic Assisted Wire Saw, *11th E. C. Photovoltaic Solar Energy Conference*, 12–16 October, Montreux, Switzerland, pp. 1061–1062.

Hauser, C., (1997) 'Procédé pour l'orientation de plusieurs monocristaux posés côte à côte sur un support de découpage en vue d'une découpe simultanée dans une machine de découpage et dispositif pour la mise en oeuvre de ce procédé', European Patent EP 0 802 029 A2.

Hauser, C., (1996) 'Process for the orientation of single crystals for cutting in a cutting machine and device for practicing this process', US Patent number 5,720,271.

Hauser, C., and P. M. Nasch, (1998) Slicing Technology and Thin Wafer, Workshop *'Dependable and Economic Silicon Materials Supply for Solar Cell Production'*, December 7–12, JRC-Ispra, Italy.

Hauser, C., and P. M. Nasch, (1999) Waste Processing in Wafer Slicing Techniques Used as Economical Factor, *SEMICON China 1999 – Technical Symposium*, March 17–18, China World Hotel, Beijing, China, pp. K1–K7 of Technical Program.

Helmreich, D., R. M. Knobel, and W. Ermer, (1988) Experiences with a Pilot System for Preparing Multicrystalline Silicon Wafers, IEEE, 1390–1394.

Huff, H. R., and R. K. Goodall, (1996) Material and Metrology Challenges for the Transition to 300 mm Wafers, *Interface* **5**(2), 31–35.

Kao, I., V. Prasad, F. P. Chiang, M. Bhagavat, S. Wei, M. Chandra, M. Constantini, P. Leyvraz, J. Talbott, and K. Gupta, (1998(a)) Modeling and Experiments on Wiresaw for Large Silicon Wafer Manufacturing, *Proc. 8th Int. Symp. on Silicon Materials Science and Technology*, in special volume on 50 Years of Silicon Materials and Technology, San Diego, May.

Kao, I., M. Bhagavat, V. Prasad, J. Talbot, M. Chandra, and K. Gupta, (1998(b)) Integrated Modeling of Wiresaw in Wafer Slicing, in *NSF Design and Manufacturing Grantees Conference*, pp. 425–426, Monterrey, Mexico, Jan. 5–8.

Kao, I., S. Wei, and F. -P. Chiang, (1998(c)) Vibration of Wiresaw Manufacturing Processes and Wafer Surface Measurement. In *NSF Design and Manufacturing Grantees*, pp. 427–428, Monterrey, Mexico, January 5–8.

Kao, I., (1998(a)) Wiresaw Technology and Research: Integrated Wiresaw Manufacturing Modeling. Rolling-Indenting in Free Abrasive Machining, private communication (summary handout).

Kao, I., (1998(b)) Wiresaw Technology and Research: Large Diameter Wafer Preparation. *Vibration Analysis and Surface Measurement, Private Communication* (summary handout).

Li, J., I. Kao, V. Prasad, (1998) Modeling Stresses of Contacts in Wire Saw Slicing of Polycrystalline and Crystalline Ingots : Application to Silicon Wafer Production, *Trans. ASME, J. Electron Packag.* **120**, 123–128.

Lundt, H., M. Kerstan, A. Huber, and P. O. Hahn, (1994) Subsurface Damage of Abraded Silicon Wafers, *Proc. 7th Int. Symp. on Silicon Materials Science and Technology 1994*, pp. 218–224, Semiconductor Silicon, eds, H. R. Huff, W. Bergholz, K. Sumino, Electronic Division, Proc. Vol. 94–10.

Oishi, H., K. Asakawa, and J. Matsuzaki, (1998) Design Concept of Multi-Wire Saw for 400 mm Diameter Silicon Ingot Slicing, *Proceedings of 'Silicon Machining'*, 1998 Spring Topical Meeting, Vol. 17, pp. 109–112, The American Society for Precision Engineering (ASPE), April 13–16, Carmel-By-The-Sea, California.

Phillips, K., Crimes, G. M., and T. R. Wilshaw, (1977) On the Mechanism of Material Removal by Free Abrasive Grinding of Glass and Fused Silica, *Wear* **41**, 327–350.

Sahoo, R. K., V. Prasad, I. Kao, J. Talbott, and K. P. Gupta, (1998) Towards an Integrated Approach for Analysis and Design of Wafer Slicing by a Wire Saw, *Trans. ASME, J. Electron Packag.* **120**, 35–40.

Schmid, F., M. B. Smith, and C. P. Khattak, (1994) Development of Shaped Wire for FAST Slicing, *12th European Photovoltaic Solar Energy Conference*, pp. 1009–1010, Amsterdam, The Netherlands, 11–15 April.

Smith, M. B., F. Schmid, C. P. Khattak, L. Teale, and J. G. Summers, (1991) Multi-Wire Fast Slicing for Photovoltaic Applications, 22nd *IEEE Photovoltaic Specialists Conf.*, Vol. 2, pp. 979–981, Las Vegas, Nevada, October 7–11.

Steffen, E., J. Schandl, J. Junge, and H. Lundt, (1994) Manufacturing Processes for Advanced Silicon ULSI Wafers, *Proc. 7th Int. Symp. on Silicon Materials Science and Technology 1994*, pp. 197–206, Semiconductor Silicon, eds, Huff H. R., W. Bergholz, K. Sumino, Electronic Division, Proc. Vol. 94–10.

Tabata, K., M. Geshi, K. Kaneko, M. Takatani, K. Hashimoto, and M. Fujisaki, (1984) Silicon Slicing for Solar Cell, in *Technical Digest of the International PVSEC-1*, Kobe, Japan, pp. 809–812.

Tönshoff, H. K., W. V. Schmieden, I. Inasaki, W. König, G. Spur, (1990) Abrasive Machining of Silicon, *Annals of the CIRP*, **39/2**, 621–633.

Tönshoff, H. K., M., Hartmann, Przywara R., and M. Klein, (1994) Defects in Silicon-Wafers – Characterisation and Prevention, In *Advancement of Intelligent Production*, Elsevier Science B.V./Jap. Soc. Prec. Eng., pp. 627–632.

Tönshoff, H. K., Karpuschewski B., Hartmann M., C. Spengler, (1997) Grinding-and-Slicing Technique as an Advanced Technology for Silicon Wafer Slicing, *Machining Sci. Technol.*, **1 (n. 1)**, 33–47, August.

Werner, P. G., and I. M. Kenter, (1988) Comparative Study of Advanced Slicing Techniques for Silicon, *Inter-Society Symposium Machining of Ceramic Materials and Components* – 2nd International Symposium Chicago-Illinois 11/28–12/2.

25 Methods and Tools for Mechanical Processing of Anisotropic Scintillating Crystals

MICHEL LEBEAU

CERN 1211 Geneva 23 Switzerland

25.1 INTRODUCTION

The electromagnetic calorimeter of the Compact Muon Solenoid (CMS) experiment (Figure 25.1) will start taking data at the CERN-Geneva Large Hadron Collider in 2007. 82 000 PbWO$_4$ (lead tungstate) scintillating crystals will be assembled in a tight cylindrical array completed by two endcaps, the hermeticity of which is mandatory for the claimed detector performance. A safe, accurate and economical solution has been developed at CERN [17, 19, 21, 22] for the mechanical processing of the crystals, which will be mass produced in Russia [5] from 2001 until 2005. For each of the 17 different shapes in the barrel part 3600 pieces will be produced. Each of the two endcaps will consist of about 10 000 identical pieces. As a result of the claimed dimensional accuracy, the crystal shape error will not contribute to more than 1 % of the detector sensitive area (Figure 25.2).

25.2 CRYSTALS

After growth by the Czochralski method every PbWO$_4$ crystal ingot is first oriented to its identified crystal lattice orientation, a reference face is ground as a start for the next five faces. The oval section (Figure 25.3) resulting from the growth conditions is a clear indication of the lattice orientation (Figure 25.4). It is used to define the position of the starting face A to (1) balance the raw material around the finished shape and (2) provide a reference for the processing of the following faces (Figure 25.5).

The strong anisotropy of PbWO$_4$ is confirmed by experimental results [11]: the fracture strength varies from 26.8 to 31.7 MPa for [010] and [001] tensile directions and is 18.7 MPa normal to (011) cleavage plane. The average

Crystal Growth Technology, Edited by H. J. Scheel and T. Fukuda
© 2003 John Wiley & Sons, Ltd.   ISBN: 0-471-49059-8

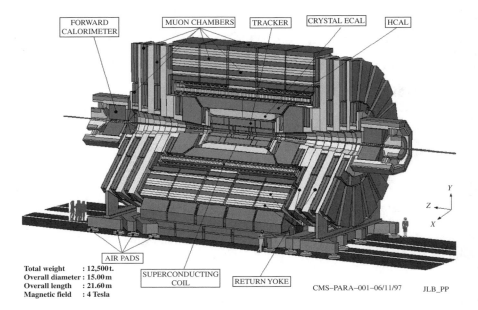

Figure 25.1   General view of the CMS experiment at CERN (2005). The electromagnetic calorimeter is the light gray layer near the centre of the assembly.

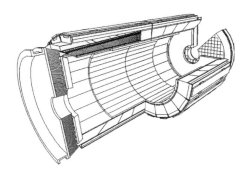

Figure 25.2   General view of the CMS electromagnetic calorimeter showing the spatial arrangement of the 82 000 PbWO$_4$ (lead tungstate) crystals.

coefficients of thermal expansion measured between 25 and 900 °C are (29.5 ± 0.6) × 10^{-6} °C^{-1} along the [001] axis (12.8 ± 0.6) × 10^{-6} °C^{-1} along the [100] and [010] axes.

   17 crystal types are needed with dimensions given in Figure 25.6. [1, 4]. Each face is cut to specification in one operation, with the same cutting orientation

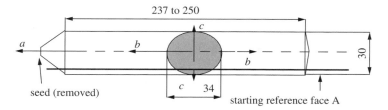

Figure 25.3   Size and crystal lattice orientation of a PbWO₄ Czochralski ingot (or boule).

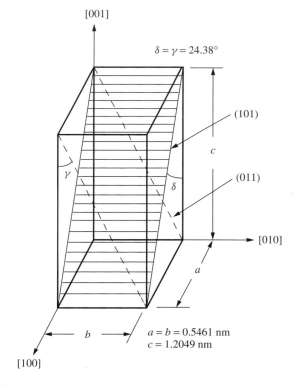

Figure 25.4   Lead tungstate (PbWO₄) crystal lattice [11].

to cleavage planes. The designed surface finish is then reached in only two quick steps, first by lapping and second by polishing. This surface finish progression minimises the number of operations – and machines – and reduces stock removal, time, and subsurface damage. The polishing grade of the four side faces is matched to produce the required light collection uniformity.

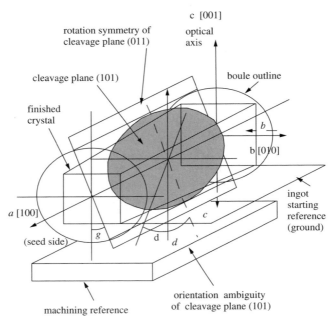

Figure 25.5 PbWO$_4$ lattice orientation with respect to finished crystal shape.

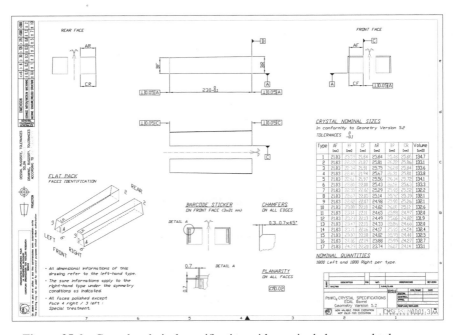

Figure 25.6 Crystal technical specification with required shapes and tolerances.

25.3 MACHINE-TOOLS AND DIAMOND CUTTING DISKS

Standard machine tools [23] have been optimised by simple accessories or modifications (Figures 25.7, and 25.8). Cutting disk thickness, feed speed versus revolution speed ratio, abrasive type and size have been optimised to reduce processing time, while maintaining dimensional and shape accuracy and reducing cutting forces, thus minimising subsurface damage, chipping or breaking risks. Special care is taken to achieve the 0.05 mm flatness in the disk manufacturing (Figure 25.9) [9], and correct it to any extent when mounting the disk on the machine spindle. All five faces are generated with the same side of the cutting disk to exclude disk-thickness variations in the settings.

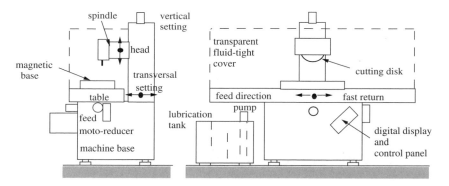

Figure 25.7 Precision planar grinding machine transformed into a crystal-cutting machine.

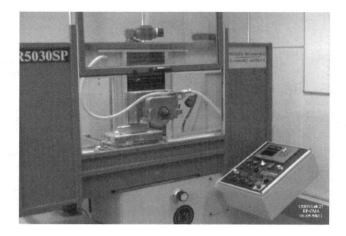

Figure 25.8 Crystal-cutting machine.

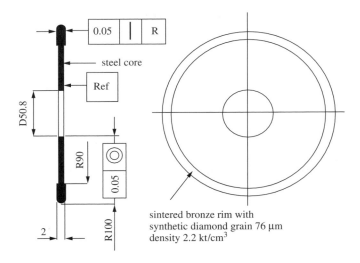

Figure 25.9   Cutting disk with special shape tolerances.

Cutting and lubricating conditions have been optimised as the result of numerous tests (Figure 25.10 and Table 25.1). The cutting depth is set to 11 mm, i.e. larger than the rim width of 10 mm so that the most off-plane diamond grains in the disk will scan the whole processed face, and the face geometry will be generated by the feed translation movement. Cutting force direction close to cleavage plane normal cannot be avoided (Figures 25.16, 25.18 and 25.20) and care is taken to reduce vibrations. The cutting disk is mounted on a dynamical balancing head (Figure 25.15 and 25.17) [10].

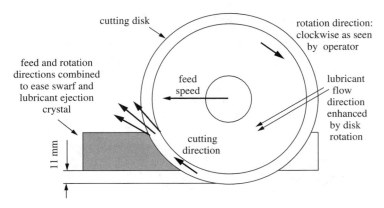

Figure 25.10   Cutting and lubricating conditions of a $PbWO_4$ crystal.

Table 25.1 Cutting and lubricating conditions

Cutting	rotation	feed speed	feed per turn	cutting force	cutting area	surface
parameter	4000 rpm	40 mm/min	10 μm/turn	axial < 10 N	50 mm^2	2 μm Ra

Lubrication	Lubricant	lub. (option)	lub. flow	nozzle dia.	lub. velocity	filtering
condition	Petrol	pH9 solution	180 cm^3/s	2 × 1 mm	2 m/s	gravity

25.4 TOOLING FOR CUTTING OPERATIONS

A quick, accurate and reproducible positioning of the crystals at every operation is performed by a rugged tooling fit for industrial conditions. Quality control is achieved by simple inspection jigs, using a traveling (setting) reference base similar to the processing reference base and magnetic fixations at every processing step.

25.4.1 TRAVELING (SETTING) REFERENCE BASE

The traveling (setting) reference base is in contact with the reference stops of the positioning tooling (points shown as * on Figure 25.11) exactly at the same

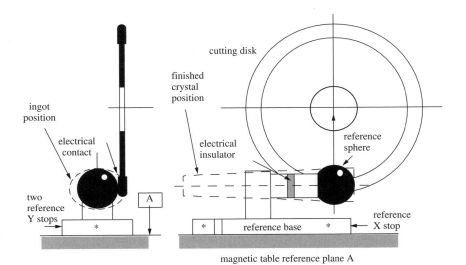

Figure 25.11 Setting disk-to-tool distance to zero with the traveling reference base (principle).

Figure 25.12   Setting disk-to-tool distance to zero with the traveling reference base.

points as the processing reference bases during cutting operations (Figure 25.12). The setting is computed to produce the angular value and the lateral dimensions of the opposite faces 3 and 4 (Figure 25.13).

25.4.2   PROCESSING REFERENCE BASE

The processing reference base consists of a glass plate permanently bound to the steel base by epoxy resin resistant to lubricant and acetone solvent. The 10-μm parallelism tolerance required between the crystal reference face A and the tool plane is maintained using the production lapping machine during maintenance periods. The crystal is bound to the glass base with high precision transfer adhesive (50 ± 5 mm thick), ref. 3M Isotac 9460 IP, resistant to lubricant and easily removed after completing the cutting operations of the five faces in 4 h hours with acetone. In Figure 25.14 a glass fake is used instead of a crystal to test the procedure.

25.4.3   POSITIONING TOOLS

The processing reference base follows the crystal at every cutting and inspection step. It is positioned on the respective tools with mechanical stops placed at identical positions. This reduces tolerance requirements on the numerous processing reference bases and preserves accuracy on a small number of positioning tools (Figures 25.15, 25.16, 25.17, 25.18, 25.19 and 25.21). Positioning tools consist of a magnetic base that is positioned on the cutting machine magnetic table via

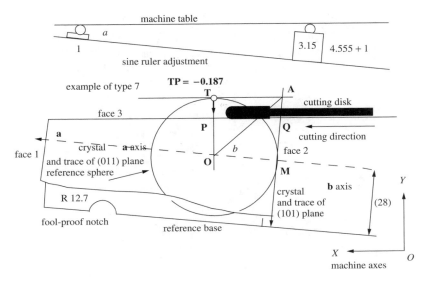

Figure 25.13 Computation of disk setting distance TP from zero to cutting position [16].

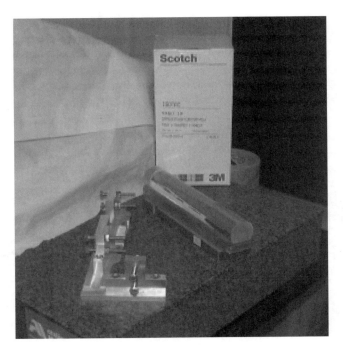

Figure 25.14 Fixation with a special jig of a crystal on a processing reference base.

Figure 25.15 Cutting operation 3 showing the crystal on its processing reference base.

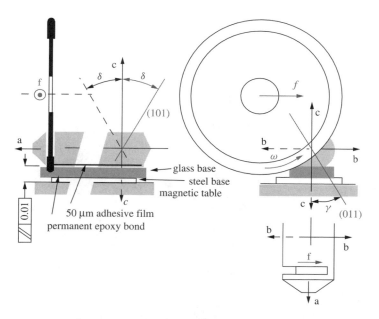

Figure 25.16 Principle of cutting operation 1.

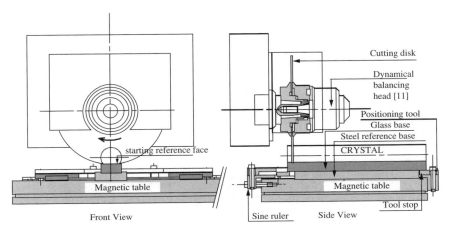

Figure 25.17   Positioning tool for cutting operation 1.

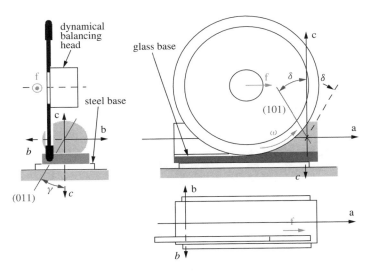

Figure 25.18   Principle of cutting operation 3.

a sine ruler, and a frame with steel spheres acting as accurate positioning stops (see * on Figure 25.11).

For the last cutting operation (Figures 25.20 and 25.21) the crystal is cantilevered on a vertical magnetic table. To prevent the crystal chip from falling and bending the cutting disk, an additional glass support is glued on face N°3 with a conventional transfer adhesive.

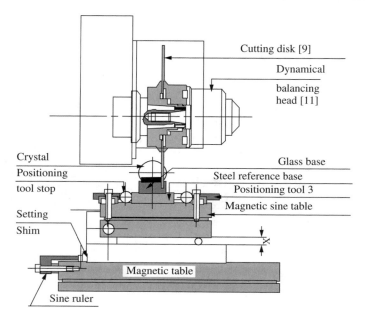

Figure 25.19 Positioning tool for cutting operation 3.

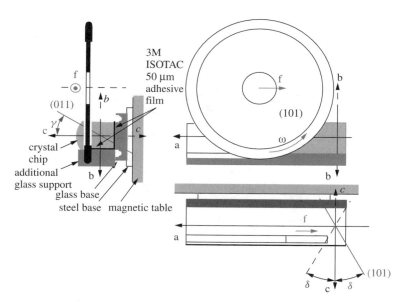

Figure 25.20 Principle of cutting operation 5.

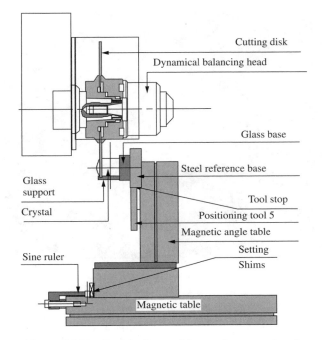

Figure 25.21 Positioning tool for cutting operation 5.

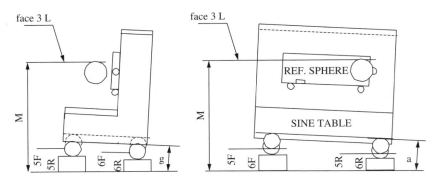

Figure 25.22 Principle of setting inspection tool for cutting operations 3 and 4.

25.4.4 INSPECTION TOOLS

The same positioning principles apply to the inspection tool, as to the processing tool (Figure 25.15), including the use of the reference sphere. The nominal value to be measured (Figures 25.24 and 25.25) is M + T3P3 + LAP where M is measured in the same position with the reference sphere (Figures 25.22 and 25.23), T3P3 is the setting displacement for cutting operation of face 3 (Figure 25.13)

Figure 25.23 Setting zero on inspection tool for cutting operation 3.

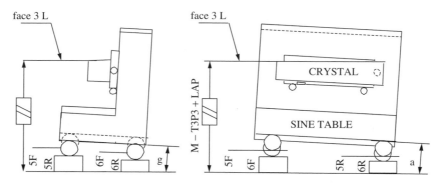

Figure 25.24 Principle of crystal inspection after cutting operation 3 and 4.

and LAP is the extra thickness left for lapping and polishing operations (see Section 25.5). To produce the nominal crystal face angles α and γ, shims of values 5F, 5R, 6F, 6R are placed for inspection of faces N° 3 and 4, and respectively 3F, 3R, 4F and 4R for face N°5.

As each face is inspected after its production in a step-by-step procedure, wrong parts can be discarded or corrected at the earliest opportunity, and machine adjustments immediately corrected. The machine operator's idle time may be profitably used during the cutting operations proper (10 mn on faces 1 and 2, 20 mn on faces 3, 4 and 5) to perform the inspection of the previously machined crystals.

Figure 25.25 Inspection of cut crystal face N° 3.

25.5 TOOLS FOR LAPPING AND POLISHING OPERATIONS

The cutting operation that provides the crystal its required geometrical and dimensional accuracy is followed by two operations only, to eliminate the subsurface damage resulting from the previous processing operation and to reach the optical surface finish required by the light collection. In both lapping and polishing the tools and processing procedures are similar, and differ only in the abrasive grain size and resulting finish grade. Lapping brings the cut surface finish of about $2\,\mu m$ Ra down to $0.2\,\mu m$ Ra, a necessary initial condition for polishing, and finally polishing reaches a minimum of $0.02\,\mu m$ Ra. It is important to reach the mentioned surface finish values in the shortest possible time, to avoid inducing new subsurface tensions. The tool (Figures 25.26 and 25.27) was designed to

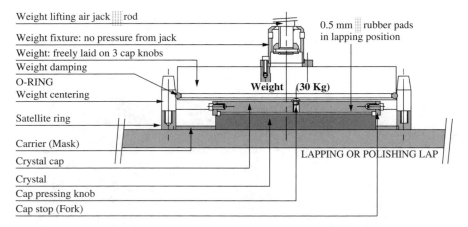

Weight lifting air jack ▓ rod
Weight fixture: no pressure from jack
Weight: freely laid on 3 cap knobs
Weight damping
O-RING
Weight centering
Satellite ring
Carrier (Mask)
Crystal cap
Crystal
Cap pressing knob
Cap stop (Fork)

0.5 mm ▓ rubber pads
in lapping position

Weight (30 Kg)

LAPPING OR POLISHING LAP

Figure 25.26   Tool for lapping or polishing crystal side faces.

Figure 25.27   Polishing side faces on the prototype machine.

exert a uniform pressure of the crystal on the lap, and use the surface to be
treated as a free-adjusting reference.

The drum-and-cartridge tool set (Figures 25.28 and 25.29) allows six crystals
to have their end faces processed in one operation. Each crystal is accurately positioned with respect to its two orthogonal side faces on the cartridge inner reference
faces. Both end faces of a crystal can be processed without changing the setting

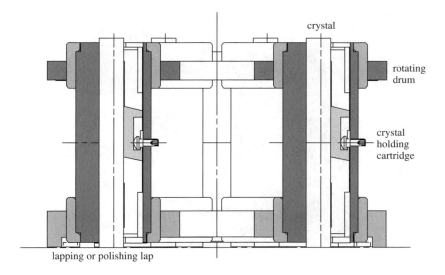

crystal

rotating
drum

crystal
holding
cartridge

lapping or polishing lap

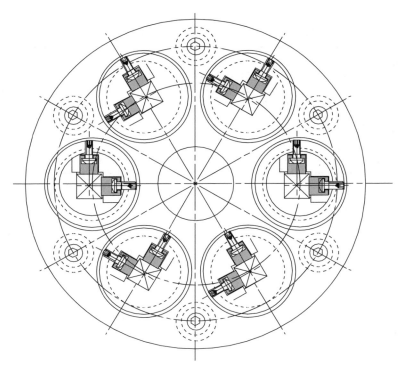

Figure 25.28 Tool for lapping and polishing crystal ends.

Figure 25.29 Polishing crystal ends on prototype polishing machine.

Table 25.2 Lapping and polishing parameters

Parameters	Lapping	Polishing
Lap material	Special resin lap	Polishing fabric on al. lap
Lap diameter/flatness	860 mm/20 μm [14]	860 mm/20 μm [14]
Lap periph. rotation	3 m/s	3 m/s
Abrasive	15 μm synthetic diamond [14]	3 μm synthetic diamond [14]
Pressure on crystal	2 N/cm^2	2 N/cm^2
Processing time	3 mn	10 mn

in the cartridge. After one face is processed, the cartridge + crystal assembly is turned upside down to process the opposite face. Cartridges are free in the vertical direction and exert the required processing pressure of the crystal end face onto the lap. Cartridges do not rotate in the drum bores. End-face lapping and polishing parameters are identical to those for side faces (Table 25.2).

25.6 OPTICAL METHOD FOR INSPECTION OF CRYSTAL RESIDUAL STRESSES

The mass production of the 82 000 PbWO$_4$ crystals complies with tight quality inspection standards. Built-in mechanical tensions – detrimental for the mechanical processing, handling during assembly and long-term reliability of the calorimeter – can be revealed by photoelasticimetry: PbWO$_4$ is birefringent monoaxial [3] and becomes optically biaxial under mechanical stresses. The application of photoelasticity to the inspection of internal tensions in transparent media is well established, in particular for materials with isotropic optical properties, for which the observed fringe pattern has a linear relation with the stress gradient [2]. We previously applied this method to determine the photoelastic constant of BGO, a scintillating crystal with a cubic lattice [18]. More generally, the fact that finished scintillators are polished to a good optical grade renders the method attractive as a simple observation in polarised light reveals internal stresses. The procedure can be part of the quality-inspection protocol, and the information fed back to the crystal growth or annealing conditions [7, 8].

When monochromatic light crosses a uniaxial medium, rays at an angle with the optical axis are refracted according to the ordinary and extraordinary indices (Figure 25.30). There is a phase difference between these two rays, which is proportional to the path length inside the medium. With the incidence increasing from a to a' the phase difference increases with the path length (a'b' > ab) and an alternation of constructive and destructive interferences is observed from m to m', etc. with polariser and analyser properly oriented. Because of the rotational symmetry of the optical properties versus. axis c, m and m' are circles on the screen.

These circles are interrupted in the polarisation direction of the polariser and of the analyser: the ordinary and extraordinary rays coming out at a point b are polarised at right angles, and do not recombine, because of the absence of either the ordinary or the extraordinary ray in these directions (Figure 25.31).

The rotational symmetry of the refractive-index ellipsoid corresponds to the lattice symmetry, with $[100] = a$ and $[010] = b$ directions interchangeable (Figures 25.4 and 25.32). In the optical-axis direction there is only one index – the ordinary one. In any other direction the ordinary index is kept but combines with the extraordinary index and may produce observable interferences.

Under certain stress conditions – no stress tensor symmetry versus crystal lattice symmetry – the rotational symmetry of the index ellipsoid is broken and the

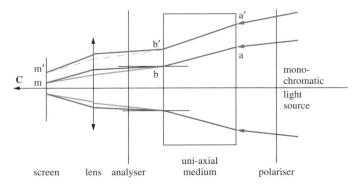

Figure 25.30 Principle of the conical interference.

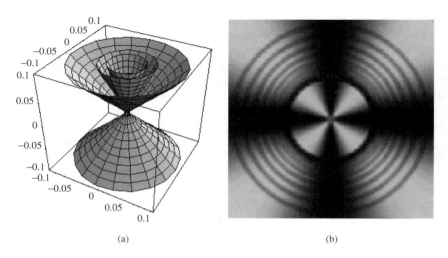

(a)   (b)

Figure 25.31 Typical symmetries of the fringe pattern of a uniaxial crystal (simulation models) [8].

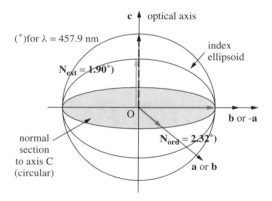

Figure 25.32 Lead tungstate (PbWO$_4$) index ellipsoid.

two axes a and b are different. There are only two central circular sections to such an ellipsoid; they correspond to a single value of the index, i.e. the ordinary index. The directions N1 and N2 normal to these sections are the two optical axes of the crystal (Figure 25.33). In such conditions it is said that a uniaxial crystal becomes biaxial. The corresponding fringe pattern degenerates from the uniaxial fringe pattern. It reveals two pseudocentres corresponding to the axes N1 and N2 (Figures 25.34 and 25.35). The phase difference – and therefore the fringe density – depends on the indices of the medium. The distance between the centres – or the angle N1; N2 – is a function of the photoelastic constant [24], and of the stress conditions. The effect is more visible on the

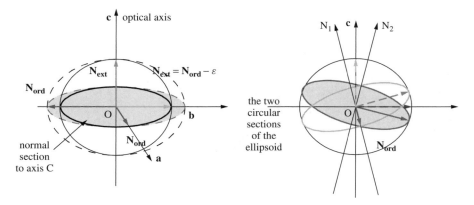

Figure 25.33 Index ellipsoid for a biaxial crystal.

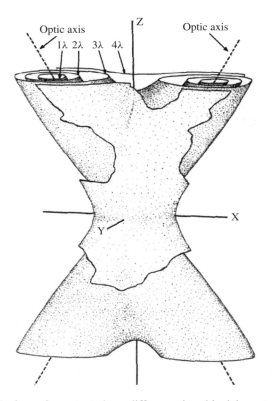

Figure 25.34 Surface of constant phase difference in a biaxial crystal (principle) [6].

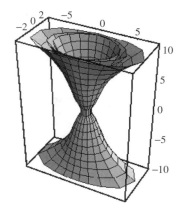

Figure 25.35 Surfaces of constant phase difference in a biaxial crystal (simulation) [7].

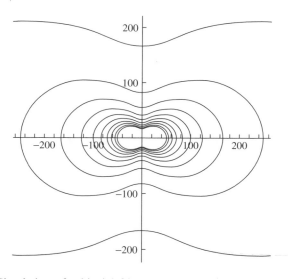

Figure 25.36 Simulation of a biaxial fringe pattern (section of Figure 25.35 by plane *xy*) [7].

fringes of the first orders, fringes of larger orders tend to become closer to circles (Figure 25.36).

The photoelastic constant of PbWO$_4$ crystals is theoretically predicted using the relationship

$$\{\Delta n\} = [\pi]\bullet[S]\bullet\{t\} \tag{25.1}$$

with $\{t\}$ the stress vector, $[S]$ the compliance matrix [15], and $[\pi]$ the piezo-optic matrix [13], which relates the strain in a variation of the refraction index n.

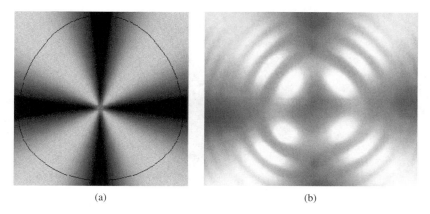

(a)   (b)

Figure 25.37 Similarity between model and image of interference in the case of flexural loading.

The corresponding interference pattern was modelled with MATHEMATICA® software and compared to experimental results. Not only simple load states as compression were predicted but also flexural states, with good correspondence between theory and observed image of interference (Figure 25.37).

The experimental setup consisted of a polariscope and an accurate loading system. As a result the photoelastic constant of PbWO$_4$ was determined: $F_\sigma = 0.1\,MPa^{-1}$ [7].

On this basis crystals can be inspected in a simple polariscope (Figure 25.38) in the flow of mass production.

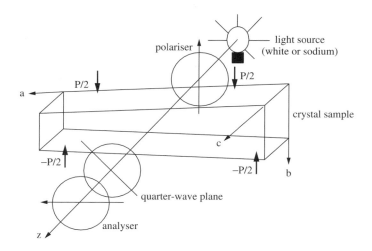

Figure 25.38 Polariscope scheme.

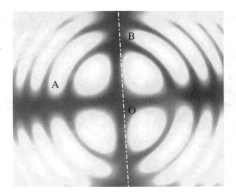

Figure 25.39   The ratio OA/OB is a linear function of the crystal stress level.

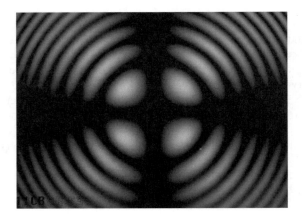

Figure 25.40   Fringe pattern of an unloaded stress-free $PbWO_4$ crystal sample showing the symmetry of the cross pattern.

The image of interference is digitised using a video camera, the data is processed to produce the value of the residual stress σ as a function of the cross-arm lengths ratio [7, 8] (Figure 25.39):

$$\sigma = ((OB/OA) - 1)/F_\sigma$$

Figures 25.40 and 25.41 illustrate typical fringe patterns observed in mass-production crystals. Crystals with low built-in stress levels display obviously distorted pattern. As an example Figure 25.40, corresponds to a compression level of the order of 1 MPa to be compared to a cleavage value of 18.7 MPa [11].

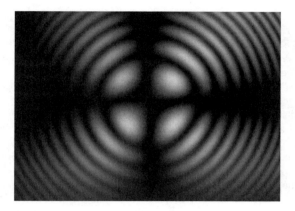

Figure 25.41 Fringe pattern of an unloaded PbWO$_4$ crystal sample with built-in stresses showing a distortion of the cross pattern.

25.7 CONCLUSIONS AND PRODUCTION FORECASTS

Most of the above processing method has already been put to work in Russia [5] for the production of the 82 000 PbWO$_4$ crystals needed for the CMS experiment. In both places cutting machines and tool prototypes were installed in 1997 and test results communicated to monitor processing progress, following the pace of crystal-growth performance. Combined with progress in growth and post-growth conditions and investigation on mechanical properties of PbWO$_4$ [11, 12, 20], they have resulted in considerable reduction of accidental breaking. Furthermore, dimensional tolerances of $+0/-100 \, \mu$m, planarity and perpendicularity of 20 μm and satisfactory surface finishes are reached in the already delivered 6000 pieces of 1999–2000's pre-production delivery from the Russian company. To achieve this goal, a fully equipped production line, consisting of five cutting machines for the cutting of the side faces, was installed in early 1999. One lapping machine and two polishing machines with their complete processing and inspection tools followed in that summer. End faces are cut with locally available internal-diameter (ID) saws adapted to the principle of this method (reference bases positioned on magnetic tables).

With one production line more, and working two shifts a day, the annual crystal growth rate of 8500 crystals planned as from 2000 should easily be absorbed, therefore completing the total delivery by the end of 2005 with a minimal capital investment in processing equipment.

REFERENCES

[1]   Auffray, A. *et al.* (1998) *Specification for Lead Tungstate Crystal Pre-production*, CMS Note/98-38

[2]   Avril, J. (1984) *Encyclopédie des Contraintes* (in French), Micromesures, 98 Bd. Gariel-Péri, 92240 Malakoff, France

[3]   Baccaro, S. *et al.* (1995) *Optical properties of lead tungstate (PbWO₄) crystal for LHC em-calorimetry*, internal note No. 1066, University of Rome 'La Sapienza', Physics Department, Rome

[4]   Bally, S. (1997) *CMS ECAL Crystal Geometry Management*, CMS Internal Note, CERN, Geneva

[5]   Bogoroditsk Technochemical Plant, 301800 Bogoroditsk, Tula region, Russia

[6]   Born, M. and Wolf, E. (1975) *Principles of Optics*, 6th edn., Pergamon Press, Oxford

[7]   Cocozzella, N. (1999) *Quality control in scintillating crystals through interference fringe pattern analysis* (in Italian) Thesis, University of Ancona

[8]   Cocozzella, N. *et al.* (1999) *Quality inspection of anisotropic scintillating crystals through measurement of interferometric fringe pattern parameters*, SCINT'99, International Conference on inorganic Scintillators and their applications, Moscow

[9]   Diamant Boart, avenue du Pont de Luttre, 74, 1190 Bruxelles, Belgium *Diamond cutting disks*

[10] Elettronica G. T., via Ruffilli 2/4, 20060 Pessano c. B. (MI), Italy *Balancing head GTR 250 M*

[11] Ishii, M. *et al.* (1996) *Mechanical properties of PbWO₄ scintillating crystals*, *Nucl. Instrum Methods A* **376** 203–207

[12] Klassen, N. (1996) *About some features of the mechanical properties of monocrystalline lead tungstate.* (in Russian), Tchernogolovka Institute of Solid State Physics, Moscow

[13] Kobelev, N. and Soyfer, I. (1996) *Elastic moduli of lead tungstate*, *Solid State Phys.*, **39 N12**, 3589–3594 (in Russian)

[14] Lamplan S. A., 7 rue des Jardins, 74240 Gaillard, France

[15] Landau, L. and Lifshitz, C. (1946) *Theory of Elasticity*, Moscow Editions, Moscow

[16] Lebeau, M. (1996) *Settings for the cutting of a PbWO₄ crystal boule to standard dimensions*, CMA Internal Note, CERN, Geneva

[17] Lebeau, M. (1997) *Principles of the cutting method proposed for the PbWO₄ crystals of the CMS Electromagnetic Calorimeter*, CMS note 1997/024, CERN, Geneva

[18] Lebeau, M. *et al.*, (1997), *Photoelasticity for the investigation of internal stresses in BGO scintillating crystals*, *Nucl. Instrum. Methods A* **397** (1997) 317–322

[19] Lebeau, M. and Le Marec, J.-C. (1985) *Coupe des cristaux de BGO. Ordonnancement des opérations*, LAPP internal note (in French), LAPP, Annecy, France

[20] Lebeau, M. and Rinaldi, D. (1997) *Photoelasticimetry for the inspection of mechanical stresses in transparent bi-refringent mono-axial crystals. Application to CeF₃ and PbWO₄*. Scint'97 International Conference on Inorganic Scintillators, Shanghai

[21] Lebeau, M. *et al.* (1997) *Cutting of five PbWO₄ crystals on the CERN prototype cutting machine*, CMS Note 1997/028, CERN, Geneva

[22] Lebeau, M. *et al.* (1997) *Cutting of five PbWO₄ crystals in industrial prototype conditions*, CMS note 1997/029, CERN, Geneva

[23] LGB, 45 allée du Mens, B. P. 6062, 69604 Villeurbanne, France, *Grinding machine LGB R 5030*

[24] Perelomova, N. and Tagieva, M. (1983) *Crystal Physics*, 1st English edn., Mir Publishers, Moscow

26 Plasma-CVM (Chemical Vaporization Machining)

YUZO MORI, KAZUYA YAMAMURA, and YASUHISA SANO

Department of Precision Science and Technology, Graduate School of Engineering, Osaka University, 2-1 Yamada-oka, Suita, Osaka 565–0871, JAPAN

26.1 INTRODUCTION

Figure 26.1 shows the relationships between removal rates and surface roughness in various machining methods. By applying the EEM (elastic emission machining) process that we developed, it is possible to realize the best flatness of all machining methods and surface roughness of the atomic order. However, there has been no effective machining method for preparing premachined surfaces for EEM.

Grinding and polishing are highly efficient machining methods with respect to removal rate, but the obtained surfaces have many defects like microcracks, dislocations, vacancies, and so on. When such machining methods are employed for preparing premachined surfaces, the damaged layer must be removed during final polishing with EEM. But the removal rate of EEM is very slow, as shown in Figure 26.1 so that conventional machining methods are not suitable for premachining the surfaces for EEM. On the other hand, plasma etching is less damaging because of the chemical processes, but its removal rate is also low. The new machining method, named plasma-CVM (chemical vaporization machining) was developed to fabricate the required shapes efficiently with no damaged layer by utilizing chemical processes [1–3].

26.2 CONCEPTS OF PLASMA-CVM

P-CVM utilizes highly reactive neutral radicals (such as halogen atoms with high electronegativity) generated in high-pressure VHF (very-high frequency) plasma. The generated radicals react with the surface atoms of the work-piece and transform them into volatile molecules. Because P-CVM is a chemical process that proceeds at the atomic level, a geometrically perfect surface, which is

Crystal Growth Technology, Edited by H. J. Scheel and T. Fukuda
© 2003 John Wiley & Sons, Ltd.   ISBN: 0-471-49059-8

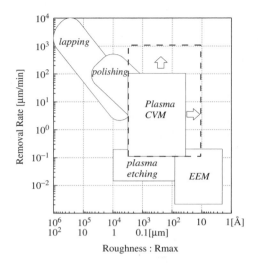

Figure 26.1 Relationships between removal rate and surface roughness of various machining method.

also free from crystallographic or structural damages, can be obtained. A similar machining phenomenon is already utilized in the plasma etching in LSI manufacturing. Plasma etching, however, has low machining efficiency and is spatially uncontrollable because it utilizes low-pressure ($1-10^{-3}$ torr) plasma. Instead, in P-CVM, a high-pressure plasma (higher than 1 atm) is utilized. The VHF high-pressure plasma is localized around the electrode and can generate a high density of radicals. The conventional process may realize a spatial resolution comparable to conventional mechanical machining. The new P-CVM allows a significantly better machining efficiency to be obtained. And by using the power supply of the VHF (150 MHz) frequency, the amplitude and energy of the ions in the plasma are reduced to 1/100 of the usual 13.56-MHz plasma, and this prevents the surface damage induced by collision of the ions to the work-piece surface. Special care is taken to prevent an occurrence of high-temperature arc-discharge plasma. We coated the rotary electrode surface with alumina to reduce the thermal effects and to eliminate the secondary-electron emission from the electrode.

26.3 APPLICATIONS OF PLASMA-CVM

In an actual P-CVM process, rotary electrodes with high velocity are utilized as shown in Figure 26.2. By placing a rotating electrode close to the work-piece surface and by applying VHF power to generate a plasma in the gap between the electrode and the work-piece surface, we can proceed with the P-CVM as if a grinding wheel performs mechanical grinding. Rotating an electrode at a high

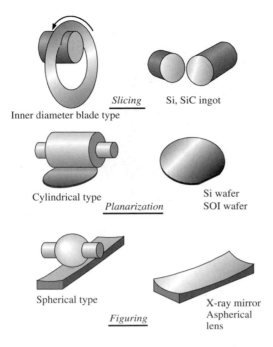

Figure 26.2   Applications of rotary electrode.

velocity enables one to achieve: (1) high supplying efficiency of reactive gas; (2) high machining efficiency due to the high power supply, mediated by the cooling of the electrode; (3) realization of surface flattening due to the high gas velocity in the horizontal direction upon the work-piece surface. The phenomenon (3) occurs due to the unique property related to the viscosity of the high-pressure plasma. By changing the electrode shape, as is shown in Figure 26.2, it is possible to carry out slicing, planarization, and figuring. The details of the applications are described in the following.

26.4   SLICING

26.4.1   SLICING MACHINE

Figure 26.3 shows a schematic view of a P-CVM slicing machine. An inner diameter blade electrode is fixed to an electrode body at an outer diameter, and tensioned by adding a strain just inside the peripheral edge. An electrode body is insulated from the chamber base by ceramic insulators and from the motor shaft by magnet coupling, and it is rotated at high velocity by the motor installed

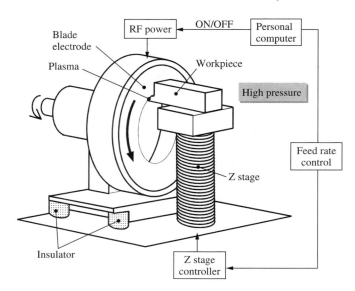

Figure 26.3   Schematic view of slicing machine.

outside the chamber. RF power is supplied to an electrode through an impedance-matching unit. The plasma is generated in the machining gap between the inner edge of a blade electrode and the workpiece.

26.4.2   SLICING RATE

The quantity of reactant SF_6 gas supplied by the viscosity increases in proportion to the rotational velocity of the electrode. Figure 26.4 shows the relationship between the slicing rate and the rotation velocity of the electrode. The SF_6 concentration is 10 %, and the supplied RF power is 1000 W.

Figure 26.5 shows the relationship between slicing rate and concentration of SF_6 for rotating velocities of the electrode of 72 m/s and 39 m/s, and the supplied RF power 1000 W. It is found that the slicing rate increases with the SF_6 concentration. For a further increase of the slicing rate, it is necessary to increase the RF power in order to generate more fluorine radicals by the decomposition of the reactant gas. The relationship between slicing rate and RF power for various machining conditions is shown in Table 26.1, and it is found that the slicing rate is almost proportional to the RF power. Table 26.2 shows the effect of the addition of oxygen for the decomposition of SF_6. It is clearly seen that the slicing rate with O_2 is 1.5 times faster than without O_2. O_2 addition seems to activate the SF_6 decomposition and to remove sulfur from SF_6 decomposition. In short, Tables 26.1 and 26.2 show that the increase in the slicing rate up to 0.7 mm/min can be realized by sufficient SF_6 supply to the machining gap.

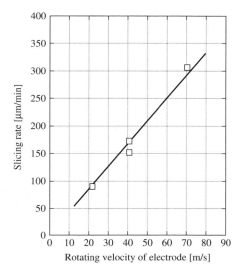

Figure 26.4 Relationship between slicing rate and rotation velocity of electrode.

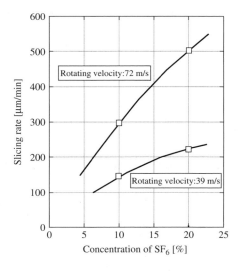

Figure 26.5 Relationship between slicing rate and concentration of SF_6.

26.4.3 KERF LOSS

In general, for the purpose of slicing with small kerf loss, it is necessary to use a thin tool. In plasma-CVM, the tool for slicing is the plasma generated in the machining gap between the electrode and the workpiece. So it is necessary to

Table 26.1 Relationship between slicing rate and RF power

	SF$_6$ concentration (%)	Rotation velocity of electrode (m/s)	RF power (W)	Slicing rate (μm/min)
A	10	39	500	70
			1000	**150**
B	20	72	1000	500
			1300	**700**

Table 26.2 Effect of the addition of the oxygen

	RF power (W)	SF$_6$ concentration (%)	O$_2$ concentration (%)	Slicing rate (μm/min)
A	600	10	0	200
			10	**280**
B	1000	10	0	300
			10	**450**
C	1000	20	0	400
			20	**600**

generate plasma only in a narrow area. For this, it is important not only to use a thin-blade electrode but also to control the plasma itself. The SF$_6$ gas restrains the discharge, and it is generally used as a gas for insulation. Therefore, even with increased SF$_6$ concentration, the plasma is only locally generated at the edge of the inner-blade electrode, and a small kerf loss is expected.

Figure 26.6 shows the relationship between kerf loss and SF$_6$ concentration for the rotation velocity of the electrode of 72 m/s, RF power of 1000 W and blade thickness of 0.1 mm. It is found that the kerf loss decreases with increased SF$_6$ concentration to a minimum kerf loss, which has been realized so far of 0.3 mm.

26.4.4 SLICING OF SILICON INGOT

Next, the result of slicing a silicon ingot (25 × 25 mm) by plasma-CVM is described. Figure 26.7 shows the relationship between slicing length and machining time. The gradient of the plotted line corresponds to the slicing rate, and it is found that stabilized machining properties have been obtained. Figure 26.8 is a photograph of a side view of a sliced ingot. The vertical dark line is the kerf loss. This figure shows that reactant gas was efficiently supplied into the deep groove by a high-speed rotary inner-blade electrode, and the performance of this system allows stable slicing in a deep groove. Figure 26.9 is a photograph of the sliced surface and a surface profile of this sample. It is found that there is a waviness

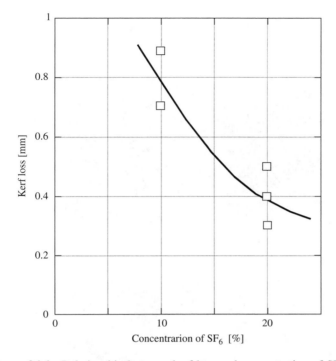

Figure 26.6   Relationship between kerf loss and concentration of SF$_6$.

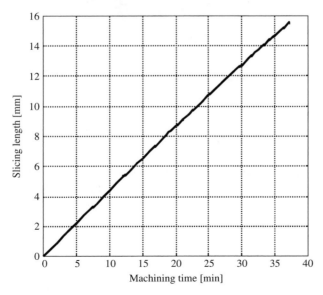

Figure 26.7   Relationship between slicing length and machining time.

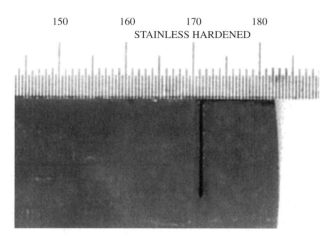

Figure 26.8   Side view of a sliced ingot.

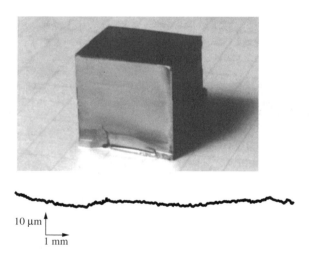

Figure 26.9   Profile of a sliced surface.

of the about 10 μm in a 20 mm region of the sliced surface. However, it seems to be possible to reduce the waviness by improved control of the workpiece speed.

26.5   PLANARIZATION

26.5.1   PLANARIZATION MACHINE

Figure 26.10 shows the planarization machine for Si wafers where the RF power ($f = 150$ MHz) is supplied to the cylindrical rotary electrode through a cavity

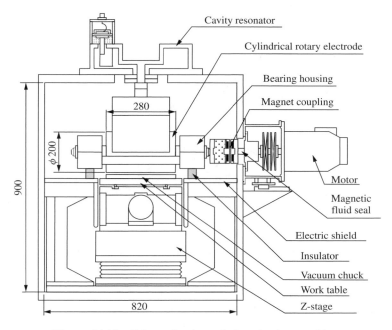

Figure 26.10 Schematic view of planarization machine.

resonator. The electrode, insulated by magnet coupling supported in Teflon, and the insulator, made of alumina, is driven by an AC spindle motor through a magnetic fluid seal. The silicon wafer is held by a vacuum chuck, and it is possible to polish a wafer of 8-inch diameter.

26.5.2   MACHINING PROPERTIES

Figure 26.11 shows the relationship between stock removal and surface roughness of the Si wafer polished by plasma-CVM. In this experiment, the premachined plane is lapped and chemically etched. A surface roughness Ra = 1 nm was obtained after removing 50 μm. Figure 26.12(a) and (b) shows the surface roughness of the Si before and after polishing with Plasma CVM, respectively, and demonstrates clearly the improvement by P-CVM polishing. The waviness of several tens of μm period is effectively eliminated. This kind of waviness called 'ripple' is generated by transferring the roughness of the polishing pad, and is an unavoidable phenomenon in mechanical machining. So, plasma-CVM is superior to conventional mechanical polishing for finishing the surface with subnanometer flatness.

So far, geometrical machining properties have been discussed. However, the physical properties of polished surfaces are also important in functional materials

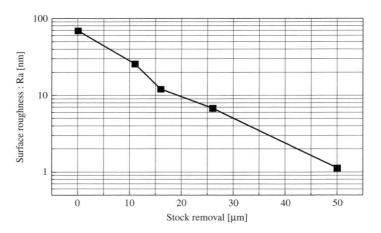

Figure 26.11   Relationship between stock removal and surface roughness of si wafer.

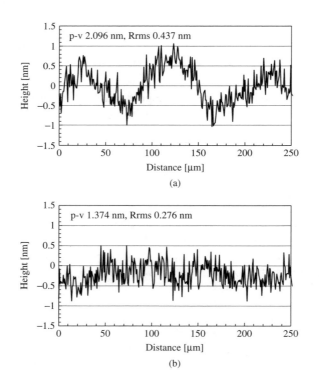

Figure 26.12   Surface roughness of Si wafer. (a) mechanical polished surface, (b) plasma-CVM polished surface.

like silicon. The defect densities of polished surfaces are measured by SPV (surface photo voltage) spectroscopy [3]. SPV spectroscopy is a sensitive method to detect electron transitions as the surface voltage changes by applying subbandgap light irradiation. When defects (dislocations, vacancies, etc.) are introduced into the semiconductor surface during machining, the electronic level induced by the defect is generated in the bandgap. The SPV spectra are shown in Figure 26.13, and the machining conditions applied for the samples (p-type Cz-Si, $\rho = 10\ \Omega$cm) are shown in Table 26.3.

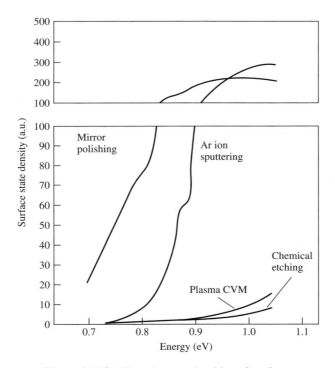

Figure 26.13 Electric state densities of surface.

Table 26.3 Machining conditions of measured specimens

Mechanical polishing + Cleaning	Powder particle: 0.1μm, SiO_2 Polishing pressure: 150 gf/cm^2 HF:$H_2O = 1:3$ (RT)
Ar$^+$ sputtering + Cleaning	Accelerating voltage: 1 kV Ion current density: $5\ \mu$A/cm^2 HF:$H_2O = 1:3$ (RT)
Plasma CVM + Cleaning	Concentration of SF_6:1 % HF:$H_2O = 1:3$ (RT)
Chemical etching + Cleaning	HF:HNO_3:$H_2O = 1:6:8$ HF:$H_2O = 1:3$ (RT)

From this figure, it is found that mechanically polished and Ar⁺-sputtered surfaces have many defects. The peak at 0.17 eV below the conduction band edge is the defect named an A-center, which has been reported as the complex defect of an oxygen atom and a vacancy [5, 6] and is the main defect of mechanically damaged crystals. On the other hand, no peak structure appears in the spectra of surfaces machined by plasma-CVM and chemical etching. Furthermore, the defect densities of these surfaces are about 1/100 of those of mechanically polished or Ar⁺-sputtered surfaces. These results show that the removal mechanism in plasma-CVM is purely chemical, and ion-bombardment damages are not introduced. The high-frequency power supply of 150 MHz seems to avoid ion-bombardment damages.

26.6 FIGURING

26.6.1 NUMERICALLY CONTROLLED PLASMA-CVM SYSTEM

The numerically controlled Plasma-CVM machine for figuring ultraprecise optical devices with aspherical shapes such as X-ray mirrors and lenses for steppers was developed and applied [7–10].

Figure 26.14 shows the outline of the numerically controlled plasma-CVM machine. This machine is composed of the high-speed rotary electrode, an X-Y-θ table, cavity resonator for RF impedance matching, and the process chamber made of aluminum alloy. The main elements are fixed on the precision granite base, so that deformation effects of the process chamber are eliminated. Furthermore,

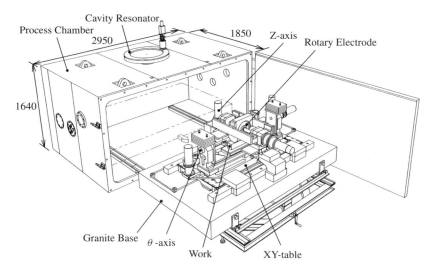

Figure 26.14   Numerically controlled plasma-CVM system.

mounting of the workpiece and maintenance of the machine are facilitated by pulling out the granite table from the process chamber.

Figure 26.15 shows a schematic view of the rotary electrode for generating the plasma. The surface of the electrode is coated by aluminum oxide to prevent arc discharges. The shaft of the electrode is supported by a hydrostatic bearing with process gases in the chamber, so that contamination of the workpiece from particles generated by wear and organic vapours from lubrication oil is minimized. The maximum rotating velocity is $5000\,min^{-1}$.

Figure 26.16 shows the XY-table for moving the work-piece. Both X-table and Y-table float on the same reference plane of the precision granite table with a hydrostatic pad, and are driven by pulling the stainless steel belts that are

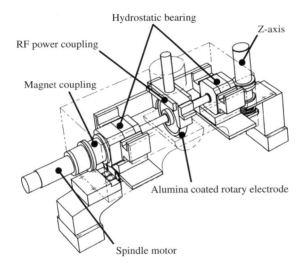

Figure 26.15   Schematic view of the rotary electrode.

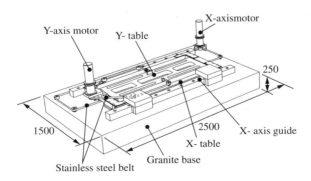

Figure 26.16   XY-table with hydrostatic bearing.

connected to each table. And stroke of the XY-table is 500 mm for the x-axis and 200 mm for the y-axis. Figure 26.17 shows the straightness of the XY-table measured by an autocollimator. It is found that straightness within 0.5 μm is achieved in the range of the full stroke.

Figure 26.18 shows the gas-circulation system for keeping the machining conditions constant. This system consists of the 0.01 μm filter, a gas purifier for eliminating reaction products by chemical reaction, a heat exchanger for keeping the temperature of the reactive gas constant, and a gas supply for keeping the concentration of the reactive gas constant. Figure 26.19 shows the gas temperature is controlled within ±0.1 °C.

In conclusion, an ultraclean environment for a precision chemical machining process is achieved by applying the hydrostatic bearing and the gas-circulation system.

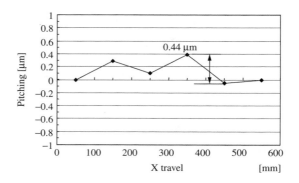

Figure 26.17   Straightness of the XY-table.

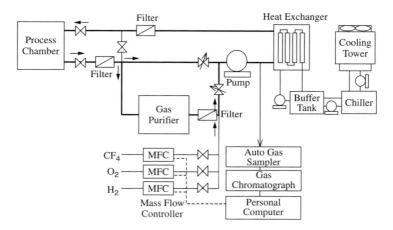

Figure 26.18   Gas-circulation system.

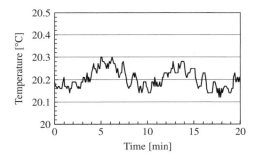

Figure 26.19 Fluctuation of the gas temperature.

26.6.2 MACHINING PROPERTIES

The shape of the removal unit depends on machining parameters such as composition of the reactive gas, RF power, rotation velocity of the electrode, machining gap between the electrode and the workpiece, curvature of the electrode, etc. Figure 26.20 shows the shape of the removed area in unit time with machining conditions shown in Table 26.4. In this case, the removed area has an oval shape, and the size is about 30 mm × 6 mm.

Next, the correlation of removal rate and table feed is described. The machining experiment was carried out by scanning the work at the various scanning speeds. The cross-sectional shape of the removal area formed each scan is shown in Figure 26.21, and machining conditions are shown in Table 26.5. From these experimental results it is found that only the removal depth increases in either condition, A or B, without changing the removal width, when the feed speed

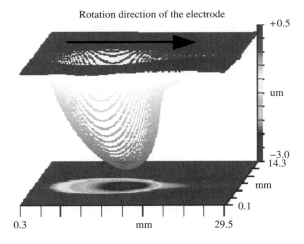

Figure 26.20 Shape of the removed area.

Table 26.4 Machining conditions

Specimen	$\phi 6$ (100) Si wafer
Gas composition	He:CF_4:O_2 = 99.8:0.01:0.01
RF power	50 W
Rotation speed of electrode	2000 rpm (31.4 m/s)
Machining gap	0.6 mm
Machining time	30 min

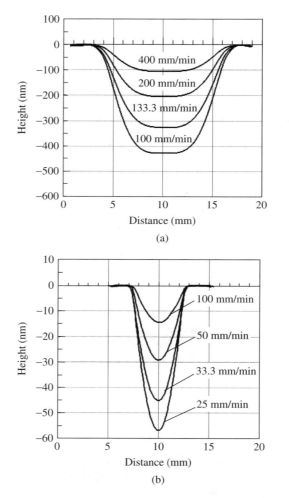

Figure 26.21 The cross-sectional shape formed by scanning: (a) condition A, (b) condition B.

Table 26.5 Machining conditions

Parameters	A	B
Specimen	Si (100) ϕ150 × 0.625^t	Si (100) 400 × 50 × 30^t
Reaction gas	He + 0.1 % CF$_4$ + 0.1 % O$_2$	He + 0.01% CF$_4$ + 0.01% O$_2$
RF power	500 W	50 W
Rotation speed of electrode	2000 rpm (31.4 m/s)	2000 rpm (31.4 m/s)
Machining gap	1000 μm	600 μm

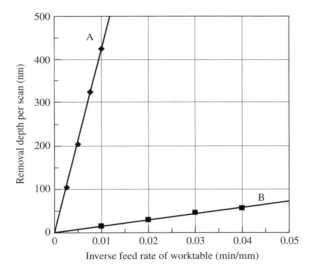

Figure 26.22 Relationship between inverse scanning rate of worktable and removal depth per scan.

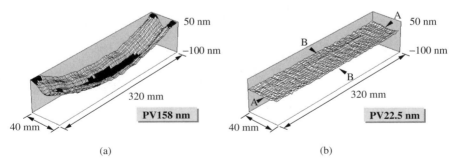

Figure 26.23 The shape-correction result of an X-ray flat mirror by plasma-CVM. (a) before correction and (b) after correction by NC-PCVM.

of the worktable decreases. The relationship between inverse scanning speed of worktable and removal depth per scanning operation is shown in Figure 26.22. From this result, it is seen that the perfect linearity has been obtained in both conditions. Therefore, it is possible to carry out highly precise figuring by the numerically controlled plasma CVM.

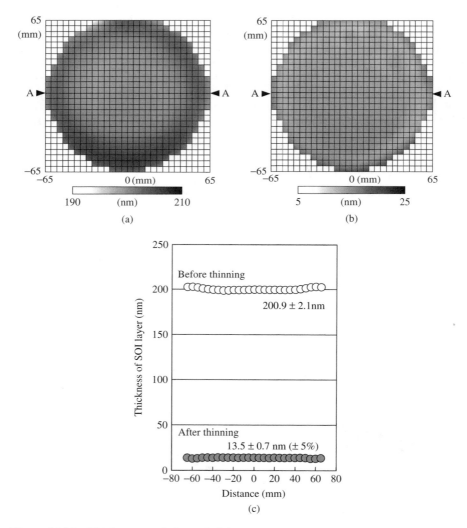

Figure 26.24 Thickness variation of SOI: (a) SOI layer thickness distribution of as-received SOI wafer (UNIBOND® wafer); (b) SOI layer thickness distribution of the thinned SOI wafer by numerical controlled plasma CVM; (c) A−A cross sections of (a) and (b).

26.6.3 FABRICATION OF THE FLAT MIRROR

Shape-correction machining of a flat mirror was carried out. This mirror is used in order to eliminate the high-order harmonics in the beam line of synchrotron radiation, the material of the mirror is single-crystal silicon, the orientation of the mirror surface is (100), and the size is 400 mm (length) × 50 mm (width) × 30 mm (thickness). Numerically controlled machining was carried out using the machining parameter shown in B of Table 26.5. The shape evaluation region in the mirror was 320 mm × 40 mm, and it was measured by a laser interferometer (Zygo : GPI-XPHR). The machining result is shown in Figure 26.23. The flatness of the mirror before correction is 158 nm (p-v), and the shape is shown in Figure 26.23(a). As a result of correction machining by numerically controlled plasma-CVM, 22.5 nm (p-v) flatness was achieved, as shown in Figure 26.23(b). Seven correction cycles of this mirror were carried out and the total thickness removed was about 1 μm.

Figure 26.24 shows the result of thinning the silicon layer of the SOI wafer. The film thickness before thinning is 200 nm, and the thickness distribution is ±4.2 nm. After film thinning to 13 nm the thickness distribution was improved to ±2 nm.

In the figuring by numerically controlled plasma-CVM, there is no degradation of the machining accuracy by the effect of heat despite the existence of the plasma heat source. This is because plasma-CVM is a noncontact machining method that controls the removal volume only during the dwell time of the plasma. On the other hand, in the mechanical machining method, the heat that arises at the machining point becomes a direct factor that causes degradation of machining accuracy, because the fluctuation of cutting depth or polishing pressure by the thermal deformation causes instability in machining properties. Therefore, plasma CVM has a remarkable advantage of the insensitivity for various disturbances.

ACKNOWLEDGEMENTS

This work was partially supported by the Japan Science and Technology Corporation and by a Grant-in-Aid for COE Research (No. 08CE2004) from the Ministry of Education, Science, Sports and Culture.

REFERENCES

[1]   Y. Mori, K. Yamamura, K. Yamauchi, K. Yoshii, T. Kataoka, K. Endo, K. Inagaki, and H. Kakiuchi, 'Plasma CVM: an ultra precision machining technique using high-pressure reactive plasma.', Nanotechnology **4** (1993) 225–229.

[2]   Y. Mori, K. Yamamura, K. Yamauchi, K. Yoshii, T. Kataoka, K. Endo, K. Inagaki, and H. Kakiuchi, 'Plasma CVM – An ultra precision machining with high pressure reactive plasma', *Technol. Rep. Osaka Univ.* **43**, No. 2156, (1993) pp 261–266.

[3]   Y. Mori, K. Yamauchi, K. Yamamura and Y. Sano, 'Development of plasma chemical vaporization machining' *Rev. Sci. Instrum.* **71** (2000) 4627–4632.

[4]   Y. Mori, K. Yamauchi, K. Endo, H. Wang, T. Ide and H. Goto, 'Sub-Band-Gap Surface Photovoltage in Finely-Polished Si Surfaces', *Technol. Rep. Osaka Univ.* **40** (1990) 293–301.

[5]   G. D Watkins and J. W Corbett, 'Defects in Irradiated Silicon I. Electron Spin Resonance of the Si-A Center', *Phys. Rev.* **121** (1961) 1001–1014.

[6]   J. W. Corbett, G. D. Watkins, R. M. Chrenko and R. S. McDonald, 'Defects in Irradiated Silicon. II. Infrared Absorption of the Si-A Center', *Phys. Rev.* **121** (1961) 1015–1022.

[7]   K. Nemoto, T. Fujii, N. Goto, H. Takino, T. Kobayashi, N. Shibata, K. Yamamura, and Y. Mori, 'Laser beam intensity profile transformation with a fabricated mirror', *Appl. Opt.* **36** (1997) 551–557.

[8]   H. Takino, N. Shibata, H. Itoh, T. Kobayashi, H. Tanaka, M. Ebi, K. Yamamura, Y. Sano and Y. Mori, 'Plasma Chemical Vaporization Machining (CVM) for Fabrication of Optics', *Jpn. J. Appl. Phys.* **37** (1998) L894–L896.

[9]   H. Takino, N. Shibata, H. Itoh, T. Kobayashi, H. Tanaka, M. Ebi, K. Yamamura, Y. Sano, and Y. Mori, 'Computer numerically controlled plasma chemical vaporization machining using a pipe electrode for optical fabrication', *App. Opt.* **37** (1998) 5198–5210.

[10] Y. Mori, K. Yamamura and Y. Sano, 'The study of fabrication of the X-ray mirror by numerically controlled plasma chemical vaporization machining: Development of the machine for the X-ray mirror fabrication', *Rev. Sci. Instrum.* **71** (2000) 4620–4626.

27 Numerically Controlled EEM (Elastic Emission Machining) System for Ultraprecision Figuring and Smoothing of Aspherical Surfaces

YUZO MORI, KAZUTO YAMAUCHI, KIKUJI HIROSE, KAZUHISA SUGIYAMA, KOHJI INAGAKI, and HIDEKAZU MIMURA

Department of Precision Science and Technology, Graduate School of Engineering, Osaka University, 2-1 Yamada-oka, Suita, Osaka 565-0871, JAPAN

27.1 INTRODUCTION

EEM (elastic emission machining) is an ultraprecise machining process employing surface-chemical activities of ultrafine powders [1–6]. The removal unit is atomic size and no crystallographic damage is introduced during machining. EEM has been said to be a possible technique to achieve quasiatomically flat surfaces. This technique has been successfully applied to polish the mirrors of the laser jailoscope employed as an orbit controller of the Japanese H-II rocket.

The EEM process is carried out with a fluid consisting of water and ultrafine powders with smaller than 0.1 μm particle size. They are softly transported and contact the workpiece surface in a flow of ultrapure water. Upon contact, chemical reactions between the workpiece surface and the particle occur with some probability. When the particles are separated and removed from the workpiece surface by the viscous drag of the water flow, atomic removal from the workpiece surface is taking place (Figure 27.1). The machining technology utilizing this kind of chemical mechanism is called EEM [3, 6].

27.2 FEATURES AND PERFORMANCES

The distinctive features of EEM process are:

First, extremely smooth surfaces of quasiatomic flatness can be obtained easily. Figure 27.2 shows WYKO measurements. Figure 27.2(a) is the profile of a

Crystal Growth Technology, Edited by H. J. Scheel and T. Fukuda
© 2003 John Wiley & Sons, Ltd.   ISBN: 0-471-49059-8

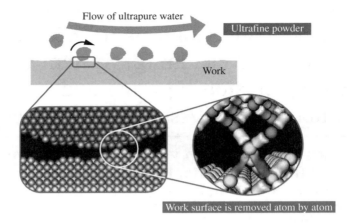

Figure 27.1   Schematic drawing of atom-removal process in EEM.

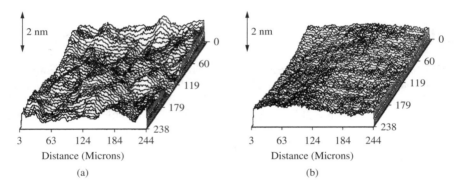

Figure 27.2   WYKO observations. (a) As-received Si wafer surface. (b) EEM surface.

commercial Si wafer surface that has a roughness of more than 3 nm. In contrast, the EEM surface shown in Figure 27.2(b) is extremely smooth with less than 1 nm peak-to-valley roughness. Figure 27.3 shows the results of STM observations on a microscopic scale, Figure 27.3(a) the as-received Si wafer surface and Figure 27.3(b) the EEM surface with quasiatomic flatness.

Secondly, the finished surfaces have no crystallographic damage. Figure 27.4 shows the surface-state densities of differently prepared Si(100) surfaces observed by surface photovoltage spectroscopy. The EEM surface has the smallest surface-state density, which is even slightly lower than that of the surface chemically etched with HF and HNO_3.

The third feature is the removal mechanism in EEM, which is important to understand. The removal rates strongly depend on the types and interactions between powders and workpiece surfaces. Figure 27.5(a) shows the variations of

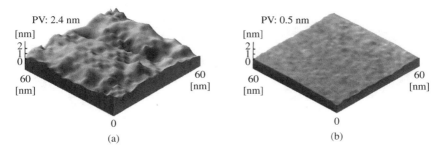

Figure 27.3 STM (scanning tunneling microscopy) observations. (a) As-received Si wafer surface. (b) EEM surface.

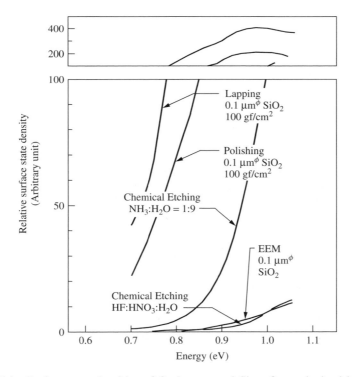

Figure 27.4 Surface state densities of finely prepared Si surfaces obtained by surface photovoltage spectroscopy.

removal rates with zirconia powder and different workpiece materials, whereas Figure 27.5(b) shows the variations of silicon removal rates with different powder materials. Such wide variations do not appear in conventional mechanical polishing. These three features of EEM indicate that the atomic-removal mechanism in EEM is chemical [2, 4–6].

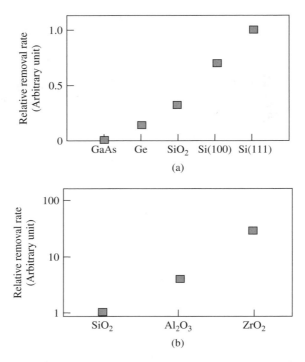

Figure 27.5 Relationship between removal rates and combinations of workpiece and powder materials. (a) Wide variations of removal rate of different workpiece materials with ZrO_2 powder. (b) Wide variations of removal rate of different powder materials with workpiece Si.

27.3   ATOM-REMOVAL MECHANISM

27.3.1   GENERAL VIEW

An atomic-removal mechanism in EEM has been proposed [1–6] as follows. In water, surface atoms both of ultrafine powder and workpiece are believed to be terminated by hydroxide species as shown in Figure 27.6(a). When they approach and contact each other, the binding structure as shown in Figure 27.6(c) might develop through an intermediate hydrogen bonding state (Figure 27.6(b)) followed by dehydration.

The interfacial structure has not been elucidated yet, but it is natural to assume the existence of this structure, because this kind of interfacial structure is the same as in usual metal oxide, e.g., $SiZrO_4$. Subsequently, the water flow drags the powder, and the backside bonds between the workpiece surface atom at the interface and the second-layer atoms are separated. This mechanism could explain the monoatomic removal process.

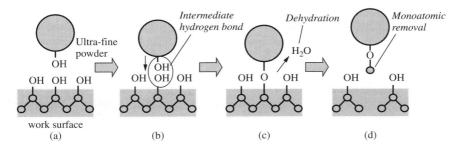

Figure 27.6 Interactions between surfaces of ultrafine powders and workpieces. (a) Surface structures before interaction. (b) Hydrogen bonding between two surfaces. (c) Formed interfacial structure after dehydration. (d) Separation of adsorbed powder with removal of workpiece atom.

27.3.2 PROCESS SIMULATION AND RESULTS

The atomic-removal process indicated in Figure 27.6 is a most fundamental process to create atomically flat surfaces in EEM, for which we carried out first-principle molecular-dynamics simulations. The atomic motion and the dependency of the force on powder material are analysed. The simulation is based on the density-functional theory and the local-density approximation [7, 8]. The norm-conserving pseudo-potential [9] is adopted. The plane-wave expansion method and repeated slab model are applied with the cut-off energy 24 Ry. The sampled k-point is (0.25,0.25,0.25). The Si (001) surface model consists of a 5-layer slab, with saturating the silicon bonds at the bottom of the slab with hydrogen atoms, as shown in Figure 27.7(a). The cluster model of ultrafine

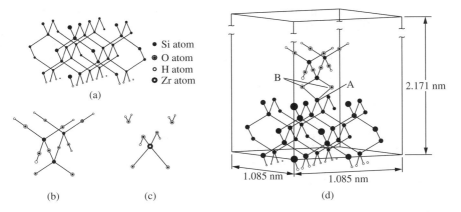

Figure 27.7 Models employed in calculations. (a) Si (001) surface with the bottom surface saturated by hydrogen. (b) Cluster model of SiO_2 [10]. (c) Cluster model of ZrO_2 [11]. (d) Supercell model in the case of SiO_2 cluster employed.

powder is cristobalite structure for SiO_2 [10], and tetragonal for ZrO_2 [11], as shown in Figure 27.7(b) and (c), respectively. The two oxygen atoms at the bottom of each powder clusters are taken to be not terminated by hydrogen atoms to form a reaction site. Powder clusters are adsorbed onto the Si surface at the top layer Si atoms to form covalent bonds to the dangling bonds of sp^3 hybridized orbitals, as shown in Figure 27.7(d).

In the actual EEM process, the powder adsorbed on the Si surface is considered to be removed by the water flow. In order to simulate this process, step-by-step pulling up of the powder cluster is performed with the step length of 0.01 nm. After each pull-up step, the topmost 3 layers consisting of 24 Si atoms on the surface and two O atoms (indicated as B in Figure 27.7(d)) of the cluster are relaxed by the molecular-dynamics method.

Figure 27.8 shows some snapshots in the simulation on each powder material. In this figure, the grey spheres represent atoms, and white clouds the electron distribution (equivalent valence charge-density surface of 0.65 electrons/a.u.^{3}). When powders are pulled up, the Si atom at the interface (indicated as A in Figure 27.7(d)) is lifted by being captured by the powder. The amount of electronic density at the back-bond between A and Si atoms in the second layer and the bond population decrease apparently, as shown in Figure 27.8(c) and (f) and in Table 27.1. Thereby the other surface atoms make no drastic motion. These results show that the back-bonds are being broken. From noninduction of defects onto the surface even employing ideal surface models that are chemically active, we deduce that it is more difficult to induce defects onto real passivated surfaces. Consequently, it is concluded that a monoatomic-removal process while introducing no defects on the machined surface is theoretically feasible.

Next, we calculate the change of total energy and the force arising when pulling up the powder, as shown in Figure 27.9. The force is evaluated by calculating the numerical difference of the energy against pull-up height. The energies in the cases of two powder materials increase as pulling up progresses and the forces also increase until 0.1 nm (SiO_2) or 0.15 nm (ZrO_2), respectively. The maximum value of the force is about 5×10^{-9} N. The lack of smoothness in the force data is considered as due to the insufficiency of the convergence of the structural optimization, but the obtained value is not so far from the exact one.

The force to remove SiO_2 powder is slightly larger compared to the case of ZrO_2 powder, which is consistent with experimental results. Although only a small difference of the force is observed, further investigation such as simulation on the developed model (adding surface terminations or considering water ambient) or on the bonding process between powder and workpiece surface (from Figure 27.6(a) to (c)) are required in order to analyse precisely the machining-rate dependence on material.

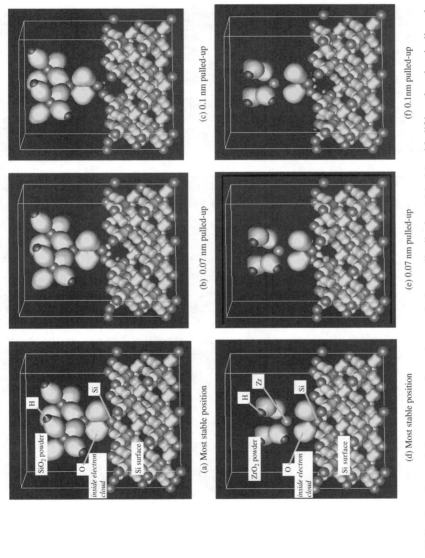

Figure 27.8 Snapshots of the atom configurations and charge distributions (a)–(c) with lifting the chemically adsorbed SiO_2 cluster ($Si_3O_{10}H_6$), and (d)–(f) with lifting the chemically adsorbed ZrO_2 cluster (ZrO_6H_6). Hydrogen atoms terminating the bottom side of the silicon surface are removed for illustration.

Table 27.1 Evaluated bond population near the interface using the calculation method of ref. [6]. Breaking of bond between the Si atom to be removed and the second layer Si atom can be seen

Bond position	SiO$_2$ powder		ZrO$_2$ powder	
	Most stable position	After lifting 0.1 nm	Most stable position	After lifting 0.1 nm
O atom at interface (B) and powder surface atom (Si or Zr)	0.43	0.41	0.38	0.34
O atom at interface (B) and work surface Si atom (A)	0.44	0.46	0.49	0.51
Work surface Si atom (A) and second layer Si atom	0.63	0.39	0.61	0.46
Work surface Si atom (not A) and second layer Si atom	0.67	0.64	0.70	0.69

27.4 NUMERICALLY CONTROLLED EEM SYSTEM

27.4.1 REQUIREMENT OF ULTRACLEAN ENVIRONMENTAL CONTROL

The removal mechanism in EEM is chemical so that the following two points have to be avoided. Firstly, organic contamination on the processed surface during EEM acts as a blocking mask against the chemical reaction between powder surface and processed surface. This affects the microroughness of the finished surface. Secondly, oxygen dissolved in water is very active and destructively oxidizes the processed surface. Especially when the oxygen-containing water interacts with the Si surface at a shear rate higher than 1 m/s, the growth rate becomes higher than 20 nm/h. Such a high oxidation rate of Si surface in water was first discovered in this study [5].

27.4.2 NUMERICALLY CONTROLLED STAGE SYSTEM

The developed EEM system is shown in Figure 27.10. All of the numerically controlled stages are closed in the ultrapure water vessel to avoid introduction of oxygen gas and organic materials such as oil vapour. Two worktables, which are equipped with x, y and Θ stages, are mounted in the system. One worktable is used for flat, cylindrical and toroidal X-ray mirrors, and the other worktable is used for axis-symmetrical optical components. Strokes of x and y stages are 600 mm and 200 mm, respectively. In the guides, bearings and drive-screws, a hydrostatic supporting system utilizing ultrapure water is employed. With these

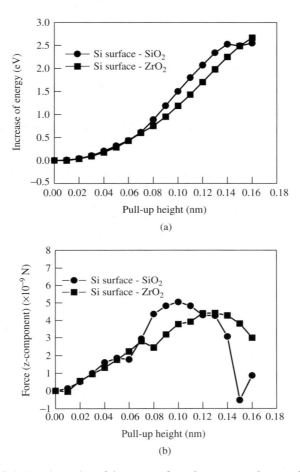

Figure 27.9 Calculated results of increase of total energy and numerically obtained force. (a) Increase of energy versus pulling-up height of particles (b) Force versus pulling-up height of powder particles.

systems, both organic contamination from the lubricant and the metallic dust produced through the mechanical friction are completely avoided.

27.4.3 EEM HEADS

Two types of EEM head are equipped in this system. One is the conventional rotating-sphere type. Another is an ultrahigh pressure jet-nozzle type.

The latter has been developed to realize extremely high shear-flow on the workpiece surface. Figure 27.11 shows the calculated results of the jet flow near the workpiece surface. The jet flow from the nozzle is directly transported near

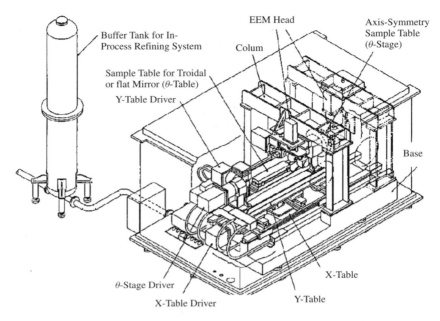

Figure 27.10 Schematic drawing of numerically controlled EEM system developed in this study.

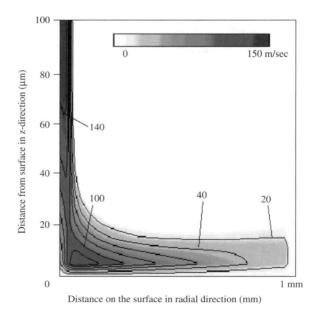

Figure 27.11 Flow-velocity distribution near the processed surface. The initial jet velocity in this calculation is 150 m/s.

to the processed surface and the flow direction is changed to be parallel to the surface by hydrostatic pressure.

The trajectory of ultrafine powder also changes its direction near the processed surface and softly intersects it. The kinetic energy of ultrafine powders at the interaction with the processed surface is calculated to be smaller than several eV, which is small enough to finish the surface without any introduction of crystal defects. The shear rate of the radial-direction flow on the processed surface is more than 10.0 m/s as shown in Figure 27.11. Such a high rate is sufficient for the effective removal of the ultrafine powders adsorbed chemically on the processed surface.

27.4.4   IN-PROCESS REFINING SYSTEM OF THE MIXTURE FLUID

An internal refining system of the mixture fluid is equipped in the EEM system. All mechanical components of the EEM system are completely closed in the water vessel so that the water injected for the hydrostatic supporting system should be eliminated from the vessel. Ultrafine powders, the reactants for the EEM process, should not be wasted so that a solid–liquid separator is provided. The separated liquid, which is pure water containing a little impurities such as metallic ions dissolved from the stainless steel, is refined to ultrapure water, which serves as the fluid for the hydrostatic supporting. The environmental cleanliness for EEM process is maintained by this system.

27.5   NUMERICAL CONTROL SYSTEM

27.5.1   CONCEPTS FOR ULTRAPRECISE FIGURING

EEM is realized within the area in which the shear rate of the water flow on the workpiece surface is higher than 1 m/s. The practical size of the area is about 0.1–5 mm diameter. The constancy of the material removal and the processed surface shape in unit time is sufficiently reproducible so that figuring is performed by scanning the EEM head in the area to be machined with a suitable speed, as shown in Figure 27.12.

27.5.2   SOFTWARE FOR CALCULATING SCANNING SPEED

The software for the calculation of scanning speed based on a RMS minimum error algorithm is developed for this system. The thickness distribution to be removed $M(r_i)$ is calculated from the differences between the premachined surface profile measured by an ultraprecision profilometer and the required surface profile. r_i shows the sampling point on the surface to be processed. The square

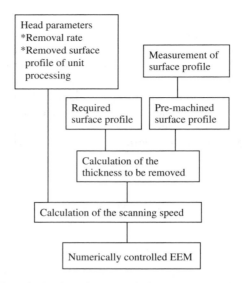

Figure 27.12   A procedure for numerically controlled EEM processing.

sum of the figure errors of the finished surface is expressed by the following equation

$$E = \sum_i^m \{M(r_i) - \sum_j^n N_j(r_i) \cdot t_j\}^2 \qquad (27.1)$$

where $N_j(r_i)$ shows the removed depth at the point r_i by the processing at the point r_j for a unit time and t_j shows the dwell time to be calculated at the point r_j. $t_j (j = 1, 2, 3, \ldots, m)$ are obtained from the following simultaneous equation

$$\frac{\partial E}{\partial t_j} = 0 \qquad (j = 1, 2, 3, \ldots n) \qquad (27.2)$$

The scanning rate of the EEM head is proportional to the inverse of t_j. The RMS error of the processed surface is minimized in this algorithm.

27.5.3   PERFORMANCES OF NUMERICALLY CONTROLLED PROCESSING

Figure 27.13 shows the spherical surface machined on the optical flat quartz glass with a flatness of $\lambda/100$. With a single trial for figuring, the obtained surface profile error is smaller than 0.01 μm, that is 0.7 % of the maximum depth. This

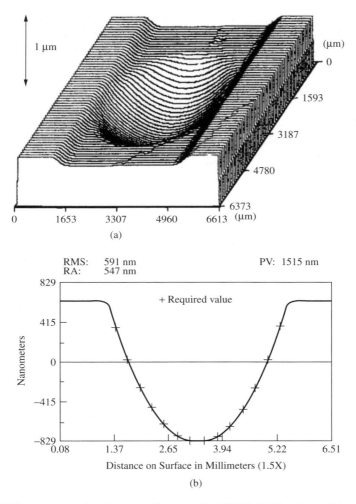

Figure 27.13   An example of numerically controlled EEM. (a) Top view of the machined area. (b) Profile of the centerline of the machined area (+ shows the required depth).

result shows that the high accuracy required for the X-ray mirrors of SR, EUV lithography, etc., can be expected from EEM machining.

27.6   CONCLUSION

A numerically controlled EEM system for figuring and smoothening of aspherical surfaces has been developed. The developed system and software for aspherical figuring already have very high performance so that the figure error of the processed surface is much smaller than 1 % of the maximum removal depth. The obtained surfaces have quasiatomic smoothness and are free from any crystallographic damages.

ACKNOWLEDGEMENT

The authors gratefully appreciate that this work was partially supported by a Grant-in-Aid for COE Research (No. 08CE2004) from the Ministry of Education, Science, Sports and Culture.

REFERENCES

[1]   H. Tuwa and Y. Aketa: Proceedings of International Conference on Production Engineering, Tokyo **2** (1974) 33.
[2]   Y. Mori, K. Yamauchi, and K. Endo: *Precis. Eng.* **9** (1987) 123–128.
[3]   Y. Mori, K. Yamauchi, and K. Endo: *Precis. Eng.* **10** (1988) 24–28.
[4]   Y. Mori, K. Yamauchi, K. Endo, T. Ide, H. Toyota, and K. Nishizawa: *J. Vac. Sci. Technol.* **A 8** (1990) 621–624.
[5]   K. Yamauchi, T. Kataoka, K. Endo, K. Inagaki, K. Sugiyama, S. Makino, and Y. Mori: *J. Japan Soc. Precis. Eng.* **64** (1998) 907–912 (in Japanese).
[6]   K. Yamauchi, K. Hirose, H. Goto, K. Sugiyama, K. Inagaki, K. Yamamura, Y. Sano, and Y. Mori: *Comput. Mater. Sci.* **14** (1999) 232–235.
[7]   W. Kohn, and L. J. Sham: *Phys. Rev.* **140** (1965) 1133.
[8]   J. P. Perdew, and A. Zunger: *Phys. Rev.* **B23** (1981) 5048.
[9]   N. Troullier, and L. Martins: *Phys. Rev.* **B43** (1991) 1993.
[10] M. Kawai, M. Tukada, and K. Tamaru: *Surf. Sci.* **111** (1981) L716.
[11] F. Babou, B. Bigot, and P. Sautet, *J. Phys. Chem.* **97** (1993) 11501.

Part 6

Epitaxy and Layer Deposition

28 Control of Epitaxial Growth Modes for High-Performance Devices

HANS J. SCHEEL
SCHEEL CONSULTING, Sonnenhof 13, CH-8808 Pfaeffikon, Switzerland

ABSTRACT

The epitaxial fabrication method, the interfacial energy between substrate and film, and the major growth parameters – thermodynamic driving force, misfit between substrate and layer, substrate misorientation, and the growth temperature – determine the growth mode and thus structural perfection, purity, homogeneity, surface flatness and interface abruptness of the layers. In addition to the classical three growth modes of Bauer (1958) we define four distinct growth modes and describe their occurrence. By systematic analysis one can derive the single optimum growth technology for a specific device structure, thereby taking into account layer and surface perfection, but also device performance, economic and ecological factors. For compounds with limited thermodynamic stability like compound semiconductors and high-temperature superconductors, the optimum layer perfection can in many cases be achieved by liquid phase epitaxy (LPE).

28.1 INTRODUCTION

Performance and lifetime of microelectronic, photonic and magnetic devices are limited by the purity, structural perfection and homogeneity of the epitaxial layers and by the flatness and abruptness of the layer surfaces and interfaces. For example, the detrimental effects of dislocations for transistors (variations of threshold voltages across the wafer) has been proven by Miyazawa *et al.* 1986, and the dependence of the efficiency of light-emitting diodes on dislocation densities has been reviewed by Lester *et al.* 1995. The control of the growth mode and thus the epitaxial fabrication method and the growth parameters determine the layer and surface perfection. In addition to the three well-known epitaxial growth modes (Volmer–Weber, Stranski–Krastanov, Frank–Van der Merwe) there are four distinct growth modes (step flow mode, columnar growth, step bunching, screw-island growth) which will be described. The occurrence of the growth modes depends on numerous parameters of which the most important are

Crystal Growth Technology, Edited by H. J. Scheel and T. Fukuda
© 2003 John Wiley & Sons, Ltd.   ISBN: 0-471-49059-8

the thermodynamic driving force and the misfit between substrate and layer. In silicon epitaxy, chemical vapour deposition is well established.

Optoelectronic devices based on compound semiconductors (GaAs, GaInAsP, GaN, CdTe) were first developed and fabricated by liquid phase epitaxy (LPE) which successively has partially been replaced by metal-organic chemical vapour deposition (MOCVD). Now still 60 % of light-emitting diodes LEDs and III–V solar cells are produced by LPE, whereas laser diodes and electronic III–V devices are predominantly fabricated on layers grown by MOCVD or by molecular beam epitaxy MBE. Also in the development of 'novel' materials for devices like SiC and GaN the epitaxial processes from the vapour are dominating so far. In contrast some of the highest-performance optoelectronic devices (red LEDs) and photovoltaic solar cells are produced by LPE. This discrepancy can be understood considering the complexity of the LPE process requiring a subtle control and optimisation of many growth parameters by specially educated and experienced engineers, whereas MOCVD equipment (together with some standard process) is readily available so that the development of new device structures is facilitated. Furthermore, LPE has several limitations, so that it cannot be applied in cases of immiscibility, of special dopants and dopant profiles, and in cases of narrow confinement (e.g. submicrometer structures). In the case of complex oxide compounds (magneto-optic compounds and high-temperature superconductors) the best layer performance and stoichiometry control can be achieved by liquid phase epitaxy, which is also the dominating fabrication technology.

In the development of the *single optimised epitaxial growth technology* for a specific device structure with given performance, economic and ecological factors also have to be taken into account. This approach is different from the general propagation of specific best epitaxy methods (by leading scientists and equipment companies) which are listed under 'General References'. Education in epitaxy technology is needed in order to recognise the optimum technology and to minimise the use of unsuitable methods with the example of high-temperature superconductors to be discussed.

28.2 SEVEN EPITAXIAL GROWTH MODES AND THE ROLE OF GROWTH PARAMETERS

In heteroepitaxy the mode of nucleation and initial growth is strongly dependent on the bonding between the substrate and deposited film as was established by Bauer 1958. Hence the surface and interfacial free energies can be used to discuss the initial phase of film formation.

If the deposit is in the form of a droplet as shown in Figure 28.1, then the equilibrium condition is defined by the Young equation

$$\gamma_{SV} = \gamma_{FS} + \gamma_{FV} \cos \theta \qquad (28.1)$$

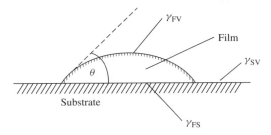

Figure 28.1 Surface and interface energies and contact angle Θ of a substrate with deposited film for the case of significant wetting ($0 < \Theta < 90°$).

where γ_{SV}, γ_{FS} and γ_{FV} are the free interfacial energies of substrate–vapour, film–substrate and film–vapour, respectively, with θ the contact or wetting angle. Significant wetting is defined by a small contact angle: at $\theta \to 0$ with $\gamma_{SV} = \gamma_{FS} + \gamma_{FV}$, we expect layer-by-layer growth as described by Frank and Van der Merwe 1949 (abbreviation F–VM). For the case of medium or little wetting, corresponding to a large contact angle and the relation $\gamma_{FS} + \gamma_{FV} > \gamma_{SV}$, the deposit forms discrete nuclei that successively grow three-dimensionally and coalesce to a compact continuous film: this is the Volmer–Weber (V–W) growth mechanism. An intermediate case, when the substrate–film interactions are stronger than the binding within the film with $\gamma_{FS} + \gamma_{FV} < \gamma_{SV}$, then first a continuous film of one or two monolayers is deposited onto which in the second phase discrete islands are formed that eventually coalesce. This epitaxial mechanism was named the Stranski–Krastanov mode (S–K, 1938). These three 'classical' growth modes defined in terms of the interface energies by Bauer 1958 and systematically described by Le Lay and Kern (1975), are shown in the upper part of Figure 28.2.

An alternative description is based on the Dupré equation (1969) for the specific interfacial free energy

$$\sigma^* = \sigma_f + \sigma_s - \beta \qquad (28.2)$$

with σ_f and σ_s the specific surface free energies of film and substrate, respectively, and β the specific adhesion free energy between substrate and film. The F–VM mode is expected for $\beta > 2\sigma_f$, and the V–W growth mode for $\beta < 2\sigma_f$.

These thermodynamic approaches can qualitatively describe certain epitaxial phenomena. However, different growth modes have been described for the same substrate–film system, thus indicating that growth methods and growth parameters influence the growth modes. Furthermore, the vast epitaxial growth experience of the past 40 years has given clear evidence of further growth features that may be described by four distinct and different epitaxial growth modes, which are shown in the lower part of Figure 28.2. These aspects are important for the structural perfection of the layers and thus for the device performance and will be described in the following.

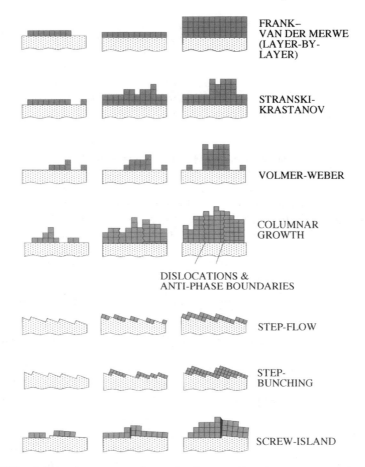

FRANK–
VAN DER MERWE
(LAYER-BY-
LAYER)

STRANSKI-
KRASTANOV

VOLMER-WEBER

COLUMNAR
GROWTH

DISLOCATIONS &
ANTI-PHASE BOUNDARIES

STEP-FLOW

STEP-
BUNCHING

SCREW-ISLAND

Figure 28.2 Schematic cross sections of substrate–film, in three successive growth stages, of the seven growth modes. The upper three classical growth modes had been defined by Bauer 1958, the lower four growth modes are defined in this work.

The discussion of both the three classical growth modes and of the newly defined four growth modes will be based on the most important growth parameters: thermodynamic driving force (supersaturation), crystallographic misfit between substrate and film, and the misorientation of the substrate (the angular deviation of the substrate surface from the ideal crystallographic plane). Table 28.1 lists these major growth parameters, along with other parameters the role of which needs further investigations and are not discussed here.

The definitions of the supersaturation include the equilibrium concentration of the growth species. This concentration is very different in epitaxy from the vapour phase and from the liquid phase. In the vapour phase the density of GaAs, for example, is reduced from 5.3 g/cm^3 (solid) to 0.0065 g/cm^3 (vapour),

Table 28.1 Major growth parameters of epitaxial film deposition by vapour phase epitaxy VPE and by liquid phase epitaxy LPE

Surface energy of substrate	γ_{SV}
Surface energy of deposited film	γ_{FV}
Interface energy substrate – film	γ_{SF}
Supersaturation, relative*	$\sigma_{LPE} = (n - n_e)/n_e$; $\sigma_{VPE} = (p - p_e)/p_e$
Supersaturation ratio*	$\alpha_{LPE} = n/n_e$; $\alpha_{VPE} = p/p_e$
Misfit between substrate and film*	$f = (a_S - a_F)/a_F$
Misorientation angle (and direction) of substrate	
Surface diffusion coefficient*	–
Growth temperature*	–
Growth species in VPE, solvents in LPE with characteristic reaction kinetics, thermodynamics, complex formation and desolvation effects, diffusion coefficients, etc.	
Stoichiometry/oxidation stage of deposited compounds	
Condensing impurities (surfactants); surface liquid or surface melting due to impurities (VLS mechanism) or due to partial decomposition	

*The relative supersaturation, the equilibrium values of concentration n_e and pressure p_e, the supersaturation ratio, the misfit, and transport phenomena like surface diffusion and bulk diffusion are determined by the growth temperature.

n and p are the actual values of concentration and pressure, and a_s and a_f are the lattice constants of substrate and film, respectively.
From this table only the effects of the most important parameters supersaturation, misfit, and misorientation on the epitaxial growth modes will be discussed.

thus by nearly three orders of magnitude. This corresponds at a typical working pressure of VPE of 10^{-3} to 10^{-4} Torr (10^{-6} to 10^{-7} bar) to a further reduction so that the concentration in VPE is 10^{-9} or less. In metal-organic chemical vapour deposition MOCVD with a working pressure of 10 Torr and a 1:20 dilution by carrier gas, the concentration of growth species is less than 10^{-6}. In comparison the concentration in liquid phase epitaxy is much higher, typically around 10^{-1}. In growth from the vapour phase generally the material flux (in number of species per cm^2 and s) is the rate-determining factor for growth.

In LPE it is the mass transport through the diffusion boundary layer δ (Nernst 1904) which at a given supersaturation is limiting the growth rate v according to

$$v = \frac{D(n - n_e)}{\delta\rho} \tag{28.3}$$

where D is the diffusion coefficient and ρ the density of the crystal. The growth rates in LPE can be enhanced by reducing the diffusion-boundary-layer thickness (by a high solution flow along the growth surface) so that surface kinetics may become the rate-limiting factor.

However, with increasing supersaturation and growth rate we observe successively step bunching, wavy macrosteps, formation of inclusions, edge nucleation and surface dendrites, hopper growth and bulk dendrites, in the transition from stable growth to growth instability, see Elwell and Scheel 1975. There is a maximum stable growth rate v_{max} that is the fastest growth rate without such instability phenomena and without formation of inclusions. This growth rate was derived from an empirical boundary-layer concept of Carlson 1958 by Scheel and Elwell 1972, see also Elwell and Scheel 1975, Chapter 6:

$$v_{max} = \left\{ \frac{0.214 D u \sigma^2 n_e^2}{Sc^{1/3} \rho_c^2 L} \right\}^{1/2} \tag{28.4}$$

with u the solution flow rate, Sc the Schmidt number $\eta/\rho_L D$ with η the dynamic viscosity and ρ_L the density of the liquid, and L the substrate diameter. By observing this concept, extremely flat surfaces with interstep distances of 6 to 14 μm can be achieved by LPE, in contrast to a widespread prejudice (Scheel 1980, Scheel et al. 1982, Chernov and Scheel 1995 for GaAs; Klemenz and Scheel 1993, Scheel et al. 1994 for high-T_c superconductors).

Whereas the supersaturation in growth from solutions is readily defined and applicable, the ratio of vapour pressures in vapour phase epitaxy leads to unrealistic supersaturation values. Stringfellow (1991) has proposed to compare the thermodynamic driving forces of epitaxial processes by means of the Gibbs free-energy differences between the reactants before growth and the crystalline product after growth. For GaAs and growth temperature 1000 K this comparison of epitaxial driving forces is shown in Figure 28.3.

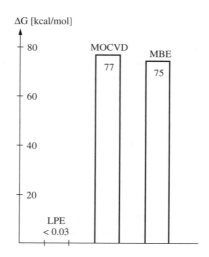

Figure 28.3 The estimated thermodynamic driving forces for LPE ($\Delta T < 6$ K), MOCVD (TMGa + arsine) and MBE (Ga + As$_4$) of GaAs at 1000 K, after Stringfellow 1991.

This figure shows that the supersaturations in vapour phase epitaxy, with examples of metal-organic chemical vapour deposition MOCVD and molecular beam epitaxy MBE, are orders of magnitude greater than in liquid phase epitaxy. In LPE the thermodynamic driving force can be very precisely adjusted by the undercooling – by precise temperature control – between negative values for etching, zero at equilibrium, and positive values for growth, as was shown by Scheel 1980. The free-energy differences demonstrated in Figure 28.3 give only approximate values of supersaturation. The effective supersaturation during the growth process can be derived from the morphology of as-grown surfaces (Scheel 1994, 1997). The distances y_0 between steps are related to the size of the two-dimensional nucleus r_s^* according to the Burton–Cabrera–Frank BCF theory 1951 in the improved treatment of Cabrera and Levine 1956 by

$$y_0 = 19r_s^* = \frac{19\gamma_m V_m}{a^2 RT\sigma} \tag{28.5}$$

with γ_m the energy per growth unit, V_m the molar volume, and a the size of the growth unit.

With the typical interstep distances of 50 nm in MBE and in MOCVD of GaAs, and with 6 μm in LPE of GaAs (Scheel 1980) the supersaturation ratio of vapour growth and LPE is 120, whereas the ratio derived from the free-energy differences is about 2500, obviously too high a value. For vapour growth an attempt should be made to establish the supersaturation at the step edge (Nishinaga 1996).

In heteroepitaxy the lattice mismatch between substrate and film, the so-called misfit as defined in Table 28.1, has a significant effect on nucleation and growth modes. Grabow and Gilmer (1988) have shown by atomistic simulations using the Lennard–Jones potential that the pure layer-by-layer F–VM growth mode requires quasi-zero misfit at growth temperature as demonstrated in Figure 28.4.

Misfit normally induces the Volmer–Weber mode except for large interface energies between substrate and deposited film, which will cause the Stranski–Krastanov mode. If structurally perfect layers or quasi-atomically flat surfaces are required, either homoepitaxy or substrates with zero misfit at growth temperature have to be applied. It is experimentally found that the supersaturations in epitaxy from the vapour phase are so high that epitaxial deposition can be achieved even at very high misfit. In LPE, however, the necessary supersaturation for epitaxial growth increases with misfit. This higher supersaturation then leads to step bunching, growth instability or even 2D-nucleation.

The misorientation of the substrate is providing misorientation steps depending on the angle and the direction of misorientation. For a given supersaturation, even for the large supersaturation in epitaxy from the vapour phase, the density of the misorientation steps can be made so high and the interstep distance so small, that 2D-nucleation and the V–W and S–K modes can be suppressed. The layers grow then in the step-flow mode and have a relatively high structural perfection because defects due to coalescence are prevented. Surfaces can, on average, be

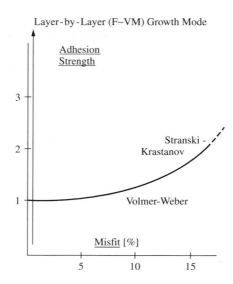

Figure 28.4 The effect of misfit on occurrence of the three classical growth modes, after Grabow and Gilmer 1988.

quite flat but of course show the high step density. In LPE this high step density will lead to step bunching and eventually growth instability if the supersaturation is not kept sufficiently small and precisely controlled. On the other hand, a low supersaturation in LPE will lead to the transition of the misoriented surface to faceting and the Frank–Van der Merwe growth mode, as was shown by Scheel 1980 and by Chernov and Scheel 1995.

In the following the seven epitaxial growth modes will be briefly described. Frank–Van der Merwe F–VM: Layer-by-layer growth *over macroscopic distances,* whereby a new layer is nucleated after completion of the layer below (for not too large substrate diameters and/or sufficient substrate rotation). In the case of perfect substrates without screw dislocations the supersaturation would increase until surface nucleation occurs, then the supersaturation would drop during the growth of the monolayer until after layer completion it increases again. Normally, however, there are continuous step sources like screw dislocations or other defects, so that the F–VM growth mode works continuously and is spreading growth steps (in ideal cases monosteps) over macroscopic distances. Interstep distances normally exceed 1 μm and can be up to 100 μm. Screw dislocations may cause the spiral-growth mechanism, described by the BCF theory, which may lead to growth hillocks with very small slopes, depending on supersaturation with angles of seconds or minutes of arc. At the boundaries of the growth hillocks the steps are annihilated by steps arriving from neighbour hillocks. The F–VM mode is best proven by optical microscopy for the interstep distances and by AFM or STM for step-height measurements, see the examples of

GaAs (Scheel 1980, Scheel *et al.* 1982) and the high-temperature superconductor $YBa_2Cu_3O_{7-x}$ (Klemenz and Scheel 1993, Scheel *et al.* 1994).

Screw dislocations, other defects, and the hillock boundaries may cause local stress fields and variations of the incorporation rates of impurities and dopants, or the local stress fields may getter or reject impurities during annealing processes in device fabrication. Thus one would expect a few local variations of carrier lifetimes within large quasi-defect-free areas of epitaxial films grown by LPE of compound semiconductors, or grown by VPE of germanium (Tikhonova 1975) or silicon at high temperatures when interstep distances above $1\,\mu$m can be observed, see Figure 28.5.

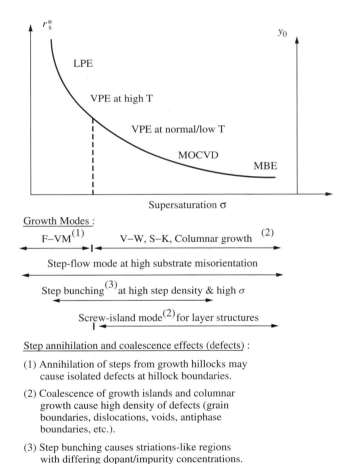

Figure 28.5   The supersaturation effects on the critical size of surface nuclei r_s^*, on the interstep distance y_0, and on the growth modes for the major epitaxial growth methods. Also, the defects caused by the different growth regimes are listed (Scheel 1997).

In MOCVD the physisorbed precursor state of an adatom, due to surface reconstruction, may lead to high surface diffusivities and to F–VM growth (Stringfellow 1994). The optimum layer homogeneity can be achieved by a one-directional movement of steps in the F–VM mode initiated by a precisely controlled small substrate misorientation or by defects at one edge of the substrate.

The Volmer–Weber V–W growth mode (1926) consists in the first phase of the formation of a large number of surface nuclei, typically 10^6 to $10^{11}\,\text{cm}^{-2}$, and their spreading: theoreticians like to call this the birth-and-spread mode, see Grabow and Gilmer 1988. These locally relatively rough surfaces, with three height levels within 100 nm, have been proven by scanning-tunnelling microscopy (Pashley 1990) and by high-resolution transmission electron microscopy using 'chemical lattice imaging' in combination with digital pattern recognition (Ourmazd 1989). In the second phase three-dimensional islands are formed followed by coalescence to a compact layer. As the growth islands may have slightly different orientations, numerous defects may be formed during coalescence as schematically shown in Figure 28.6.

The angular deviation of the growth islands from the substrate orientation is dependent on the misfit and may reach $2°$. The annihilation of steps coming from different directions will then cause defects along the annihilation lines and especially at the triple points where steps from three growth islands meet, as is shown in Figure 28.6. The number of 'triple-point defects', often in the form of screw dislocations, is the same as the number of surface nuclei and growth islands, for example of GaN layers and of spirals in the spiral-island growth mode discussed below.

Continued growth of the layer, after the initial V–W growth, occurs by columnar growth, unless there is a healing procedure involving surface diffusion, for instance by growth interruption or by an annealing phase in MBE.

This V–W mode is quite common in VPE, MOCVD, MBE due to the inherent high thermodynamic driving force, even in homoepitaxy. Initial V–W growth followed by columnar growth is the common feature in epitaxy of GaN, diamond, and high-T_c superconductors where, due to the thermodynamic stability limits of the compounds the high growth and annealing temperatures, sufficient for surface diffusion and healing, cannot be applied. As an example we may cite the Akasaki–Amano group with columnar growth of GaN on AlN buffer layers (Hiramatsu et al. 1991).

In LPE the V–W mode is normally not observed and would be expected only in cases of very high supersaturation when this is required for substrates with large misfit or with low interfacial energy with the film to be deposited.

The Stranski–Krastanov S–K mode can be regarded as intermediate between the F–VM and V–W modes: Due to large substrate–film interface energies first one or two compact monolayers are formed onto which by surface nucleation 10^6 to 10^{11} nuclei and growth islands per cm^2 are formed in analogy to the Volmer–Weber growth mode. This phenomenon can be explained by a

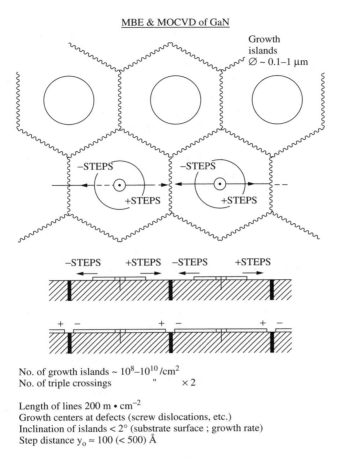

MBE & MOCVD of GaN

No. of growth islands ~ 10^8–10^{10} /cm^2
No. of triple crossings   "   × 2

Length of lines 200 m • cm^{-2}
Growth centers at defects (screw dislocations, etc.)
Inclination of islands < 2° (substrate surface ; growth rate)
Step distance $y_0 \approx 100$ (< 500) Å

Figure 28.6 Schematic top view (upper picture) and cross section of growth islands in the V–M, S–K and screw-island growth modes, along with the directions of steps. Defects may occur along the lines of step annihilation and especially at the triple points where three steps meet. The example is shown for GaN with 10^8 to 10^{10} growth islands per cm^2, after Scheel 1998.

significant misfit as was discussed above in connection with Figure 28.4. The S–K mode with the initial compact layer formation followed by the island nucleation stage has been demonstrated by Sasaki and coworkers (Nabetani *et al.* 1993) and by Nishinaga *et al.* 1994 who used transmission electron microscopy, RHEED, HRSEM and AFM in order to demonstrate the appearance of nuclei only after deposition of one to two monolayers of InAs onto GaAs by MBE at around 480 °C.

The step-flow mode is clearly distinct from layer-by-layer growth in the F–VM mode. Unidirectional step flow is induced by substrate misorientation, which is frequently applied to prevent island formation, their coalescence, and the following columnar growth in epitaxy from the vapour phase as treated above. In this way the defects caused by coalescence can be minimised, and thus the performances of optoelectronic devices fabricated on VPE-, MOCVD- and MBE-grown layers could be significantly increased. The fine-stepped surface is on average relatively flat, but behaves quite differently with respect to surface reactivity and surface states compared to the quasi-atomically flat equilibrium surfaces grown by the F–VM mode, by LPE. In LPE the supersaturation should be limited when applying misoriented substrates in order to prevent step bunching and surface ripples as first stages towards growth instability. Localised step flow and small interstep distances of less than 100 nm are frequently observed on films growing by the V–W or S–K mode from the vapour phase. This step flow is not unidirectional and spreads from screw dislocations or other defects on the top of growth islands.

Step bunching is observed when a high density of steps moves with large step velocities over the growth surface. By fluctuations, higher steps catch up with lower steps and then move together as double, triple steps, etc., or in general as macrosteps that can exceed thicknesses of 1000 monosteps. This is a traffic-flow problem that was theoretically analysed by Lighthill and Whitham 1955. Cabrera and Vermilyea and independently Frank in 1958 applied this theory to macrostep formation in their kinematic wave theory. The terrace-macrostep (or thread-riser) morphology causes different incorporation rates of impurities and dopants due to locally varying growth rates: the resulting striations have been demonstrated by photo- or cathodoluminescence and by etching by Kajimura et al. 1977, Raul and Roccasecca 1973, Nishizawa et al. 1986 and by Nishinaga et al. 1989. It was shown experimentally by Scheel 1980 and theoretically by Chernov and Scheel 1995 how at low supersaturation the step bunching can be suppressed by the transition to faceting, to a growth surface with mono- or double steps depending on the Burgers vector of the step-generating defects. These small steps with typical interstep distances of 1 to $20\,\mu m$ propagate over macroscopic dimensions (F–VM mode) and lead to very homogeneous layers and quasi-atomically flat surfaces. This surface flatness was proven by differential interference contrast microscopy after Nomarski (Scheel 1980) and in the first investigation of an epitaxial surface by scanning tunnelling microscopy STM by Scheel et al. 1982.

Bauser (1994) and coworkers have published beautiful photographs of surface growth features, unfortunately without details of the important growth conditions, so that the interpretation is difficult.

The seventh growth mode, the spiral-island mode or screw-island mode, was discovered during STM surface investigations of epitaxial layers of the high-temperature superconductor $YBa_2Cu_3O_{7-x}$ (YBCO) first by Hawley et al. 1991 and then by Gerber et al. 1991. A high density of screw islands was discovered on YBCO layers grown by sputter deposition and by MOCVD. The origin of these screw islands (occasionally also called Tokyo towers or towers of Hanoi) was suggested by Scheel 1994: Coalescence of the large number of initial growth islands (10^8 to $10^{10}\,\mathrm{cm}^{-2}$) may lead to screw dislocations due to the layer structure of YBCO. Similar formation of screw dislocations had been described for organic (Kozlovskii 1958) and inorganic layer structures (Baronnet 1973). Further evidence is based on the same orders of magnitude of first-nucleated islands and screw islands, and on the fact that the number of screw islands is reduced by misorientation steps (Scheel 1994). For semiconductors with a layer structure one would also expect this screw-island growth mode at not too high growth temperatures. Continuous growth by the screw-island mode leads, in analogy to the V–W and S–K modes, to coalescence and to columnar growth whereby the density of screws may be reduced during further growth at low supersaturation and at high growth temperatures.

So far the surface morphologies characteristic for growth modes and growth methods have been described by step heights and interstep distances neglecting the statistical distribution. For the overall surface flatness the relative roughness parameter R is defined (Scheel 1987) which takes into account all surface areas parallel to the surface and all surface areas vertical to the surface:

$$R = \frac{\Sigma(\text{Horizontal areas}) - \Sigma(\text{Vertical areas})}{\Sigma(\text{Horizontal areas})} < 1 \qquad (28.6)$$

Surface steps, respectively, the sum of vertical areas provide the active sites in catalytic reactions, in corrosion, and in crystal growth. In the case of the high-T_c superconductor $YBa_2Cu_3O_{7-x}$ with $\{001\}$ equilibrium faces the sum of vertical areas and thus the R value determines the rates of oxidation and of oxygen losses. Other definitions of roughness like rms, the maximum peak-to-valley value, the average peak height value etc. and also the energetic roughness definition of Burton et al. (1951) are either not practical or not meaningful in the applications given above and in a comparison of the epitaxial growth modes. Even the R value has to be complemented by other data like interstep distances in order to achieve a sufficient description of the surface activity as can be seen from Figure 28.7.

The ten examples of cross sections of surfaces show the different surface morphologies with different R values, and the number of nines (N) gives an indication of the surface flatness. The ideal surface would of course have $R = 1$.

28.3 CONTROL OF GROWTH MODES

The appearance of the epitaxial growth modes is closely connected with the growth method and its inherent range of supersaturation as this is schematically shown in Figure 28.5. Here the sizes of the critical nuclei r_s^* and the interstep distances y_0 are plotted as functions of supersaturation for the epitaxial growth methods. Not shown in Figure 28.5 is the misfit between substrate and film, which would shift the ranges of the V–W, S–K and the screw-island growth modes to the left, as was discussed above. Also not shown in Figure 28.5 is the growth

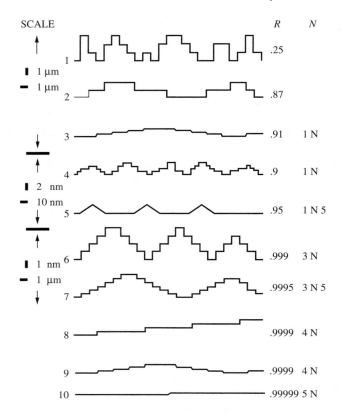

Figure 28.7 Schematic presentation of surface structures and their relative roughness factors and the number of nines (*N*) indicating surface quality. Note the different scales for surfaces 1 to 3, 4 and 5, and 6 to 10.

temperature: high growth temperatures shift the curve up and the ranges of the vapour-phase-epitaxy methods to the left, due to enhanced surface diffusivities and due to lower supersaturation in epitaxy from the vapour phase. From this general concept, a specific and realistic diagram could be designed for epitaxial deposition of specific elements and compounds as a base for the interpretation of experimental results and for process optimisation. Such a diagram would also be helpful to realise the effects of the other growth parameters of Table 28.1, which are not treated here.

For compounds of limited thermodynamic stability like GaAs, GaN, SiC, diamond, high-T_c superconductors, the growth modes are predetermined by choice of epitaxy from the vapour phase (MOCVD, MBE, etc.) or from the liquid phase (LPE) through the supersaturation inherent to the growth method. Since the majority of structural growth defects are caused by coalescence of growth islands, we have to avoid the V–W, S–K and the screw-island growth modes when there is interest in high structural perfection and excellent surface and interface flatness of the epitaxial layers. In epitaxy from the vapour phase this is only possible through the step-flow mode by using substrates of a misorientation precisely adjusted to the specific supersaturation and growth temperature.

An alternative approach to increase the structural perfection of VPE-grown layers is based on epitaxial lateral overgrowth ELO (Nishinaga and Uen 1994) and on pendeo-epitaxy (Zheleva 1998). Lateral overgrowth is initiated from columns grown through channels (ELO) or from columns obtained by etching (pendeo-epitaxy). The propagation of line defects or planar defects to the successively grown upper layer is hindered so that top layers of improved structural perfection can be achieved. This approach is of special interest for GaN, although such processing steps normally are not required for LPE due to the inherent high structural perfection of LPE-grown layers.

By far the best structural perfection and quasi-atomically flat surfaces can be achieved by the Frank–Van der Merwe growth mode, which requires substrates of very low misfit and a very low supersaturation, see Figure 28.8.

Thus, the F–VM mode can only be achieved by a near-equilibrium process like LPE. However, the complexity of LPE using solvents in a multiparameter process is great so that proper process control and optimisation is rarely achieved. Often LPE-grown surfaces show macrosteps, surface ripples and meniscus lines: detrimental defects that can be prevented by careful selection and surface preparation of the substrate, by the purity of the process atmosphere, and by the precise adjustment of the growth conditions (Scheel 1980, Chernov and Scheel 1995). Trained crystal growth and epitaxy engineers with special experience would be needed to chose the optimum and economic epitaxial growth process for the fabrication of high-quality layers and surfaces. In the field of compound semiconductors, the highest-brilliance red light-emitting diodes LEDs and clear green LEDs have been achieved by LPE of simple layer structures (Nishizawa and Suto 1994) whereas complex quantum-well structures are required for brilliant MOCVD-grown LEDs. Figure 28.8 schematically shows the summary of the role of the parameters supersaturation and misfit, the ranges of the epitaxial growth modes, and the relatively small parameter range in the corner, where the Frank–Van der Merwe growth mode allows layers of highest structural perfection with quasi-atomically flat surfaces to be obtained.

The majority of LPE-grown layers and multilayers of compound semiconductors for light-emitting diodes and for solar cells are fabricated by slider systems, even for mass production (Nishizawa et al. 1986, Mauk et al. 2000) except for S. Kamath, who developed a vertical dipping technology for photovoltaic solar cells for space applications. A large variety of slider systems were developed in the 1970s, horizontal sliding, rotary sliding, etc. However, all sliding systems have severe disadvantages that are schematically shown in Figure 28.9. The sliding action itself leads to abrasion of graphite particles, and to scratching of the raised edges of the substrates formed during growth. Furthermore, the gap between the substrate and the slider with the solutions is very critical: if the

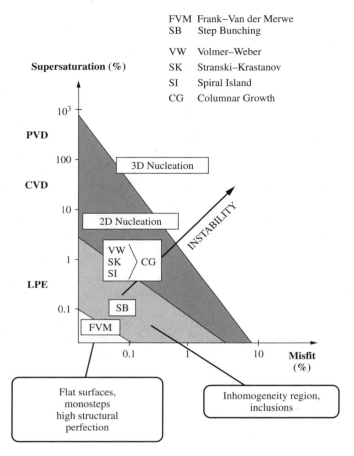

Figure 28.8 The effects of both supersaturation (and the growth method) and misfit on the nucleation and growth regimes. Only at very low supersaturation in LPE using low-misfit substrates can really flat surfaces be expected.

gap is too wide then the trapped solution fraction will mix with the following solution, if too narrow then the grown layers will be scratched, thus leading to decreased yield of devices.

Anyhow, the total thickness of grown layers is geometrically limited, so that for efficient mass production by LPE a slider-free technology should be applied. This was developed by Scheel 1977 with a double-screw graphite device, where different solutions wash over the substrates, without getting mixed, and deposit layers upon cooling. Superlattice structures with 100 *p-n*-GaAs layers could be prepared, and by transition to faceting atomically flat surfaces could be achieved (Scheel 1980, Scheel *et al.* 1982) in a laboratory-scale process. For mass production by LPE another topology is required, which for example will

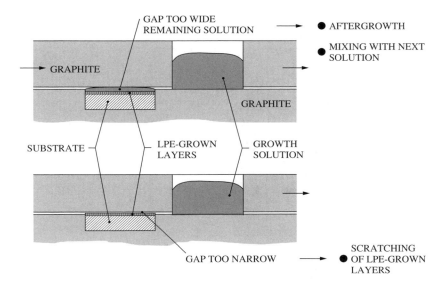

Figure 28.9   The major problems with slider systems in LPE, after Scheel 1987.

allow the production of $0.26 \, \text{km}^2$ of 11-layer structures per month in a quasi-continuous process, which is relatively ecologic, energy-efficient and economic (Scheel, unpublished). Contamination of the free surface of the Ga solution can lead to wetting problems, which, however, may be overcome by low oxygen partial pressure of the process atmosphere at high growth temperature. Such low $P(O_2)$ also has the advantages of a low carrier concentration ($3.7 \times 10^{12} \, \text{cm}^{-3}$), high mobility of GaAs ($244\,000 \, \text{cm}^2/\text{V/s}$ at $77 \, \text{K}$) and an increased lifetime of semiconductor lasers (Kan *et al.* 1975).

In the case of high-temperature superconductors HTSC with critical temperatures above the boiling point of liquid nitrogen at $77 \, \text{K}$ (which were discovered by Wu *et al.* 1987) initially more than a thousand groups attempted to achieve homogeneous high-quality layers by epitaxy from the vapour phase. However, for a reliable planar HTSC tunnel-device technology, quasi-atomically flat surfaces are required due to the short coherence lengths of the HTSC compounds (Braginski 1991). As was discussed above, such flat surfaces of thermodynamically not very stable compounds can only be achieved by the near-equilibrium technique LPE, which at optimised conditions yielded facets with interstep distances of more than $10 \, \mu\text{m}$ as was shown by Klemenz and Scheel 1993, Scheel *et al.* 1994, Scheel 1998. Figure 28.10 shows the application ranges and the lateral and vertical resolutions of microscopes (after Komatsu and Miyashita 1974) along with the step heights and interstep distances of VPE- and LPE-grown HTSC surfaces and also the required value range for HTSC tunnel-device applications.

Somewhat similar is the situation with GaN and its solid solutions the layers of which have been fabricated so far by MOCVD and MBE and show

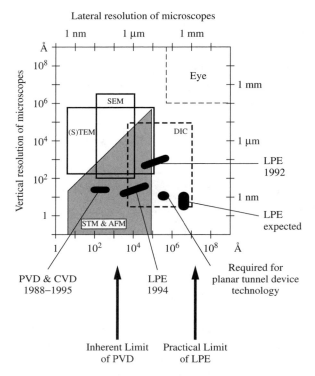

Figure 28.10 Surfaces (step heights and interstep distances) of HTSC layers(YBCO, NdBCO) grown by vapour phase epitaxy (PVD, CVD) and by LPE, introduced into a diagram of lateral and vertical detection limits of microscopes (after Komatsu and Miyashita 1993). The vertical heavy arrows indicate the inherent and practical limits of physical vapour deposition PVD and of liquid phase epitaxy, respectively, after Scheel 1998.

SEM: scanning electron microscope; (S)TEM: scanning transmission electron microscope, TBI: two-beam interferometer; MBI: multiple-beam interferometer; REM: reflection electron microscope; DIM: differential interference contrast (Nomarski) microscope; PCM: phase contrast microscope; STM: scanning tunneling microscope; AFM: atomic force microscope.

dislocation densities of 10^9 to 10^{11} cm^{-2}, even with buffer layers. Eventually also for GaN (and SiC) LPE will become the economic approach to reduce dislocations (and other defects like pipes) by orders of magnitude. Initial LPE of GaN has been reported by Klemenz and Scheel (2000) and by the group of Sasaki and Mori (Yano *et al.* 1999).

These examples show that education of crystal-growth and epitaxy engineers is required so that in future the single optimum technology can be applied from the beginning and thus allows accelerated developments at reduced development funds (Klemenz and Scheel 1996).

28.4 CONCLUSIONS

The epitaxial growth phenomena with metallic, with semiconducting, and with superconducting oxide films showed that there are seven distinct epitaxial growth modes, instead of the three classical V–W, S–K and F–VM modes. A unifying attempt was made to define the conditions under which these 7 growth modes occur. Thereby it was realised that the growth parameters supersaturation, misfit of the substrate, and misorientation of the substrate are of similar importance as the surface and interface energies by which the three classical growth modes had been defined by Bauer in 1958. There are further parameters listed in Table 28.1 like surface diffusivity, surfactants, oxidation stage of surface species, types of precursors in MOCVD, VPE and MOMBE, partial pressures of reactive species and of impurity gases, etc., which will have an effect on the growth modes and the perfection of the layers.

The most critical factor for the structural perfection of epitaxial layers is coalescence of the initially formed growth islands in the V–W, S–K, and screw-island growth modes. It was shown how by substrate misorientation a high step density and the step flow mode can be achieved in epitaxy from the vapour phase so that coalescence can be suppressed. Still higher perfection of epitaxial layers can be obtained by liquid phase epitaxy, where by proper choice and preparation of substrates with very low misfit such a small supersaturation can be adjusted so that the pure F–VM growth mode, with spreading of growth steps (with large interstep distances) over macroscopic distances, can be achieved. Such LPE-grown layers are not only structurally very perfect, but also have quasi-atomically flat surfaces and interfaces. Since LPE can be scaled up to real mass production and allows relatively high growth rates, it is the most economic epitaxial growth process for compounds with limited thermodynamic stability (compound semiconductors, oxide superconductors) with minimum ecological problems. Therefore LPE would allow real mass production of highest-efficiency light-emitting diodes and photovoltaic solar cells based on the systems Ga-Al-In-arsenides-phosphides and Ga-Al-In-nitrides. Furthermore, LPE would be the single method by which flat layers of oxide superconductors could be produced. The disadvantage of LPE is its chemical and process complexity, which requires specialised education and experience.

In conclusion, one could say that MBE, MOCVD and related techniques are powerful for quick development of novel layer, superlattice, and device structures and therefore are preferred at universities for PhD theses. LPE will regain the major role for mass production of most highest-performance optoelectronic devices, whereas MOCVD and MBE will be used for production of devices that cannot be fabricated by LPE. Thus the various epitaxial technologies are complementary, and in future the well-educated epitaxy engineers will have the knowledge to chose the one optimum technology for a specific material and device.

GENERAL REFERENCES

A. A. Chernov: Modern Crystallography III: Crystal Growth, Springer, Berlin 1984.

D. Elwell and H. J. Scheel: Crystal Growth from High-Temperature Solutions, Academic Press, London – New York 1975, reprint foreseen 2003 by Elsevier.

Encyclopedia of Materials: Science and Technology. Elsevier Science Publishers Oxford 2001: Several epitaxy reviews.

M. M. Faktor and I. Garrett: Growth of Crystals from the Vapour. Chapman and Hall, London 1974.

Handbook of Crystal Growth, editor D. T. J. Hurle, Volume 3a and 3b. North-Holland/Elsevier, Amsterdam 1994: Several reviews.

M. A. Herman and H. Sitter: Molecular Beam Epitaxy. Springer, Berlin-Heidelberg 1989.

M. L. Hitchman and K. F. Jensen: Chemical Vapor Deposition – Principles and Applications. Academic Press, London 1993.

J. W. Matthews, editor: Epitaxial Growth Part A, Part B. Academic Press, New York 1975.

D. C. Paine and J. C. Bravman: Laser Ablation for Material Synthesis., Mater. Res. Soc. Symp. Proc. No. 191, 1990.

M. Razeghi: The MOCVD Challenge. Vol. 1: A Survey of GaInAsP – InP for Photonic and Electronic Applications. Inst. of Physics, Adam Hilger, Bristol 1989.

G. B. Stringfellow: Organometallic Vapor Phase Epitaxy: Theory and Practice. Academic Press, San Diego 1989.

W. A. Tiller: The Science of Crystallization: Microscopic Interfacial Phenomena, Cambridge University Press 1991, Chapter 5.

REFERENCES

A. Baronnet, J. Cryst. Growth **19** (1973) 193.

E. Bauer, Z. Kristallogr. **110** (1958) 395.

A. I. Braginski, Physica C **185–189** (1991) 391.

W. K. Burton, N. Cabrera and F. C. Frank, Philos. Trans. **A243** (1951) 299.

N. Cabrera and D. A. Vermilyea in Growth and Perfection of Crystals, editors R. H. Doremus, B. W. Roberts and D. Turnbull, Wiley, New York 1958, 393.

N. Cabrera and M. M. Levine, Philos. Mag. **1** (1956) 450.

A. E. Carlson, Ph.D. thesis, University of Utah 1958.

A. A. Chernov and H. J. Scheel, J. Cryst. Growth **149** (1995) 187.

F. C. Frank in Growth and Perfection of Crystals, editors R. H. Doremus, B. W. Roberts and D. Turnbull, Wiley, New York 1958, 411.

F. C. Frank and J. H. Van der Merwe, Proc. Roy. Soc. London **A198** (1949) 216.

C. Gerber, D. Anselmetti, J. G. Bednorz, J. Mannhart and D. G. Schlom, Nature **350** (1991) 279.

M. H. Grabow and G. H. Gilmer, Surf. Sci. **194** (1988) 333.

M. Hawley, I. D. Raistrick, J. G. Beery and R. J. Houlton, Science (1991) 1537.

K. Hiramatsu, S. Itoh, H. Amano, I. Akasaki, N. Kuwano, T. Shiraishi and K. Oki, J. Cryst. Growth **115** (1991) 628.

T. Kajimura, K. Aiki and J. Umeda, Appl. Phys. Lett. **30** (1977) 526.

C. Klemenz and H. J. Scheel, J. Cryst. Growth **129** (1993) 421.

C. Klemenz and H. J. Scheel, *Physica C* **265** (1996) 126.

C. Klemenz and H. J. Scheel, *J. Cryst. Growth* **211** (2000) 62.

H. Komatsu and S. Miyashita, *Jpn. J. Appl. Phys.* **32** (1993) 1478.

M. I. Kozlovskii, *Sov. Phys. – Crystallogr.* **3** (1958) 236.

G. Le Lay and R. Kern, *J. Cryst. Growth* **44** (1978) 197.

S. D. Lester, F. A. Ponce, M. G. Craford and D. A. Steigerwald, *Appl. Phys. Lett.* **66** (1995) 1249.

M. J. Lighthill and G. B. Whitham, *Proc. Roy. Soc.* **A299** (1955) 281, 317.

M. G. Mauk, Z. A. Shellenberger, P. E. Sims, W. Bloothoofd, J. B. McNeely, S. R. Collins, P. I. Rabinowitz, R. B. Hall, L. C. DiNetta and A. M. Barnett, *J. Cryst. Growth* **211** (2000) 411.

S. Miyazawa *et al.*, *IEEE Trans. Electron Devices* **ED-33** (1986) 227.

Y. Nabetani, T. Ishikawa, S. Noda and A. Sasaki, 12th Rec. Alloy Semicond., Phys. and Electronics Symp. 1993, 223–228.

T. Nishinaga, private communication 1996.

T. Nishinaga, C. Sasaoka and K. Pak, *Jpn. J. Appl. Phys.* **28** (1989) 836.

T. Nishinaga and W. Y. Uen, in Control of Semiconductor Interfaces, editors I. Ohdomari, M. Oshima and A. Hiraki, Elsevier Science 1994, 87.

T. Nishinaga, I. Ichimura and T. Suzuki, in Control of Semiconductor Interfaces, editors I. Ohdomari, M. Oshima and A. Hiraki, Elsevier Science B.V. 1994, 63–67.

J. Nishizawa and K. Suto, private communication 1994.

J. Nishizawa, Y. Okuno and K. Suto, in JARECT Vol. 19, Semiconductor Technologies, editor J. Nishizawa, OHMSHA and North-Holland (1986)17.

A. Ourmazd, *J. Cryst. Growth* **98** (1989) 72.

M. D. Pashley, *J. Cryst. Growth* **99** (1990) 473.

R. H. Saul and D. D. Roccasecca, *J. Appl. Phys.* **44** (1973) 1983.

H. J. Scheel, *J. Cryst. Growth* **42** (1977) 301.

H. J. Scheel, *Appl. Phys. Lett.* **37** (1980) 70.

H. J. Scheel, Proceedings E-MRS Meeting June 1987, Les Editions de Physique (Paris) **XVI** (1987) 175.

H. J. Scheel and D. Elwell, *J. Cryst. Growth* **12** (1972) 153.

H. J. Scheel, G. Binnig and H. Rohrer, *J. Cryst. Growth* **60** (1982) 199.

H. J. Scheel, C. Klemenz, F. K. Reinhart, H. P. Lang and H.-J. Güntherodt, *Appl. Phys. Lett.* **65** (1994) 901.

H. J. Scheel in Advances in Superconductivity VI (Proceedings of the 6th Internat. Sympos. on Superconductivity ISS'93 Hiroshima Oct. 26–29, 1993), editors T. Fujita and Y. Shiohara, Springer, Tokyo 1994, 29.

H. J. Scheel in Proceedings Internat. Sympos. on Laser and Nonlinear Optical Materials Singapore Nov. 3–5, 1997, editor T. Sasaki, Data Storage Institute, Singapore, 1997, 10–18.

H. J. Scheel in First International School on Crystal Growth Technology ISCGT-1, Beatenberg/Switzerland Sept. 5–16, 1998, Book of Lecture Notes, editor H. J. Scheel, p. 480–494.

I. N. Stranski and L. Krastanov, *Acad. Wiss. Math.-Naturw. Klasse IIb* **146** (1938) 797.

G. B. Stringfellow, *J. Cryst. Growth* **115** (1991) 1.

G. B. Stringfellow in Handbook of Crystal Growth, editor D. T. J. Hurle, Elsevier, Amsterdam 1994, Chapter 12, 494.

A. A. Tikhonova, *Sov. Phys.-Crystallogr.* **20** (1975) 375.

M. Volmer and A. Weber, *Z. Phys. Chem.* **119** (1926) 277.

M. K. Wu, J. R. Ashburn, C. J. Torng, P. H. Hor, R. L. Meng, L. Gao, Z. J. Huang, Y. Q. Wang and C. W. Chu, *Phys. Rev. Lett.* **58** (1987) 908.

M. Yano, M. Okamoto, Y. K. Yap, M. Yoshimura, Y. Mori and T. Sasaki, *Jpn. J. Appl. Phys.* **38** (1999) L1121.

T. Zheleva, *MRS Internet J. Nitride Semiconductor Res.* **4S1** (1998) G3. 38.

29 High-Rate Deposition of Amorphous Silicon Films by Atmospheric-Pressure Plasma Chemical Vapor Deposition

YUZO MORI [a]**, HIROAKI KAKIUCHI**[a]**,
KUMAYASU YOSHII**[a]**, and KIYOSHI YASUTAKE**[b]

[a] *Department of Precision Science and Technology, Graduate School of Engineering, Osaka University, 2-1 Yamada-Oka, Suita, Osaka 565-0871, JAPAN*
[b] *Department of Material and Life Science, Graduate School of Engineering, Osaka University, 2-1 Yamada-Oka, Suita, Osaka 565-0871, JAPAN*

ABSTRACT

A new high-rate deposition technique has been developed, in which the atmospheric-pressure plasma generated by the rotary electrode was utilized to create a high-density of radicals. A 150-MHz VHF (very high frequency) power supply was employed to get a high-density plasma and to avoid ion bombardment of the film. By virtue of these advantages, high-quality films could be made at unprecedented high deposition rates. Amorphous silicon films were prepared in gas mixtures containing helium, hydrogen and silane, and their electrical and optical properties were investigated. The results showed that the optical sensitivity (σ_{ph}/σ_d) of 10^6 was obtained at extremely high deposition rate of 0.3 μm/s.

29.1 INTRODUCTION

Hydrogenated amorphous silicon films have been applied to functional devices such as solar cells and thin-film transistors. They have mainly been prepared by plasma-CVD (chemical vapor deposition) [1, 2]. In general, plasma-CVD processes are performed under relatively low pressure (less than 100 Pa) with a parallel plate, and capacitively coupled radio-frequency (rf) plasma. Silane (SiH$_4$) is typically used as a source gas with inert dilution or hydrogen gases. However, high-rate deposition (1 μm/s) of high-quality films cannot be realized by the conventional plasma-CVD method.

Crystal Growth Technology, Edited by H. J. Scheel and T. Fukuda
© 2003 John Wiley & Sons, Ltd. ISBN: 0-471-49059-8

In order to make a-Si:H films at high deposition rate, atmospheric pressure plasma-CVD technique with rotary electrode has been developed [3, 4]. In this chapter, the concept of this technique is explained first. Then experimental conditions, deposition rate and film properties are described.

29.2 ATMOSPHERIC-PRESSURE PLASMA CVD

Atmospheric-pressure plasma-CVD is a new technique for fabricating high-quality silicon films at very high deposition rate. The advantageous aspects of this technique are described in the following.

29.2.1 ATMOSPHERIC PRESSURE, VHF PLASMA

High-pressure reactive gas (1×10^2–5×10^3 Pa) is diluted by plenty of inert gas such as helium or argon. Hence, a stable plasma is easily generated at such high pressure, and the deposition rate is remarkably increased. Furthermore, the kinetic energy of ionic species is reduced due to the increased collision frequency of ions and other species. This reduces the ion damage of the films. Furthermore, the oscillation amplitude of ionic species is reduced due to the high-frequency electric field. Lower oscillation amplitude increases the density of ions, which do not collide against the electrode surface and results in a higher power density in the plasma region.

On the other hand, the following problems must be solved for applying this technique: (1) a fixed electrode might be thermally damaged because the gas temperature of the atmospheric-pressure plasma is much higher than that of the conventional low-pressure plasma, (2) plenty of particles, which is directly connected with the degradation of the film quality, is generated in the plasma because of the high density of reactive gases. Utilizing a rotary electrode, however, can solve these problems.

29.2.2 UTILIZATION OF ROTARY ELECTRODE

In Figure 29.1, the reactive gas is homogeneously introduced at a high speed into the plasma area between the rotary electrode and the substrate. Also, it is possible to sufficiently cool the electrode surface, so that a high electric power can be applied without thermal damage of the electrode. Therefore, the deposition rate and homogeneity of the films are remarkably improved, and particles generated in the plasma are easily removed from the plasma area. Furthermore, it is possible to make a large-area film by scanning the substrate.

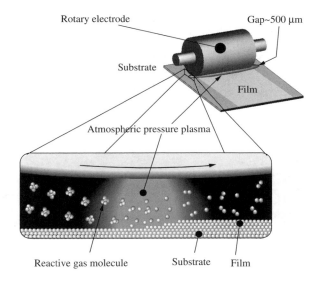

Figure 29.1   Schematic of the atmospheric-pressure plasma-CVD with rotary electrode.

29.3   EXPERIMENTAL

Figure 29.2 shows a schematic of the atmospheric-pressure plasma-CVD system used in this study. A cylindrical rotary electrode with 300 mm diameter and 100 mm width is placed in the reaction chamber. Magnetic coupling is used for conveying the torque of the high-speed rotation spindle drive. The substrate is vacuum chucked by the substrate holder, which can be heated up to 300 °C. The 150-MHz VHF power is supplied to the rotary electrode through the impedance-matching unit. The vacuum system consists of a turbomolecular pump and a scroll-type dry pump.

The gas-circulation system for recycling the reactive gases is connected to the chamber for removing particles generated by the chemical reactions in the plasma. The reactive gases are absorbed through the particle-collecting duct located behind the electrode. The gas circulation rate is 1260 l/min.

In a series of experiments, a-Si:H films were deposited on Corning #7059 glass substrates or on float zone (FZ) silicon wafers. After the substrate was loaded into the reaction chamber, it was evacuated to a pressure of 1.5×10^{-4} Pa and then filled with helium, hydrogen and silane up to 1 atm. The hydrogen-to-helium ratio (H_2/He) was fixed at 1 %, and the SiH_4/He ratio was varied as a parameter. The substrate temperature of 200 °C was monitored by a thermocouple.

The thicknesses of the a-Si:H films were determined directly from the cross-sectional SEM image. The deposition rate was evaluated from the film thickness

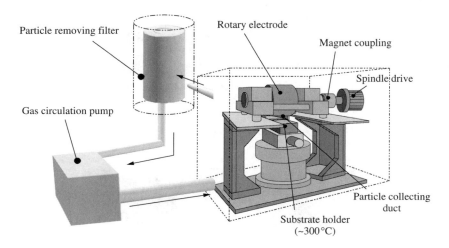

Particle removing filter

Rotary electrode

Magnet coupling

Spindle drive

Gas circulation pump

Substrate holder
(~300°C)

Particle collecting
duct

Figure 29.2 Atmospheric-pressure plasma-CVD system.

and the growth time. The conductivity was measured at room temperature in the ohmic region with a coplanar cell using aluminum electrodes of 1 mm spacing. The photoconductivity was measured under AM1.5, $100\,\mathrm{m\,W/cm^2}$ illumination. The optical gap (E_{opt}) was calculated from the $(\alpha h\nu)^{1/3}$ plot [5], where α is the absorption coefficient of the a-Si:H films. The hydrogen content (C_{H}) was calculated from the IR absorption around 1800 to $2200\,\mathrm{cm^{-1}}$ due to silicon–hydrogen bonds.

29.4   RESULTS AND DISCUSSION

29.4.1   DEPOSITION RATE

Figure 29.3 represents the input-power dependence of the deposition rate, in which the silane-to-helium ratio is varied from 0.1 % to 5 %. The maximum deposition rate is $1.6\,\mu\mathrm{m/s}$ at $SiH_4/He = 5\,\%$ and the input power of 2000 W. In this deposition condition, a-Si:H film $0.5\,\mu\mathrm{m}$ thick over a $100\,\mathrm{mm} \times 140\,\mathrm{mm}$ region can be deposited in 6 s with a substrate scanning speed of 19 mm/s.

From this it follows that a deposition rate over 100 times faster than in the conventional plasma-CVD is realized.

29.4.2   ELECTRICAL AND OPTICAL PROPERTIES

Figure 29.4 shows the deposition-rate dependence of conductivity of the a-Si:H films. The conductivity of a-Si:H films formed by conventional plasma-CVD is also shown as comparison. From Figure 29.4, σ_{d} is in the range of

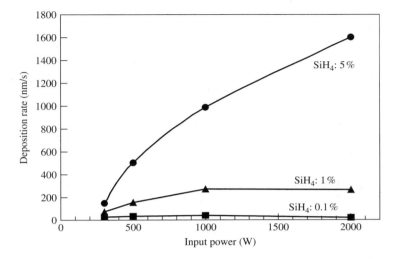

Figure 29.3 Input-power dependence of the deposition rate.

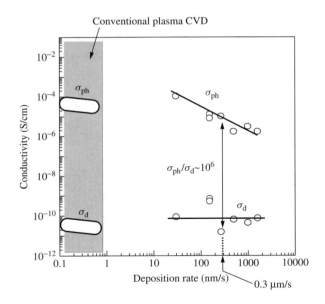

Figure 29.4 Deposition-rate dependence of the photo and dark conductivity.

$10^{-11}-10^{-10}$ S/cm, while σ_{ph} tends to decrease with increasing deposition rate. Yet, the optical sensitivity (σ_{ph}/σ_d) over 10^4 is obtained even at the maximum deposition rate. In particular, an optical sensitivity of about 10^6 is obtained at the deposition rate of 0.3 μm/s. Amorphous-silicon films formed by this technique are found to have an excellent conductive characteristic equivalent to those formed by the conventional plasma-CVD method [6].

Figure 29.5 shows the variation of the optical gap and the hydrogen content with the deposition rate. As the deposition rate increases, C_H tends to increase. The tendency of E_{opt} is opposite to that of the conventional plasma-CVD. The reason for this variation of E_{opt} may be insufficient structural relaxation on the growth surface due to the extremely high deposition rate compared to the conventional plasma CVD. C_H is in the comparable range of 10–25 % to that of the conventional method, in spite of the high deposition rate condition. It is assumed that highly decomposed radicals such as SiH_2, SiH and Si mainly contribute to film growth.

The deposition-rate dependence of the ratio of hydrogen configuration (SiH_2/SiH) in the film is shown in Figure 29.6. The SiH_2/SiH ratio is a parameter indicating a degree of disorder of the film structure. In device-grade a-Si:H films, the SiH_2/SiH ratio should be lower than 0.1. In Figure 29.6, although the SiH_2/SiH ratio is slightly higher than that of the conventional technique, a sufficiently low SiH_2/SiH ratio is realized with the extremely high deposition rate. The film-growth mechanism in the atmospheric-pressure plasma-CVD is assumed to be absolutely different from that of the conventional technique.

From these results, it is clear that amorphous-silicon films fabricated by atmospheric-pressure plasma-CVD have excellent electrical and optical characteristics, which are equivalent to those grown by the conventional plasma-CVD.

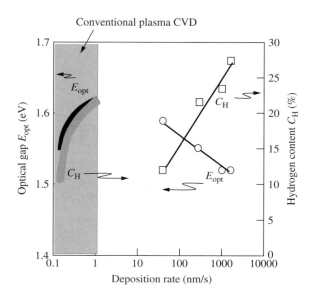

Figure 29.5   Deposition-rate dependence of the optical gap and hydrogen content.

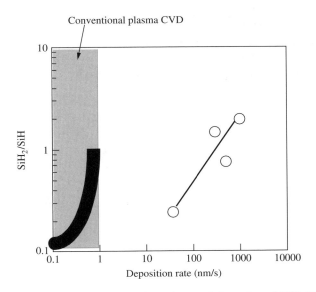

Figure 29.6   Deposition-rate dependence of the value of SiH_2/SiH.

29.5   CONCLUSION

High-rate deposition of amorphous-silicon films was achieved by the atmospheric-pressure plasma-CVD. The results obtained are as follows:

(1)   Extremely high deposition rate of $1.6\,\mu m/s$ was achieved at $SiH_4/He = 5\,\%$, $H_2/He = 1\,\%$ and input power of $2000\,W$.
(2)   Hydrogenated amorphous-silicon films with $\sigma_{ph}/\sigma_d = 10^6$ were realized at the very high deposition rate of $0.3\,\mu m/s$.

ACKNOWLEDGEMENTS

This work was supported by a Grant-in-Aid for COE Research (No.08CE2004) from the Ministry of Education, Science, Sports and Culture.

REFERENCES

[1]   D. E. Carlson, and C. R. Wronski: *Appl. Phys. Lett.* **28** (1976) 671.
[2]   Y. Tawada, K. Tsuge, M. Kondo, H. Okamoto, and Y. Hamakawa: *J. Appl. Phys.* **53** (1982) 5273.
[3]   Y. Mori, K. Yoshii, T. Kataoka, K. Hirose, K. Yasutake, K. Endo, Y. Domoto, H. Tarui, S. Kiyama, and H. Kakiuchi: 1996 The Japan-China Bilateral Symposium on Advanced Manufacturing Engineering (1996) 84.

[4] Y. Mori, K. Yoshii, H. Kakiuchi and K. Yasutake: *Rev. Sci. Instrum.* **71[8]** (2000) 3173.

[5] Y. Hishikawa, N. Nakamura, S. Tsuda, S. Nakano, Y. Kishi and Y. Kuwano: Jpn. *J. Appl. Phys.* **30** (1991) 1008.

[6] Y. Hishikawa, S. Tsuda, K. Wakisaka and Y. Kuwano: *J. Appl. Phys.* **73** (1993) 4227.

Index

Entries in italics refer to figures, entries in bold refer to tables.

Crystal Growth Technology, Edited by H. J. Scheel and T. Fukuda
© 2003 John Wiley & Sons, Ltd. ISBN: 0-471-49059-8